Historical & Cultural Astronomy

More information about this series at http://www.springer.com/series/15156

Harm J. Habing

The Birth of Modern Astronomy

Springer

Harm J. Habing
Leiden Observatory
Leiden, The Netherlands

Extended English translation of the original Dutch edition published by
Veen Media, The Netherlands, 2014

ISSN 2509-310X ISSN 2509-3118 (electronic)
Historical & Cultural Astronomy
ISBN 978-3-319-99081-1 ISBN 978-3-319-99082-8 (eBook)
https://doi.org/10.1007/978-3-319-99082-8

Library of Congress Control Number: 2018956729

This Springer imprint is published by the registered company Springer Nature Switzerland AG
The registered company address is: Gewerbestrasse 11, 6330 Cham, Switzerland

Preface

Astronomy is booming. Every month there are reports of some spectacular discovery in newspapers, TV broadcasts and other means of distributing important news all over the world. It has been this way for more than two decades. And, believe me, several of these loudly announced discoveries were spectacular indeed.

All this news is coming from huge, high-tech instruments operating in space and on all continents, even on Antarctica. Space telescopes have been successfully launched by the USA, by Russia, by (the rest of) Europe, Canada, Japan, India, and China. More than ten optical telescopes with mirrors of 8–10 meters in size were brought into use in the 1990s, although only a few years earlier it was generally assumed that such mirrors were too large and too massive. A radio telescope consisting of over 50 smaller telescopes has been erected in a desert in Chile at 5000 meters altitude by a cooperation of countries from the Americas, Europe, and Asia. South Africa and Australia are now engaged in a joint enterprise to build the largest radio telescope ever in two parts, one in each country. There is an observatory at the South Pole profiting from the thin and dry atmosphere to detect radiation that does not arrive at the surface of the Earth anywhere else in the world. The biggest radio telescope with a diameter of 500 meters has opened in the last few years in China!

How come? How did astronomy develop from a minor and somewhat sleepy discipline in 1945 that was taught by a very small staff at universities and with research going on in a few larger observatories? How did this modest enterprise grow to a large, costly activity all over the world and find general appreciation everywhere? That is the subject of this book.

In 1958, at the age of 20, I became involved in astronomical research and I stayed there for over 50 years. I was a witness to all these developments and a humble participant. Over the last 5 years I have tried to reconstruct all that happened. It brought back many memories, not only successes but also unpleasant battles and mishaps. But the dominant feeling was one of happy surprise. The sky, the Milky

Way, the stars, the planets, the comets, and even the supernovae, they all have had an important place in civilizations all over the world and over thousands of years. To me, the last 70 years have brought a joyful and impressive apotheose. That is what I have wanted to describe.

Leiden, The Netherlands
June 2018

Harm J. Habing

Acknowledgments

I would like to thank Ewine van Dishoeck for stimulating me to translate and expand the existing Dutch text. At Springer Verlag, New York, Harry Blom and Butler Burton gave my proposal a warm welcome. Ramon Khanna at Springer, Heidelberg, was always available for questions and for leniency in postponing the submission date, not once, but quite a few times. The Netherlands Organisation for Scientific Research, NWO, gave me a subsidy for the translation, and I very much appreciated the enthusiastic support I received from Ronald Stark. Leiden Observatory, through Evelijn Gerstel and Alexandra Schouten-Voskamp, was helpful at critical moments. Johan Bleeker and Huib van Langevelde read parts of the text and their comments were very valuable.

I thank Marilyn Hedges for her competence as a translator and for our very enjoyable, cheerful collaboration. I remember several stimulating discussions about such subjects as the linguistic differences and similarities between Dutch and English. Lisa Pothoven very competently dealt with all the figures, restoring the sometimes poor copies that I had made.

Lastly, my thanks go to my wife, Connie, who has been such an enormous support to me throughout all the years that it took to write the book that I wanted to write for reasons that I did not fully understand.

Contents

Prologue

Why This Book?

Astronomy attracts a lot of attention today. Leaf through any popular science magazine or the science supplement of a newspaper and you are bound to come across news about what is a relatively small field of science that nonetheless embraces the very biggest phenomena. The interest in the field corresponds with the exceptional growth in research since 1945, a growth that has been prompted by the insight that more radiation reaches us from the world outside our galaxy than just light. In terms of physics, light is electromagnetic radiation at a wavelength of between 300 and 800 nm; 1 nanometer (nm) is one-billionth part of a meter. There is also electromagnetic radiation at other wavelengths, radiation that we know by different names, such as X-radiation, infrared radiation, radio waves.

Figure 1 shows the different kinds of electromagnetic radiation, arranged according to wavelength, from long (right) to short (left). Fortunately for our health, much of this radiation is held in check by the Earth's atmosphere. In other words, if the atmosphere were not there, we would not be here.

Electromagnetic radiation can be thought of as waves traveling through space at the velocity of light but also as a stream of photons, or light particles. Each photon has energy that in Fig. 1 increases from right to left: radio photons have the least energy and gamma photons the most. The two descriptions (waves vs. photons) are fully compatible with one another and one uses whichever is most convenient. Light, radio and infrared radiation, is usually described by waves, and wavelength is then the most important parameter. X-radiation and gamma rays are usually described as streams of photons (traveling, to be sure, at the velocity of light) for which the photon energy is the most convenient parameter.

Stars radiate mainly in visible light and in the UV. Red giant stars emit little or no UV but are strong in the infrared. A few objects are known to be bright at all wavelengths. The most outstanding of these is the Crab nebula. In this book the Crab nebula is as important as the Rosetta stone was for deciphering the Egyptian hieroglyphs in the nineteenth century (Fig. 2).

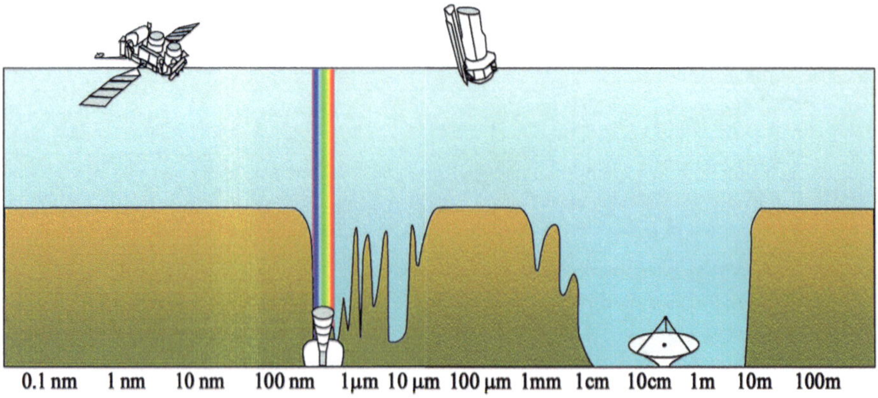

Fig. 1 All the types of electromagnetic radiation arranged according to wavelength. Radio waves have the longest wavelengths and gamma rays have the shortest. The upper diagram indicates the transparency of the atmosphere to electromagnetic radiation, and the lower diagram explains what is going on. Notice in the upper figure that visible radiation (light), indicated by the colors of the spectrum, is no more than a very narrow part of the whole spectrum. Only "visible light" radiation and radio radiation with wavelengths between 1 cm and 10 m reach the surface of the Earth. Source: NASA

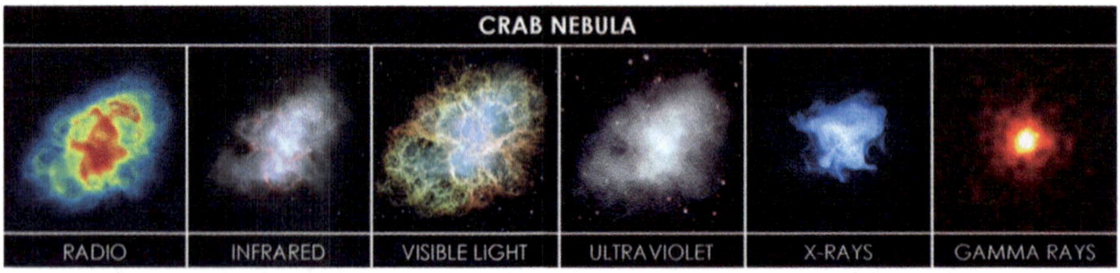

Fig. 2 Images of the Crab nebula at all wavelengths that have been detected. Compilation of images by (1) NRAO/AUI and M. Bietenhol; (2) NRAO/AUI and J.M. Uson, T.J. Cornwell; (3) NASA/JPL-Caltech/R. Gehrz; (4) NASA, ESA, J. Hester and A. Loll; NASA/Swift/E. Hoversten; (5) NASA/CXC/SAO/F.Seward et al.; (6) ASA/DOE/Fermi LAT/R. Buehler

After August 1945, when the war was finally over, the first amazing results came from physicists in England and Australia, who had been trained as radar experts during the war. They chose to look for natural sources of radio radiation in the sky; during the war they had occasionally picked up hints of such sources. The Sun was their first goal but soon it became clear that there was much more to be recorded. The celestial sources they detected were definitely astronomical objects, the physics of which could not be explained by conventional astronomy. Slowly but surely, the more traditional astronomers also became involved, and radio astronomy was born as a new field of research. One of the consequences was that "old" astronomy now became known as "optical astronomy."

The unexpected findings from radio astronomy encouraged other physicists to look for extraterrestrial sources of infrared and X-radiation. The start was difficult, but ultimately they were not disappointed: at these two wavelengths, too, great surprises were in store.

The aim of this book is to describe how this broadening of astronomical research progressed from 1945 to 2015. I will split the book into two parts. The first part, 1945–1985, may be summarized by "Anything goes," the title of a well-known musical in the 1930s. In this period astronomical research was a kind of "wild west" show: scientists (always physicists, rarely astronomers) measured celestial radiation in a wavelength band that had never before been studied by astronomers and this led to totally unexpected discoveries; sources were detected that had previously been the subject of pure speculation, almost topics of science fiction: "black holes", for example. These objects became food for spectacular movies or for famous pop songs. This first period was adventurous but also chaotic. Within a few years, in the 1960s in particular, there were at least four major breakthroughs that affected all aspects of astronomy.

I will divide the time between 1945 and 1985 into three periods, each of which has main themes: "technical progress," "stars," and "galaxies and the universe."

Sometime between 1980 and 1990, a major change occurred in the way research was conducted. The most spectacular objects had been discovered at every wavelength and astronomers started to study these objects systematically. Clearly defined goals could now be formulated. Huge instruments were built and systematic studies followed, often by very large teams of scientists, more often than not consisting of researchers from all the continents of the world. Ultimately, even, including astronomers from Antarctica! I choose to describe these developments by tracing the history of the main types of object (star formation, supernovae, galaxies, etc.).

It is remarkable that, even with the impact of fundamental new discoveries in other wavelength areas, traditional optical astronomy continued to play a prominent role. Admittedly this was only possible due to major improvements in detection techniques (in particular, the replacement of photographic plates by CCDs) that were applied in new, much larger and better-located telescopes or in telescopes in space.

All the important new insights that came about after 1945 resulted from totally new instruments and from improvements in existing instruments. Theoretical physics as it then existed was adequate to explain the revolutionary new observations up until 1980: we are talking here about the general theory of relativity from around 1916 and quantum theory from the 1920s and 1930s. But from around 1980, questions arose that present-day physics is unable to answer. The three most important questions are: What is the nature of dark matter? What is the nature of dark energy? Was there an extremely brief process called "inflation" right after the Big Bang? Nonetheless, a cohesive body of knowledge has been gained about how this Universe developed from the Big Bang to what we see today, a Universe of stars and galaxies. This is a remarkable achievement.

This book will not discuss planets, comets and other objects within our solar system. With space research, so much is known today that it could not all be fitted in within the pages available; another book would be needed. The Sun and the planets will be discussed, but only in comparison with other stars and their planets.

Peace and Destruction in 1945

In August 1945 the war was over, both in Europe and in the Pacific Ocean. American soldiers returned home to carry on where they had left off before 1941. Australia became a center for radio astronomy, as I will discuss in the next chapter. The Soviet Union, China, and Japan were in no state to pay attention to high-level research. Europe had suffered enormous blows. England may well have been the victor, but the country was drained, both financially and in terms of manpower. On the continent, people started to pick up the pieces. Until 1939, Germany had been a leading player in natural science research and now the country lay in ruins. This is graphically indicated by the fate of Germany's most famous astronomical research journal *Astronomische Nachrichten*. The journal was launched in 1821; in 1939 it contained 350 articles, in 1944 a mere 25, and in 1945 and 1946 none at all; in 1950 it published 47 articles. The journal was never to recover its former standing, partly because after 1945 it became an East German publication; in 2000 a total of 44 articles were published.

American Supremacy

There was one clear conclusion to be drawn: the USA was superior in natural science research. In the field of astronomy, this supremacy manifested itself in several different ways. The first of these was in the instruments used, by far the most important of these being the telescopes. There are two properties of telescopes that determine how significant they are. The first is the size of the diameter of the mirror: the bigger the mirror, the more light it receives and the more sensitive it is. The second is the location of the telescope: it should be located on a site that is as high and as dry as possible. In 1945, there were ten telescopes with a diameter of a meter or more. The six bigger ones were in the USA and Canada, all at good locations. The four smaller telescopes, on the other hand, two in Italy and two in Germany, were all at deplorable locations. The exceptional qualities of the summits of volcanoes on the Canary Islands only became accessible to the rest of Europe after the fall of the Franco regime in 1975.

A second example of the dominance of the USA is in its role as the "hub" where all the news was collected and distributed. If anything new was seen in the sky, such as a new comet, telegrams were sent from Germany to observatories all over the world so that people could observe the object with the minimum loss of time. The last telegram was sent in 1944 from Berlin. This news service started up again shortly after the war, this time from the Harvard observatory in the USA.

Thirdly, given its size, the USA had much greater potential for research than the divided Europe, and not only that, the Americans had benefited from the immigration of eminent—mainly Jewish or half-Jewish—scientists from Europe, such as H. Bethe, G. Herzberg, A. Penzias, B. Rossi, M. Schwarzschild, and E. Salpeter. It is only in the last 20 years that major contributions have come from Europe and from countries such as Russia, Japan, South Africa, and, recently, from China. Europe's contribution has been primarily through the foundation of organizations such as the European Southern Observatory (ESO) and the European Space Agency (ESA).

Discoveries in Times of War

However atrocious war may be, it does bring some benefits: forced technical developments, for example, such as the enormous improvements that were made in radio techniques both by the Germans and the allied forces. German radar installations that were constructed along the whole stretch of the "Atlantik Wall" (along the coast from Norway to Spain), the so-called Würzburg radar antennae with a diameter of 7.50 meters, were used enthusiastically in developing radio astronomy in the Netherlands, England, France, and even in neutral Sweden. The cardinal discovery—that a very strong radio source, Cygnus A, coincided with a very small galaxy—was only made after astronomers in Cambridge had measured the position precisely using two German radar antennae.

A second "new astronomy"—infrared astronomy—started in the USA with a detector based on lead sulphide. This was a German military discovery that was taken to the USA as spoils of war. During the last two years of the war and immediately behind the front line, a group of American scientists under the code name ALSOS were looking at secret technical developments in German laboratories. The objective behind this operation was to find out how far the Germans had progressed with constructing an atom bomb, but the ALSOS scientists also conducted a more general, and not strictly necessary, exploration, during which they made a number of other discoveries. One of these was that lead sulphide (PbS) was a better detector for infrared radiation than researchers on the allied side knew.

The third item is a secret American operation, "Operation Overcast," whereby shortly after the German capitulation dozens of engineers and technicians who had worked on the V-2 rockets were picked up and taken to New Mexico. Among them was the highly sought after Wernher von Braun. A large number of V-2s were also transported. Operation Overcast meant that the American space program began after 1945 with a flying start. Von Braun was such an unscrupulous opportunist that I cannot help but include two animated photos showing him in the midst of his friends. See Figs. 3 and 4.

There were also important technical developments on the allied side, the most impressive being the building of the atomic bomb, regretted by many who were involved in it. A less problematic development was the first electronic computer "Colossus." Colossus was built under great pressure and at record speed in the secret research center at Bletchley Park (England) to decode messages exchanged between the German submarines in the Atlantic Ocean and the headquarters of the German war navy.

Looking back, the war years were a turning point in the progress of astronomy research. The profession of astronomer changed because, among other things, all

Fig. 3 German rocket experts who were taken to the USA after victory was declared; 35 of them are not on this photo from 1947 or 1948 because they were involved in a test flight of a V-2 elsewhere in the USA. Von Braun is in the front row, the seventh person from the right. Source: NASA

Fig. 4 Von Braun in 1941 with his friends of the time. Source: German Federal Archive

"new astronomies" brought its own researchers, who were in most cases experimental physicists. The increasing productivity manifested itself in the number of publications. The table below shows the number of publications in *Astrophysical Journal* and *Astronomical Journal*, two American journals, together with the English magazine *Monthly Notice*. Each figure is the sum of 3 consecutive years.

Years	1935	1950	1960	1970	1980	1990	2000
No. of publications	1,010	1,257	1,713	2,936	5,592	7,393	11,669
Compared with 1960	0.60	0.73	1.0	1.71	3.26	4.32	6.81

Looking Back: What Did We Know in 1945 about Stars and Galaxies?

Stars

In 1549, Copernicus proposed that the Sun was the center of the Universe, and the Earth was one of the planets that orbited around the Sun. In around 1610, Galilei and Kepler provided so much evidence for Copernicus' proposal that it was no longer a matter of serious scientific debate. It was then that people started to understand how large the Sun was compared to the Earth, which led to the hypothesis that the stars were (very) distant Suns. This hypothesis was generally accepted, but it was not until 1837 that it was proven by observations. Stars, like the Sun, were spheres of gas, held together by gravity and radiating enormous amounts of energy.

In the eighteenth and nineteenth centuries, the concept of "energy" took form very gradually, ultimately reaching its final definition in around 1840. It became a leading principle in physics through the law that if a system is isolated, its total content of energy cannot change. Application of this principle to the Sun dictated that the energy radiated by the Sun had to come from an internal and finite source. At some time in the future the supply of energy would be exhausted and then the Sun would stop shining.

Then came the question of what stock of energy the Sun contained. A first proposal at the end of the nineteenth century by the English Lord Kelvin and his German friend Von Helmholtz identified a reserve sufficient for perhaps a million years. This was not long enough by far for the geologists and the Darwinists. Finally, after Einstein had proven theoretically that matter could be converted into energy ($E = mc^2$), there was now the possibility that the formation of the nucleus of helium from four nuclei of hydrogen provided the energy for solar radiation: the helium nucleus had a mass that is a few percent smaller than the sum of four hydrogen nuclei. If this conversion took place, the energy of the Sun was no longer a problematic issue. Finally, in 1939 two young physicists, both from Germany, but during the war on opposite sides, Von Weiszäcker and Bethe, independently proposed a series of nuclear reactions that produced the desired result.

Let us take a look at the observation of stars. The brightness of a star is indicated by its magnitude. A tradition that is more than two millennia old implies that a star with a higher magnitude is fainter; a star (or nebula) with a magnitude of 15 is a hundred times *fainter* than a star of magnitude 10. This is not logical or even rational, but disputing it is like Don Quixote tilting at windmills.

Stars have colors: a star can emit more blue light than red, or vice versa. To determine the color of a star, you measure the magnitude B of the star as seen through a blue filter and the magnitude V through a yellow ("visual") filter. The color of the star is the difference between the magnitudes B-V. Color is an important concept.

Now take a large number of stars at the same distance from the Earth and measure their magnitude and color. Put them in a diagram with the color of the star horizontally and the magnitude vertically. The result is the Hertzsprung–Russell diagram

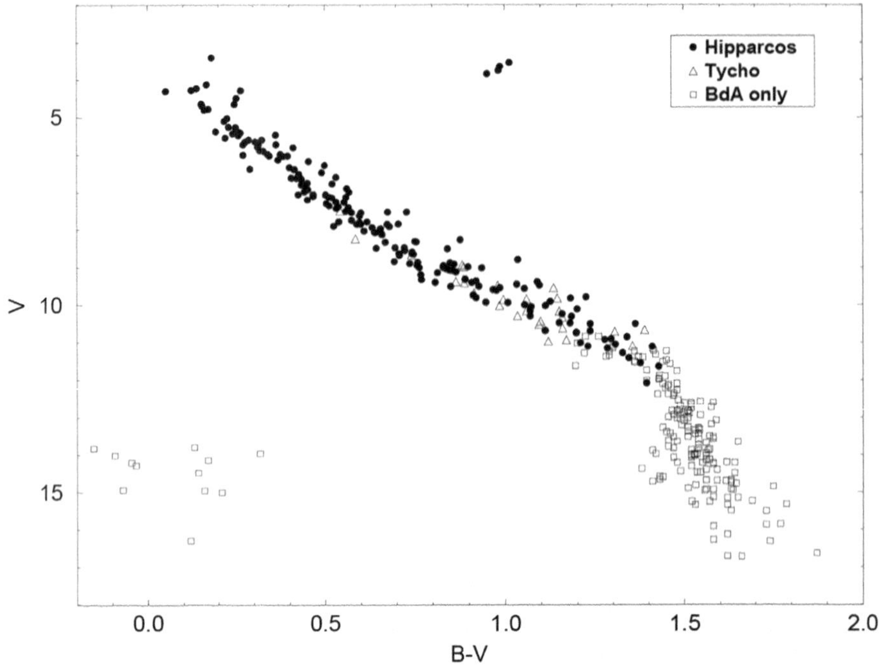

Fig. 5 The Hertzsprung-Russell diagram of stars in several nearby stellar clusters in the Milky Way. The color (or surface temperature) of the stars is indicated by the horizontal scale and in the vertical direction is the magnitude all stars would have if they were put at the same distance. The curve underneath is the so-called main sequence; stars will stay there the longest part of their life. When stars get old they leave the main sequence and become brighter and redder; the point where the cluster stars leave the main sequence depends on the age of the cluster: older clusters turn off from the main sequence more to the right (Perryman et al. 1998).

(generally abbreviated to "HR diagram"), an extremely important tool to compare and understand stars. When the first HR diagrams were drawn, in around 1910, all stars seemed to be in a narrow band from blue and bright to red and faint. This band is referred to as the "main sequence." The Sun fits well in such a diagram: it is a perfectly ordinary star in the main sequence. A present-day example of such an HR diagram can be seen in Fig. 5.

An important discovery showed that the blue, bright stars were much heavier than the red, weak stars. This led to a surprisingly simple conclusion: the mass of a star dictated where the star was positioned on the main sequence. To put things into perspective: a blue main sequence star is approximately 30 times more massive than the Sun, and the weakest red stars have a mass that is about one-tenth that of the Sun.

An explanation of the main sequence was provided in the 1920s by English astrophysicist Eddington. He calculated stellar models in which a particular amount of energy was produced in the center of the star in the form of light particles or photons (how it happened nobody knew at the time). He also worked out how that light found its way to the star's atmosphere and was radiated outward from there. Eddington showed that high-mass models were large, bright, and blue, whereas

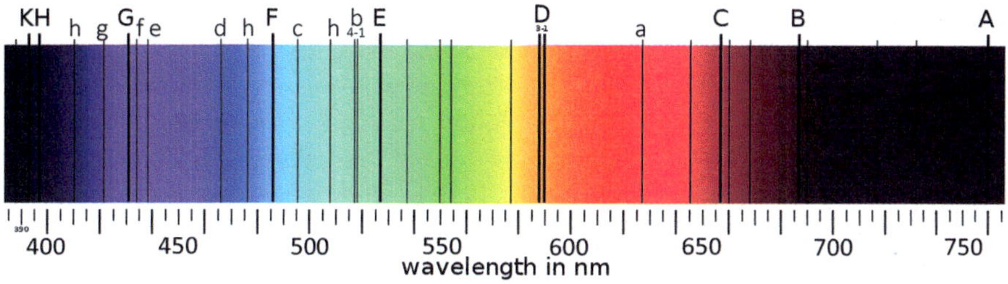

Fig. 6 The solar spectrum and the spectral lines that Fraunhofer identified in 1814. Source: Wikimedia Commons

low-mass models were small, faint, and red: these models matched the main sequence remarkably well.

Stars have an atmosphere, which is where the light we receive is emitted. This light can be split into a rainbow with colors from blue to red (see Fig. 6). If you look carefully at this "spectrum" (the official name of the rainbow) you will see it contains black lines—first recognized by 27-year-old Joseph Fraunhofer in 1814. The lines come about because atoms in the atmosphere absorb light selectively at specific wavelengths—the two lines labeled D, for example, are produced by sodium atoms. By studying these lines minutely, it became possible (but only 100 years after Fraunhofer's discovery!) to calculate temperature and pressure in the atmosphere as well as how many atoms there are and of which elements.

The composition of the solar atmosphere appeared to be as follows: of every hundred atoms, ninety were hydrogen atoms, nine were helium atoms and the remaining one was of all other elements together (carbon, nitrogen, oxygen, iron, etc.). There were good arguments to assume that this was the composition the Sun was born with. The composition of the Earth, however, is very different because hydrogen is a minor element and helium is almost completely absent. Cecilia Payne, who was the first person to publish on this difference between Sun and Earth in her Harvard dissertation in 1925, was strongly advised against this conclusion: common sense dictated that this could not be true. But within 10 years everyone was convinced she was right. Payne was also able to prove that in all the main sequence stars studied the atomic composition was the same as in the Sun, even though there were large differences between the spectra of different stars. The existence of "cosmic abundances" as they are called, was an unexpected and surprising finding.

Eddington's explanation that the properties of main sequence stars could be understood as the consequence of their difference in weight (mass) was valid for stars on the main sequence. But in the HR diagram above, there are also a few stars at the bottom left ("white dwarfs") and above right ("red giants"). What were these objects?

The white dwarfs were discovered in the nineteenth century, but in around 1900 they seemed to have one inexplicable characteristic. They were about as large as the Earth but they contained about as much matter as the Sun. The consequence was that 1 cubic centimeter of matter in a white dwarf contained 2000 kilograms. Such a density of matter had never been seen before and it gave the physicists a headache. What could the characteristics of this matter be? The answer came in around 1925: matter of such density could be and must be described using the new laws of quantum mechanics! White dwarfs were the first large objects for which the older physics could not give any explanation, but quantum mechanics could. This had consequences, as was shown by a young Indian student, Subrahmanyan Chandrasekhar, who was travelling to England by boat in 1930 to study astronomy there under the master Eddington. During the journey, he thought about the characteristics of this unusual matter and concluded that white dwarfs could only exist if they were not too massive; he estimated that there was an upper limit of around 1.4 times the mass of the Sun. However, Eddington in Cambridge believed none of this. Chandrasekhar was put out, but he persevered, and eventually he was proved right (in 1983 he received the Nobel Prize for his work): all white dwarfs of which the mass has been measured are lighter than 1.4 times the mass of the Sun. White dwarfs are fundamentally different types of objects from main sequence stars. They must have been formed under great pressure, but how that happened nobody knew.

And the red giants? Nobody could explain these objects and in 1945 their existence was still a mystery.

Novae and supernovae. There is a striking story about Tycho Brahe, the great Danish astronomer, who comes out of his workshop one evening in 1572 and sees a bright star in the sky, which was never there before. He thinks it is a mirage and that he is seeing something that is not actually there. He approaches some farmers and asks them to confirm that he, Tycho Brahe, has not gone out of his mind. The farmers do not understand the excitement of this high noble lord, but they do confirm: yes, they see that star too. Brahe's astonishment is easy to understand, for it came from a deep, thousands of years old belief that heaven was immutable, an eternal majesty; change was typical of the Earth inhabited by imperfect man, not for the divine space beyond Earth. But now, on that evening in 1572, this old, comfortable thought was convincingly destroyed. The new star slowly but surely weakened and, after a few months, disappeared from view. A few decades later Kepler also saw such a short-lived "new" star (a "stella nova"), after which it took until 1987 before the phenomenon could be seen again with the naked eye—see later in this book.

New stars, "novae," were observed more often when telescopes became available. In particular, their number increased after the introduction of photography in around 1880. With large telescopes such as those in California, novae were found in the Andromeda nebula; in 1885, a very bright nova was even seen for a short time. It was this exceptionally bright nova that in 1934 set Baade and Zwicky thinking; these were two young European astronomers, one connected to the Mount Wilson observatory at Los Angeles and the other to the California Institute of Technology. They realized that that the very bright nova in Andromeda had been much brighter than the other novae in that galaxy. They wondered: "Was this an ordinary nova? Could there be more of such "super-novae"? They looked up old data, among other things about the new stars seen by Brahe and Kepler, and concluded that there was indeed a separate class of "supernovae" that reached a brightness many times greater than that of other novae. Novae were probably a big explosion on a star, but, Baade and Zwicky speculated, supernovae heralded the definitive end, the total explosion and destruction of a star.

This history became even more interesting when in around 1940 some astronomers remembered that in the nineteenth century a sinologist had reported on a "bright guest star" that had been noticed by Chinese astronomers in the year 1054. The star had been visible for a few months and then slowly disappeared from sight: with reference to the work of Baade and Zwicky, this guest star had to have been a supernova. What caused some excitement was that the position indicated by the Chinese was in line with that of a nebula in the constellation of Taurus, known as the Crab nebula. It was a unique object, which looked very different from all the others. Surprisingly, it had been discovered that this nebula became larger every year. If one calculated backward to the point in time when the nebula was as small as a star, the answer was: 900 years ago. This was a decisive result: the Crab nebula had to be the remnant of the guest star, the supernova of 1054. The Crab nebula will play a key role in the development of radio astronomy in the following chapters.

Milky Way

The large, luminous band of the Milky Way has always intrigued humanity—in the north and in the south. In very ancient cultures throughout the world, the Milky Way became a firm item in myths about gods. Astronomical insight emerged when Galilei directed his homemade telescope at the Milky Way in around 1610 and

Fig. 7 The central part of the Milky Way seen above the large telescopes of the European Southern Observatory. The red line is a ray of light from a laser near the telescope. High in the Earth's atmosphere the ray is bent by turbulent winds, and it is this that gives information on how to distort the mirror of the telescope to compensate for the turbulence ("active optics"). Source: ESO

decided that it was made up entirely of stars. In retrospect, he was largely, but not entirely, right. Other astronomers, who also acquired a telescope, saw not only a multitude of stars but also some luminous clouds, such as those in Orion. It was assumed that such nebulae were stellar farrowing rooms, which turned out to be an inspired idea centuries later, after spectroscopy had made its appearance.

Dark spots in the Milky Way can be observed without a telescope, as is shown by the presence of the so-called Coal Sack in the logs of the first European ships that traveled south of the equator. When the first photographs of the Milky Way were taken, around 1900, many more of these dark clouds were noticed. What were they? Peep holes in the Milky Way, or clouds of some dark material that halted the light of stars behind the cloud? E.E. Barnard, 1857–1923, a famous photographer at Yerkes Observatory near Chicago, discussed some "vacant areas" in the Milky Way in 1908, but 10 years later he was convinced that these areas were actually dark clouds (Fig. 8).

Fig. 8 A dark cloud not associated with a bright nebula. Source: Astronomy Picture of the Day, 8 October 2017, FORS Team, 8.2-meter VLT Antu, ESO

What made these clouds so dark? Atomic and molecular gases were thought of first, but these absorbed energy in selective, small wavelength intervals whereas the extinction occurred at all wavelengths. The partial solution came in 1930 when Truempler, a Swiss astronomer who had emigrated to the USA, proved that everywhere in the Milky Way, even between the clouds, a thin gas was present that took away a little bit of the stellar light, and the greater the distance to the star, the less light you see. The darkness also appeared to be stronger in blue light than in red light. After some discussion it was concluded that very small particles of matter (often called "dust") and smaller than the wavelength of light, scattered the rays of light and thus weakened the light of distant stars. The explanation was satisfactory, but created a new problem: How and where were such particles formed? One possibility could be excluded: the particles could not form in the interstellar space; the extreme low density of atoms there prevented such a formation. It was not until 1968 that the answer was found.

Galaxies

Let us now look at the galaxies, starting with the Milky Way, that luminous band that stretches across the whole sky from north to south. The band is made up of stars, light-emitting nebulae, and dark, black nebulae. These black nebulae block the light from stars located deep in the Milky Way seen from the Earth: as a result, we are only able to observe those objects that are close to us. However, in 1918, American astronomer Harlow Shapley produced an indirect but strong argument that the Milky Way has a distant nucleus in the direction of the constellation of Sagittarius, and Dutch astronomer J.H. Oort showed in 1927 that the stars close to us orbit in a circle around that nucleus: the Milky Way was a flat stellar system rotating around an axis far away in the direction of Sagittarius.

For centuries people had been asking whether the Milky Way was in fact the whole Universe that contained everything we could see. Or were there other galaxies just like our Milky Way? In the 1920's Edwin Hubble produced convincing evidence that several nebulae, including the Andromeda nebula (M31) were so distant that they had to be a Milky Way of their own, outside ours; M31 had even the same size as our Milky Way. Objects like these were henceforth referred to as "galaxies." Such galaxies existed in all kinds and sizes—and this is what will occupy us in this book. The larger galaxies (of which ours is one) can contain a hundred billion stars—and there are billions of such galaxies. I will refer to our galaxy as the "Milky Way."

Hubble photographed many galaxies and organized them according to their shape; see Fig. 9. This was an obvious system and it served as the standard for more than 80 years. Yet this system did not acquire the significance that the HR diagram had had. The HR diagram led the research toward a proper understanding of what stars are and what makes up their internal structure. Hubble's tuning fork did not generate any such insights; the reason being that Hubble's tuning fork is largely based on morphology and thus on literally superficial criteria.

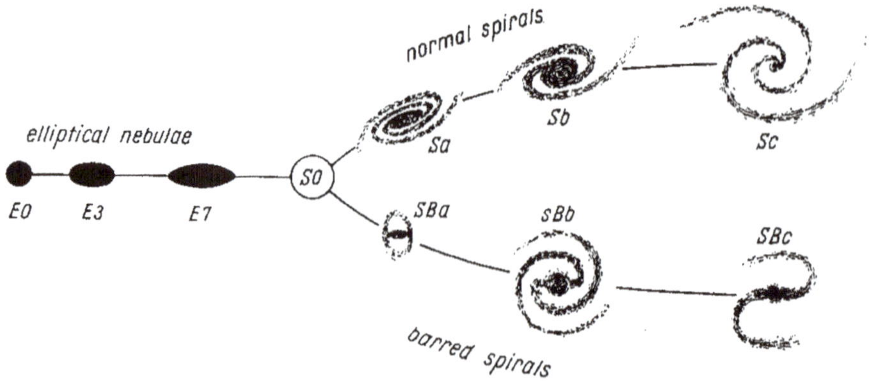

Fig. 9 A diagram showing the galaxies organized according to their shape. Source: Hubble, 1936 (Realm of the Nebulae)

Fig. 10 The elliptical galaxy NGC 205, a companion of the Andromeda nebula. Gas clouds can be seen dimly in the interior. Their presence is unusual. Source: Astronomy Picture of the Day, 8 September 2006. Johannes Schedler (Panther Observatory)

I will limit myself to a division of all galaxies into four main shapes: elliptical, disc, irregular, and dwarf systems. The last two categories do not feature in Hubble's tuning fork.

Elliptical galaxies, see Fig. 10, have a regular, uniform light distribution, and their outline can be described fairly well as an ellipse.

Fig. 11 A rotating thin disk viewed from above: the M101 galaxy M101. The spiral arms are very prominent and the area has a blue tint due to clear blue and young stars. Around the center there is a red area made up of old stars. The sharp, dark-brown threads in the inner areas are of the same matter as the dark clouds within the Milky Way. Source: Astronomy Picture of the Day, 2 March 2006. NASA, ESA, CFHT, NOAO; K. Kuntz (GSFC), F. Bresolin (U. Hawaii), J. Trauger (JPL), J. Mould (NOAO), Y.-H. Chu (U. Illinois)

In second place are the spiral galaxies, thin, flat disks generally surrounded by a weaker elliptical-shaped light. See the Figs. 11, 12, and 13. If you look diagonally from above, you can see the spiral arms (Figs. 11 and 12). If you take a direct side-on view of a spiral galaxy, you will see a black line (the dark clouds in the disk) embedded in a rather elliptical star division: Figure 13. In addition to the disk there is generally a central bulge.

The third category is made up of the irregular galaxies. See Fig. 14, which shows the Small Magellanic Cloud. There is no regular structure to be seen anywhere. The Small Cloud is much smaller than the Milky Way or the Andromeda nebula, and is a companion of our Milky Way. The magenta objects are light-emitting gas clouds around newly formed stars. Just as in the Milky Way, the clouds have this color because of the radiation from the bright spectral line Balmer-alpha of hydrogen. The red objects with a whitish nucleus are nearby stars inside the Milky Way.

These categories were not the end of the story. In 1938, Harold Shapley discovered another type on a photo taken at Harvard's southern station. He saw a collection of weak stars that apparently belonged together. This had to be a galaxy, but one

Fig. 12 Spiral galaxy M81 seen obliquely from above. There is a strong similarity with the Fig. 11. The area of the spiral arms is blue as a result of the young, hot stars. The central area is red as a result of the many old stars found there. Source: Astronomy Picture of the Day, 7 July 2006. Giovanni Benintende

Fig. 13 A side-on view of a spiral galaxy. The colored band running through the center of the galaxy is explained as a thin layer of dark clouds—just as in the Milky Way. Source: Astronomy Picture of the Day, 2012 May 26. Composite Image Data—Subaru Telescope (NAOJ), Hubble Legacy Archive, Michael Joner, David Laney (West Mountain Observatory, BYU); Processing—Robert Gendler

Fig. 14 The Small Magellanic Cloud. Source: Astronomy picture of the day, 17 June 2005. Josch Hambsch, Robert Gendler

with so few stars per unit of area that you could see through them. Shapley's galaxy was the first of a new category, the "dwarf galaxies"; see Fig. 15. These dwarf galaxies are difficult to find, and new ones are continually being found, even today. Detecting them is so difficult because it is all too easy to see through them. A sketch like these is clearer than a photo.

Large: Groups of Galaxies

Galaxies often appear in groups called clusters. These can be small clusters with a small number of galaxies or very large with hundreds of them, like the cluster in Virgo. Figure 16 shows part of the Virgo cluster.

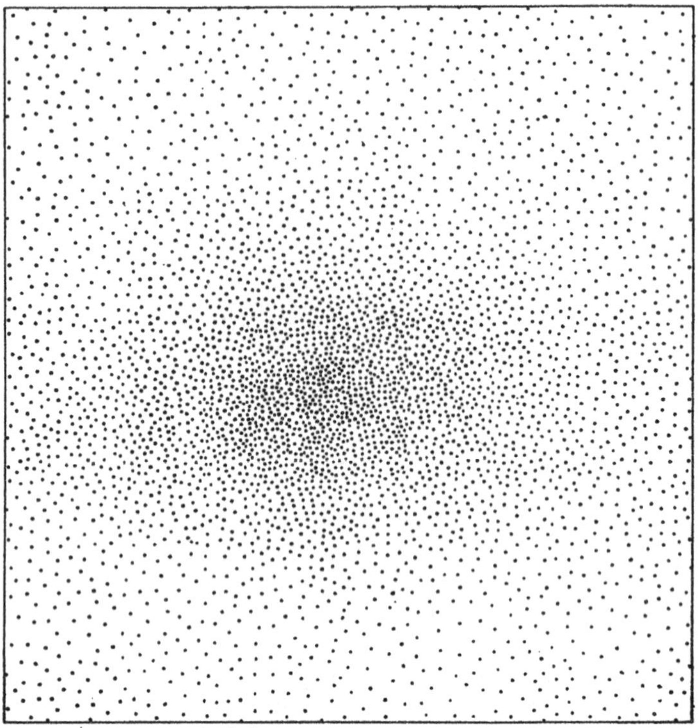

Fig. 15 A sketch of the stars in the dwarf galaxy in the constellation of Sculptor (Shapley 1938)

Fig. 16 An area in the constellation of Virgo photographed with the Hubble space telescope. There is a cluster of galaxies, but there are also many bright, mainly blue, spots. These are not galaxies, but rather individual stars within the Milky Way. Source: Astronomy Picture of the Day, 22 April 2012. Rogelio Bernal Andreo

Clusters of galaxies were fine examples to which the theory of statistical mechanics could be applied, a theory that was developed particularly in the late nineteenth century and that questioned how a cluster of objects behaved that attracted one another through gravity. In 1933, Zwicky—we have come across him previously—wrote an article in German in a relatively unknown Swiss magazine, in which, in addition to a number of other cosmological issues, he also determined the total mass of the nearby Coma cluster by adding together the masses of all the galaxies present there. On the basis of the physical theory, he then calculated what the (quadratic) average velocity of the systems should be, and came up with 80 km/s. But the observed average was 1000 km/s!

This result was totally unexpected and Zwicky could not find any explanation. He, therefore, suggested that the cluster contained a lot of invisible matter, 400 times more than the matter emitted by light. There was no other indication whatsoever that such "dark matter" existed and Zwicky's assumption was not taken seriously. What helped was that Zwicky was a somewhat antisocial and difficult person and a loner. Several renowned astronomers looked at the problem again with new data and decided that Zwicky's measurements and reasoning were unobjectionable. In the 1970s, the same problem, "there is more matter than you can see," was found in very different observations but it still took a while before others dared to confirm the existence of dark matter. And so it happened: it took the astronomical community 50 years before Zwicky's conclusion was accepted.

Biggest: The Universe

And finally, there is the Universe, space and its contents occupied by galaxies as far as we can see. The Universe had become highly interesting to theoretical physics following the publication in 1916 of Einstein's "Algemeine Relativitätstheorie" ("General Theory of Relativity"). It incorporated Newton's 1687 theory of gravity and expanded upon it. In developing the theory, Einstein had found a model of the Universe that was fully consistent with his theory—Newton had not been able to find such a model that was consistent with his theory: it would collapse. Newton, a very religious person, had drawn the conclusion that God interfered so that the collapse would not happen, an explanation that was not appreciated in the world of physics.

In the years around 1916, Einstein spent several weeks a year with his friend Paul Ehrenfest in Leiden. The First World War was raging, but Holland had remained neutral and was a haven of tranquility to Einstein who was a confirmed pacifist and absolutely opposed to the war. In Leiden, he also had discussions with Lorentz, his scientific godfather, for whom he had enormous respect. He discussed his model with Willem de Sitter, director of the Observatory at Leiden University. De Sitter was originally a mathematician and it was no problem for him to understand Einstein's new theory. He told Einstein that his model had a major flaw: it would collapse, a problem similar to that encountered by Newton. Einstein reflected on de Sitter's criticism and finally agreed. His theory contained only a single force, gravitational attraction, and it needed a pressure to keep the galaxies apart. He therefore introduced a new force and in his mathematical equations he symbolized it with the Greek capital Λ. The trick worked, but Einstein had a bad conscience: it was poor science to introduce a totally new concept to solve existing problems without having additional arguments in favor of one's hypothesis.

With hindsight, the de Sitter–Einstein problem arose because both scientists were absolutely convinced that the Universe was unchanging, eternal. This was not what most religions, including Christianity, had always accepted as a fundamental truth: that there had been a beginning of it all. The Earth, the Sun—they all had been created at some point in the past.

Still, the idea of a changeable Universe made its reappearance when theorists simplified Einstein's mathematical equations on the assumption that the Universe is the same everywhere and in all directions (in technical language: homogeneous and isotropic; the so-called cosmological principle). In 1922, Friedmann, a Russian astronomer, applied this assumption and derived simpler mathematical equations that clearly showed that a changeable Universe was physically possible. His work appeared in a Soviet publication and in the confusing early days of the USSR it went unnoticed in the West. Friedmann died in 1925, at only 37 years old. A few years later, a young Belgian priest, Georges Lemaître, rediscovered the same equations, not knowing of Friedmann's work. Lemaître's article, published in a French-language Belgian magazine, was also ignored at first. But a few years later when Eddington along with his student McVittie in Cambridge was again studying the problem, he was alerted to a reprint of Lemaître's article that lay unread somewhere in his office. Eddington generously recognized the contribution made by Lemaître, whose reputation was now established. The result was in any event that in theory the Universe was no longer regarded as eternal and immutable.

In 1929, Hubble made an unexpected and remarkable discovery, only a few years after he had proven that some nebulae (NGC6822 and M33) were far outside the boundaries of the Milky Way. He had recorded the spectra of many galaxies and saw that the spectral lines all had longer wavelengths than those of the same atoms measured in the laboratory. This could easily be explained by the so-called Doppler Effect, an effect predicted in 1842 and since seen in many stars. When a star approached the Sun, the atomic lines in its spectrum had a shorter wavelength (i.e., they were "blue-shifted"), but the same lines had longer wavelengths (they were "red-shifted") when the stars moved away. Observers noticed that about as many stars had blue-shifted lines as had redshifts. But in galaxies this was surprisingly different: all spectra showed only red-shifted lines; only in a very few nearby galaxies were blue shifts found. The conclusion was therefore that all galaxies were moving away from us. Hubble's diagrams showed that the velocity "V" at which the galaxies move away from us was proportional to the distance "d"; in algebraic form: $V = H \times d$. The equation became known as the "Hubble equation" and the constant "H" as the "Hubble constant" (Fig. 17).

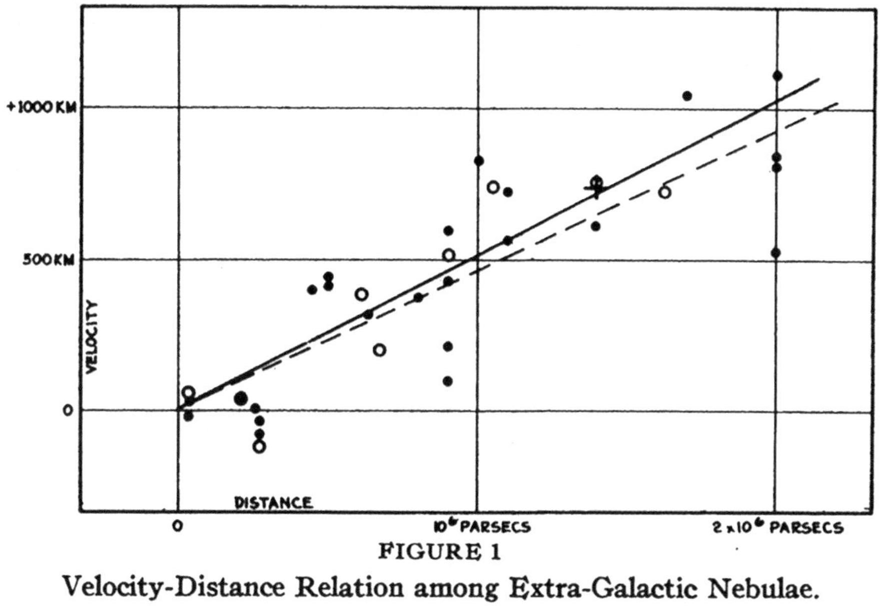

Fig. 17 The famous figure that Hubble published in 1929, which proved that the Universe had a definite beginning and that gave mankind a fact to wonder at. The dots are galaxies. The distance as determined through the medium of variable stars is given horizontally. Vertically is the velocity of recession as measured from the Doppler shift of the lines in the spectra (Hubble 1929)

The conclusion that the Universe was expanding was quite remarkable. The opposite conclusion, drawn most effectively by Lemaître in 1931, was even more remarkable, namely that in the past the galaxies had been packed much more closely together: in "a primeval atom", according to Lemaître. The atom exploded and formed the Universe that we observe. At first Einstein was not amused. He wrote to Lemaître (in French): "Your mathematics is OK, but your physics is abominable." One should not be surprised: the idea that the Universe had a definite beginning contradicted the basic—but tacit—assumption in science and it was a complete reversal of what Einstein and de Sitter had discussed in around 1916. Hubble's observations were considered convincing. The idea that the Universe had a definite beginning was generally considered to be remarkable and it has remained so ever since.

This turnaround in scientific attitude had one consequence to be remembered: since the Universe was expanding, there was no longer any need in Einstein's theory for a pressure term symbolized by Λ. Einstein happily withdrew the symbol and is supposed to have said, according to Gamow, "Introducing Λ has been my biggest blunder." Sixty years later, luckily after Einstein had died, Λ was reintroduced, as we will see in Chap. 16.

Without the Λ, the equations derived by Friedmann and by Lemaître, and now known as the Friedmann-Lemaître-Robertson-Walker equations, were proven to have three different solutions for the mathematical structure of the Universe. In each solution the expansion of the Universe must slow down, thanks to gravity. If there were enough matter in the Universe, and therefore gravity was strong, then the expansion would stop at some time in the future; the Universe would turn around and begin to decrease in size until it had regressed to the primeval atom again. If there were not enough matter, the Universe would expand forever; the expansion would slow down but not enough to stop. Between these two possibilities was the "critical" solution: the expansion would reach velocity zero at time infinity. The critical solution is sometimes referred to as the Einstein–de Sitter Universe, since in 1932 these two proposed the model as a "toy model" for further use.

The question: "In which Universe are we now living?" was for a very long time the only important question for cosmological research. Even in 1961, Hubble's successor, Alan Sandage, wrote a widely read paper entitled: "The ability of the 200-inch telescope to discriminate between selected world models." Little progress was made in cosmological research for more than 30 years; the breakthrough came in 1965 and the victory was completed in 2000. Read on! (Fig. 18)

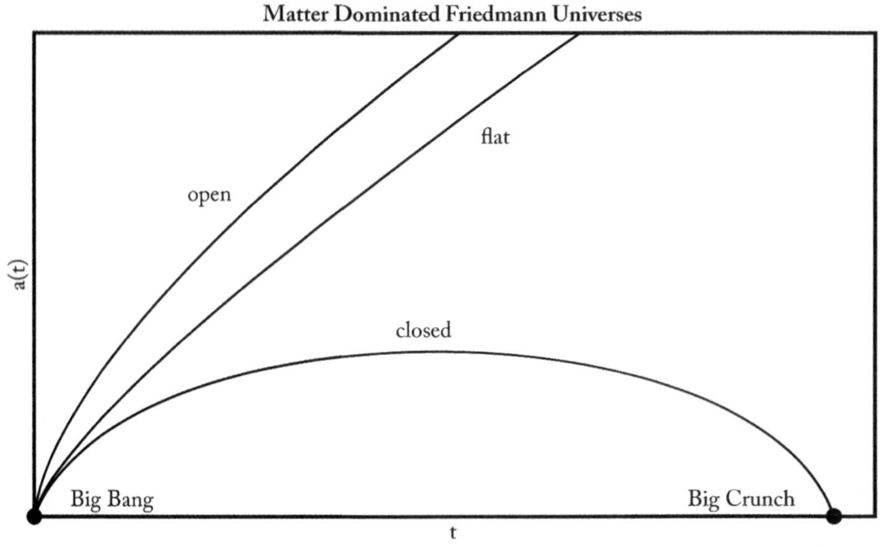

Fig. 18 The expansion of the Universe from theory. Horizontally is the time, starting at the Big Bang. Vertically is a yardstick of size a. Give it a definite value now. Then it will have been smaller in the past and it will be larger in the future. How fast it grows and how it will behave in past and in future is mathematically determined by the Friedmann-Lemaître-Robertson-Walker equation. To solve the equation one must know the present (average) matter density ρ in the Universe and the present rate of expansion H (the Hubble constant). Measurements of sufficient accuracy were obtained around the year 2000; until then all possibilities had to be left open. The two major possibilities were a 'closed Universe' (that would contract after maximum expansion and ending in the 'Big Crunch) or an 'open Universe' that would expand forever. The border curve between open and closed was called the 'critical Universe'. It would expand forever, but approach zero expansion at time infinity. The geometry of the Universe was different in the three cases. Only in the critical Universe the geometry was 'flat', i.e the geometrical relations followed Euclid's rules

With an eye to later discussions, a new parameter is introduced here, Ω, which is defined as the ratio between the actual density in the Universe, ρ, and the critical ρ_{crit} so that $\Omega = \rho/\rho_{crit}$. If $\Omega > 1$, the Universe is closed, and if $\Omega < 1$, then it is open. The critical, or Einstein–de Sitter model assumes $\Omega = 1$. For several decades observers argued that there is too little matter to close the Universe (thus: $\Omega < 1$), but eventually (in around 2000), the Universe will be shown to be critical, $\Omega = 1$.

This is also a good moment to introduce the parameter z that describes the redshift. The definition is $z = (\lambda - \lambda_0)/\lambda_0$. Here λ is the wavelength of a spectral line measured by the observer and λ_0 is the wavelength at which the line was emitted, or the wavelength one would measure in a laboratory. "No redshift" means $z = 0$. For small values of z, the velocity at which a galaxy moves away from us is given by $v = z \cdot c$, with c the speed of light. Later we will see redshifts z larger than 1. Simplistically this would imply that v is larger than the speed of light. The theory of relativity prevents this from happening and shows that the relation between z and c is more complex and not as simple as above.

Readers with some understanding of differential equations and of integrals will find very instructive lemmas in Wikipedia. Start with the lemma on the "Hubble constant" or on the "Friedmann-Lemaître-Robertson-Walker equation."

The Hubble Constant: A Warning

The Hubble equation and the Hubble constant have been introduced above. Hubble himself found for H the value of 550 km per second and per megaparsec. Let us express this in words: a galaxy that is at a distance of 1 million parsec (3.3 million light-years) is speeding away from us at a velocity of 550 km/s; at a distance of 2 million parsec the velocity is 1100 km/s, etc. At a congress of the International Astronomical Union in Rome in 1952, Baade reported that the value of H was at least two times smaller. This caused something of a sensation but it was not yet the end of the story; the discussion carried on. In around 1960, there were two possible values for H, each of which had its supporters. One group "proved" that $H = 100$ km/s/Mpc, the other "proved" $H = 50$ km/s/Mpc. The dilemma was eventually resolved in around 2000. The value of $H = 72$ km/s/Mpc was the result of a long program of observations by a large group of astronomers who had spent many hours using the Hubble Space Telescope. Even in 2016, there is still an uncertainty of ±5 km/s/Mpc in the value of H. The 50-year-long discussion on such an elementary measurement as that of the Hubble constant shows how difficult it was to make progress in cosmology.

Let us make a small detour, with an eye to the future. Imagine that there are two galaxies—A and B—at a short distance d from one another. The velocity v at which B distances itself from A is $d \times H$. But now look at the past: a time d/v ago their distance was 0: we arrived at the beginning of the Universe. Now $d/v = 1/H$ and thus $1/H$ is a simple estimate of the age of the Universe. If $H = 75$ km/s/Mpc, its age is

13.7 Gjr, a value that is about right when compared with modern values obtained by more detailed arguments.

Despite this standstill, the discovery of a clear beginning of the Universe had a major impact on the culture of the entire West. Philosophers, for example, were affected by this sense of finiteness. It was a conclusion of similar importance to Copernicus' statement that the Earth revolved around the Sun. Remarkably enough, the Roman Catholic Church kept out of the discussion about the Big Bang and thus prevented a repetition of the Galileo tragedy (and blunder). It is said that this was mainly thanks to the priest Lemaître who had good access to the Pope and to the Roman Curia.

A Comment in Retrospect

In conclusion, a comment on the awareness of the history of the Universe, an awareness that was barely present in 1945 but that was later to become so important.

In 1945, the mission of astronomy, or so it seems to me now, was to describe the Universe without thinking about its possible evolution. All the attention was focused on the Universe today. Nonetheless, there were a few scientists who did give careful thought to the concepts of time and history. One of these was the great German astrophysicist Unsöld (certainly not a Nazi), who wrote an impressive article in 1944 (that was not published until 1948) in which he investigated what we already knew about temporality in the cosmos. The article is a fine example of a systematic and in-depth exploration of a fundamental problem: the history of the Universe. It was a question that was being thought about, but for the time being it was impossible to look farther back than just a few percent of the age of the Earth.

References

Baade, W., Zwicky, F.: Proc. Natl. Acad. Sci. **20**, 254 (1934)
Barnard, E.E.: Astron. Nachr. **177**, 231 (1908)
Barnard, E.E.: Astrophys. J. **49**, 1 (1919)
Hubble, E.P.: Astrophys. J. **62**, 409
Hubble, E.P.: Astrophys. J. **63**, 236
Hubble, E.P.: Astrophys. J. **69**, 103
Hubble, E.P.: Proc. Natl. Acad. Sci. **15**, 168 (1929)
Shapley, H.: Nature **142**, 715 (1938)
Unsöld, A.: Z.A. **24**, 278 (1948)
Zwicky, F.: Acta. Helv. Phys. **6**, 110 (1933)

Part I
1945–1984: Anything Goes

In the years 1945–1985 the Columbuses of the sky sailed out in new directions and came back with big surprises; most were totally unexpected discoveries and a few had been suggested long ago by fantasists you would not trust. These were the pioneering years, to which the title of a 1930s musical is applicable 'Anything Goes'. The heyday was between 1960 and 1975, when the most spectacular findings were made, some within a few years of each other.

I have split these 40 years in three periods: 1945–1960: the running start; 1960–1975 the great discoveries; 1975–1985: reflection and preparation for further study. In each period there is a chapter on instruments, a second on newly forming stars, ordinary stars and their remants and a third chapter on galaxies and the Universe.

Chapter 1
1945–1960: New Optical Instruments and the Beginning of Radio Astronomy

1.1 Optical Astronomy

Even before the War it was clear that the biggest and best-equipped telescopes were in North America, and specifically in the US. After the War this contrast between the US and other countries became even greater, as a comparison with European countries will prove.

There are three main quality criteria to judge how effective a telescope is. The first is its location: this is what determines the percentage of photons (light particles) that pass through the atmosphere to reach the telescope. The best locations are those that are both high and dry. Secondly, a large mirror will receive more photons than a small one and often it is the faintest signals that are the most interesting. Thirdly, you need to have the best possible 'receiver', the device that counts the number of incoming photons. Until around 1980 this device was a photographic plate. It has to be said that the largest European telescopes in 1945 were at deplorable locations and were also smaller than the American ones. Not only this, Europe was also falling behind in detector developments.

The gap became painfully obvious when in 1948 two telescopes were commissioned in the US that were both the biggest of their kind: the 200-inch Hale telescope and the 48-inch Schmidt telescope. A meagre consolation for Europe was that even in the US these telescopes could only be used by a 'fortunate few'.

The 200-inch Hale Telescope and the 48-inch Schmidt

The biggest and most famous telescope had a diameter of 5 meters (200 inches) and was located on Mount Palomar in Southern California. It was officially commissioned in 1948 and belonged administratively to the Mount Wilson observatory north of Los Angeles. This telescope was also the fourth largest to be constructed

H. J. Habing, *The Birth of Modern Astronomy*, Historical & Cultural Astronomy,
https://doi.org/10.1007/978-3-319-99082-8_1

with private means that George Ellery Hale (1868–1938) had managed to acquire: the first of these, at the time the largest lens telescope (40 inch diameter) ever, was constructed in 1892 in the new Yerkes observatory to the north of Chicago; the second—in the much better climate of Southern California—was the 60-inch in 1908, and thirdly the 100-inch in 1919, followed finally by the 200-inch. Hale had acquired the funds for this last telescope from the Rockefeller Foundation as early as in 1928, but the War caused a serious delay in the construction. Hale did not live long enough to attend the opening in 1948, nor was he to know that the telescope would be named after him (Figs. 1.1 and 1.2).

The 200-inch was thought to be the biggest telescope that could ever be constructed. The glass weighed 40 tons and needed three diesel trucks to transport it up the mountain. A bigger telescope would be even heavier: a mirror twice as large would contain 2^3 (in other words 8) times more glass. Pouring such a glass disk, making sure the melted glass cooled down without breaking, suspending the polished mirror in a smooth mechanical construction (directing the telescope precisely): this seemed an impossible task. These arguments were confirmed in 1975 when the Soviets constructed a telescope with a diameter of 6 meters. This never became a real competitor of the 200-inch as you can read if you look up item BTA-6 in Wikipedia.

Hale had managed to persuade financiers to fund the new telescope, one of his arguments being that Hubble's research on the most distant galaxies could go much

Fig. 1.1 George Ellery Hale (1868–1938) was the founder of three major observatories (Yerkes, Mount Wilson, Mount Palomar) and found four times the funds to build the largest telescope in the world: the 1 m refractor of Yerkes Observatory (1897); the 60-, 100-, 200-inch reflectors in 1908, 1917, 1948; the first two at Mount Wilson and the third at Mount Palomar. Source: Caltech Archives/Mount Wilson Observatory

Fig. 1.2 The figure shows the open dome of the '200-inch' telescope during inauguration in 1948. The human figures that can be seen at the bottom right show clearly the size of the instrument. To the left, in the center, we can see the yoke that rotates on an inclined axis. A large cylinder runs from the yoke diagonally through the figure that supports the axis of the cage in which the mirror is suspended. The yoke and the axis of the cage can rotate over a wide angle while maintaining the required precision, although the person making the observations still has to do some fine-tuning. Source: Caltech Archives/Palomar Observatory

further with the 200-inch than with the 100-inch telescope. He argued that the 200-inch would make it possible to observe the very limits of the Universe. We can say with hindsight that although the 200-inch met all the requirements and was therefore technically a success, it never fulfilled the hope of reaching the furthest limits of the Universe. The definitive breakthrough in cosmology came between 1964 and 2000 from a not so large radio telescope, several space telescopes and small telescopes suspended under a balloon.

The 200-inch had a parabolically curved reflector: this reflector, an aluminium layer on a polished surface of glass, gives the best image at the focal point, but the quality of the image deteriorates when you look outside the focal point. A parabolic telescope produces undistorted images in an area of at most tens of arc minutes, which is too small an area to see other nearby objects that may belong to the object under investigation. A second telescope was therefore needed on Mount Palomar that might not be able to look so deep into space, but that could produce very sharp images, even up to a degree or more from the main object. The decision was taken to have a 48-inch (1.2 m) telescope constructed based on the design by the genius German/Estonian optical expert Bernhard Schmidt (see Box 1.1). This telescope met the requirements, but unfortunately Schmidt himself did not live to see the success of the 48-inch; he had already been dead for 10 years by the time the telescope was commissioned.

Box 1.1 Bernard Schmidt and his Telescope

Bernhard Schmidt (1879–1935), was born in a Swedish-speaking community on a small island off the coast of Estonia. At home his family spoke German. Schmidt had an inquisitive mind; as a young boy, he liked to do chemistry experiments, but one particular experiment carried out when he was 15, had dire consequences. He was experimenting with a cylinder filled with buckshot when something went wrong, costing him an arm.

After some wanderings, he settled in Germany and started to earn a wage by polishing lenses. What a remarkable idea, a one-armed lens polisher! For a number of years this was the way he earned a living but during the great depression of the 1920s he was forced to accept a poorly paid job as an instrument technician in the observatory in Hamburg. In 1930 he designed a wide-field reflecting telescope. He based this instrument on a spherical mirror, rather than the usual parabolic mirror, and placed a specially constructed lens at the center of the curvature of the mirror. This gave the telescope a sharp wide-field image of several square degrees. During a boat trip back from an eclipse expedition, Schmidt discussed his ideas with the then young astronomer Walter Baade, who had just arrived in Hamburg. After returning home, Schmidt built the telescope he had envisaged, with a mirror that had a diameter of 44 centimeters. The telescope did indeed give the large wide field that he had intended.

Not long afterwards, Baade was given an appointment at the Mount Wilson observatory in California. He worked together with Fritz Zwicky, a Swiss physicist employed by the California Institute of Technology in nearby Pasadena. Together they discovered the existence of 'supernovae', stars that could become thousands of times brighter in less than a single day and slowly extinguished again after this 'explosion'. It was a sensational discovery. To prove their theory, they needed data on many more supernovae than had been observed up to that point. However, supernovae are very rare. If you wanted to see them flare, you had to continuously monitor a large number of galaxies and immediately spring into action if a supernova flared in one of them. Baade and Zwicky came up with the idea of using a telescope with as wide a field of vision as possible that could photograph a whole group of galaxies at one time. Then Baade remembered the discussion he had had with Bernard Schmidt. Zwicky organised the funding needed to construct an 18-inch Schmidt telescope, which was commissioned in 1936 on Mount Wilson. It met all expectations, firmly establishing Schmidt's reputation. Sadly, he had died a short time before this.

After 1945, a number of observatories were equipped with a Schmidt telescope. The German Democratic Republic confirmed its technical ability by locating the world's largest Schmidt (with a 2-meter mirror) in Thüringen, unfortunately at such a disastrous location that the telescope barely made any contribution at all to astronomy research. The test telescope that Schmidt built was used by the German war navy in the Second World War to photograph ports from an aircraft. The telescope, like its fatherland, did not survive the war (Fig. 1.3).

Fig. 1.3 Schmidt in one of his workshops. Source: D. Reimers, Hamburg Observatory

Photo-Electric Light Measurements in Place of Photochemical Measurements

In around 1880, photography started to make its mark on astronomy, in the form of a glass plate covered with an emulsion of silver chloride or a similar compound. When a light particle (photon) impacted a silver chloride molecule, the molecule could fragment. After exposure, the remains of the molecule were removed by means of a chemical process. It was then possible to see where light had fallen and where not. The contribution from photography had been an enormous advance in 1880 but ultimately it proved to have its limitations. No two photographic plates ever had the same sensitivity; each plate had to be calibrated separately, a time-consuming procedure. Next, the photographic plate was basically not sensitive enough: only 1 in 1000 photons caused a photochemical reaction. There were other ways of detecting photons, one of which I will mention that played an extremely important role between 1945 and 1980. When light falls on certain metals, electrons are released from the surface (the photo-electric effect); these electrons cause an electrical current whose strength can be measured precisely. At the Berlin observatory in 1914, P. Guthrick had the light from a star fall on such a sensitive surface. The object was an extremely bright star, magnitude 3, seen through a 30-cm telescope. The star caused a measurable electrical current and the strength of the current enabled Guthrick to calculate the magnitude of the star with an unprecedented precision of 1%, much more precise than is possible with a photographic plate. But even though this 'photo-electric' detector was much more precise, it was also much less sensitive than the photographic plate. In 1945, 30 years after Guthrick's measurement, American astronomers Joel Stebbins and Albert Whitford published a table with very precise measurements of the magnitude of 238 stars, but the large majority of these stars were brighter than magnitude 7. And even in 1945 astronomical research needed precise measurements of stars that were at least 10,000 times fainter.

What was required was a way of enhancing the photoelectrical effect. An instrument that was capable of this was the 'photo-multiplier' that was devised shortly before the Second World War. When a photon collided with a sensitive surface and released an electron, this was accelerated by an electrical field; it then collided with a subsequent surface and released more electrons; these in turn collided with the following surface, and so on. A cascade of electrons was thus set off by a single photon. In 1946, American astronomer Gerald Kron wrote that RCA (Radio Corporation of America) had launched a 'photon multiplier', the 1P21, onto the market. This name would remain legendary for several decades within astronomy. In 1950, Stebbins gave the George Darwin lecture in London and reported that he had measured the magnitude of stars at magnitude 19 and the precision was within 1% of a magnitude, a precision that was believed to be impossible just a few years earlier.

Even after the 1P21 was commissioned, photo-electric photometry still had a significant handicap, namely that it was very time-consuming. Each star had to be measured separately and because each measurement took at least several minutes, and sometimes even up to 15 minutes, only a relatively few stars could be measured in the course of a night. This caused problems when, for example, astronomers wanted to measure the magnitudes of all the stars in a compact cluster of stars with thousands of members. As a workable compromise, from then on a photographic plate of the star clusters was made and the magnitudes of a limited number of individual stars were measured photo-electrically and used as the standard for calibrating the photographic plate.

The 'Palomar Observatory Sky Survey': A Public Treasure Trove

Once it was commissioned in 1948, the large 48-inch Schmidt telescope on Mount Palomar was used primarily to make a systematic photographic map of the sky visible above South California. Blue-sensitive and red-sensitive plates were taken showing stars up to magnitude 22: 12 magnitudes or 60,000 times fainter than the faintest stars in the illustrious 'Bonner Durchmusterung' of around a century earlier. After recording the last of the almost 1900 plates on Mount Palomar, a large number of prints were made of each plate and these prints were shipped to observatories that could pay the costs of copying, packaging and transferring them. Although the costs (the amortization of the telescope, the manpower) were not passed on, the copies were too expensive for many institutes. Observatories in Western Europe, Australia, Japan and South Africa could afford to buy a set, but behind the Iron Curtain, there was just one copy per country.

This 'Palomar Observatory Sky Survey' (POSS) proved to be highly significant. It was a rich archive in which every researcher could look to his heart's content for his favorite objects. Elsewhere in the world, catalogues were made of clusters of galaxies, special galaxies, dark clouds, planetary nebulae, white dwarves, weak red stars, etc. We will come across the results of some of these systematic searches further on. The POSS was later supplemented with photographs of the southern sky using a British-Australian Schmidt telescope in Australia and one at the European Southern Observatory in Chile. In the nineties, all the plates were digitized and made freely accessible via internet. They can be found, for example, on http://cdsweb.u-strasbg.fr, by clicking on 'Aladdin'. After the POSS survey, a new exploration of the sky followed 50 years later; again this was an American initiative. We will come across this 'Sloan Digital Sky Survey' at the end of the book.

1.2 The Rise of Radio Astronomy

At the start of the 1930s, Bell Telephone Company in the US instructed their radio engineer Karl Jansky to trace the sources of the interference in radio telephony. The possibilities considered included storms and human activity such as the construction of electric tram routes. Jansky confirmed these suspicions but also discovered that part of the interference came from the Milky Way; he even found that the strongest radiation came from the region of the center of the Milky Way. Jansky's discovery made the front page of the New York Times on 5 May 1933 but this did not prevent his employer, 'Ma Bell', from scaling down his research. And because astronomers showed no interest either, radio astronomy seemed to be dead in the water. But fortunately a radio amateur, Grote Reber, took up the challenge a few years later. He was a 25-year-old bachelor who was still living with his mother when he built a parabolic antenna in her garden in Wheaton, Illinois, in 1937. Using receivers that he also built, he charted radio waves from the cosmos at a wavelength of 2 metres and in 1940 and again in 1944 published a celestial map showing the contours and radio waves in the highly reputable Astrophysical Journal. When the editor of this journal had received Reber's manuscript he was in doubt. None of the editors understood anything about radio techniques. Was this paper important or was it produced by a crackpot? A committee of the editorial office went to visit Reber, saw his instrument and heard him talk about his measurements and then decided: this was good science. And so the journal published the manuscript. They were right. After the war both articles played a significant role in the revival of radio astronomy but Reber himself no longer contributed very much to these developments.

Although Reber's publications were widely read, they received no reaction from any of the established astronomers, with one exception: Dutch astronomer J.H. Oort, as we will see. The stresses and strains of the war probably contributed to this lack of interest, even though Reber's articles did contain some explosive material: radiation from the nucleus of the Milky Way was the first direct detection of that nucleus (with the exception of Jansky's forgotten result). Not only this, the radiation had a characteristic that current knowledge was unable to explain: at longer wavelengths radiation was stronger, not weaker, as was expected from all known radiation mechanisms. This fact alone would be reason enough for further research. It may be that these measurements with a radio antenna were a bit too remote for most astronomers, but it could also be that people thought that radio radiation is so low in energy that detecting it will lead us to minute, low-energy processes. The reasoning probably went something along the lines of: 'There will be an explanation, but then that will be the end of it. It has no significance for "real" astronomy.' E.G. ('Taffy') Bowen, Director of Radiophysics of CSIRO (Commonwealth Scientific and Industrial Research Organisation) in Australia, where radio astronomy got off to a flying start in 1945, recalled a lecture on 'The future of astronomy in Australia' by the future 'Astronomer Royal' in England, Sir Richard van der Riet Woolley (1906–1986). When asked what the significance of radio astronomy would be in 10 years' time, Woolley replied: 'It will be forgotten'. Many 'traditional' astronomers probably shared his opinion.

Let us return to radio astronomy. Shortly before the outbreak of the Second World War radar had been discovered in England and Germany as a means of picking up enemy aircraft long before they became visible. You transmit a pulse of radio waves and, if aircraft are approaching, the metal of the planes reflects the pulse, and the reflected signal is then received by the radar system. During the War this new technology was developed further by both sides, including the Australians, with large teams of specialist engineers and technicians being put together for this purpose. After the War their task was more or less over. The victors banned any further radar research in Germany, but on the allied side, in England and Australia, radar experts were permitted to direct their antennas to the Sun (at times during the war the Sun disrupted radar communications) and also to the rest of the universe. This led to the establishment of the first radio observatories in Sydney, Jodrell Bank (England) and Cambridge (England). It was these radar specialists who continued the pioneering work of Jansky and Reber. Originally they knew nothing about astronomy, but they were inquisitive and were also keen to use their expertise to meet the challenge of finding out where radiation comes from. In 1945 radio astronomy, that had been destined to die, underwent a revival.

It is remarkable that over a period of just a few years a new branch of astronomy developed that gave an enormous impetus to the whole field—and this was achieved by people who knew practically nothing about astronomy. Box 1.2 describes the 'wild' early years in Australia. Similar stories are told by English radio astronomers. Graham Smith, one of the most important pioneers on the English side, wrote about his pioneering period: 'It took a long time before we really understood that we had become astronomers.' It may have taken a long time, but ultimately two of these pioneers, Martin Ryle (1918–1984) and Francis Graham Smith (born in 1923), became successively 'Astronomer Royal' in England, the country's highest honor for an astronomer, and both were given a noble title. Ryle was also awarded the Nobel Prize in 1974 for his contribution to the development of radio astronomy.

Box 1.2 The Second World War and Radio Astronomy in Australia

During the Second World War, in Australia, just as in other war-torn countries, intensive efforts were made to improve the still very new radar systems. In Australia this research was carried out in the division for Radio Physics of the Council for Scientific and Industrial Research (CSIR). Immediately after the end of the War, the chairman of CSIR decided not to disband this radar department, but to give the staff the opportunity to engage with new research themes. This meant maintaining an excellently equipped laboratory with around 200 staff of radio engineers and technicians who were also experts in research. Around 30% of the staff bade farewell to radio physics at the end of 1945 and returned to their pre-war occupations, but 70% remained. For those who remained, a new goal had to be found. This was initially solar radiation, but later it encompassed the whole of radio astronomy. E.G. ('Taffy') Bowen, born and educated in England and head of Radio Physics at CSIR between 1946 and 1971, recalled that there were no astronomers among the scientific staff, and even more surprising, none of the staff had ever attended a single lecture in astronomy.

Bowen mentions a detail that typifies the first chaotic years after the War: in Australia nobody knew what to do with the enormous stockpile of radar and radio communications equipment. As a stopgap they started loading it onto an American aircraft carrier to dump it several kilometres out in the Pacific Ocean. Fortunately, Radio Physics managed to rescue a large part of this equipment in time.

One of the most important pioneers on the Australian side was John Bolton (1922–1993). He was born in Sheffield (England), graduated in 1942 in physics from Cambridge and shortly afterwards started to work at the secret 'Telecommunication Research Establishment', where he met many of the later pioneers of radio astronomy. In 1944, at the age of 22, he was assigned as a radar expert on the aircraft carrier 'HMS Unicorn' in the Pacific Ocean. Once the war was at an end, Bolton decided to stay in Australia and joined the Radio Physics.

Bolton's direct boss was 'Joe' Pawsey (1908–1962), deputy head of Radar Physics. Pawsey was a highly respected researcher with a paternal charm, but he intervened strongly when Bolton and his colleague Gordon Stanley used their telescope to look for cosmic radio sources, rather than looking at the Sun, as Pawsey had directed them to do. They were recalled from their outpost in Dover Heights near to Sydney to the laboratory where they had to prepare equipment for an eclipse expedition to Brazil. The following is a fragment from the autobiographical notes made by Bolton in 1982. 'Towards the end of February 1947, Pawsey came into the room where Gordon and I were working and told us that the expedition to Brazil was not going to take place. He then said, 'If you can think of anything to do with all the equipment, you

(continued)

can have it.' As he reached the door he turned around and, almost as an afterthought, and in typical Pawsey fashion, said, 'If you can think of anything to do with Gordon Stanley, you can have him too.' The opportunity was too good to miss; we spent the afternoon loading everything we had built on a truck, together with tools, spares and test equipment and early the following morning we were unloading in the former Army blockhouse on the edge of the cliff at Dover heights. By 11 a.m. we had built a one-valve (….) receiver for the broadcast band, since a Test Match between England and Australia was due to start at that time. We then started to install the solar receivers. The day we had everything in working order was when the largest bipolar spot for some years appeared on the limb of the Sun….' (end of quote). The incident has something rather boyish about it, but it did get results. Shortly afterwards Bolton and Stanley showed with their equipment that the radio source known as Taurus A (or Tau-A) coincided with the well-known Crab nebula. This nebula was to play a key role in integrating optical and radio astronomy (Fig. 1.4).

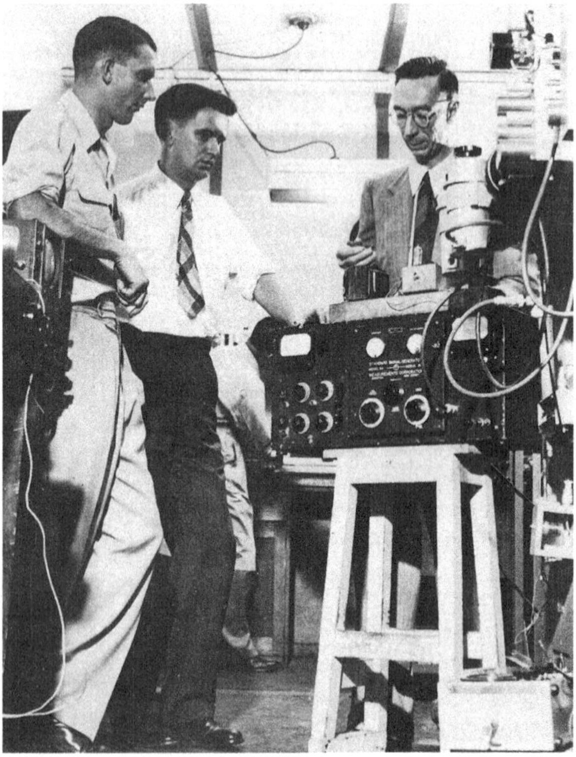

Fig. 1.4 A photo showing from left to right Bolton, Stanley and Pawsey at around the same time as described in the quote in the text. Notice how young everyone is. Source: 'The Early Years of Radio astronomy' by Woody Sullivan

The Existence of 'Radio Stars'

In the first radio maps, such as those produced by Reber, the contours on the chart seem like a weathered old mountainous area. It reminded the observer of drawings of the Milky Way and this led to the supposition that the radiation was a phenomenon produced in the wide-open space between the stars, 'interstellar space'. This space, as one knew, contained tenuous clouds of gas and 'interstellar dust particles' (see the Prologue) and thus it was obvious that attempts were made to explain the radiation by processes occurring in this gas and dust (Greenstein et al. 1946). The attempt was not successful although 10 years later such radiation was detected but then at wavelengths expressed in centimeters (20 cm e.g.) and not in meters (4 m e.g.), the radiation studied in all of radio astronomy in these early postwar years.

It therefore came as a surprise when in 1946 English radar specialists, Hey et al. (1946) found a small radio source in the direction of the constellation of Swan (Cygnus). As an aside: the report came from a military establishment; at the end of their short paper the authors thank their Director General for permission to publish this communication. The difference between military radar experts and radio astronomers was not yet well-marked.

The existence of a discrete source was based on the fluctuations in the signal. It appears that the radio astronomers involved assumed that such a small discrete source was a somewhat peculiar star, a 'radio star'. The assumption was based on wartime experience: German and Allied radar experts had discovered that the Sun was an irregular source of strong radio waves with wavelengths of a few meters. The Cygnus 'radio star' proved not to be unique. In 1948 Ryle and Smith in Cambridge announced the discovery of another discrete source, this time in the constellation of Cassiopeia. The 'star' was even stronger than the one in Cygnus. They also found some weaker sources (Ryle and Smith 1948).

A little explanation will be useful. Radiation and light are the same wave phenomenon, the only difference being the length of the waves: blue light has a wavelength of half of one-millionth of a meter, in 1946 short-wave radio radiation had a wavelength of 1 meter. This difference becomes important when making an image of a source of light or radiation. As early as the nineteenth century it was calculated that the smallest recognisable detail in the sky has a size of $\lambda/D \times 57.3°$; D is the diameter or mid-line of the telescope and λ is the wavelength—see the Box 1.3. The smallest detail viewed using a telescope with a 30-cm diameter lens and at a wavelength of blue light is smaller than 1 arc second (1/3600 of a degree) but for a radio telescope in 1957 (like the 25-meter Dwingeloo instrument used at a wavelength of 21 cm) the smallest detail is about 2000 times larger, half a degree, comparable to the size of the moon.

Box 1.3 Image Definition

What distinguishes light, infrared light and radio radiation from one another is simply a difference in wavelength. Many of the parameters that are derived from light waves apply equally to radio and infrared waves.

(continued)

Assume a light source at such a great distance that the rays of light are practically parallel. Take a telescope with a diameter of D meters. In 1835, the Englishman Airy, the then 'Astronomer Royal', calculated that the light from a star in the focal point of the telescope produces a disk with faint rings around it: see Fig. 1.5. Around the disk is a darker ring, with a diameter of $\lambda/D \times 57.3°$. Imagine that there is a second light source that is separated from the first by a small angle α. If the angle α is smaller than the diameter of the ring, this second source cannot be distinguished from the first. This boils down to the statement that the smallest visible detail is the size of this ring.

Ideally a perfect astronomical mirror should show a star as an Airy disk. In practice, however, things were different; the sharpest image of a star was always much greater than the Airy disk because turbulence in the Earth's atmosphere smeared out the image. The atmosphere has no effect on radio waves and the Airy disk is important—very important.

We can take as an example the 25-meter Dwingeloo telescope that was mainly used at a wavelength of 21 cm. This means that, according to the above formula, the diameter of the Airy ring is 0.60°. As a student working in Dwingeloo, I guided the telescope across a strong source (Cassiopeia A of Cas A, was suitable) many times and saw that the signal started to increase when the telescope was a quarter of a degree away from the position of the source, reached a maximum at the position of the source and then weakened and disappeared when the telescope was around a quarter of a degree past the source. Half a degree is roughly the measurement of the Moon and the Sun: with the Dwingeloo telescope, details smaller than the Moon were lost or blurred. Even the largest single-unit radio telescope, the one in Arecibo (Puerto Rico) with a diameter of 300 m, has at 18 cm (one of the most frequently used wavelengths) an Airy disk of 2 arc minutes. Because radio astronomers, just like their optical colleagues, want to see details that are hundreds of times smaller than the size of the Moon, this limited 'angular resolution' became a major problem for radio astronomers. It was not until the late 1960s that this problem was resolved thanks to interferometry. More on this will follow in the coming chapters.

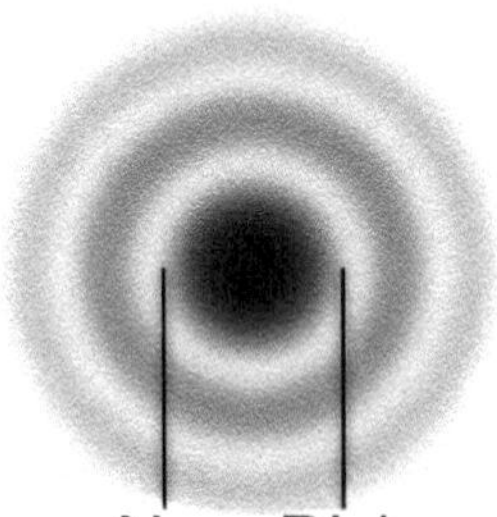

Fig. 1.5 A very small source at an astronomical distance in the focal plane appears as a bright disk (the 'Airy disk') surrounded by a black ring. Source: Wikimedia Commons

Let us go back to the beginnings of radio astronomy. Another wartime experience proved to be a large gold nugget. In practicing the detection of incoming airplanes by radar the experts discovered that the signal they received oscillated in strength, not by much, but easily noticed. The oscillations had the form of a (mathematical) sine and were strictly periodic. The effect was explained as caused by detecting the sum of two signals from the airplane, not just one. The first was the direct reflection of the signal by the airplane. The second was weaker and somewhat delayed; it had to be the signal reflected by the surface of the sea: the antenna stood on a cliff some tens of meters above sea level. It was well known that two strictly periodic signals 'interfere' when they are added up in a receiver; they strengthen and extinguish each other in a sinusoidal way, strictly periodical (see Fig. 1.6). The analysis of this interference showed that a much-improved measurement could be obtained of the direction to the incoming airplane.

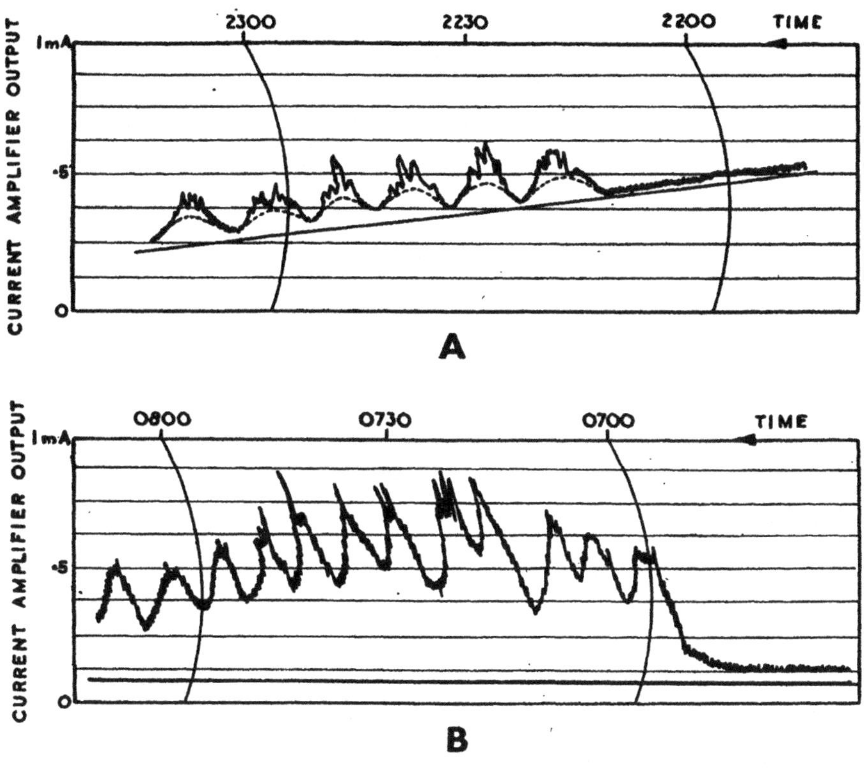

Fig. 1.6 A 'modulated' signal from source Cyg A (above) and from the Sun (below) as measured in the autumn of 1947 with the 'sea interferometer' on Dover Heights near to Sydney. As the signal strength increases, it is also modulated: see the minima indicated by the arrows. Source: Bolton and Stanley 1948

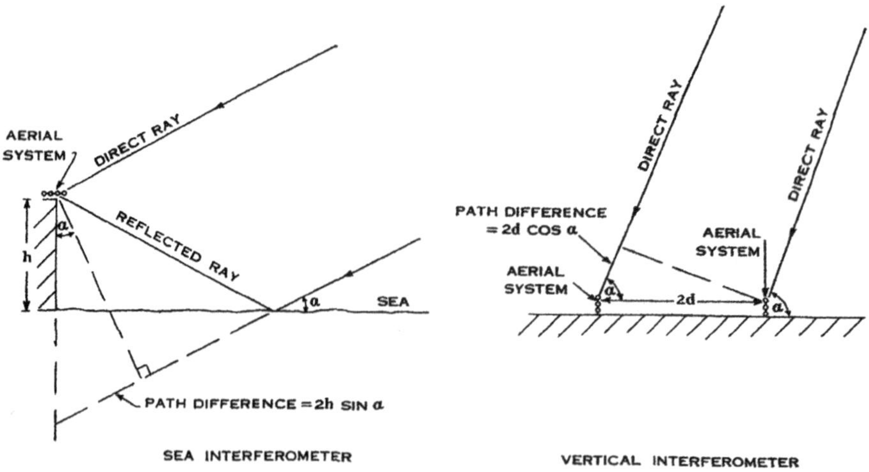

Fig. 1.7 The left-hand image is a sketch of the 'sea interferometer'. The antenna is directed towards a source close to the horizon. The antenna receives the signal from the source twice: once directly, and the second time after it is reflected from the surface of the sea. The second signal has a somewhat longer path. When the two signals are combined in the antenna, this gives a modulation as seen in Fig. 1.6. The right-hand image shows that two telescopes can be used if they are so well connected that a signal from one telescope is transmitted to the other telescope while maintaining the phase. Source: Stanley, Slee, 1950

In 1947 McCready, Pawsey and Payne-Scott used this same set-up to observe the Sun when it rose above the horizon. Interference of the two signals was clearly present and they could conclude that the solar radiation came from localized spots on the Sun and not from the whole disk. This success inspired Bolton and Gordon in Australia to observe the source in the constellation of Cygnus that was detected in England. Again there was interference: see Fig. 1.6. After the analysis of the signal, it appeared that the source was smaller than 8 arc minutes in agreement with the expectation by Hey et al. Bolton and Stanley also derived a much more accurate direction to the source. They obtained a photograph of the sky in that direction but they found only a few stars and no obvious candidate 'radio star' (Bolton and Stanley 1949) (Fig. 1.7).

A short while later Ryle in Cambridge (UK), also found interference in signals from radio sources. He did not have a high cliff and the sea as mirror, but he put two telescopes one behind the other and combined the signal from one telescope with that of the other telescope; see the right-hand image in Fig. 1.7. Once again there was interference and he could show that the radio source in Cassiopeia was also a small 'radio star': 'Cas A'.

The addition of the two signals in a receiver leads to the desired interference only when the two paths from antenna are equally long: these have to differ much less than a wavelength. This is a rigorous demand, one that is difficult to meet when the telescopes are hundreds to thousands of kilometers apart—as we will see in later chapters.

So, 'radio stars', small radio sources, had been found, but were they stars, or something quite different? Was it possible to recognize radio stars on a photographic plate and if so what were we actually seeing? To find out, the positions of the radio stars had to be measured much more precisely and that happened over the next few years.

The First Identifications of 'Radio Stars' with Optically Known Objects

The first victory came in 1949 and it came from 'down under', from Sydney. This is where Bolton and Stanley succeeded in determining the size and position of three radio stars. Two were galaxies and the third, called Taurus A or Tau A, coincided with the Crab Nebula. This object deserves some discussion.

The Crab nebula was number 1 in a list of 1781 by Charles Messier, a French comet-hunter. Several times after he had discovered a new comet, it turned out to be a nebula. He became so irritated that he published in 1781 a list of 100 of these obnoxious nebulae. The objects in this list became so worthwhile that Messier is only remembered because of this list. Later, in the middle of the nineteenth century lord Rosse observed M1. Rosse was a somewhat eccentric English amateur astronomer who had built the largest telescope of the nineteenth century on the grounds of his castle in Ireland. A refurbished version can still be seen there. The Lord made the first drawing of M1 and recognized a Crab in the image; hence the name although photographs made later did not confirm the 'crabbiness'. In 1846 E. Biot, a French railway engineer *and* a Sinologist, published a translation of Chinese annals in which a very bright 'guest star' is mentioned that appeared in 1054 AD at the position of M1. After Baade and Zwicky had proposed the existence of 'super-novae' in 1936 it soon (1939) was realized that the Chinese 'guest star' had been a supernova.

A remarkable characteristic of the nebula helped the identification: the nebula was found to be slightly bigger in each subsequent year. If you calculated backwards, the nebula would become a vanishing dot about 900 years ago. This evidence led in 1940 to the widely accepted view that the Crab Nebula was the remnant of the supernova from 1054. Since then it became the 'Open Sesame' for the treasures of radio astronomy and, remarkably and much later, also for X-ray and gamma-ray astronomy (Fig. 1.8).

Closer study of the Crab nebula around 1940 had revealed that the nebula was made up of two components, as mentioned already in the prologue. First there were the conspicuously colored filaments. Spectroscopy showed that these filaments looked very much like those of other, known nebulae in the Milky Way. But there was a second component: look at the blue light. It showed no fine structure and the spectrum was featureless. This 'amorphous' structure will play an important role in further events.

Fig. 1.8 A color photo of the Crab Nebula, taken with the Hubble Space Telescope. Source: NASA, ESA and Allison Loll/Jeff Hester (Arizona State University)

In 1949 Bolton, Stanley and Slee also identified two other 'radio stars' with well-known galaxies. The first was the radio source Centaurus A that coincided with an unusual galaxy in the southern hemisphere that had hardly been studied because of the lack of telescopes: NGC 5128 (Fig. 1.9). NGC 5128 had a puzzling appearance and did not fit in with Hubble's tuning fork; something unusual could be going on here.

Secondly, the radio source Virgo A was found to be the radio counterpart of M87, an elliptical galaxy from the cluster in the constellation of Virgo. At first there was nothing unusual about M87 until Baade and Minkowski remembered that in 1918

Fig. 1.9 The elliptical NGC 5128 galaxy appeared to coincide with the strong radio source Centaurus A. Source: Astronomy Picture of the Day, 6 August 2003, Marina Rejkuba (ESO-Garching) et al., ISAAC, VLT ANTU telescope, ESO Paranal Obs

H.D. Curtis had seen that a 'jet'—or jet stream—emerged from the center of M87. The jet stream could only be seen if a plate was illuminated briefly; with a longer exposure the jet was bleached out. The jet stream originated in the center of M87: was there something strange going on? This question remained unanswered for a long time (Fig. 1.10).

In 1949 three 'radio stars' had been identified with known objects, none of which turned out to be a real star, but were a completely different type of object. The word 'radio star' was clearly misleading; the objects had nothing to do with stars as they were known and so the word disappeared from the literature.

Fig. 1.10 Figure shows the jet stream and the nucleus of M87, a very large elliptical galaxy that coincided with radio source Virgo A. Source: Astronomy Picture of the Day, 2011 August 28, J. A. Biretta et al., Hubble Heritage Team (STScI/AURA), NASA

The 21-cm Line: Another New Field of Research

A further application of radio astronomy came about in 1951 when a prediction from 7 years previously (1944) became reality: at a wavelength of 21 cm strong radiation was received from interstellar space. At the basis of this success was a 'traditional' or even 'old-fashioned' astronomer with no experience of the world of radar: J.H. Oort from the Observatory at Leiden University. At the end of 1940 the German occupiers closed Leiden University. Shortly before and by a stroke of luck, Oort had received a copy of the 'Astrophysical Journal' with the first of the two papers on radio astronomy by Grote Reber. The USA was not yet at war with Germany and US mail was still arriving in occupied Europe. In 1942 Oort withdrew with his family to a small house in the Veluwe region in the center of the Netherlands and bided his time. He read Reber's articles and in spring 1944 Oort told Henk van de Hulst, a promising student of Marcel Minnaert at the University of Utrecht: 'Mr Van de Hulst, I've read that interesting article by Reber in the Astrophysical Journal in 1940. We ought to have a conference on the subject. Would you have a good look at the article? Radio astronomy could really become enormously important if there was a line in the radio spectrum. Then we could apply the method of differential galactic rotation as we do in optical astronomy.'[1] Van de Hulst quickly found in the literature that hydrogen atoms emit radio photons at a wavelength of 21 cm. The likelihood of a hydrogen atom doing this was very small; atoms could wait millions of years before emitting a photon. But Van de Hulst was aware that there was a lot of space between the stars so that, even if the gas was very rarefied, there would still be enormous numbers of hydrogen atoms, and he estimated that it might be possible to detect the spectral line. He presented his result in 1944 at an illegal meeting in Leiden of the Netherlands Astronomers' Club, the association of professional astronomers, and he wrote a short article about it that was published after the War (1945) in the Dutch Journal of Physics.

Oort knew that he had a bite—the idea was too good to leave it to other people; he had to reel in the prey himself. Fortunately, time was on his side. The Netherlands was impoverished as a result of the war and as all universities in the Netherlands (and in most European countries) were institutes financed solely by the central government, they were poor as well. But there was also a new fervor. The ZWO organization (Pure Scientific Research) was founded to award government funding for scientific projects throughout the country that needed more money than universities themselves had available. Oort grasped his opportunity. An institution was set up to measure 'Radiation from the Sun and the Milky Way' (the two most important objects that radar specialists knew of in 1945), the first goal of which was to detect the predicted spectral line at 21 cm. Oort employed a young physicist who started very slowly and when at some moment the equipment under construction was destroyed by fire the young man was sacked and replaced by an electronic engineer, C.A. Muller, who had just finished his studies. Oort, though, received an alarming telephone call from Van de Hulst who was at Harvard, where he had been informed that the physicists Ewen and Purcell were building a receiver for the same purpose. Later it appeared that astronomers were working towards the same goal in Australia too—that Dutch article from 1945 had reached a broad audience. In the summer of 1951 the three groups demonstrated in rapid succession the existence of the interstellar 21-cm line; (1) Harvard; (2) Leiden; (3) Sydney. This was without doubt one of the great discoveries in astronomy since 1945. In a move that amazes today's scientists, Purcell offered Oort to delay the publication of their result until Leiden had also had success. And so it happened that in one issue of Nature three publications appeared next to each other and announcing the detection of the 21 cm line. The following anecdote shows how much our attitude has changed (Box 1.4).

---

[1] Cited from the opening speech by Van de Hulst at the first IAU symposium on radio astronomy, Manchester, 1955.

Box 1.4 Then and Now

Once a week I row in a coxed four. We row for a while, stop for a break and then turn around. On one occasion, during the break, the rower in front of me, Dick Rollema, turned to me and said: 'Yesterday I realized something strange that I just have to tell you. When I was studying at the Technical University in Delft, I was a member of a student club of radio amateurs. The club regularly organized evening lectures that were often given by former members. In the spring of 1951 I was there on one of these evenings when a former member, Lex Muller, came to talk about his time working at the Leiden Observatory. After graduating, he went to work there because Oort wanted to detect radiation from space. It wasn't just about measuring the noise, but discovering a spectral line: if you detuned the receiver, then at around 1420 MHz (21 cm) the noise should increase strongly for a brief while and then diminish again. This blip in the noise was caused by hydrogen atoms in space.

Muller explained that it was no mean feat to build a receiver that was both sensitive and reliable at such a high frequency. Nonetheless, it was possible, thanks to the excellent training he had had at Delft University. And, what do you know, last week the experiment was successful! Muller took a sheet of recording paper out of his briefcase. On this paper we could see the same blip twice, once negative and once positive, just as you would expect with the measurement technique that had been used (see below). That was quite some achievement. And we—engineers from Delft—were the ones who did it. Were the astronomers happy with it? They certainly were, because this was where things really took off for them. This technology allowed them to view the whole of the Milky Way. We didn't realize that they hadn't been able to do that before, but we realized that it was important that they could do it now (Fig. 1.11).

'Some questions were asked, the speaker was thanked and he was given a book token (for a modest amount since the costs came out of the member contributions). In the café around the corner we drank a beer. And,' said Dick in the boat: 'it was only yesterday that I realized that we were probably the first outsiders to learn of a spectacular astronomical discovery, one of the most significant after the War. That evening there was no spectacle, no embargo, no nervous spokesperson organizing a press conference. They had made an interesting discovery, and that was it. My goodness, the world has certainly changed!'

Lex Muller died in 2004. I wrote the above story in a letter of condolence to his widow. She wrote back, saying: 'Yes, that's just like Lex. He wasn't one for a lot of fuss.'

Fig. 1.11 21-cm detection. Source: Van de Hulst et al. (1954)

Synchrotron Radiation

What physical process produced the radio radiation? The question had been asked since the days when Jansky detected his 'cosmic static' and more than 10 years later, in 1945, it had not yet been answered. In 1937 Whipple and Greenstein, two young American astronomers, showed that radiation from stars fell short of what had been measured by a large factor. They also excluded the possibility that the radio radiation was emitted by the small particles discovered a few years before in the space between the stars. The question of the radiation mechanism became quasi-insoluble shortly after 1945 when it was proven that the radiation was much stronger at longer wavelengths. All radiation processes known in astronomy predicted that the radiation should be fainter at longer wavelengths. The contradiction meant a crisis.

In history, finding the solution of a difficult scientific question is sometimes like an expedition along a winding path with a new, temporarily different view at every curve. In this crisis the path began with a paper by Ryle in 1949. He started from the discovery of the first radio star, Cygnus A. He implicitly assumed that the object was a real star, and then remembered the theory that stars are sources of the cosmic rays, the highly energetic particles that strike the Earth's atmosphere and are believed to be generally present in the space between the stars. Could these cosmic rays be the cause of the radio radiation from the supposed star Cygnus A? Two well-known Swedish physicists took up the bait. But a diversion was needed.

In 1947 American physicists were studying the acceleration of electrons in a so-called synchrotron. They discovered that the electrons started to radiate a bluish light when they reached a velocity that is a sizeable fraction of the velocity of light. Scanning the literature, they found that in 1944 several Soviet physicists had predicted the blue light: the electrons moved in circles around the magnetic lines because of the so-called Lorentz force. Circular motion meant a continuous change

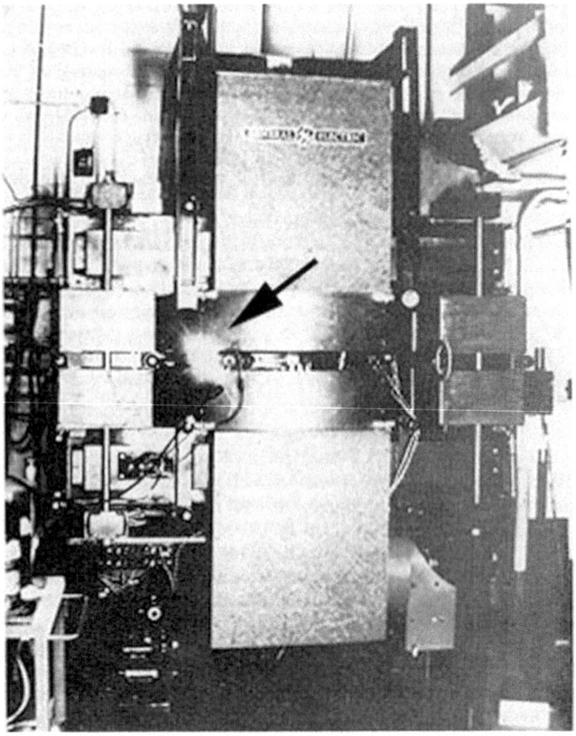

Fig. 1.12 The picture shows a particle accelerator that produces radiation as a by-product. The arrow indicates where the radiation leaks from the accelerator. Source: Wikimedia Commons

in the direction of the velocity, which implies acceleration. Accelerated electrons emit radiation. In this case some relativistic, but well-established calculations were needed that gave a good explanation of what was observed (Fig. 1.12).

Two Swedish scientists, Alfvén and Herlofson, both solar and plasma physicists, pursued Ryle's speculation and in 1950 they suggested that synchrotron radiation could be produced in the atmospheres of some stars and that this was how galactic radio radiation was produced. A paper by the German solar physicist Kiepenheuer quickly followed this paper. Kiepenheuer rejected the idea of 'radio stars' and showed that the familiar cosmic rays had sufficient energy to produce the galactic radio radiation. Whatever produced the cosmic rays also produced the radio radiation.

Kiepenheuer's suggestion was generally well received, but the idea was so far from mainstream astronomy that an independent confirmation of the theory was needed. It came in 1953 when the Soviet astronomer Shklovsky proposed that the amorphous blue structure, the second component in the Crab nebula (see above), was in actual fact synchrotron radiation and he suggested how this could be tested: the light should be highly polarized. In a separate Box 1.5 I explain what this means. Shklovsky's suggestion was taken up by two observers, Vashakidze and Dombrovsky. They quickly confirmed that, indeed, the blue light of the Crab nebula was strongly polarized.

Box 1.5 An explanation of the concept of 'polarization'
In the source of radiation, a billion times a billion individual electrons each briefly emit a wave from time to time, independent of each other, and we receive all these waves at the same time. Each wave has a preferred direction but the preferred direction of one wave is completely different from that of another and because billions of waves are received simultaneously there is no generally preferred direction. But this is different when all the electrons are aligned in one way or another. In this case the waves are also slightly aligned, and then when the radiation is received, it has a net preferred direction. It can be measured and it is called the 'polarization of radiation'. In the case of the Crab, the polarization measured showed that the electrons that were radiated were aligned to some extent. Synchrotron radiation is emitted by electrons that are aligned by a magnetic field. This was the reason why Shklovsky had suggested that this must be the radiation mechanism in the Crab.

In 1953 the Cold War was raging and the exciting results in the Soviet Union attracted little or no attention in the West. This changed when Dutch astronomer Oort visited the Soviet Union at the invitation of the powerful Soviet Academy of Sciences, of which he was a member, just as he was a member of many other academies. Oort met Shklovsky and was highly impressed by the results Shlovsky had achieved. As soon as Oort returned home he convinced a new employee at the observatory, Theodoor Walraven, a highly talented instrument maker, to build a polarimeter and to repeat the observations made by Vashakidze and Dombrovsky. Oort and Walraven confirmed the results of their Soviet colleagues within just a few months.[2] In their publication they discussed several implications. For example, that in a galaxy like M87 the radio radiation came from the center and had nothing to do with the stars in that object. This well-known galaxy had a totally new property.

In their discussion, Oort and Walraven also drew the conclusion that the synchrotron-radiating electrons would exhaust their energy in at most 200 years. But the Crab Nebula was 900 years old! The nebula must therefore have a permanent source of new, high-energy electrons: an infernal machine. Where was this machine? The authors came up with an answer that was nonetheless unconvincing. In 1967 a different and definitive answer was found.

The successful explanation of radiation from the Crab Nebula was the much needed, independent confirmation of Kiepenheuer's suggestion. From then on the paradigm was established that the radiation discovered by Jansky and confirmed by Reber was synchrotron radiation. The consequences proved to be enormous.

---

[2] Incidentally, the observations by Oort and Walraven were the last significant observations made at the Leiden Observatory that is located in the center of the city; the telescopes were too small and the street lighting too bright.

1954 Breakthrough: Baade and Minkowski Identified the Cygnus A Radio Source

And now we come to 1954 and the article that would change astronomy fundamentally. From then on radio astronomy was incorporated into mainstream astronomy, giving two branches of astronomy: optical astronomy and radio astronomy. The authors were Baade and Minkowski, who both worked at Mount Palomar. We have already come across Baade; Rudolf Minkowski[3] was a spectroscopist. He was a German astronomer with Jewish roots who had fled to the US.

Baade and Minkowski had already made recordings in 1951 with the 200-inch mirror on Mount Palomar of the area around the radio sources Cygnus A and Cassiopeia A. The radio positions were not precise enough to identify one of the objects photographed with the radio source. But in 1951 Graham Smith in Cambridge (UK) managed to measure the positions of radio sources with the Ryle interferometer much more precisely, thereby settling the matter. The photographic plate showed that Cassiopeia A coincided with a ring of faint, barely recognizable nebulae. Minkowski managed to record a spectrum of several nebulae (a beautiful piece of craftsmanship) and these showed that the nebulae moved at very high velocity through space: they constituted the remainder of a supernova. Cas A was thus related to Tau A, the Crab nebula. Later recordings, including with X-ray satellites, confirmed that Cas A is the remains of a supernova from the seventeenth century. Strangely enough, nothing can be found about this supernova in the scientific literature of the time, although by then many astronomers were already using their telescopes to scan the night sky.

The identification of Cas A was interesting, but the identification of Cygnus A caused a sensation. In 1953 two radio astronomers discovered with an interferometer at Jodrell Bank that the source consisted of two parts: see Fig. 1.13. In between the two radio sources Baade and Minkowski saw a faint spot of magnitude 16 that could be a nebula or a seriously misformed galaxy; see Fig. 1.15 When Minkowski recorded a spectrum he measured from the redshift of the spectral line that the object was moving away from us at a velocity of almost 17,000 km/s: the spot

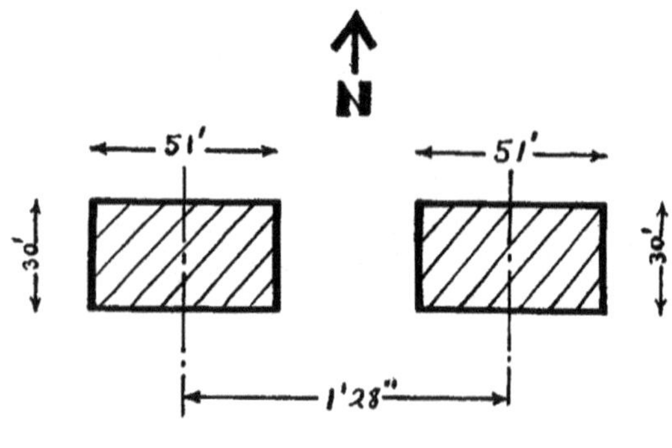

Fig. 1.13 Cygnus A as a double radio source, as found by Jennison and Das Gupta (1953)

[3]Rudolf Minkowski was closely related to Hermann Minkowski, who is well known because he expressed Einstein's special theory of relativity in a mathematically simpler form.

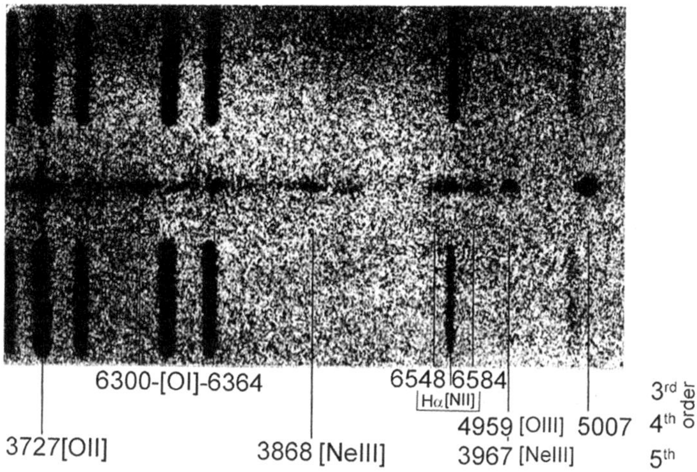

Fig. 1.14 The spectrum of Cyg A. The thick lines that are the most noticeable are from a light source in the spectrograph and they serve to measure the wavelengths of the lines of Cyg A. Between the strong lines a long horizontal line with thicker areas can be seen. This is the spectrum of Cyg A (Baade and Minkowski 1954)

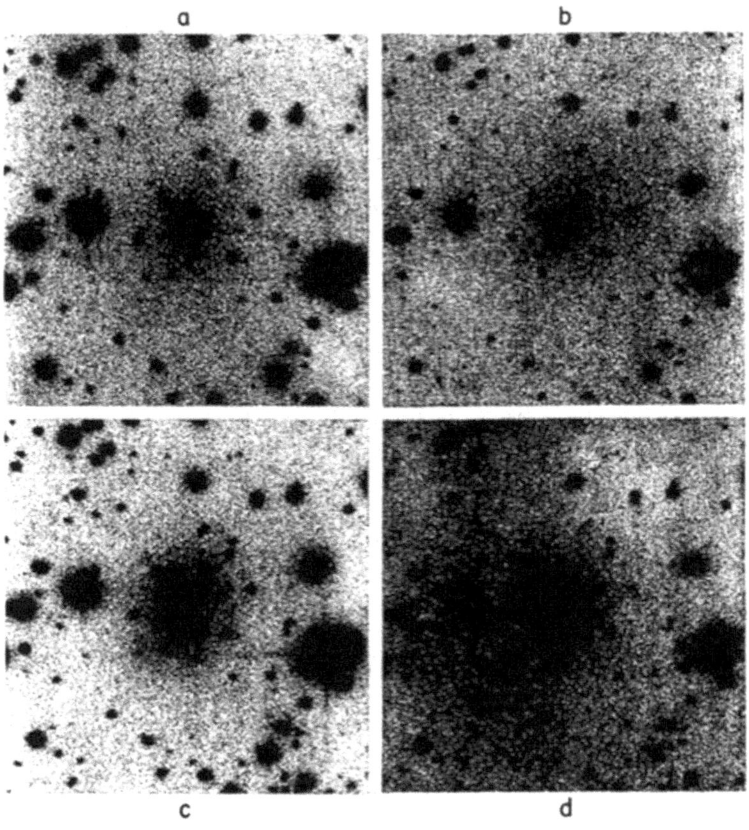

Fig. 1.15 Four slightly different photographs of the galaxy located between the double radio source (see Fig. 1.13) that is the optical counterpart of radio source Cyg A. The figure is difficult to interpret! (Baade and Minkowski 1954)

had to be a galaxy! The identification of this galaxy caused a sensation: the strongest radio source in the sky coincided with a small, faint optical object! It turned the world upside down. How much further could you look in space if you studied a much fainter radio source? In the previous decades the redshift of around 600 galaxies had been measured, and only a handful had a higher redshift than Cyg A.

The energy contained in each of the two radio sources around Cyg A was comparable with the energy confined in M87. It was natural to speculate that this small galaxy supplied the energy via a jet stream to the two radio sources on both sides of the galaxy. Probably this small galaxy also contained such an infernal machine as M87. Cygnus A, Virgo A and Centaurus A belonged to a new and exciting category of galaxies that were given the name 'radio galaxies'.

In their article, Baade and Minkowski included a footnote in which they recalled an almost forgotten article from 1943 by C.J. Seyfert (1911–1960). He discussed a small group of galaxies, the spectrum of which contained very broad emission lines including hydrogen. Seyfert ascribed this breadth to gas that moved at great velocity in the center of these galaxies. These galaxies appeared to be the first examples of a class of galaxies henceforth named 'Seyfert galaxies' that would in the future give rise to a great deal of research with telescopes at all conceivable wavelengths, from radio waves up to and including hard X-ray radiation.

In 1956 Baade added something to this work. When he heard of the explanation of the radiation from the Crab Nebula and the polarization of the blue light, he wondered whether the light from the jet stream, the 'jet' in M87, was also polarized (see Fig. 1.10 earlier in this chapter). Back home he took a number of photos with the 200-inch telescope and saw that the jet stream was indeed polarized: this meant that the light from the jet stream was probably also synchrotron radiation. The jet stream pointed to the center of M87—a new argument for thinking that a 'monster' might be present there, too, an 'infernal machine' much stronger than that in the Crab Nebula, since M87 was thousands of times further away. The English-American astrophysicist Geoffrey Burbidge calculated the energy of relativistic electrons (electrons with a velocity approaching the velocity of light) and the magnetic energy in M87 at least 10^{49} J. That was much more than was emitted in a supernova explosion. Burbidge and everyone else was highly surprised, but nobody had an explanation for what could have produced this colossal amount of energy.

1955: Radio Astronomy Gives a New Vision of the Universe

In 1955 the International Astronomical Union organized the first symposium on radio astronomy in Manchester. Although most participants were radio experts, there were also a number of *bona fide* optical astronomers: Baade, de Vaucouleurs, Edmondsson, Minkowski, Oort, Unsöld, van de Hulst. It has been said previously that 'old-fashioned' astronomers expected little from radio astronomy: radiation contained little energy and would probably be generated in an unknown and unimportant process. It now appeared that this radiation was a small leak in an enormous reservoir of energy, a reservoir that until then had never been seen and that was very probably at the core of a galaxy (see M87 and Cygnus A). It was from around this time that the term 'the violent universe' started to be used. Radio astronomy had provided a completely new vision of the universe.

1.3 The Legacy of the V2: X-Radiation from the Sun

In the Prologue the story has been told about the US who in 1945 and in a secret operation, took home many German rocket experts and a large number of V2-s. Scientists of the upper-atmosphere and of the Sun were very content because they realized that they now had the chance to send instruments much higher than ever before with balloons.

They were not the first to be aware of the V2 as an instrument for solar research. During the war a renowned German solar physicist, Kiepenheuer, had proposed to Von Braun to equip a V2 with an UV-spectrograph, to launch it towards the zenith and photograph the solar spectrum at wavelengths that had never been explored. Von Braun agreed, and the spectrograph was delivered, but by that time the war was almost over and in the chaos and disorder of the day the experiment never took place.

Already in 1946 a V2 was launched that carried a spectrograph and several followed and this time not in Germany. Apparently there were failures among these first examples of rocket research, but ultimately reliable measurements were made. The Sun emitted more hard UV-radiation than expected, and it proved to be a source of X-radiation. The results were interesting and useful but not very spectacular. The solar X-radiation was weak when compared to the output in the UV. If all stars were like the Sun their X-radiation could not be detected. A new discipline like X-ray astronomy, it was thought, had no future—a wrong conclusion as we will see in Chap. 3.

Figure 1.16 shows the launch of a V2 with a smaller rocket on top.

Sputnik and the Space Age

The technical success of the V2 inspired each of the allied winners, US, USSR, UK and France, to start its own rocket program—and of course this was classified information. In the USA Army, Navy and Air Force each had its own development programme and did its own technical research. Slowly they made progress. About what happened in the USSR nobody in the West knew, apparently not even the Intelligence organizations in the US; but they certainly could not be farther advanced than the US. Imagine the shock wave that went through the American population when on 4 October 1957 the USSR put a satellite in space, which they called the Sputnik; every few seconds it gave a 'peep', and each peep told the Americans: see what the USSR is capable of.

Fig. 1.16 A rocket launch from Cape Canaveral, July 1950. The small cylinder above the broad first stage is a WAC Corporal rocket developed in America; the rocket launcher itself is a German V-2. Source: Astronomy Picture of the Day, 1 October 2008, GRIN, NASA

In the US the shock was enormous. Self-flagellation followed: what had the nation done wrong. Where had it failed? Quick analyses were made and published. Culprits were assigned. Predictably, 'the schools and the universities had been too soft'. And to make things worse, at 3 November 1957 the Soviets launched Sputnik 2 to celebrate the 50th anniversary of the revolution that ultimately brought this paradise for farmers and workers, the Soviet Union.

To make things worse there was no US answer, not by the Air Force, the Army nor the Navy. Finally there was a hastily assembled two-stage rocket on Cape Canaveral that was to be launched on December 6, 1957. Journalists from over the world had come to watch. What a difference with the Russians who did these things secretly! At time zero ignition started, the rocket rose to a height of a few meters, tumbled over and went up in flames. The press was unrelenting. British newspapers reported about the 'Flopnik' and the 'Kaputnik'. And Time Magazine chose the Soviet leader, Nikita Krushchev as 'man of the year'.

But the last is the best. On 31 January 1958 the Army launched secretly but successfully Juno-1, a four-stage rocket that put a satellite in orbit, Explorer 1. It was a success and one of the key people in this success had been Von Braun. The launch of Explorer 2 failed but on March 26 Explorer-3 was put in orbit and it brought some revenge: it detected radiation belts around the Earth, the now famous 'Van Allen belts'. The Soviets could have discovered them, but failed to do so. After all the hysteria the Administration took the right decisions. On March 5 1958 President Eisenhower signed the law that created NASA, a civilian organization that from now on should lead American space research although the military were given some research possibilities of their own. Looking back on NASA's history, one has to admit that it has been very successful, in spite of some serious accidents.

1.4 The Remarkable Beginning of Infrared Astronomy

Astronomical research at infrared wavelengths has always had a connection with astronomy. It was an astronomer who detected the radiation in 1800. William Herschel, an English self-made astronomer, experimented with lenses of different colours and had noticed that when he tested a red lens his hand felt more heat than when he used a blue lens. He then used a prism to get a spectrum like a rainbow and put a thermometer at various colours. He concluded that the temperature of the thermometer rose faster and higher when he held the thermometer in the red part of the spectrum. He then put the thermometer outside the red edge of the visible spectrum and saw the thermometer react even more rapidly: there clearly was radiation outside the visible spectrum, radiation that the eye did not see.

Modern infrared astronomy began in 1945 in post-war Germany. ALSOS, a group of American military, and among them mainly physicists, followed the fighting at short distance to collect information how far Germany had advanced in developing the atomic bomb. This was many months before the US Air Force would drop the bomb on Hiroshima and when the world learned of its existence. The ALSOS mission served therefore a secret military purpose. One of the ALSOS members was Gerard Kuiper, a prominent American astronomer, born in the Netherlands but emigrated to the USA after his PhD. In one way or the other Kuiper learned that German physicists had discovered that a combination of lead and sulphur, PbS, made a superior detector of infrared light.[4] This discovery may well have been a German military secret.

Back in the USA, Kuiper, together with Wilson and Cashman built a spectrometer to measure the infrared radiation of planets and of cool stars. The infrared spectrum of Jupiter and Saturn had strong molecular absorption bands. For a long time the PbS cells remained the main detector of astronomical infrared radiation.

[4] In the paper that describes the construction of the spectrometer there is a somewhat mysterious sentence: 'During his work in Germany, Mr. Kuiper had been impressed by the astronomical possibilities opened up by the new PbS cells'.

References

Baade, W., Minkowski, R.: Astrophys. J. **119**, 206 (1954)
Bolton, J.G., Stanley, G.J.: Aust. J. Scient. Res. A. **2,** 139 (1949)
Greenstein, J.L., Henyey, L.G., Keenan, P.C.: Nature. **157**, 805 (1946)
Hey, J.S., Parsons, S.J., Phillips, J.W.: Nature. **158,** 234 (1946)
Jennison, R.C., Das Gupta, M.K.: Nature. **172**, 996 (1953)
Kron, G.: Astrophys. J. **103**, 326 (1946)
McCready, L.L., Pawsey, J.L., Payne-Scott, R.: Proc. R. Soc. Lond. A Math. Phys. Sci. **190**, 357 (1947)
Ryle, M., Smith, F.G.: Nature. **162**, 462 (1948)
Stebbins, J., Whitford, A.E.: Astrophys. J. **102**, 318 (1945)
Van de Hulst, H. C., Muller, C. A., Oort, J. H. Bull.: Astron. Inst. Netherlands, **12,** 117 (1954)

Chapter 2
1945–1959: Stars and Galaxies

This chapter has two parts: one about stars and one about galaxies.

2.1 Stars

How the energy of the solar radiation was produced was understood in 1939: at or close to the solar center the temperature and pressure were high enough to force the nuclei of four hydrogen atoms to merge into the nucleus of the helium atom. A small amount of matter was left over and this was converted into photons. At some future point in time all the hydrogen nuclei would have been merged and that would mean the end of the Sun: in other words, the Sun had a finite lifetime and the same would be true for all other stars. But anything that perishes must have been created: how had this happened for the Sun and the Earth? Looking to higher powers for an answer was not an attractive proposition in terms of physics: the existence of stars was a scientific problem on the plate of astronomers.

There was already a theory about how the Sun came into being, or rather, a kind of theory: a plausible proposition but one that had not been proven in any detail. This was the scenario put forward by Kant and Laplace that has already been mentioned, according to which the Sun and its planets were formed from a cloud of gas. But did this same scenario also hold true for other stars? Nobody knew whether other stars also had planets, and in 1945 there were no gas clouds known that were contracting and likely to produce a star. Thousands of objects similar to the Orion nebula had been discovered, but the gas in these nebulae was now expanding, not contracting. Dark clouds could be birthplaces of stars, but nothing was known about these clouds, other than that they existed.

This lack of observations meant that star formation was above all a matter for theoreticians, as the specialist literature of the time shows. And unfortunately, theory was useless. It was nothing more than a few very general propositions, such as the Jeans criterion, that served little purpose: its usefulness came much later.

H. J. Habing, *The Birth of Modern Astronomy*, Historical & Cultural Astronomy,
https://doi.org/10.1007/978-3-319-99082-8_2

In 1945 the lack of knowledge was much deeper. As we will see later, almost nothing was known about how old stars died. There was, as it were, no 'historical awareness', It was as if the Universe had eternal life. I can give two examples. Adriaan Blaauw obtained his doctorate in 1946 with distinction in Groningen based on a study of a group of bright blue stars, that he was able to demonstrate would disintegrate within a short period. However, as Blaauw himself wrote much later, he did not realise that his conclusion meant that the group of stars had been formed only a short while previously. It was the famous Soviet astronomer Ambartsumian to whom this thought occurred several years later. And another example: in 1945 the American astronomer Alfred Joy introduced a new type of variable star, the 'TTauri stars'. Joy noted that the stars occurred in the region of dark clouds, but he did not speculate on the question of whether the stars had indeed been created there a short time before, as later proved to be the case.

In 1956 the topic of star formation was put firmly on the agenda with an article about stars that must be very young, much younger than had previously been thought. A young American astronomer, Merle Walker reported on photo-electric measurements of the brightness and colour of stars in a known star cluster (see Fig. 2.1).

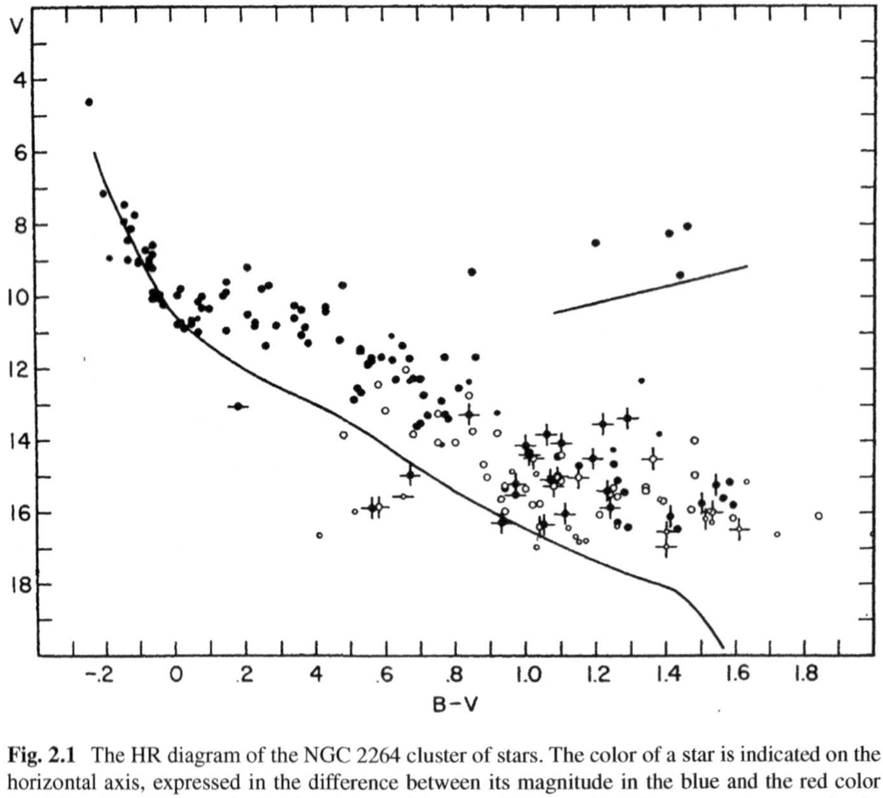

Fig. 2.1 The HR diagram of the NGC 2264 cluster of stars. The color of a star is indicated on the horizontal axis, expressed in the difference between its magnitude in the blue and the red color band. The bluer the star, the brighter it is, and the more light it emits (all stars are close together in space and thus they are all equidistant from us). The curve drawn indicates the main sequence—see the Prologue. The fainter stars (color B-V greater than 0) are all to the right, beside the main sequence: they are en route to the main sequence but have not yet arrived there. (Walker 1956)

Walker was able to show that the stars were colder than stars in the main sequence, but that they emitted just as much light. He drew the conclusion that these stars were en route to the main sequence, but had not yet arrived there. They would later be referred to as 'Pre-main-sequence stars', PMS stars. Walker's article made a strong impression.

And then there was another observation that might well have something to do with star formation: in 1951/1952 the Mexican Guillermo Haro and the American George Herbig both observed several small nebulae in the vicinity of dark clouds, and specifically in the area of T Tau stars (Herbig 1951; Haro 1952). The nebulae were peculiar because of unusual strength of spectral lines. Were they TTau stars in the making? More 'Herbig-Haro' or 'HH objects' were found in later years. The question was: What were they? The answer was not found until 1975.

In 1960 the enigma of star formation was no longer a purely theoretical question as it had been in 1945. Stars had been found that were still on their way to the main sequence and unusual nebulae had been discovered in places where perhaps new stars were formed. It was something but still only very little (Box 2.1).

Box 2.1 Kelvin-Helmholtz Time

In the middle of the nineteenth century physicists discovered the concept of 'energy' and the fundamental law that energy is never lost: the principle of the conservation of energy. Applied to the Sun, this meant that the energy emitted by the Sun must originate somewhere inside the Sun. But what was the source of this energy? The best answer available at that time came from Irish-Scottish physicist William Thomson (1824–1907; after 1892 Lord Kelvin) and his colleague and friend Prussian Hermann von Helmholtz (1821–1894). They looked for this source of energy in gravity. According to fairly elementary reasoning, the total amount of gravitational energy (U) for a star is indicated by $U = 3/5\ G\ M^2/R$, where G is the (known) constant of gravity, M is the mass of the star and R is the radius. It follows that the gravitational energy of the Sun is approximately 3.8×10^{41} J or 3.8×10^{41} Ws. As a light source, the Sun emits $L = 3.9 \times 10^{26}$ W. An estimate of the total lifetime of the Sun is then U/L, referred to as the 'Kelvin-Helmholtz time'. In the case of the Sun, this is 30 million years.

During the lifetime of Kelvin and Helmholtz, the second half of the nineteenth century, this was initially considered a long existence for the Sun and therefore also for the Earth. But within a few years, geologists, and later Darwinists too, came up with a fundamental objection: 30 million years was much too short to explain all the geological processes or the evolution of life. Unfortunately, the discussion was hampered—particularly in England—by the conservative and influential Lord Kelvin: he told the Darwinists and the geologists to come up with a better estimate, because the Sun did not give them enough time for what they said they needed. It was only when, after 1900, the age of the Earth was calculated from the radioactivity of rock that it was generally accepted that the Earth, and therefore also the Sun, was older than the 'Kelvin-Helmholtz' age.

(continued)

Eventually, in 1939, researchers learned that the energy of the Sun and other main sequence stars comes from the fusion of hydrogen to helium and the Sun had a lifetime of many billions of years. This same insight led to the construction of the hydrogen bomb.

The Kelvin-Helmholtz age, $t_{KH} = U/L$, disappeared from astronomy, but it returned when people started to think about the origin of stars. The Kelvin-Helmholtz age, appeared to be a good estimate of the length of the contraction of a cloud needed to trigger nuclear fusion at its core. And it was immediately realized that t_{KH} is much shorter for a high-mass star. For a star with a mass of 10 times the Sun, then R is approximately 10 times greater, but L is 10,000× greater, which means that the first stage in the formation of a high-mass star—the contraction—is almost 1000 times faster than the contraction of a light star.

Stars on the Main Sequence

In 1945 it was understood that stars had to die: they emitted energy continuously and because their supply of energy was finite, it had to come to an end at some point in time. The age of the Earth had been reliably and precisely determined: it was around 4 billion years old, and everyone assumed that the Sun was about the same age, even if it had been formed a little earlier than the Earth.

The Chemical Composition of Stellar Atmospheres

The findings of English PhD candidate Cecilia Payne in Harvard were discussed in the Prologue. In the early 1920s she deduced from lines in spectra of stars that the composition of the atmosphere of all stars was the same: 90% of all atoms were hydrogen, 9% helium and the atoms of all other substances added together constituted the remaining 1%. Her results were later confirmed by others which led to this composition of the gas in star atmospheres being given the name 'cosmic abundances'. This happy simplicity was destroyed in 1951 when an analysis of the spectra of a few stars showed that they contained much less calcium and iron atoms per hydrogen atom than the cosmic abundance required (Chamberlain and Aller 1951). After this, more such stars were discovered and it became clear that the number of atoms heavier than helium could vary from star to star, although a higher content than in the Sun was an exception.

The variation of the abundances got a much wider meaning in 1950 when the American astronomer Nancy Roman showed that 'low-metallicity' stars had velocities differing strongly from those of the other stars in the solar neighbourhood. This simple fact had considerable consequences; see the discussion later in this chapter.

The discovery that the abundances varied from star to star was a remarkable finding that needed clarification. Start with the simple, but very difficult question: how did the composition of the cosmic abundances come about? The answer ought to come from detailed knowledge about the fusion of atoms, and the basis for this was laid in 1939 by Bethe and Von Weizsäcker; see the Prologue. Fusion of atoms only takes place at high temperatures and under great pressure: atomic nuclei can only form another nucleus when they collide violently with one another. Enough pressure and a sufficiently high temperature (several millions of degrees kelvin or more) were only present in all of space within a few minutes after the Big Bang and much, much later, in the center of stars.

Between 1945 and 1955 the Russian American Georg Gamow investigated whether all atoms were formed shortly after the Big Bang. He concluded that only hydrogen and helium were produced and none of the other heavier atoms. Gamow was unable to explain the existence of the heavier atoms. Gamow predicted also the existence of a very cold photon gas, a remnant of the Big Bang —but that will be discussed in Chap. 5.

The percentage of heavy atoms was not the same in all stars as Chamberlain and Aller had shown. That made it unlikely that the heavy atoms had all been formed shortly after the bigbang; they must have been formed much later and deep inside stars. A simple and convincing piece of evidence proved this. In 1952 the American astronomer Merrill discovered atoms of the element technetium in some stars. These atoms disintegrate within a short space of time. The fact that they were discovered long after the Big Bang meant that they had been formed recently in the star in which they were discovered.

The definitive evidence that atoms heavier than helium could be produced in the centers of stars came in 1957 in one of the most important articles in the history or astronomy after 1945. This article was written by Geoffrey and Margaret Burbidge, William Fowler and Fred Hoyle, three Brits and one American (Burbidge et al. 1957). It is generally known as B^2FH. In an intensive collaboration lasting several months the four looked at all the processes that created new, heavier atoms from smaller ones in the interior of stars. The conclusion was that during their lifetime, and specifically at the end of it, stars could produce all the atoms found and also in the right proportions.

B^2FH settled the matter once and for all: atoms heavier than helium were formed in the centers of stars. Only the two lightest atoms—hydrogen and helium—were formed within a 1000 seconds after the Big Bang. A star that had produced new atoms distributed these new atoms at the end of its life, when it exploded as a supernova. The next generation of stars would be created from the matter expelled that contained more heavier atoms, This process would be repeated indefinitely.

How Stars change as they become Older: the Solution to the Mystery of the Red Giants

In 1939 Bethe and Von Weizsäcker discovered independently of one another how, given a high enough pressure and temperature, the nuclei of four hydrogen atoms would merge spontaneously to form the nucleus of one helium atom—see the Prologue. In the 1920s Eddington had constructed stellar models with one missing piece: he did not know what process produced the energy. Now that process was known and the production of photons could be described using mathematical equations. This meant that the mathematical equations of all processes in a star were known and realistic models could be constructed. There was no general solution to this set of equations but if you knew the star's size and its atomic composition, you could calculate the pressure, temperature, etc., by working from the center of the star outwards in small steps. It was an awful job, but it was possible. In 1942 Schönberg and Chandrasekhar decided on the basis of these kinds of calculations that a star could burn at most 10% of all the hydrogen, which meant that the Sun had a lifetime of 7 billion years (Schönberg and Chandrasekhar 1942).

Even though the mathematical laws were known, Schönberg and Chandrasekhar had reached an impasse. The longer the fusion lasted, the more helium was produced. Schönberg and Chandrasekhar could not decide what happened next. Would the helium and hydrogen become separated in the interior, with the helium forming a ball at the center surrounded by hydrogen?

Many papers were written about inhomogeneous stars. But the decisive result appears to have been obtained in 1952 by Martin Schwarzschild with his student Bev Oke (Oke and Schwarzschild 1952). They assumed a stellar model that consisted of a highly compressed core of helium surrounded by a mantle of hydrogen. Nuclear fusion took place in a thin layer around the helium core. The results were surprising. Firstly, the model produced much light. A model, as heavy as the Sun, could emit hundreds of times more than the Sun. Secondly the hydrogen mantle would become a 1000 times more extended. So much light and such an enormous size: this was the model of a red giant—the two objects shown in the Prologue in Fig. 5, upper center—that Eddington had failed to make a model of in the 1920s. Now it had been achieved part by intuition and part by chance. Red giants, it appeared, were hybrid objects, a thick mantle of hydrogen surrounding a small, highly compressed ball of helium.

For Schwarzschild and Oke it was clear. A star might appear more or less the same for a long time, but at a certain point in time it had produced a maximum amount of helium and then it started to shine more strongly and at the same time took on a red colour. In the HR diagram a star stayed in its permanent location for a long time, but when the time was right, it rose rapidly in luminosity and became very red.

Schwarzschild's calculations in Princeton were confirmed a short time later by observations that were made completely independently on the other side of the US. Shortly after the 5-meter telescope on Mount Palomar was commissioned, three young staff astronomers made photographic plates of the M92 cluster, a distant cluster of millions of stars (see Fig. 2.2) They recorded the magnitude of as many stars as they could, using two filters of different colours; they called the magnitude seen through a blue filter 'P' and that viewed through a yellow filter 'V'. They entered these measurements in a so-called color-magnitude diagram: the color of the star was plotted horizontally ('color index, P-V') and the 'V' magnitude vertically. Because all the stars were equidistant from us, the brightnesses measured were the true magnitude plus a number that was the same for all stars. This was equivalent to an HR diagram, explained in the Prologue.

Fig. 2.2 Cluster M92. Source: NOAO/AURA/NSF

The progress that the 5-meter telescope had brought about was apparent from the faintest stars that were measured: stars up to magnitude V = 19.5. This was a magnitude of just 3.5 or 25 times fainter than in an excellent study made in 1939 by O. Hachenberg with a 1.2-meter diameter telescope near to Berlin. This difference of a magnitude of 3.5 appeared to be of crucial importance when in 1952 Sandage took these data to Princeton to discuss them with Schwarzschild.

Schwarzschild had already come to the conclusion that heavier stars must age faster than lighter stars: heavier stars may have more fuel, but they emit so much more light, that they live for a far shorter time. Stars higher up the main sequence were heavier and aged more rapidly than lighter stars and left their place in the HR diagram earlier to disappear diagonally to the upper right. Sandage and Schwarzschild now assumed that the stars at the bottom of the diagram, between the two dashed lines, were unchanging main sequence stars. But all stars above the upper dashed line had started to move diagonally upwards to the right (Figs. 2.3 and 2.4).

And there was something else. During his calculations with Oke, Schwarzschild had seen that once stars left their place on the main sequence, they did so very quickly.

He even ventured to hypothesize that all stars above the main sequence differed very little in age from the brightest main sequence stars: all the stars that had left the main sequence, all stars on the giant branch (see caption of Fig. 2.3), were more or less equally old. These two hypotheses explained the place of almost all stars in the diagram. It was a breakthrough, a victory.

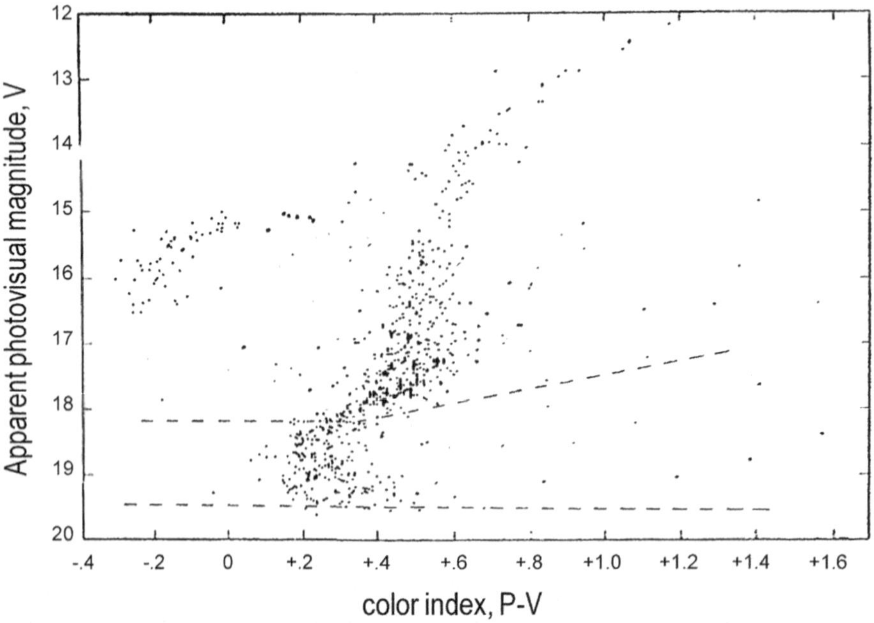

Fig. 2.3 HR diagram of the M92 cluster with the 5-meter telescope on Mount Palomar. The dots indicate the values of the individual stars in this cluster: color (horizontal: left, blue stars; right, red stars) and magnitude (vertical: the magnitude of the stars; the smaller the magnitude, the brighter the star). The bottom dashed line was the observation limit: stars below this line were too faint for the 5-meter telescope; this limit is around magnitude 19.5. The stars between the two dashed lines were main sequence stars. Above the upper dashed lines were stars with magnitudes and colors that changed with ageing: they have become redder and much brighter. The giant branch is the sequence of stars that runs diagonally to the upper right. Initially, the thin sequence of stars at half height in the left half of the diagram ('blue horizontal branch', abbreviated to BHB) could not be explained (Arp et al. 1952)

Nevertheless, the success was not complete: there were still some unanswered questions. Why did the giant branch come to an end? And where did the sequence of the 'blue horizontal branch' fit into the story?

This question was answered when English astronomer Fred Hoyle went to Princeton. He had paid an extended visit to William Fowler at the California Institute of Technology in Pasadena, California, where the two men had worked on the synthesis of elements. Later, their work led to the B^2FH article mentioned earlier. Now in the summer of 1954, Hoyle was on his way home to Cambridge, England, and he stopped off at Princeton. The visit was very successful. Together with Schwarzschild, he found the answer to the question of when stars would stop progressing upward along the giant branch: when the growing mass of helium became very large, the pressure and temperature in the center of the core became so great that a new nuclear reaction started in which three helium nuclei fused to one carbon nucleus and again a small amount of remaining material was converted into radiation: $3\ {}^4He \rightarrow 1\ {}^{12}C$, a process that is known as the 'triple alpha' process because for

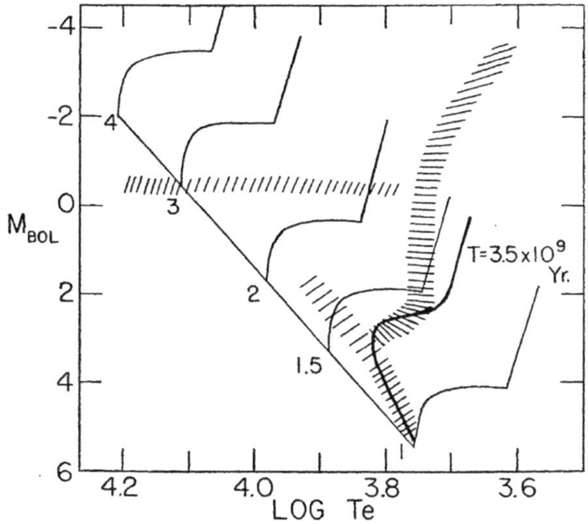

Fig. 2.4 A theoretical HR diagram. Horizontally is the temperature or the logarithm of the temperature from high to low; the vertical axis shows the amount of light emitted, logarithmically expressed in absolute magnitude. The straight line from lower right to upper left is the main sequence with the position taken up by a stars of 1.0, 1.5, 2, 3 and 4 times the mass of the Sun. The curved lines starting at the main sequence indicate the path that stars take when the hydrogen at the center of the star becomes exhausted and is replaced by freshly produced helium atoms. The dashed curves are population I and population II stars as observed by Baade in M31 (Sandage and Schwarzschild 1952)

historical reasons the helium nucleus ^{4}He is known as an 'alpha particle'. Hoyle and Schwarzschild were able to show that the onset of this process caused the red giant star to shrink to a fairly compact star on the 'blue horizontal branch' (Hoyle and Schwarzschild 1955) (Fig. 2.5).

This meant that thanks to Schwarzschild, Oke, Sandage and Hoyle, a physical explanation had been found for the position of every star in the HR diagram for M92. It was an impressive achievement. This insight has become the basis of our present-day knowledge about the ageing of stars and it was for the time being the endpoint of the research. However, the basis was still not complete: what would happen if all the helium in the core of the BHB star had been converted into carbon?

What happened after the composition of the core changed from helium to carbon is a question that could not be answered then, but the new insight that had been obtained could be put to good use, as was shown by the ever-energetic Alan Sandage. Just go back to Fig. 2.3 and look at the upper dashed line that marks the border between stars that are on the way to becoming red dwarfs and those that are less far advanced; this border is the 'turn-off' point. If you could calculate the age of the stars at this juncture, you would know the age of the star cluster, because what happened thereafter was so fast that the little time it took could be ignored. It was a new and interesting discovery. Together with photometrist H.L. Johnson, Sandage

Fig. 2.5 Martin Schwarzschild (1912–1997), an eminent astrophysicist and teacher. He was born in Germany as the son of Karl Schwarzschild, one of the predominant astrophysicists in the early nineteenth century. As a Jew, Martin had to flee Germany and built his career in the US. In 1947 he became professor of astronomy in Princeton. Source: AIP Emilio Segrè Visual Archives

set to work and determined the turn-off point of a number of star clusters, from which he was then able to determine their age. The age of M92, for example, was estimated at 5 Gj (5000 million years) and that of the well-known cluster of 'h and χ Persei' (see Fig. 2.6) was shown to be 80 times younger: 65 million years. This was an impressive achievement: a means had been found of measuring the history of our Milky Way. And even though the way forward proved much more difficult than expected, a solid start had been made.

In 1958 Schwarzschild published a textbook on the structure of stars that has since become acclaimed. This book is interesting historically because it marks the turning point in how stellar models are calculated. Mathematical differential equations were traditionally solved with calculations using pen and paper and at times a simple mechanical calculator. These calculations were mind-numbingly slow. The mid-fifties saw the advent of many more powerful electronic computers, which brought about huge improvements. Ever-faster computers have made the study of the development of stars into a cornerstone of modern astronomy.

Closely Coupled Binary Stars and Novae; Flows of Matter and Disks

In 1954 a discovery was made that caused quite a stir, although at the time nobody knew just how far-reaching the effects would be. There is a saying in Arabic: people saw the nose in the corner of the tent, but nobody had any idea how big the camel

Fig. 2.6 Image of the binary star cluster h + χ Persei. Both clusters are easily recognizable. There are also five yellow-colored stars: 'red' giants that are part of the cluster. Most stars in this image do not belong to the cluster, but are in front of or behind it. Source: Astronomy Picture of the Day, 2009 January 3. Roth Ritter

was that was standing outside. Here the nose of this proverbial camel was an article by Joy, an astronomer at Mount Wilson, about measurements of the star AE Aqr made over a period of more than 10 years. AE Aqr was an 'eruptive variable' star: it would suddenly become much brighter and would then slowly return to its former level, and at some unpredictable point in time it would explode again. Bright and dark lines appeared in the spectra. The wavelength of each line changed but the variations were strictly regular. This can be seen in Fig. 2.7.

This regular change in wavelengths was nothing new and the explanation for these changes was simple: AE Aqr was a 'spectroscopic binary star': although you only saw one star with your telescope, there were two orbiting around one another. AE Aqr was, however, unusual because the two stars were so close to one another: they completed their orbits in only 17 hours (within a single day!).

Stars that are so close together must have an effect on one another's shape. So, what did they look like? In general, there is no answer to this question, but in 1941 Dutch-American astronomer Gerard Kuiper revived an old theory that went some way to explaining the distortions (Kuiper 1941). I will address the old theory here in more detail because it will play an important role in the coming decades, including in X-ray astronomy.

Kuiper went back to the work of French mathematician Edouard Roche (1820–1883). Roche supposed that both stars were reduced in size until each was a small point. The two points orbited in circles around their center of gravity.

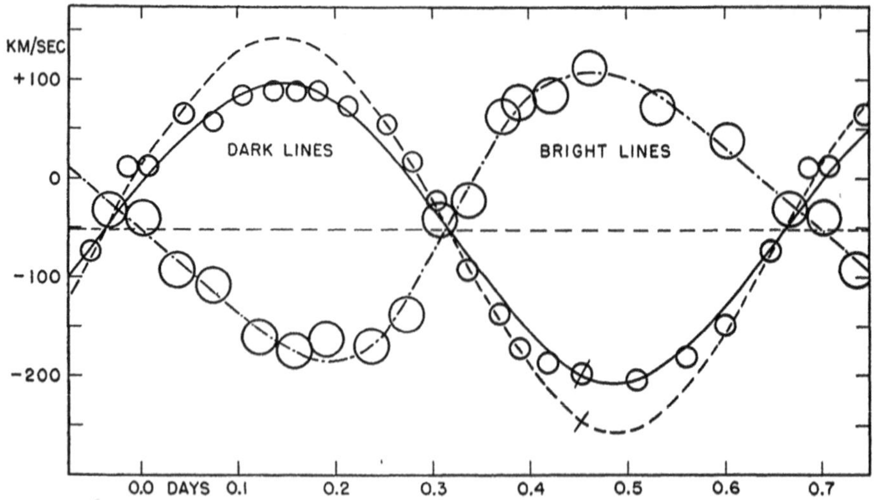

Fig. 2.7 The wavelengths of the darker and brighter lines in the spectrum of the star AE Aqr plotted (vertically) against the time (horizontal). The regular changes are explained as a consequence of the Doppler effect: AE Aqr comprises two stars orbiting around one another. The spectrum of one star is made up of bright lines, and that of the other star of dark lines. We see that in the course of a day (17 hours) first one star comes towards us and the other star moves further away; then it turns around (Joy 1954)

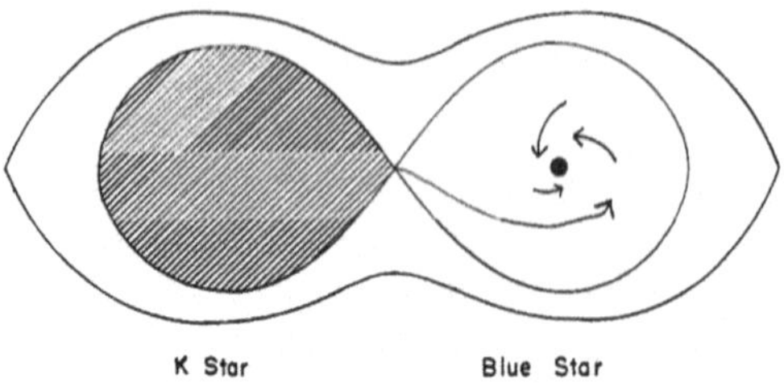

Fig. 2.8 A model for the AE Aqr star based on spectroscopic observations. You can recognize the horizontal figure 8, the Roche surface (Crawford and Kraft 1956)

Roche then calculated so-called 'equipotential surfaces': surfaces over which you could move without any effort. If you set a marble (or a satellite) on this surface in motion, it would never stop. In more detail for the curious and adventurous: equipotential surfaces are perpendicular to the sum of the gravitational and centrifugal forces; when one moves over the surface, the force is always perpendicular to the orbit and thus has no effect on the motion. There are an infinite number of such surfaces, but there is one special one and it is usually referred to as the 'Roche surface'. The surface looks like two 'teardrops', the sharp points of which are touching. You can also think of it as a figure eight turned on its axis. Each loop of the eight then forms a surface that constitutes the perimeter of the teardrop. See Fig. 2.8.

The point where the two teardrops touch one another is for historic reasons referred to as 'L1', the 'first point of Legendre'. At this point both stars exercise the same force of attraction, and a particle that reaches point L1 from one star can enter the region of the other star without any effort: L1 is a gateway between the two stars.

Kuiper now had the following idea. Suppose that one of these stars is a very small object, for example a white dwarf, but the other is an old star that is expanding. Ultimately the expanding star will be so big that it fills the Roche surface. Then matter from the expanding star will pass through the point L1 and enter the teardrop of the white dwarf. More will be said shortly about what happens next.

Two young American astronomers, J. Crawford and R. Kraft, continued Joy's work on AE Aqr. They added some of their own measurements and produced Fig. 2.8. On the left, a cool red dwarf star can be seen that had expanded to the point that its matter reached the point L1 (in other words: the star filled the Roche surface) and to the right a white dwarf can be seen. The mass of both stars was roughly the same and yet the ratio of the size of the two stars was indicated correctly: white dwarfs are actually very small! Material flowed via point L1 into the volume of the white dwarf and would eventually end up on this dwarf.

AE Aqr had been known as an 'eruptive variable', which brings us to the novae: stars that unexpectedly become much brighter—up to ten magnitudes (10,000 times)—within one or two nights and then over a period of a few months slowly fade, reverting to their previous level. This kind of eruption was sometimes repeated after many years (decades); in dwarf novae it recurred very frequently, although less violently. Novae had been observed for centuries, but so far there had been no explanation for this phenomenon.

In December 1934 a nova appeared in the constellation Hercules and reached a magnitude of 1.5, making it one of the brightest stars in the sky. The brightness slowly reduced and ultimately what remained was a variable star of the 15th magnitude catalogued under the name of DQ Her. In 1954, American astronomer Merle Walker—we came across him earlier in this chapter—discovered that DQ Her was a binary star where the two stars rotated in less than 5 hours (!) around their common center of gravity. Walker was able to show that DQ Her comprised a red dwarf star and a white dwarf. A short while later Kraft added something new to this image: the observations could be better explained if the light of the second star did not come from the white dwarf itself but from a light-emitting disk of gas around the white dwarf. Figure 2.9 shows four successive appearances of the double star.

Kraft demonstrated that material from the red dwarf had passed via the gateway in L1 to the area of the white dwarf. As a result of the rotation of the two stars, this material had ended up in a disk around the white dwarf, and while circulating, the material moved slowly inwards, in the same way that water circulates in a wash basin before disappearing down the plug hole. Having finally arrived on the white dwarf, the material would form a layer on the very hot surface. If the layer became thick, nuclear fusion could suddenly occur and there would be an explosion: the nova. Kraft's argument was convincing. A few years later he was even able to conclude that all novae were binary stars, like DQ Her.

Gas disks proved to play an important role in subsequent decades. However, they were not new in astronomy. Just as the planets form a disk around the Sun, Huygens discovered a disk around the planet Saturn in 1655. Disks form, as will be shown, around very young stars, white dwarfs, neutron stars and black holes: what these are will become apparent in Chap. 4. There is also a disk around the black hole in the center of our Milky Way. But these insights came later.

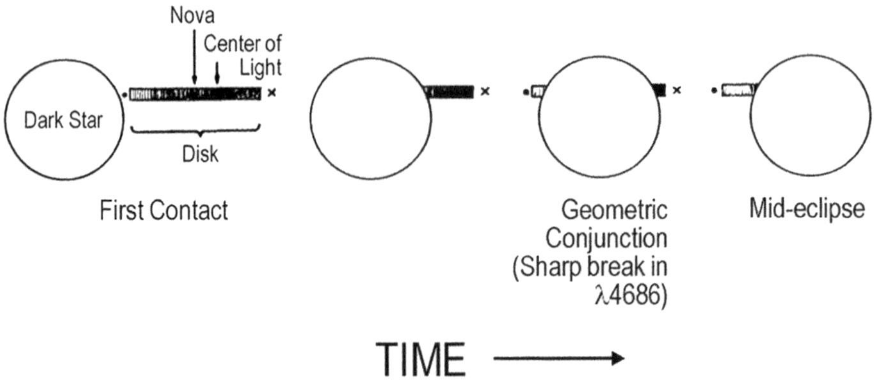

Fig. 2.9 Four successive appearances in which the binary star DQ Her is presented in less active periods. The binary consists of a normal, but intrinsically faint star and a white dwarf. Matter flowing from the normal star to the white dwarf arrives in a disk of hot gas surrounding the white dwarf. This disk is the most luminous part of the double star and determines its photometric behaviour. From time to time matter of the disk is added to the white dwarf and causes the nova explosion (Kraft 1959)

2.2 Galaxies

Research on galaxies after 1945 underwent an unexpected development. Chapter 1 mentioned the sensational developments in radio galaxies, but there were also spectacular results, from different radio observations, relating to our own Milky Way.

The Concept of 'Stellar Populations'

Walter Baade was a German astronomer at the Mount Wilson observatory—we have met him in the Prologue. He still had his German passport when Hitler declared war on the US and therefore Baade became an 'enemy alien'. He was not detained but had a very limited area around his home in Pasadena in which he could move freely. Luckily Mount Wilson Observatory fell within the limits and so he could still use the 100-inch telescope, the largest in the world. For Baade this had two advantages. First, most of his immediate colleagues were involved in War efforts and could not compete with him over observing time. Second, the streetlights in Los Angeles remained dark at night because of the (very remote) possibility of a Japanese attack by air. The sky from the observatory was much darker at night than it had been for many years.

In the fall of 1942 Baade took pictures of the central parts of M31 and of two companions of M31. He used red-sensitive photographic plates, something relatively new: photographic plates had always been most sensitive to blue light. Baade's photography was of very good quality. Adding already existing data from blue photographic plates, Baade could draw for the first time a Hertzsprung-Russell diagram of M31. It was surprising. The diagram was clearly the composition of two very different diagrams: one that compared well with HR diagrams of globular clusters and another that looked like a diagram of the stars near to the Sun. Baade drew the conclusion that M31 contained two different 'populations' (he named them Population I and II) and that each was distributed in a different way through the stars in M31. The Andromeda nebula was a composite system.

After the war had ended, Baade's report made a deep impression on the astronomical world. Baade had noted in his paper that a long time ago different populations had already been *de facto* recognized in the Milky Way without using the word. Many now embraced the word 'population' and several new populations were defined and given subtle names like population Ia, Ib etc. Remember, that at this time very little was known about stellar evolution; the nature of red giants had not yet been explained and a deeper understanding was lacking of the differences between stars of the different populations. When in 1950 Nancy Roman found that 'weak-line' and 'strong-line' stars had a different distribution in (local) space, this was seen as a clear sign of the coexistence of different populations. In the same light one could look at the discovery in 1951 by Chamberlain and Allen of stars with a chemical composition different from the 'cosmic abundances' (see above). Shortly thereafter Schwarzschild and co-workers gave a physical understanding of the HR diagram (see Chap. 1) and therefore of the differences between the two populations.

The concept of galaxies composed of various populations quickly became paradigmatic, so much that a special conference on this subject was convened in 1957 at the Vatican and with participants from over the world. In his review Baade emphasized that his study of the populations in various types of galaxies led him to two important conclusions that retained their significance during the following decades: (1) all galaxies, whatever their type, contain old stars, sometimes in very small numbers; therefore they equally old, whether they are elliptical or spiral galaxies; this implies that the different morphologies (spiral or elliptical) are *not* explained by a difference in age; (2) elliptical galaxies must have formed all their stars in one burst very early in the Universe.

The Milky Way

Interstellar Magnetic Fields

It has been noted in Chap. 1 that the magnitude of a star could be measured so much more precisely by photo-electric techniques than through photography. An unexpected result was the discovery in 1949 by two American observers independently from each other (J.S. Hall 1949; W.A. Hiltner 1949) of the polarisation of the light from distant stars. A remarkable conclusion was that the direction of the polarisation was by and large always in the same direction—see Fig. 2.10. The large distance of the stars and the constancy of the polarisation direction suggested strongly that polarisation was an interstellar effect. A little bit of the light from a star on its

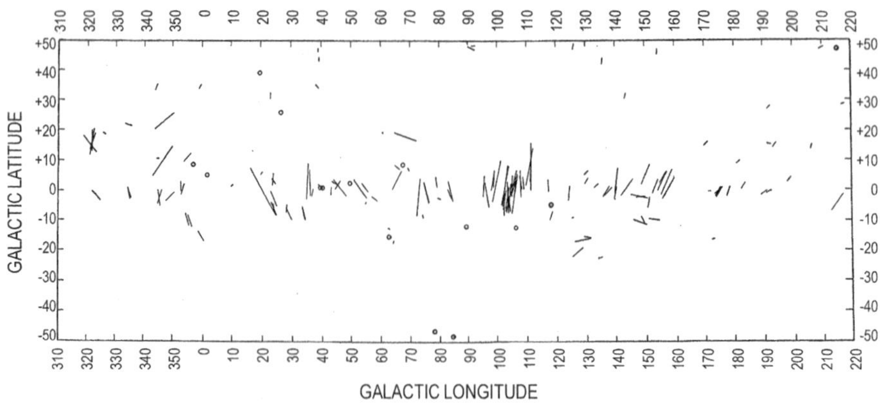

Fig. 2.10 The direction of the interstellar polarization over a long stretch along the plane of the Milky Way. The polarization of the individual stars are strongly aligned. This is best explained by a line-up of the interstellar particles by an interstellar magnetic field (Hall and Mikesel 1949)

way to us was led astray by interstellar dust particles. If these particles were lined up, one expected the light to become more and more polarized. Indeed, a few years later it was shown that an interstellar magnetic field could line up the interstellar dust particles well enough to explain the polarisation (Davis and Greenstein 1951). The existence of an interstellar magnetic field was generally accepted.

The discovery caused some sensation because a physical force never considered before was present in interstellar space. Magnetic forces could have considerable influence on the interstellar gas and perhaps contribute to the formation of spiral arms. The three-dimensional structure of the magnetic field made it difficult to study the detailed influence and the observations gave little insight into how the magnetic field was shaped. Too little was known about the field and there was little hope that much more information would become available in the foreseeable future. Estimates showed that on average the magnetic field was not as strong as the gravitational field and thus there was no need to overhaul all that was known about the interstellar space, but the effect of magnetic fields should not be overlooked. It gave the investigators something of a headache.

The radio radiation in the Milky Way was explained by synchrotron radiation, produced by relativistic electrons moving in a magnetic field. Analysis of the net polarization of the radio radiation offered a new way to study the structure of the magnetic field inside the Milky Way. After 1960, several such studies were made but they did not provide a breakthrough. The magnetic field remained an elusive topic.

Continuum Radio Waves

In the 1930s Jansky had already recorded radio waves from the Milky Way with the strongest radiation coming from the center of the Milky Way. Reber had confirmed Jansky's work but after 1945, when the interest of astronomers had been aroused,

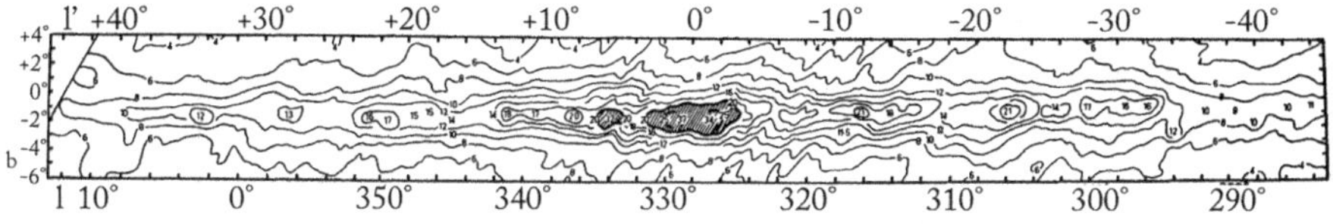

Fig. 2.11 Map of the Milky Way at a wavelength of 3.5 m. The galactic center is the middle of the map, where 0° can be seen in the frame at the top. Because the measurements were made in Australia, the map cuts off at the left-hand side (Mills et al. 1958). The horizontal scale is galactic longitude. On top is the new longitude scale with the zero point at the galactic center; below is the old longitude scale; the difference between the two scales is 327 degrees

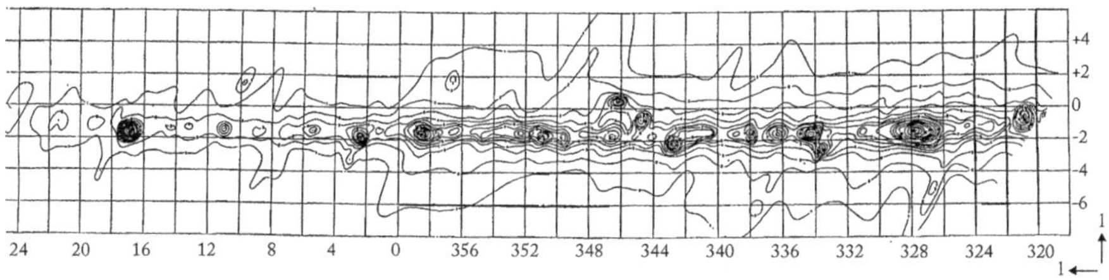

Fig. 2.12 Map of the Milky Way at a wavelength of 22 cm.; the horizontal scale is in longitude, old scale– see Fig. 2.11. The map shows a strip along the Milky Way that is noticeably shorter than the map in Fig. 2.11. The center of the Milky Way is the second strong source from the right. In its original form the map was made by G. Westerhout. He used the Dwingeloo radio telescope, where the center of the Milky Way comes just above the horizon. The map is therefore cut off on the right-hand side (Mills 1959)

the same work was repeated, this time with better instruments. In Figs. 2.11 and 2.12 the Milky Way can be seen as a ribbon of radio waves. The first of these two figures was made at a wavelength of 3.5 m, the other at 0.22 m, and there are considerable differences between the two.

First look at the map made at a wavelength of 3.5 m. The radiation forms a long, thin mountain range with a number of peaks. In the middle of the map there is a high ridge with the center of the Milky Way as one of the highest peaks. This recording did not provide very much insight into that center. Clearly, one had to wait for new observations that were able to provide sharper details. The other peaks could be individual sources, but the clarity of the figure was not yet good enough to determine this for sure. Nonetheless, information from two strong radio sources had already shown that these were the remains of supernovae: the Crab Nebula and Cassiopeia A (see Chap. 1). The other sources could also equally well be the remains of supernovae. The idea gained ground that with better radio maps it might be possible to find all the remnants of supernovae in our Milky Way and thus obtain a clearer idea of that still shadowy subject, the total explosion of a star, a supernova (see the prologue).

But even though the map had peaks, in between there was a plateau that had to be synchrotron radiation from the thin interstellar medium in the Milky Way. Synchrotron radiation has been discussed in Chap. 1.

The other map (Fig. 2.12) was made in 1957 at a wavelength of 22 cm. The map covers a much smaller part of the Milky Way. But it also shows a high mountain range with peaks. The highest peak is almost at the right-hand end and it is the center of the Milky Way. The mountain tops here are more pronounced than at longer wavelengths. The radiation in this map was almost certainly not synchrotron radiation; it would be too weak at this wavelength. The radiation came from hot gas in which the hydrogen atoms are ionised, in other words the nucleus, the proton, is separated from the electron. This kind of gas is encountered in light-emitting nebulae, such as those in Orion. All the peaks, except those at the center of the Milky Way are probably also these kinds of nebulae that had never previously been seen in such large numbers: the light from these nebulae was blocked by dark clouds but the radio radiation carried on unobstructed. A number of such hidden nebulae, for example W3, W44 and W51 (W stands for Westerhout), would cause some surprises in later research. Here, too, the positive conclusion was that radio astronomy provided the means to finally obtain a complete overview of all nebulae in the Milky Way.

The Leiden-Sydney Map of the Hydrogen Clouds in Our Milky Way

Chapter 1 reported on the discovery of the 21cm line in the summer of 1951. The American discoverers were physicists who were happy to have detected this radiation that had been predicted but had never been recorded. In Leiden and in Sydney, however, the discovery was only the beginning of a great adventure. Some explanation is needed here.

Hydrogen atoms emit photons at a wavelength of 21.10611... cm. In the Milky Way, one assumed, the clouds of hydrogen moved in circles around the galactic center and each cloud had a certain velocity. Part of that velocity, called the radial velocity, was along the line of sight from us to the cloud. The Doppler effect would then cause the observer to detect this cloud at a wavelength slightly different from 21.10611... cm. The difference should be less than a few percent, but large enough to be recorded. Figure 2.13 shows a measurement in some direction in the galactic plane. Three 'hills' can be recognized that each correspond with a separate hydrogen cloud. From the Doppler-shift one can derive the distance to the cloud.

When the 21 cm-line profile was measured in more and more directions along the galactic plane a very systematic variation in the form of the profile became apparent. This was explained by a simple model in which the clouds rotated in circles around the galactic center at a velocity that depended on the distance from the centre. The measurement of how this velocity varied with distance (called the 'rotation curve') made it possible to draw two strong conclusions.

First, it made it possible to derive the distance from the Sun to each hill in a line profile. That led to a map of the distribution of clouds of atomic gas in the Milky Way—see Fig. 2.14. Second, circular rotation implies a simple relation between the velocity of rotation and the gravitational attraction at that distance—similar to the way that the attraction by the Sun determines the velocity of rotation of each planet around the Sun. Knowing the gravitational attraction and how it varied across

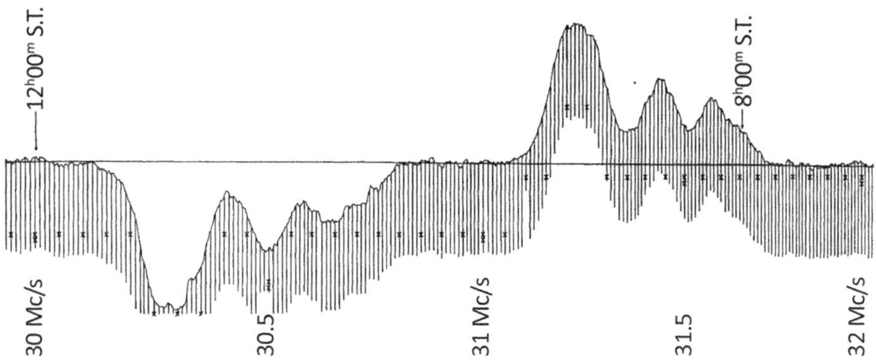

Fig. 2.13 The profile of the 21 cm line in some direction in the galactic plane. The profile has been measured twice, once positive and once negative

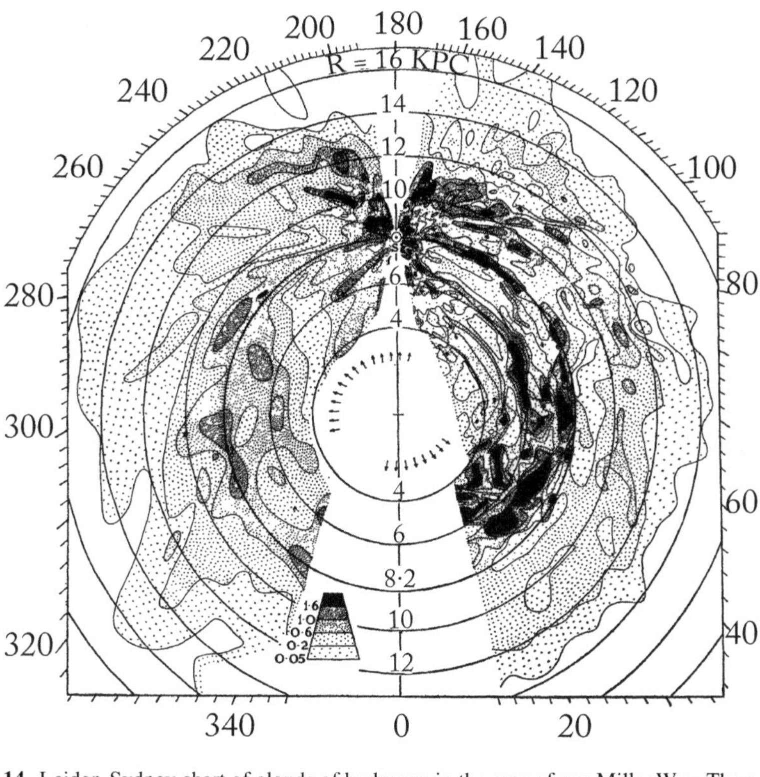

Fig. 2.14 Leiden-Sydney chart of clouds of hydrogen in the area of our Milky Way. The position of the Sun is indicated by a small circle around a dot. The center of the Milky Way is in the center of all circles (Oort et al. 1958)

the galactic plane made it possible to calculate the distribution of mass in the Milky Way. With knowledge of the rotation curve, it was possible to derive a mass model for the Milky Way—see figure farther down. All this was truly new and exciting.

As described in the Prologue, the 1930s had brought the disappointing conclusion that absorption of stellar light by small particles prevented us from ever seeing the center of the Milky Way. Now radio astronomy allowed the observer to escape from the confines of optical observations.

The 'Leiden-Sydney chart' of the distribution of hydrogen clouds in our Milky Way was the result of this Australian-Dutch collaboration. See Fig. 2.14.

Figure 2.14 was incomplete. There were open spaces, particularly around the center of the Milky Way. There were also some elongated structures that pointed like an accusing finger to the Sun—and thus at us, the arrogant humans who thought that we could explain the world. These structures (they appear also in other astronomical diagrams) are sometimes referred to as 'the finger of God'. In spite of these shortcomings, the chart showed convincingly that in our Milky Way the hydrogen gas is organised in spiral branches, and *that* was a great result. The first fundamentally better chart of the spiral arms did not come until 20 years later from optical (!) observations by a French duo, Y.M. Georgelin and Y.P. Georgelin.

The Galactic Center

Jansky and Reber had already shown that the galactic center is a strong source of radiation, long before the significance of this discovery had penetrated through to the astronomers. An example of a more detailed chart of the center was made at a wavelength of 2 cm; see Fig. 2.15. The most intense source just to the right of $l^{\rm II} = 0°$ had to be the center, which was confirmed by many of the studies over the subsequent 50 years. The center is not precisely at ($l^{\rm II} = 0°$, $b^{\rm II} = 0°$) as a result of incomplete information when the system of coordinates was defined.

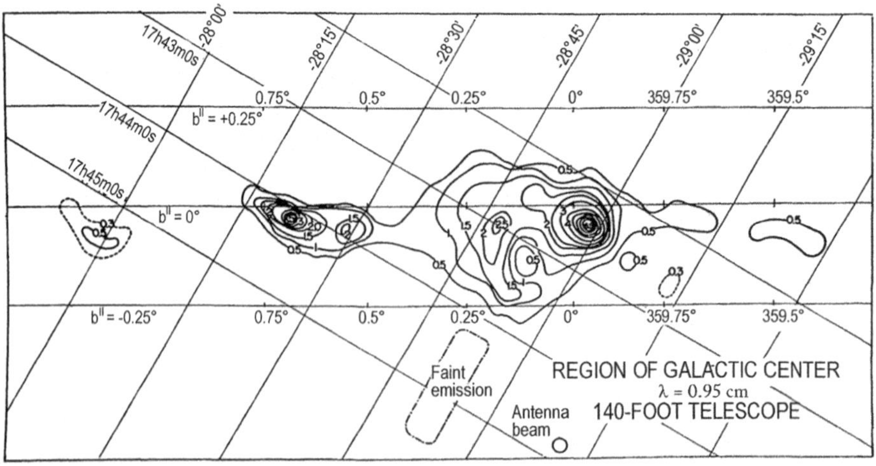

Fig. 2.15 The radio source A at a wavelength of 0.95 cm as measured by Drake (1959). The horizontal axis is the galactic longitude; the vertical axis is the galactic latitude. A grid based on right ascension and declination is also shown (Lequeux 1967)

After the completion of the Dwingeloo radio telescope in 1957, Oort started immediately to make observations of the 21 cm line in the direction of the center of the Milky Way, hoping to disentangle the gas motions there. But the work was not very successful. The measurements could not be explained clearly in spite of Oort's determination and the efforts of more than one good PhD candidate (Box 2.2).

Box 2.2 Oort

Jan Hendrik Oort (1900–1992) was born in Franeker, but grew up in Leiden where his father was a doctor in a psychiatric institution. In 1917 he went to Groningen to study maths and physics, and, encouraged by J.C. Kapteyn who was internationally one of the leading astronomers of his time, he decided to specialise in astronomy. In his first year, the small and slight Oort won the Dutch Varsity rowing competition as coxswain for the 'old four' of Aegir. Later he used this conquest to put into perspective the praise that was directed at him: 'I feel just as I did at Varsity; I was only sitting at the helm, the real work was done by other people.' Judging by my own experiences with Oort, this was definitely not false modesty.

In 1921 Oort took his doctoral exam and left shortly afterwards for the United States where he spent 2 years at Yale University. He returned to the Netherlands in 1924, but not to Groningen because De Sitter, Director of the Leiden Observatory, offered him an assistant position in Leiden, where he was to remain for the rest of his life. In 1926 he became observator, in 1930 a lecturer and in 1935 after the death of De Sitter, he was made professor by special appointment. In 1945 the Director of the Sterrewacht, Hertzsprung, retired and Oort succeeded him.

Oort gained international fame in 1927 with his research on the orbits of stars in the Milky Way. In 1918 Shapley at Harvard had shown that the center of the Milky Way should be at a great distance from us in the constellation of Sagittarius. Oort provided independent and convincing evidence that the stars in the Milky Way belong to a disk rotating around this center. He demonstrated that the stars we see in the direction of the center seem to pass us 'en masse', while in the opposite direction we in our turn overtake the stars. Oort explained this with 'differential rotation': if all the stars in the vicinity of the Sun orbit around the distant center of the Milky Way, then stars in the direction of the center will orbit faster and stars in the opposite direction will move more slowly. The solar system is a good example: in the orbit around the Sun, the Earth is overtaken by Venus but the Earth in turn overtakes Mars.

An anecdote about Oort's prestige. During the sixties, I spent some weeks as an observer in Dwingeloo, a task that was carried out by master students from Groningen and Leiden. On Friday afternoon you were taken from the telescope by company car to Hoogeveen station. But this particular time I heard that I would be going by the bus 'with which the senior and clearly important gentlemen from The Hague would be brought to the train'.

(continued)

The gentlemen arrived and an older man sat next to me: later it turned out that this was Dr A.J. Piekaar, Director-General for Science Policy at the Ministry of Education and a very powerful official. He clearly enjoyed sitting next to a young student whom he could ask all sorts of questions that he could not ask the people he usually talked to. At the end of the journey I asked him how it was that so much money was available for something as useless as radio astronomy. Was the subsidy weighed against other plans? And what were the arguments for making this choice? Piekaar smiled and said: 'The answer is simple. We give Oort (almost) all the money he asks for because we know that high-level research will be carried out with that money.' Many years later I was part of the subsidy allocation circus of the Dutch government, ZWO, the equivalent of the National Science Foundation in the US. But it never went as simply as Piekaar explained it. Nonetheless, I still believe Piekaar's answer. Oort was the exception, including within ZWO.

Oort may well have been friendly and good-natured, but he was by no means soft-hearted. He was able to ask very penetrating questions in a gentle way. He also had the talent of being able, at the end of a conference, to pose a simple question that most listeners recognized but hadn't yet formulated for themselves. On the other hand, he could admit his own limitations, untroubled and honestly. I have often heard him say admiringly: 'He is a really clever man,' or 'That's brilliantly reasoned; it's way beyond me.' His total dedication to astronomy and his 'diligence' (the word he used about himself) did not prevent the Oort family (two sons and one daughter) from making a happy impression. This was primarily due to Oort's wife Mieke, and Oort was grateful to her for this, as he said from time to time.

Oort was a wizard. He gave lectures and spoke with a mild tone of voice; he wrote in illegible handwriting on the board and generally stood in front of what he had written. But still, everyone looked on with fascination. I think that everyone understood that they were witnessing something extraordinary; they were being shown something that was hidden deep in the Cosmos. There were those who said: 'Oort's brain is built like the Universe. He only needs to look within himself and he sees the answer.' Oort had vision; he always looked further than everyone else. One effect of this was that there was almost always a natural distance between him and other people. He was friendly, but never overfamiliar.

We have had a great number of great astronomers in the Netherlands, more than the size of the country might warrant. I myself regard Oort as the greatest in the impressive list that includes Kapteyn, de Sitter, Hertzsprung, Pannekoek, Minnaert, Oort, Blaauw, van de Hulst and de Jager. He was one of the greatest, not only in the Netherlands but also internationally. When Oort passed away, the next day I received a request from the French radio to say a few words in his memory. And that for someone who at secondary school always failed his French, but, fortunately, was married to a French teacher! (Fig. 2.16)

Fig. 2.16 Oort 'wracking his brains' (his own words) at his desk in the Old Observatory in Leiden, 1960

M31, the Andromeda Nebula

In 1958 and 1959 de Vaucouleurs published maps of the two nearby galaxies M31 (the Andromeda nebula) and M33 (de Vaucouleurs published maps obtained by photoelectric measurements 1958, 1959). These did not produce important new insights, but they should be seen as the firsts of a new type of maps superior in quality to photographs. Many more such maps at all kinds of wavelengths (X-ray, radio etc.) would follow later and the intercomparison would be profitable.

The spiral nebula M31, also known as the Andromeda nebula, is the nearest large galaxy. In 1957 the brand new Dwingeloo telescope was used to measure 21m line radiation from M31. The 'rotation curve' of M31 was eventually found, in the same way as had been done for the Milky Way. In 1939 Babcock determined this curve by using spectra of light-emitting nebulae, but the radio measurements were of better quality (Fig. 2.17).

The curve of the Milky Way and that of M31 are similar. Look at the 'M31 adopted curve' and 'Galactic System final model' curves and start at the left in the center of M31 and our Milky Way. Observe that the curves reach a maximum and then slowly reduce. When the curves reach the right end of the diagram (at 75,000 light-years), the rotation velocity is still high. This was unexpected. On photos, M31 had a radius of around 20 kpc (60,000 light-years): beyond 20 kpc no stars were seen in M31. If there were no stars and thus no matter beyond 20 kpc then the rotation curve should decrease steeply. As this was not what happened, M31 still contained a lot of matter outside of the visible field. Van de Hulst, Raimond and Van Woerden

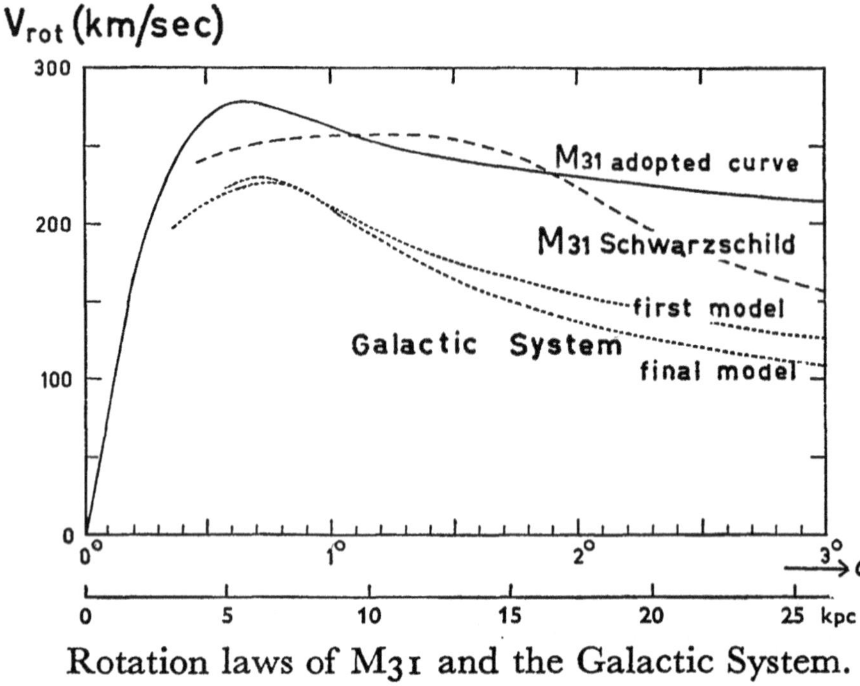

Fig. 2.17 The 'rotation curves' of our Milky Way and of the Andromeda nebula. It shows the velocity at which clouds of hydrogen gas circle around in our Milky Way and in M31. Both curves are based on 21 cm line profiles (van de Hulst et al. 1957)

noticed this and suggested that they needed to make a better search for matter in M31 that was so far outside that galaxy. Babcock had said the same in 1939. Twenty years later, as we will see, it was measurements such as these that led to the hypothesis of the existence of 'invisible matter', also known as 'dark matter'.

Clusters of Galaxies and Superclusters

In several articles in the 1950s, de Vaucouleurs showed what had been known more or less for several decades: if you plotted the galaxies brighter than the 10th magnitude they formed a long strip almost perpendicular to the Milky Way (Fig. 2.18). The impression is especially strong in the northern sky. The same plot reappeared when plots were made with a somewhat different selected sample of nearby galaxies. De Vaucouleurs proposed that all bright galaxies form a flattened system that he called the 'Local Supercluster'. The Local Group (the Milky Way and M31 plus a bunch of satellites) is near the outer rim of the supercluster and therefore this is seen best in only the northern hemisphere.

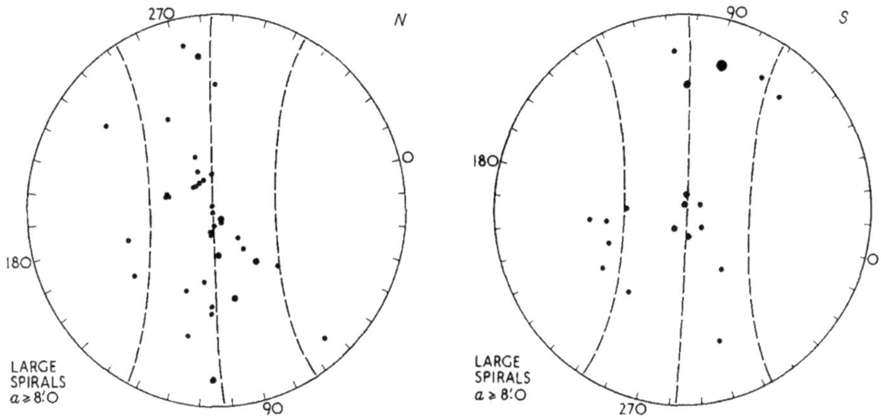

Fig. 2.18 The distribution of bright galaxies (brighter than the 10th magnitude) in the northern and in the southern sky with the Milky Way as the equator. Left is the northern sky, right the southern sky (de Vaucouleurs 1956)

De Vaucouleurs' proposal was rather speculative because this foundation is rather weak, but it has remained in the background until more evidence would be found. Whether true or not, it is an example of attempts to come to grips with the overall distribution of the galaxies and to solve the question of the structure of the Universe.

The large-scale distribution in space led to studies made in the early fifties by statisticians and astronomers from Berkeley (Cal). They counted the number of galaxies on a set of photographs. The statisticians, J. Neyman and E. Scott, and the astronomer C.D Shane came to the main conclusion that practically all galaxies belonged to a cluster although some of the clusters were very small. The clusters could also occur in collections of clusters; superclusters that were referred to as 'clusters of the second order'. A separate study by Shane and Wirtanen in 1954 led to the same conclusion, although this time derived in a more intuitive way by describing a contour map of the distribution of the galaxies.

The authors of these studies were definitely very competent scientists and yet, the conclusions were uncertain and unsatisfactory. The careful phrasing of the results suggests that there was no clear conclusion and that the truth of the matter escaped the authors. In retrospect this is not a surprise. The attempt was to derive a three-dimensional structure from a two-dimensional map with a very unsharp measurement of the third dimension, the distances of the galaxies, for which only the apparent magnitude of the galaxies could be used. Clearly the magnitude of a galaxy was not a reliable indicator of its distance. As we will see in later chapters a much clearer notion of the distribution is found after a large campaign to measure radial velocities from spectra: these were good enough proxies for measuring the distances.

The most important and lasting contribution to knowledge of the clusters of galaxies came from the systematic search for such clusters on the plates of the POSS, the Palomar Observatory Sky Survey. In 1958 George Abell published his catalogue of 2,712 'rich' clusters of galaxies. The word 'rich' implied that the cluster had at least 50 galaxies about as bright as the third brightest galaxy of the cluster. The largest redshift of the clusters was 63,000 km/s, 0.21 times the velocity of light. It took observers several decades before clusters were detected with much higher redshifts.

In 1951 Spitzer and Baade published a paper of lasting importance on clusters of galaxies. They noticed that spiral galaxies do not occur in dense clusters of galaxies and appeared to have been replaced by S0 galaxies. They argue that the S0 galaxies resemble disk or spiral galaxies but lack the interstellar matter and the spiral arms. They show that collisions between galaxies are much more frequent in clusters and state that collisions have robbed the spiral galaxies of the gas and so converted them into S0 galaxies. This is a subject that will be discussed many times again.

The Hubble Constant and the Age of the Universe

In the Prologue Hubble's 1929 discovery has been discussed of the expansion of the Universe with the reverse conclusion that the Universe had a definite beginning. In the following decades Hubble spent a lot of time working to give this sensational discovery a firmer observational basis. At the same time, theoreticians explored the basic theory that culminates in the Friedmann-Lemaitre-Robertson-Walker equation and the conclusions it led to. For further progress to be made, observations had to provide the value of two quantities: the expansion rate of the Universe, that is the precise value of the Hubble constant H, and the precise density of matter in the Universe, ρ, or—as an alternative for ρ—the deceleration of the expansion, a quantity usually called q. Hubble had determined the constant to be around 500 km/s/Mpc but could not find a value of q. The density also proved to be an elusive quantity. Unfortunately, both theory and observations had reached a dead end. Because the age of the Universe can be estimated by the 'Hubble time', $1/H$, the age of the Universe was 2 billion years, about half the age of the Earth. Here was a contradiction, that was not resolved for a long time.

This low tide in cosmological research continued after 1945. The only telescope that could bring important news would be the 200-inch. I quote from a contribution made in 1951 by Baade during a meeting on the occasion of the opening of an observatory in Michigan: 'Altogether, there are good reasons to believe that solution of the cosmological problem is much more difficult than was thought some 15 years ago and that it may well be beyond our present powers'(Baade, 1951).

In the following year, Baade announced at the General Assembly of the International Astronomical Union in Rome that Hubble had used (unknowingly) a wrong kind of variable star for his determination of H. Consequently the value of H was at least a factor of 2 smaller. It was sensational news that reached the newspapers in the Netherlands, and probably in many countries, with the statement that 'Baade had made the Universe twice as large and twice as old'.

Hubble died in 1953, at the age of 63. In 1956, Humason, Mayall and Sandage summarized what had been achieved in the subject of redshifts of distant galaxies. They published a catalog of over 800 redshift measurements, most of them obtained at Mount Wilson and estimated that about 63% of all bright galaxies in the northern sky had been observed. The largest redshift was $z = 0.23$, corresponding to a velocity of 63,000 km/s. The value of H was discussed and found to be 180 km/s/Mpc. Two year later, Sandage reduced its value to 75 km/s/Mpc. The present-day value is 72 km/s/Mpc; see Chaps. 16 and 17. The Hubble time was now 13.9 billion years, a value much more in line with ages estimated for globular clusters than the age previously estimated.

Gamow's Prediction of a Remnant of the Big Bang

Let's concentrate on the Universe immediately after the massive explosion when there were not yet any stars and galaxies and the Universe only consisted of photons and elementary particles, such as quarks. Theoretical physics had things to say about that Universe and from 1945 to 1956 George Gamow was the person who did the telling. As mentioned earlier, Gamow tried to find an explanation for the extremely divergent abundances of chemical elements, work that would later be treated more extensively and more convincingly in the famous 'B^2FH' article. But an apparent detail in Gamow's argument that did not appear in B^2FH later proved to be far ahead of its time. Gamow estimated how many photons had been created in the Big Bang and calculated that many of these were still present. He decided that those remaining could be measured as radiation, which was equally strong in all directions, and had a temperature of a few degrees above absolute zero–see Fig. 2.15. Gamow's idea was right, but it was so at odds with contemporary discussions that it was passed over. It was only in 1965 that Gamow's publications were remembered after two radio astronomers had discovered this remnant of radiation by chance—a discovery that earned them the Nobel prize—see Chap. 5.

'Steady State Theory' and Conflict

During the war many young British physicists were deployed on military research. Refugees from 'das Dritte Reich' (the Third Empire) who had come to England were initially interned as 'enemy aliens'; these included two young physicists, Hermann Bondi and Thomas Gold, both Austrian and Jewish. Fortunately, the British authorities recognised pretty quickly that these two were more useful in the lab than in an internship camp. They were assigned to Fred Hoyle, a physicist who had studied in Cambridge. At the end of 1940 all three had a place in the 'Admiralty Signals Establishment', a naval research institute. The three men became friends.

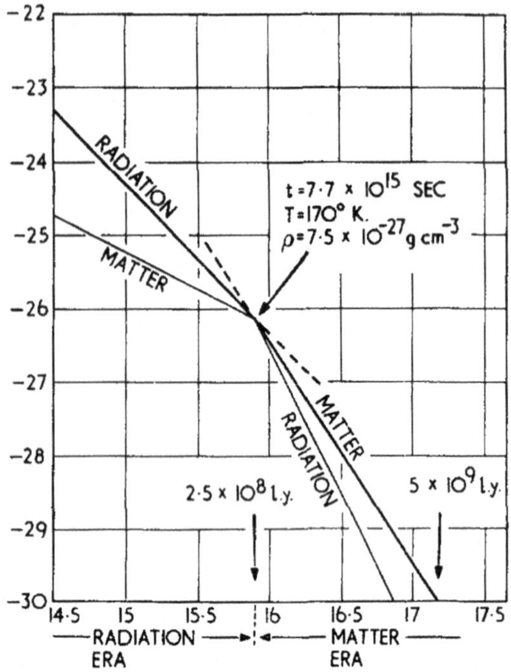

Fig. 2.19 The earliest evolution of the Universe. Horizontally is the logarithm of the time in seconds since the Big Bang. It starts at an age 14.5, and thus after $10^{14.5}$ s = 10 million years. The logarithm of the overall density in the Universe is given vertically. Matter and density develop differently in the course of time (Gamow 1956)

Bondi and Gold rented a house near to their lab and regularly had Hoyle staying with them, particularly if the weather was too bad for Hoyle to cycle home. By day they worked for the navy and victory, and in the evenings they worked for science. They discussed all manner of things and discovered a shared interest in cosmology. The expansion of the Universe with that absurd starting point was a recurring subject of discussion.

After the War, all three were given jobs at the University of Cambridge and their friendship continued. One evening in 1946 they went to see an English horror film (‘Dead of night’) that ends with images of the start of the film. While walking home, Gold suddenly said that, if new matter is constantly being created, the expansion of the Universe could continue indefinitely while everything looked the same. Gold’s idea was the start of their attack on the absurd concept that the Universe had had a starting point. They developed this idea further and in 1948 two major articles were published in which they formulated an alternative to the expanding Universe.

Bondi and Gold wrote a somewhat philosophically inclined article in which they introduced the ‘perfect cosmological principle’. This was the assumption that the Universe had always and everywhere looked the same. The name ‘steady state Universe’ underlined this. The most difficult point in the new theory was that new matter was constantly being produced out of nothing. This went against all existing principles of physics. However, if this creation of new matter proceeded slowly enough, this did not mean a contradiction of the known facts. Hoyle wrote the sec-

Fig. 2.20 From left to right Gold, Bondi, Hoyle in this photograph taken around 1950. Source: St John's College, Cambridge

ond article about how the general theory of relativity had to be adapted if new matter was continuously being created. Hoyle was also the first person to use the term 'Big Bang', which he did in a radio broadcast on the BBC: (paraphrased) 'conservative astronomers believe that the world started with a "Big Bang" and do not understand that there is a much simpler explanation for the facts, namely ours'. (Fig. 2.20).

There was a simple, but difficult to carry out test to decide whether we live in a Big Bang or in a steady state Universe. Take a galaxy at a distance of 10 billion light-years. The light that we now receive left that galaxy 10 billion years ago and shows the galaxy as it was 10 billion years ago. The Big Bang theory said that the Universe was a lot younger at that time and that the galaxy may have looked very different then. In 'steady state' theory, the galaxy looked the same as ours: Big Bang theory predicts that the Universe is evolving whereas steady state says that it does not.

This test may have been simple, but unfortunately galaxies at a distance of 10 billion light-years were much further away than even the biggest telescope could find. It was therefore not possible to make a choice between the two theories. Nonetheless, I believe that most astronomers were convinced of the Big Bang theory but their choice did not rest on hard facts but on emotional preference, and Hoyle was someone to sense that faultlessly and to demonstrate it solidly. On this particular issue of 'is the steady state theory correct or not', he found himself in a major conflict with his close colleague in Cambridge, Martin Ryle, one of the great pioneers of radio astronomy.

Under Ryles' leadership the radio astronomy institute at Cambridge University systematically searched the northern sky for all radio sources. In 1950 the first catalogue (1C) was published with 50 sources and in 1955 the 2C appeared with 1926 sources. The 2C contained so many more sources because the new receiver was much more sensitive. It could be assumed that these sources were all galaxies far from our Milky Way. Ryle counted the number of sources in his 2C catalogue and drew the conclusion that there had previously been more radio galaxies than there were at the present time, and that therefore steady state theory was incorrect. On 6 May 1955 Ryle gave the famed Halley lecture in Oxford at which he said: 'There seems to be no possibility of explaining the observations using steady state theory.'

Given Hoyle's flamboyant personality, this was a declaration of war. And within 2 years he received unexpected support from Australia. There, Mills and Slee, hardened radio astronomers, carried out a survey of the southern sky at the wavelength of 3.5 m (Mills and Slee 1957). The surveys made by Cambridge and Sydney shared a large portion of the sky and there were major differences. Mills and Slee wrote plainly, and with good arguments, that the 2C catalogue from Cambridge contained many sources that did not exist at all. There were by no means too many radio galaxies at a great distance.

The Cambridge radio astronomers found the conclusion difficult to accept, but did not dispute it; they smelled a rat. As a last resort, the radio astronomers carried out a new survey, the 3C that contained far fewer sources than the 2C and was so reliable that it became a standard work and repaired the reputation of the radio group in Cambridge. They were forgiven their one blunder.

But the Ryle-Hoyle controversy continued, reaching a peak of irritation and a nadir in friendly working relations. Using the 3C-catalogue Ryle made a new analysis and thought that this time he would be proved right. And so Hoyle received an invitation to attend a press conference in London, together with his wife. The conference was organized by Mullard Ltd., an electronics company and an important sponsor of Ryle's radio observatory. The Hoyles arrived and were received politely, after which Barbara Hoyle was given a seat among the journalists but Hoyle was seated on the podium. Ryle then came onto the stage and gave a presentation that ended with his saying that he now had 'definitive evidence against the steady state model' and the 'the Universe had once been much more densely populated with radio galaxies than now—'and therefore it is not in a steady state'. It was apparently great news. The headline of the Evening Standard newspaper was 'The Bible was Right' (meaning: the Universe is not infinitely old, it was once created) and the next day other newspapers said something similar. The general conclusion was that Hoyle was beaten. But the collegiality had been dealt a blow—it was widely regretted that a scientific dispute had been fought out in this way. A couple of days later Ryle called Hoyle and apologised—as far as he was concerned this whole affair had got out of hand and had become a highly regrettable incident. But Hoyle could not forgive him and the situation was at best a ceasefire. One of the effects was that there was a kind of iron curtain in Cambridge between the Cavendish lab, Ryle's institute, and the Observatory, Hoyle's base. Even students from one institute did not visit the other. The situation was generally referred to as 'dismal'.

Many analyses have been made of this 'battle of the Titans'. It was an illustration of the human condition. Two men with big egos, both highly competent, find themselves in a conflict and can see no way out or more probably, they do not want to see any way out. They were men with very different, even contradictory backgrounds.

Hoyle was an 'upstart'; he was born in Bingley in West Yorkshire where his father had a clothing business. The family had suffered strongly during the crisis period of the 1920s and 1930s. Fred was the first in the family to have a university education. He was brilliant and he was wild. As a lecturer he could be very inspiring, but he was also often absent. There were some students who were very much against this, while others could easily live with Hoyle's habits. As an expression of his brilliance and originality, Hoyle wrote books, predominantly highly successful 'science fiction' novels such as the 'Black Cloud'. Some of them have been made into TV series.

Martin Ryle was different in almost all those aspects that make a man what he is. He was the son of a professor of medicine in Oxford who was also physician to the king. Ryle was reserved and strongly focused on his work. He had built up his own institute and was one of the most important founders of radio astronomy. He was a careful supervisor of students and as head of the laboratory was well respected by his colleagues. Hoyle and Ryle were both members of the Royal Society; they were both raised to the nobility ('Sir Fred' and 'Sir Martin'). Ryle received the Nobel Prize for his contribution to the development of radio astronomy. Hoyle did not receive a Nobel Prize but his friend William Fowler did and that for the work that he conducted with Hoyle. External experts had the firm impression that Fowler and Hoyle had contributed equally and that both deserved the Nobel Prize.

The controversy between Hoyle and Ryle lasted for years. Neither of the two titans ever admitted he was wrong. When 'background radiation' was discovered in 1965 (see Chap. 5), the 'steady state' theory became untenable: so Hoyle was wrong. But did that mean that Ryle was right? Opinions are still divided. I quote American radio astronomer Ken Kellermann in his 'Helen B. Warner lecture' dating from 1972. He concluded that '.... the message that there were far too many weak radio sources was strongly exaggerated...'. In short, both men were wrong.

References

Arp, H.C., Baum, W.A., Sandage, A.R.: Astron. J. **57**, 4 (1952)
Baade, W.: Publications Observatory of Michigan. **10**, 7 (1951)
Burbidge, E.M., Burbidge, G.R., Fowler, W.A., Hoyle, F.: Rev. Mod. Phys. **29**, 547 (1957)
Chamberlain, J.W., Aller, L.H.: Astrophys. J. **114**, 52 (1951)
Crawford, J.A., Kraft, R.P.: Astrophys. J. **123**, 44 (1956)
Davis Jr., L., Greenstein, J.L.: Astrophys. J. **114**, 206 (1951)
de Vaucouleurs, G.: Vistas in Astronomy. **2**, 1584 (1956)
de Vaucouleurs, G.: Astrophys. J. **128**, 465 (1958)
de Vaucouleurs, G.: Astron. J. **130**, 728 (1959)
Gamow, G.: Vistas Astron. **2**, 1726 (1956)
Hall, J.S.: Publ. US Naval Observatory. **17**, 275 (1949)
Hall, J.S., Mikesel, A.H.: Astron. J. **54**, 187 (1949)
Haro, G.: Astrophys. J. **115**, 572 (1952)
Herbig, G.: Astrophys. J. **113**, 697 (1951)
Hiltner, W.A.: Science. **109**, 165 (1949)
Hoyle, F., Schwarzschild, M.: Astrophys. J. Suppl. **2**, 1 (1955)
Joy, A.H.: Astrophys. J. **120**, 377 (1954)

Kraft, R.P.: Astrophys. J. **130**, 110 (1959)
Kuiper, G.K.: Astrophys. J. **93**, 133 (1941)
Lequeux, J.: I.A.U. Symposium. **31**, 393 (1967)
Mills, B.Y.: Publ. Astron. Soc. Pac. **71**, 267 (1959)
Mills, B.Y., Slee, O.B.: Aust. J. Phys. **10**, 162 (1957)
Mills, B.Y., Hill, E.R., Slee, O.B.: Observatory. **78**, 116 (1958)
Oke, J.B., Schwarzschild, M.: Astrophys. J. **116**, 317 (1952)
Oort, J.H., Kerr, F.J., Westerhout, G.: Mon. Not. R. Astron. Soc. **118**, 379 (1958)
Sandage, A.R., Schwarzschild, M.: Astrophys. J. **116**, 463 (1952)
Schönberg, M., Chandrasekhar, S.: Astrophys. J. **96**, 161 (1942)
van de Hulst, H.C., Raimond, E., van Woerden, H.: Bull. Astron. Inst. Netherlands. **14**, 1 (1957)
Walker, M.F.: Astrophys. J. Suppl. **2**, 365 (1956)

Chapter 3
1960–1974: Developing New Instruments

The 15 years from 1960 to 1974 were more important for modern astronomy than the other 15-year periods in this book. Discoveries were made that changed fundaments of astronomy: newly born stars, masers, neutron stars, black holes, quasars, the cosmic background radiation and the presence of more 'dark matter' than there was luminous matter. Infrared astronomy and X-ray astronomy became established. Space research opened new avenues. It were amazing times and the astronomers were fully aware of what happened.

In the years from 1960 to 1974 astronomy needed and received an increasing array of new and large-scale instruments, and developed into 'big science' involving huge sums of money. I could put this in financial terms, but it is probably more convincing to use the newly constructed observatories and their telescopes, and the space projects carried out by astronomers to illustrate the extent of the investments. Anyone looking at the figures in this chapter will understand that these construction projects called for massive investments, and that the annual staffing and the maintenance costs increased enormously. Take a simple example of the change in style: before 1960, astronomers working in the Netherlands would come to the annual Netherlands Astronomers' Conference by bicycle. These conferences were held in youth hostels or very simple conference centres. However, after 1960 this Spartan style disappeared; everyone was carried along on the general tide of rising prosperity. Conferences were held in hotels or resorts and more and more researchers attended conferences on other continents.

3.1 Optical Astronomy

Plans for New and Bigger Telescopes

The commissioning of the world's biggest telescope, the 5-meter instrument in California in 1948, made it abundantly clear to the European astronomers just how far they were lagging behind. The War had impoverished Europe but as a reaction to

H. J. Habing, *The Birth of Modern Astronomy*, Historical & Cultural Astronomy,
https://doi.org/10.1007/978-3-319-99082-8_3

this there was now a preparedness to work together on a scale that had not previously been possible. In 1952 Walter Baade, who had been a staff member at Mount Wilson and Palomar since 1931, spent several months in Leiden. He had retained his German nationality and his interest in Europe was still very much alive. In discussions with Oort, the plan was conceived to construct an observatory in South Africa with in any event a telescope 3 meters in diameter. It should be a collaborative enterprise of many European countries. Oort's excellent reputation brought together a dozen leading astronomers from Belgium, Germany, France, the Netherlands, Sweden and the UK. In the stately Senate Room at Leiden University they signed a solemn statement setting out their intentions.

It was after this that their problems started. They had no great difficulty persuading officials and even ministers of the importance of their plans, but getting them to sign a joint contract was a different matter altogether, although there was already a good example of European collaboration in the form of CERN in Geneva. In all the subsequent negotiations on the South African telescope, Germany continued to be a reliable partner. The English were initially very positive but at a late stage in the negotiations they started to procrastinate, and pulled out when a plan was developed to construct two large telescopes, one in England and a second together with the Australians in Australia. Right from the outset, the discussions with France were difficult; governments in the country came and went, and there was also the revolt in Algeria to be dealt with. The frequently reiterated death knell was: 'Relax; we're all in favor of this observatory, but not right now.' On 1 March 1960 Oort wrote to his French colleague André Danjon: 'Mon cher ami, je viens de passer une demie nuit sans sommeil avec mes soucis concernant l'ESO…' (Dear friend, I have been kept awake half the night because of my concerns about ESO) and on 3 March Danjon wrote a long letter back in which he explained how many discussions he had held and at what level, but he had come away from these discussions with nothing more than a half-promise. The letter ended: 'Toute personne approchant du gouvernement se rend compte qu'il est obsédé par l'Afrique du Nord et non par l'Afrique du Sud' (Anyone close to the government is obsessed with North Africa, and not with South Africa). But then 2 years later, on 5 October 1962, Danjon wrote to his German colleague Heckman: 'Mon cher Collègue, Un mot seulement pour vous confirmer que les représentants des cinq pays ont signé la Convention aujourd'huid à midi! Alleluia!' (Esteemed colleague, just a word to confirm that the representatives of the five countries signed the covenant this morning at noon. Halleluja!).

ESO, the dream of European astronomers, had become a reality. German cosmologist Heckman was appointed as the first Director-General, and Adriaan Blaauw from Groningen became Scientific Director. Studies had been carried out over many years to determine the air quality at different locations in South Africa in order to decide on a site, and the partners were on the verge of reaching a decision when the news reached them that Americans had found locations in the Chilean Andes that were far superior to anything that South Africa had to offer. In November 1963 ESO opted for Cerro La Silla, a mountain top in Chile, hundreds of kilometers to the north of Santiago, on the edge of the Andes. It was an unexpected deviation, but one that has never been regretted.

The mountain was bought from the Chilean government; a road was laid, there was successful drilling for underground water, and construction started on the domes of the telescopes and a range of outbuildings. In March 1969, the observatory was opened in the presence of the President of Chile, Don Eduardo Frei, and of official representatives of the participating countries. Unfortunately, the observatory was far from complete. There were already two smaller telescopes, but the large 3.5-meter instrument, which was the objective of the whole undertaking, was still in parts in the workplace in Hamburg, Heckmann's territory. When Blaauw succeeded Heckmann as Director-General a year after the opening, his first task was to speed up the construction of the 3.5-meter telescope. He decided to move the construction work to Geneva, to one of CERN's buildings, where he believed there was more expertise available for carrying out large-scale scientific projects. He was proved right, but it still took a further 7 years before this large telescope was finally commissioned. This was 25 years after the plans were originally conceived by Baade and Oort (Fig. 3.1).

While ESO was being set up, a similar collaboration was formed in the US. The aim was to build two large telescopes, one in the US and another in the Andes, both of which would be accessible for American astronomers who had no access to Mount Palomar and Mount Wilson. In the 1950s, an association was created of around 30 (American) universities: AURA, the Association of Universities for

Fig. 3.1 The European Observatory on La Silla (Chile) at sunset around the year 2000. Source: ESO

Fig. 3.2 Picture of the Kitt Peak observatory in 1984. Source: NOAO/AURA/NSF

Research in Astronomy. AURA built an observatory on Kitt Peak, a mountain peak in Arizona where a 4-meter telescope was commissioned on 1 January 1976. Figure 3.2 shows the Kitt Peak observatory in 1984. At the same time, AURA also built an observatory with a 4-meter telescope in the southern hemisphere on Cerro Tololo in Chile, not far from La Silla.

And the big English telescope? It was partly a sorry farce and partly a success. The UK constructed a 2.5-meter telescope, which was named after the country's greatest scientific hero, Isaac Newton. In 1967 when the telescope was ready, the financiers decided that the telescope should be located in England. The astronomers were distraught because they realized that such a telescope could only be effective in clear, dry air—a situation that was rare in the dear fatherland. It took until 1984 before the telescope was eventually transferred to La Palma and the clean air there. This fiasco was offset by the successful story of the 3.9-meter Anglo-Australian telescope, commissioned in 1974. Although the weather conditions in Australia were not as excellent as in the Andes, a great deal of valuable research was and is carried out with the 'AAT'.

Chapter 1 already explained that for technical reasons it was not possible to construct telescopes much larger than 5 meters. In 1973, American astronomer Art Code, a successful inventor and engineer of instruments, summed up the possibilities of constructing large telescopes but he failed to mention the actual solution that appeared a short time later: the possibility of keeping thin mirrors in the right shape using computers. This solution will be discussed in a later chapter.

Below the Seeing Limit

Atmospheric irregularities passing through the field of view of a telescope erase details and blur the image. The recorded image of a star is a blot of arc-sec size and we never see the stellar surface, nor the true nucleus of a galaxy. The desire to remove the seeing effects needs no further explanation. In the period 1960–1974 two different methods were proposed that successfully removed at least part of the blurring effect.

R. Hanbury Brown proposed the first method. He was a British radar expert during the War and a pioneer of radio astronomy afterwards. If the flux of a star is recorded, it can be seen to fluctuate slightly due to variations in the Earth's atmosphere. Brown, together with the physicist Twiss, discovered that if two optical telescopes recorded the flux of a given radio source independently from each other, the fluctuations in the two recordings were not random but 'correlated': if detector A recorded a momentary flux value above the mean value, then there was a definite probability that detector B would also record such an enhanced flux. This 'correlation' seemed at first highly improbable and it was disputed by fellow physicists on fundamental principles. Ultimately, Brown and Twiss were proven to be right and the 'Hanbury Brown-Twiss effect' was confirmed as correlating with quantum theory. Brown showed that one could deduce the diameter of a star from the strength of the correlation. He moved to Australia and constructed an optical interferometer with two telescopes on a circular track with a diameter of 188 meters. He used this instrument to measure the size of several dozens of stars, data that have since been confirmed. Brown had, however, no successors.

In around 1970, French astronomer Antoine Labeyrie developed another method to obtain sharper images. He photographed stars very briefly but many times in succession and saw on the photographic plate a random distribution of very small 'grains' or 'speckles'; each of these is a sharp but very weak image of the source. He then showed that the original, undisturbed image could be restored from these speckles by a statistical method; the reconstruction involved a Fourier transform. Labeyrie's approach was put into practice successfully, showing that some stars that we see as a unit are in fact two stars that are separate but very close together.

Hanbury Brown's and Labeyrie's work had considerable influence on the development of optical interferometers some 20 or more years later when new recording techniques and faster computers made it much easier to process very brief observations. In the short run, the immediate return in astrophysical knowledge was very limited; the results mainly confirmed what was already known from accurate spectral photometry. However, it certainly helped in gaining confidence that the model stars of theory closely resembled the real thing.

3.2 Radio Astronomy

Radio waves have the same characteristics as lightwaves, except that the wavelength is much longer: 2 meters versus 500 nm, a ratio of 4 million to 1. This has all kinds of important consequences, for example for the sharpness of the image, as we saw in Chap. 1. Another consequence of this difference is that a light particle, or photon, has four million times more energy than a radio photon. In the very first years of radio astronomy, a number of eminent optical astronomers concluded that radio astronomy could have no actual significance for real (in other words, optical) astronomy, but after the discovery of giant reservoirs of energy in the 'radio galaxies', these eminent men were proven wrong (see Chap. 1).

In around 1960 it was realized that there were at least three technical limitations that radio astronomers needed to remove. Firstly, the images that the radio telescopes produced needed to become much sharper. Secondly, the telescopes needed to become more sensitive so that fainter radio sources could be detected. Thirdly, more receivers were needed for making measurements at increasingly shorter wavelengths.

A solution for the problem of the sharpness of the images was to make larger radio mirrors, as has already been mentioned. However, this was of very limited use: radio telescopes that could recognize details of 1 arc minute, for example, would need to have a diameter of several kilometers! Surprisingly enough, there appeared to be a very different and elegant solution. Ryle and Hewish, both from Cambridge (England), published their invention in 1960 (Ryle and Hewish 1960).

To explain what they proposed let us return to radio astronomy in 1947 in Australia. The fresh, new radio astronomers Pawsey, Payne-Scott, McCready used the sea interferometer (see Chap. 1) to follow the radiation of the Sun as it rose above the horizon. For their analysis they wrote down a mathematical equation that mimicked the rising signal. To their surprise, they noted that the equation was well known in mathematics as a 'Fourier transformation'. They jokingly remarked that if they could make their measurements from a large number of cliffs at different heights, they could use a mathematical technique called 'reverse Fourier transformation' to obtain a much sharper image of the Sun. But as there was only one single cliff available ('Dover Heights' near Sydney), they left it at that; they did not develop this coincidence any further.

To understand their statement better, think of cryptography. If you want to send a secret written message to someone far away, you code the text and then send the coded message. The receiver knows your code and by decoding the message he can read the text. Now take the image of the Sun. Pawsey and Payne-Scott realized that they actually measured one point of the *coded* (=transformed) image of the Sun. If there had been many cliffs, they could have measured more points of the transformed image and decoded the image (mathematically: reversed the Fourier transformation).

In the 1950s, Ryle took two small telescopes and combined these into one interferometer. He made sure that the signal arriving in each telescope was transported to the receiver by cables of exactly equal length—this is an absolute necessity for the method to succeed. In the receiver the two signals were added. Like Pawsey and Payne-Scott, Ryle realized that this measurement gave him one point of the coded image of the sky within his antenna beam and that this code was a 'Fourier transformation'. Subsequently he put many times one of the two telescopes at a different location and added the two signals in his receiver. After he had measured at enough different points he entered all signals in a computer, applied the reverse code ('reverse Fourier transformation') and obtained a detailed image of the part of the sky within his antenna beam. If he had wanted a measurement made with a radio telescope with a diameter of, say, 10 kilometers, he would have made many interferometer measurements until the largest interferometer distance was 10 km. *Many interferometer measurements with two small telescopes replaced the one measurement obtained by a single, large telescope*. That the method worked in practice was shown in a map with many small radio sources some of which could be identified with optical objects. He named his technique 'aperture synthesis', because by adding up many different measurements you obtained the same result as you would have achieved with a telescope with a 'large aperture'. Unsurprisingly, Ryle received the Nobel prize for physics in 1972, especially for his 'invention of aperture synthesis'.

There were two practical downsides of the method: one, it was very time-consuming to calculate a Fourier (reverse) transformation using pen and paper. What was needed in practice was a fast, programmable computer. In 1947 no such instrument existed, but in 1960 it did. Second, if you had only two telescopes—enough for one interferometer—it was a seriously time-consuming business to measure a large number of points of the transformed image. You could speed up the process by using many more telescopes (an 'array'), with each of them operating as an interferometer with each of the others. This is illustrated by Fig. 6.5 in Chap. 6, of the 'Very Large Array', the American 'synthesis telescope' built in 1980.

The greatest distance between the two telescopes determines the minutest detail in the final image that the synthesis telescope can measure. Now suppose there is a theoretician who proposes that at the heart of a radiogalaxy like M87 (Virgo A) is a 'supermassive black hole' (Salpeter 1964), a really small object that at the distance of M87 has a size much, much less than 1 arcsecond. Can radio astronomers prove that much (most?) of the radio radiation is from such a small object in the center of M87? An interferometer with a 'base line' (the distance between the two telescopes) of N times λ is given by 57.3°/N (see earlier) and to make the outcome equal to 1 arcsec it follows that N = 200,000. The astronomer decides to make the observation at a wavelength of 1 m and thus the base line has to be 200 km (=200,000 m). The technical obstacle now becomes the requirement that one knows the distance between the two telescopes better than, say, one-tenth of a wavelength, or 10 cm.

Geodetical measurements were not good enough to achieve such an accuracy, but there was a solution: use the measurements themselves. Record the arrival time of each signal with an accuracy better than one billionth of a second. Add the two signals in the receiver by giving one of them a little delay and check whether the sum of the two signals is as expected. If not, change the delay time slightly and repeat the same procedure. Continue until the expected signal ('the fringes') appears.

In the 1960s various groups in Canada, the UK, the US and in Sweden experimented with 'Long Baseline Interferometry' steadily increasing the baseline of successful observations. First they used telephone lines or radio emitters to send the data (signal strength and arrival time) to the central station and were successful with baselines of a million wavelengths—several hundreds of kilometers. There they reached a natural end: the telephone lines and the radio connection with the receiver became unreliable. Then, in 1967, this hurdle was overcome. The two telescopes were operated as two separate telescope and the signal was recorded on a magnetic tape together with the signal of an atomic clock. That was good enough: the arrival times had been measured so accurately that by 'shifting the arrival times a bit backwards and forwards' the experimenters found the precise time when 'fringes appeared'. The new technique was called 'Very Long Baseline Interferometry' and became known as 'VLBI'. In 1967 two groups in the US and a combined Canadian-UK group found 'fringes' over distances of many thousands of kilometers. The history of these experiments is well described in Cohen (1969).

An important by-product of VLBI was that once 'fringes' had been seen one could derive the distance between the two telescopes with an accuracy of a fraction of the wavelength. In the early 1970s I was an accidental observer when a successful experiment was carried out between Massachusetts and Sweden, and it was concluded that the distance between these two locations had increased by a couple of centimeters over the previous few years: the amount that the geological plates of the American and the European continent had moved apart. VLBI could play a useful role for geology.

Let us look now at other requirements for the construction of a radio telescope. Firstly, consider the location. A radio telescope does not suffer any interference from the atmosphere and so can easily be built at sea level. However, you do have to locate it away from the civilized world because of the many sources of interference; ideally such telescopes should be sited in lonely, isolated spots. Then, each telescope in the array has to have a large surface area: the bigger the surface, the weaker the signal that can be recorded. But just how weak a barely measurable signal is will also be determined by the quality of the detector or receiver at the focal point of the telescope. Every receiver itself produces noise, which cannot be distinguished from the radiation that is being sought, and this noise becomes stronger as the receiver heats up. To suppress the noise, the receiver has to be cooled down. At first this was done using liquid nitrogen dioxide, but later an even lower temperature was needed and liquid helium was used: a rather expensive commodity.

There is an important practical difference between detecting radio photons and light photons. Each light photon is detected by recording its energy. Radio photons have so little energy that a different method is called for. Detection depends on resonance in the antenna that is caused by the incoming radio wave, which produces an alternating current in the antenna with the same frequency as the radio wave. The receiver increases the current and records the signal. This necessary procedure limits the wavelengths of the radio radiation: a given receiver can only detect radiation in a small range of wavelengths. To make measurements at a different wavelength, another receiver has to be built. Observing at another wavelength is much more difficult in radio astronomy than in optical astronomy.

In conclusion, we can say that in the period from 1960 to 1974 the receivers in radio telescopes became increasingly noise-free and they could detect objects at ever shorter wavelengths (ever higher frequencies).

And now for the single antennae, that after 1960 consisted of parabolic radio reflectors. Larger reflectors capture more radiation and so are able to detect weaker sources. Moreover, there are a lot of sources, such as pulsars, that are too small for aperture synthesis: single telescopes remained very useful. The 25-meter Dwingeloo telescope and the 75-meter Jodrell Bank telescope in Manchester both date from 1957. In 1961 a radio telescope was commissioned in Parkes (New South Wales) with a diameter of 64 meters (see Fig. 3.5). This telescope determined the position and structure of the radio source 3C273 that, following optical identification, was found to be the first quasar ever discovered (see Chap. 5). Figure 3.3 shows the remarkable 305-meter telescope in Arecibo in Puerto Rico, for decades the largest single telescope in the world, commissioned in 1963. A larger, similar telescope has recently been opened in China. The Arecibo telescope was constructed in a valley and can only look to the zenith and a small distance beyond this. As a result of its enormous surface area, the telescope detected numerous very weak pulsars, such as the binary pulsar with which the American Jo Taylor was able to demonstrate the effect of gravitational radiation. This indirect but convincing proof gained him the Nobel Prize in 1993 (see Chap. 7). Figure 3.4 shows the 100-meter telescope in the Eiffel that was commissioned in 1971. For more than 50 years it was the ultimate in performance, exactly what can be expected from the Germans: an impeccably carried out project. It has measured such phenomena as magnetic fields in other galaxies (Fig. 3.5).

Fig. 3.3 The 305 m radio telescope in a valley near to Arecibo in Puerto Rico. Source: NAIC—Arecibo Observatory, NSF

Fig. 3.4 The 100 m radio telescope in the Eifel

Fig. 3.5 The 61 m radio telescope in Parkes, Australia. Source: John Sarkissian (CSIRO Parkes Observatory)

In 1963 an interstellar line of the OH molecule was discovered at a wavelength of 18 cm (Weinreb et al. 1963); in 1969 three other interstellar lines were detected (ammonia, NH_3, water, H_2O, and formaldehyde, H_2CO), all at cm wavelengths—see for more details Chap. 4. A new receiver had to be built for the detection of each line. These detections started a hunt for more lines of more molecules because it was rightly expected that they would be important for astronomy. All of these lines were no longer at cm-wavelengths but at mm-wavelengths and thus at very high frequencies. Building radio receivers for such high frequencies was at that time a challenge for engineers and required state-of-the-art techniques. All the new lines were detected in the US, mainly at the National Radio Astronomical Observatory. Because absorption of mm-radiation by water vapor in the atmosphere is significant, the mm-radio telescopes were no longer built at sea level, but at high and dry locations.

3.3 Infrared Astronomy

In radio astronomy radiation is studied at a wavelength greater than circa 1 mm. Optical astronomy reaches its limit when the wavelength is around 900 nm or 0.9 μm. Electromagnetic waves between 1 μm and 1 mm are referred to as being 'infrared' and 'sub-millimeter'. In general, these waves do not reach the Earth's surface due to absorption in the atmosphere, which is mainly by water vapor. What a disaster to have this happen in the final kilometers of a journey of so many light-years! There are a few atmospheric 'windows', narrow wavelength bands in which the infrared radiation can reach the surface of the Earth. In the 1960s, particularly in the US, people started to scratch away at this 'watershed' and gradually different groups managed to make successful observations in these windows, in particular at 2.2 and 10 μm (0.0022 and 0.010 mm). The telescopes used for these measurements were located on high and dry mountain tops. At wavelengths greater than 20 μm there were no more windows and the observations had to be made above the atmosphere and thus from planes or from balloons and rockets; satellites were not yet in the picture.

The first detectors were made of lead sulphide, based on technical knowledge that the ALSOS mission had come across in German laboratories and taken as war booty (see Chap. 1). But there were also new detectors, made of indium antimonide, for example, that could be used for detection at a wavelength between 1 and 6.7 μm. When a photon of the right energy reached the surface of the detector, an electron in the material was released. By counting these electrons, you knew how many photons had reached the surface. Detectors based on the combination of two particular atoms, indium and antimonium, fitted well within a new field of physics, solid state physics. In the US many of the searches for materials useful for infrared detection were financed by the American Department of Defense (DoD), which meant that only those astronomers were able to build infrared detectors that were considered sufficiently trustworthy by the American military; in the US only a few astronomers were chosen and outside the US no one!

In 1961, a new infrared detector was invented that consisted of a piece of germanium to which a small amount of gallium had been added ('gallium-doped germanium', Ga:Ge[1]). This detector made precise recordings of temperature changes caused by the reception of photons. A particularly good aspect of this discovery was that the instrument could detect photons at all wavelengths between 1 μm and 1 mm—and not in a limited spectral area such as the indium-antimonide detectors mentioned above. The Ga:Ge detector only worked efficiently if it was cooled to −270 °C (three degrees above absolute zero) using liquid helium. This extreme, deep cooling posed difficult technical demands and the final instrument was a very complex device. The detector was invented by an ingenious instrument builder, who was also a tough Texan with whom you had better not get into an argument, as I discovered during my involvement in the IRAS infrared satellite.

Further, using optical telescopes for infrared observations was problematic. The temperature of the telescope body was always in the region of 300 K and it therefore radiated strongly in the infrared; this radiation was much stronger than the astronomical signal. Because it was not feasible to cool the (open) telescope (the air would freeze on the reflector), it was decided to alternate the telescope rapidly—within 1 second—looking at the astronomical object and the adjacent space, and to measure the difference. This could be achieved by making the secondary reflector oscillate. This represented an intervention in the construction of a telescope. A better solution was to construct separate telescopes for infrared astronomy. These came on the scene after 1977.

In the late 1960s Neugebauer and Leighton from the California Institute of Technology carried out a survey for infrared sources at a wavelength of 2.2 μm using a telescope and a detector system, both of which they built themselves. It was the first astronomical infrared survey ever and it was of high quality. Over 5,600 sources were detected and most of them turned out to be very red stars already appearing in existing catalogues. A few exceptional sources (NML Cyg and NML Tau, among others) were new detections of stars with unusual characteristics. They will be discussed in Chap. 4. Observers from Berkeley discovered spectral lines in the infrared from ionized argon (ArIII of Ar^{2+}) and sulphur (SIV or S^{3+}). The lines were discovered in the gas of planetary nebulae. Planetary nebulae consist of a very hot star surrounded by a low-density cloud of ionized gas. The high level of ionization was explained by the high temperature of the star. In the period from 1960 to 1974 infrared research brought some real news. Discoveries were made in the field of star formation, on the disintegration of red giants, and infrared radiation was measured coming from the center of the Milky Way and from some galaxies. In 1975 infrared astronomy held great promise for the rest of astronomy.

[1]As a citizen of the European Union, I have always found it remarkable that a piece of germanium only becomes successful after the addition of a little gallium.

3.4 X-Ray and Gamma Astronomy

X-Ray Astronomy

From 1945, American researchers used confiscated V2 rockets to measure the ultra-violet light of the Sun from above the atmosphere of the Earth. Gradually they managed to measure at shorter and shorter wavelengths and ultimately solar X-rays were detected. The amount of solar radiation was very low as the Sun is a weak source of X-rays. It therefore seemed pointless to try to detect X-rays from other stars. As a result, only very few research groups were interested in trying this and focused their weak resources on the solar system. One such group, headed by Herbert Friedman, was housed in the Naval Research Laboratory (NRL), the research laboratory of the American navy, and another—the most successful group—was an organisation known as American Science and Engineering Inc. (AS&E), a company that developed X-ray apparatus and supplied equipment to NASA. The company had close links with the Massachusetts Institute of Technology (MIT) in Cambridge (Massachusetts); one of AS&E's most important advisors was Bruno Rossi, a professor from MIT and a respected researcher of cosmic radiation. He had moved to the US before 1940 because of Mussolini's anti-Jewish laws. Another Italian American, Riccardo Giacconi, led the AS&E group. Eventually, in 2002, Giacconi was awarded the Nobel Prize for the discovery and development of X-ray astronomy.

The low X-ray flux from the Sun led to the expectation that X-ray astronomy was doomed. Surprisingly, it was not. On 18 June 1962 at around midnight, the AS&E group launched a rocket carrying several Geiger counters to measure the fluorescence from the Moon, a possible effect of X-ray radiation of the Sun. Their surprise was huge when, during the flight, the Moon was seen but its signal was fainter than that of an X-ray source in the constellation of Scorpio (Giacconi et al. 1962) (see Figs. 3.6 and 3.7). The discoverers were able to show that the source was outside the solar system. They named it Sco X-1. Others confirmed the discovery and it dawned upon the astronomical community that X-ray astronomy had its own future alongside optical, radio (1945) and infrared astronomy (1959). Three years later, a further 10 non-solar X-ray sources were known and by 1967 this number had risen to 30. Most, but not all, were in the Milky Way.

Friedman's NRL group proved in 1964 that the good old Crab nebula was a source of X-ray radiation. The proof was ingenious: a rocket was launched minutes before the Crab nebula would disappear behind the Moon. Rocket observations last only a few minutes, thereafter the rocket start falling down. But the timing was accurate and in the few minutes available the detector saw the X-ray source disappear behind the Moon at exactly the same time as the Crab (Bowyer and Byram 1964). Other well-known radio sources also appeared to emit a high level of X-ray radiation: M87 (Virgo A), NGC 5128 (Centaurus A) and 3C 273 (the first quasar).

It was not until four years later that Sco X-1 was identified with an optically visible object, a discovery that caused something of a sensation: see Chap. 4.

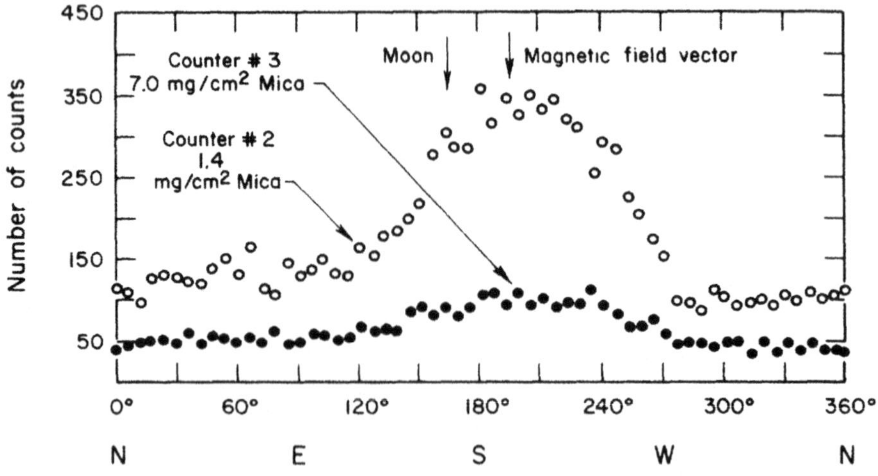

Fig. 3.6 The number of clicks made by the Geiger counters on board a rocket during a flight on a June night in 1962. The number of clicks rose until the camera seemed to reach a maximum when the Moon was in view, the planned goal of this flight. However, when the Moon was lost from view, the number of clicks did not fall back to the former level but became even higher. An unknown X-ray source was found next to the Moon and without any relation to any object in our solar system: an extra-solar source (Giacconi et al. 1962)

All these results were obtained from rocket and balloon flights. This led to the assumption that using a satellite for X-ray radiation would generate a lot of new information. NASA took up the challenge and in December 1970 launched a small satellite from a platform off the coast of Kenya. The satellite was named 'UHURU', a word that means 'freedom' in Swahili. UHURU discovered more than 300 X-ray sources. The number of identifications with an optical object also increased rapidly. X-ray astronomy had gained a place in the canon of astronomy.

Figure 3.8 from 1968 shows a scan along the Milky Way that is repeated after some time. The degree of agreement can be seen, as well as the progress.

X-ray radiation is fundamentally the same as light, but X-ray photons are so much harder (that is, more energy-rich) that the detection and imaging of X-ray radiation differs strongly from the detection of (visible) light. The hardest X-ray photons have a million times more energy than the softest ones. A million is a big range: the hardest photon of *visible* light has only 2.5 times more energy than the softest one. Separate X-ray satellites have therefore been used for 'soft' and 'hard' X-ray radiation.

The detection of X-ray photons is not the problem: the photons have so much energy that they easily cause an observable effect. The biggest problem is to find where they come from and what precisely their energy is. The first is needed to gain a precise image of the source of the radiation and the second is needed to discover any spectral line.

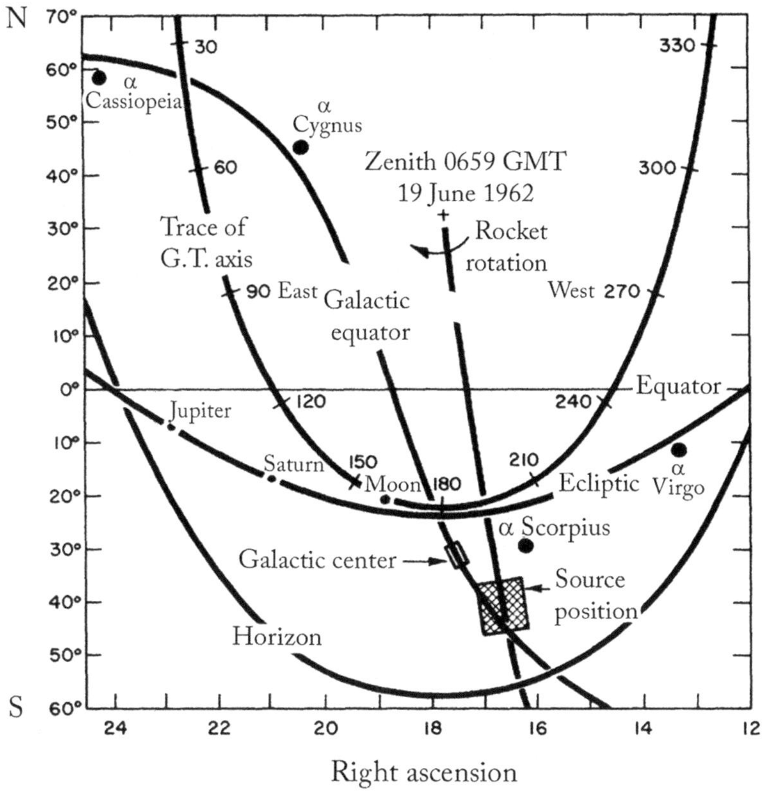

Fig. 3.7 The path through the heavens of the Aerobee rocket with which extra-solar X-ray radiation was discovered. The square shows how large the area is within which the source is found (Giacconi et al. 1962)

I will restrict myself now to the first problem: producing a good image. Most X-ray photons pass diagonally through a glass mirror: a reflector telescope is useless. Low-energy X-ray photons reflect on metal if they arrive at a shallow angle, something known as the 'grazing incidence'. German engineer Hans Wolter designed a reflector telescope in 1952 that made use of this phenomenon—see Chap. 9.

There is no surface known that will reflect hard photons. To decide from which direction the photons came one went back to the 'camera obscura': on a sunny day, darken the room by closing the heavy curtains at the windows and then make a small hole in the fabric. A weak but easily recognizable image of the opposite side of the street will appear on the wall opposite the window, albeit upside down. The image is weak because the hole only allows very few photons to pass. To make the image stronger you could make the hole bigger, but then the image would become more blurred.

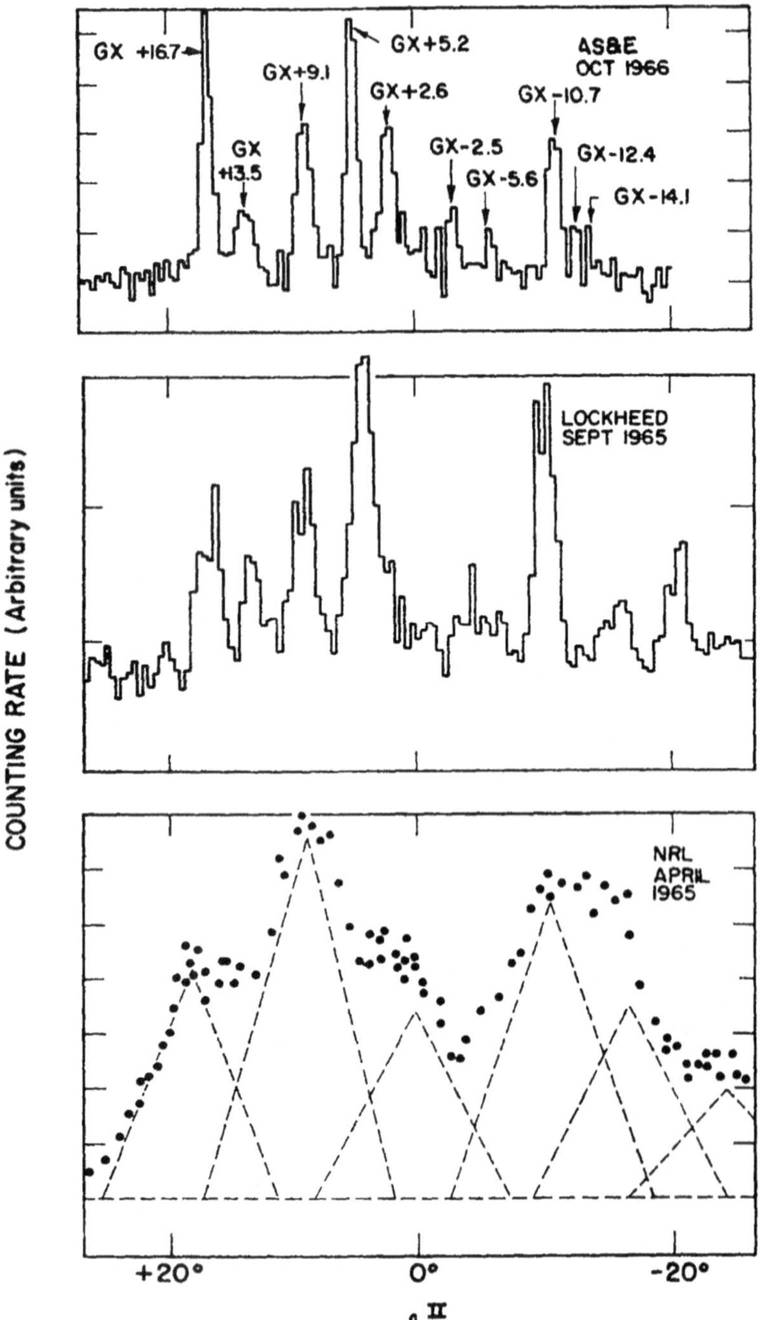

Fig. 3.8 Three scans along the Milky Way equator measured by different groups. The date of the measurement is indicated in the top right-hand corner (Giacconi et al. 1968)

A natural camera obscura. On a sunny day look at the ground under a deciduous tree: you will see dappled spots where the rays of the Sun reach the ground through the leaves. Look more carefully and you will see that the speckles are round: they are images of the Sun! The small open space between the leaves gives a camera obscura effect on the ground. With a partial solar eclipse you will see that the images are no longer round, but have a chunk bitten out of them. This happened once in the sixties during the national Dutch Astronomers' Conference when Marcel Minnaert, the author of the famous book *Physics of the Open Air*, instructed all the participants (the author was one of them) to go outside and observe this phenomenon.

And now to the hard X-ray photons: only a camera obscura could produce an image, but even then it would be a very weak image. In 1965 Japanese astronomer Minoru Oda devised a clever solution to overcome this problem: imagine making two holes in the wall of the camera. That will give you twice as many photons, but also two weak images that partly overlap one another. Oda then reasoned that it was a minor problem for a computer to reconstruct the true image from the measured intensity variation. Oda went further and proposed making a large number of holes: even more photons would then be received and, as he showed, the computer could still reconstruct the true image from the overlapping images. This was possible because the computer knew the positions of all the holes and therefore also the code with which the true image was converted into the observed image. The mathematical technique for decoding the image was already known; it shared some similarity with the previously mentioned Fourier transformation from radio astronomy. The opaque plate with the holes was referred to as a 'coded mask'. This mask was the basic element in all X-ray satellites at photon energies where the Wolter telescope was no longer effective.

Gamma Radiation

Photons with energy above 100 keV are for historical reasons called gamma photons.[2] Gamma radiation from the Sun had been discovered with V2 rockets. In the 1950s it became clear that the Milky Way should also produce gamma rays: the long wavelength radio radiation from the galaxy proved the presence of relativistic

[2] In the final years leading up to 1900, experiments were carried out to study the emission of radioactive matter in greater detail. There proved to be three types of emissions: alpha radiation was made up of positively charged nuclei of helium atoms; beta radiation comprised negatively charged electrons and what remained was gamma radiation, that proved to be electromagnetic radiation with photon energies even higher than those of X-ray radiation.

electrons that were part of the Cosmic Radiation. Relativistic protons were another component of the Cosmic Radiation and collisions of these protons with interstellar atoms should produce gamma radiation. One of the basic problems in actually detecting the gamma photons was that they are so rare: because of their enormous energy, they make a huge impression when they arrive but they arrive only rarely. Gamma radiation from the Milky Way was first detected in 1968 after Americans Clark, Garmire and Kraushaar had analysed 36 weeks of measurement data from the OSO-3 satellite. The radiation was strongest in the galactic plane: their instrument recorded about 2 gamma photons per 3 hours, 16 per day (Clark et al. 1968).

Around 1970 a rumour went around that American spy satellites had detected bursts of gamma radiation from outer space. The rumours were confirmed by announcement in 1973 by Klebesadel, Strong and Olson (1973) that on 16 occasions a brief explosion of gamma radiation had been detected, at all kinds of different times, with different satellites and apparently from very different directions. The observations had been made with 'Vega-satellites', American surveillance equipment, and the authors had links with the Los Alamos military research lab. The data were published after it became clear to the military that the Gamma Ray Bursts (they were soon called GRBs) came from space and were not produced on Earth during atom bomb tests that were banned under a US-USSR agreement. These astronomical gamma explosions probably released more energy than a supernova, an idea that caused a huge sensation. Within several months, many articles followed putting forward very different hypotheses about these GRBs. These articles were of course fully speculative because nobody knew where the explosions were produced (far away or close by?). GRBs were registered time and again in subsequent years by different satellites. But that was all. What the sources of these explosions were remained a mystery for a further 20 years.

3.5 Computers and Automation

In 1959 I was involved as a master student in the research with the new Dwingeloo radio telescope. I found myself in a small but lively group of researchers from Leiden and Groningen. It was a privilege. The connection with this chapter is my experience that it took so much mindless routine work before you could start to interpret the result of a measurement—and I am convinced that this was just the same elsewhere.

Let me explain what would happen. The researcher received by post the 'recording strips', long reams of paper with a curve drawn by a thin pen with red ink while the paper rolled slowly out of the recorder. The first step was to read the graph. Two men (it was never a woman) sat opposite one another and one read the height of the red line out loud, and the other recorded the findings. After a half hour you had a table with two columns, the first indicating the wavelength and the second the height of the red curve. This deadly boring work corresponded with what was later called 'digitisation', which would be done a few years later electronically within the receiver. I am surprised that I did not anticipate this development; afterwards it was so obvious.

Once the readings were made, there were all kinds of arithmetical calculations to be carried out. This was taken over by the first 'big' computer, the pride of Groningen University. It was called ZEBRA, an acronym for the Dutch Zeer Eenvoudig Binair Reken Automaat (Very Simple Binary Calculating Machine). It just about managed to carry out all the necessary arithmetical calculations, but only after the figures had been fed in via a ribbon of punched tape, a long, narrow strip of paper that could with the greatest of ease tie itself into knots or—even worse—tear. Punching the tape was the most boring work and it became downright unpleasant if you discovered you had punched in a wrong number somewhere. Then the punched tape had to be 'read' by hand until you found where the mistake was. The incorrect holes had to be masked with sticky tape and the correct holes had to be punched in. I found myself wondering, was it for this that I had passed my bachelor's exams? My father told me not to complain: 'We've all had to do dumb work when we were young.'

It was not until 1970 that the measurement data were recorded electronically. The data no longer had to be read out and the mathematical calculations were made automatically. Automation saved researchers a huge amount of time; not only that, it also gave them an opportunity to mumble on about the good old times.

3.6 Observations at High Altitude

The atmosphere shrouds the Earth like a blanket. It makes life possible, but it hampers a lot of astronomical research. There are four ways of combatting this: make observations from an aircraft (at a maximum altitude of 13 km), a balloon (at 40 km), a rocket (at 200 km) or a satellite (above some 500 km). Each option has its pro's and con's, but ultimately a satellite is the best solution.

A plane can fly high enough to measure infrared radiation at any wavelength, but not high enough to detect X-rays or gamma rays. The observer is present, so the telescope is operated optimally. Even so, the observations can take several hours. The view ahead is restricted, as the observer can look only to the side. There is an outward and a return journey, and generally the view is optimal on only one of these. Then, a plane is never completely still; there are a lot of annoying vibrations. An example of successful pioneering work can be seen in the flights by American astronomer Low in a NASA Lear Jet, in which he discovered that many sources (from planetary nebulae to galaxies) emit a lot more infrared radiation than predicted; the strong peak at a wavelength of around 100 μm (one-tenth of a millimeter) was a total surprise.

Balloons, too, have made important contributions to infrared research, in particular at wavelengths between 30 and 1000 μm (0.03 and 1.0 mm), and to the research in the near ultraviolet (220–320 nm). Balloons could float for hours at the highest possible altitudes (40 km), but there was no observer on board and the basket tended to sway about most of the time. In the 1960s and 1970s computers were still rather primitive and there was very limited interaction with the ground. There was no possibility to change the observation plan during the flight. The data were recorded on board on a tape recorder, which of course had to be recovered after the flight. In spite of these difficulties, numerous research groups (American, British, French, German, Dutch) have launched balloons carrying detection equipment, with which they had some success.

Rocket flights reached the greatest height and so could serve to record astronomical X-ray and gamma photons. The greatest limitation of rocket flight was the short duration, which might be as little as tens of minutes. Just as with balloons, telemetry was very limited and the measuring instruments had to be able to carry out the observations largely autonomously. Returning the measuring equipment to Earth was also a risky operation, as the history of Don Morton will show in the next chapter. As well as American rockets, there have been countless launches with the British Skylark rockets. Only a small percentage of the hundreds of rockets launched were intended for astronomical research outside the solar system.

The aircraft, balloon and rocket observations from this period were mainly of an exploratory nature, and in a number of cases they led to unexpected discoveries. It was pioneering work and adventurous—there are many stories to be told! The results stimulated and justified the desire for quantitatively better measurements and a more complete overview. Astronomers therefore frequently found themselves begging for a satellite—and sometimes their pleas were rewarded.

Satellites are undoubtedly the best platforms at all wavelengths: they move far above the atmosphere and can in some cases be operational for years. However, there are a few significant problems. The first is that the observation equipment cannot be adjusted or replaced, while operating the instrument from the Earth was problematic: the on-board computer in the 1960–1974 period offered very little assistance. The second and third problems were the high costs of a satellite project and the lengthy duration of the preparations. It is not surprising that the first satellites that proved valuable for astronomical research were not launched until the end of the sixties, some 10 years after Sputnik.

In terms of the investments needed, balloons were cheapest, then rockets (that could only be used once), followed by aircraft and finally satellites. If the cost is calculated per hour of observation, the satellite is the cheapest: it can be used for many years to make measurements, 24 hours a day, 7 days a week.

Between 1960 and 1974, 540 satellites were launched, 237 of them by the Soviet Union, 256 by the US and 36 by other countries, often using American launch rockets (Bleeker et al. 2001). No more than a tenth of these were intended for research outside our solar system. NASA launched most astronomical satellites for research in extreme ultraviolet, and for X-ray and gamma radiation. NASA launched one European, one British and one Dutch satellite to detect X-ray and UV radiation from outside the solar system. The Soviet Union twice put a UV spectrograph (Orion 1 & 2) on board the space station where cosmonauts made the observations. Japan launched a couple of satellites with their own rockets but not for astronomical purposes.

Let us return to the developments in space research in Western Europe. The bigger countries (Germany, England, France and Italy) each set up their own space programmes after Sputnik but from the outset all the parties strove for European collaboration, in particular in scientific research. CERN in Geneva had already convincingly demonstrated the value of scientific collaboration. In 1964 the European Space Research Organisation (ESRO) was set up and was given a test centre in Noordwijk in the Netherlands. Plans for research were made that soon proved to be too ambitious because as yet people were not hampered by any significant knowledge. Three projects were started, each for the launch of a large space telescope, christened TD1, TD2 and TD3. Ultimately, TD2 and TD3 were combined into one project that in the final event was never implemented. TD1 was launched in March 1972 with an American rocket and provided astronomical information about ultraviolet light from thousands of stars; not big news, but a useful start.

Besides ESRO, there was also ELDO, the European Launcher Development Organisation, where researchers were trying to construct a three-stage rocket with a British rocket, the Blue Streak, as the first stage. The project was a disaster, with nine failed attempts at a launch from Australia and a further two from a base in French Guyana. The programme was brought to an end without success. Clearly, Europe, including Western Europe and excluding the Soviet Union was about 10 years behind the US.

Here I want to pay my respect to Henk van de Hulst, who was instrumental in making space research useful for science, above all for astronomy. He did this on a worldwide scale through his chairmanship of the COSPAR committee, in Europe by his leading role in establishing ESA and in the Netherlands by helping to create SRON, the Dutch organisation for space research. Apart from that he predicted in 1944 the existence of the 21 cm line and he wrote a classical book on the scattering of radiation by small particles, a book that had wide applications, also outside astronomy. He was a great man.

References

Bleeker, J.A.M., Geiss, J., Huber, M.C.E.: The Century of Space Science. Kluwer Academic Publishers, Dordrecht, The Netherlands (2001)

Bowyer, S., Byram, E.T.: Science. **146**, 912 (1964)

Clark, G.W., Garmire, G.P., Kraushaar, W.L.: Astrophys. J. **153**, L203 (1968)

Cohen, M.: Annu. Rev. Astron. Astrophys. **7**, 619 (1969)

Giacconi, R., Gursky, H., et al.: Phys. Rev. Lett. **9**, 439 (1962)

Giacconi, R., Gursky, H., van Speybroeck, L.P.: Annu. Rev. Astron. Astrophys. **6**, 373 (1968)

Klebesadel, R.W., Strong, I.B., Olson, R.A.: Astrophys. J. **182**, L85 (1973)

Pawsey, J.L., Payne-Scott, R., McCready, L.L.: Nature. **157**, 158 (1947)

Ryle, M., Hewish, A.: Mon. Not. R. Astron. Soc. **120**, 220 (1960)

Salpeter, E.: Astrophys. J. **140**, 796 (1964)

Weinreb, S., Barrett, A.H., et al.: Nature. **200**, 829 (1963)

Chapter 4
1960–1974: Stars

4.1 How Nature Creates Stars

Dark Clouds and Molecules

In 1960 there was a strong conviction that new stars were born in dark[1] clouds. These dark clouds were always near to light-emitting nebulae that were illuminated by very bright stars. The stars had so short a lifespan that they were still in the place where they were born. This conviction was endorsed in 1954 by the discovery by Walker of immature stars near to dark clouds, a discovery that has been discussed in Chap. 2. Unfortunately, very little was known about these dark clouds, other than that they contained small grains of dust or grit that occluded all the light from outside. It was impossible to say how many particles of dust there were, nor was it known whether the clouds contained any gas. Radio telescopes had been used to search for the 21 cm line of hydrogen atoms, but nothing was found. Nonetheless, astronomers continued to believe that the clouds should contain some gas, and if gas was present, since all the stars consisted mainly of hydrogen, this gas should also consist mainly of hydrogen.

Since the gas did not consist of hydrogen atoms, the hydrogen must be in molecular form: H_2 molecules, two atoms bound to one another. On Earth, for example, hydrogen is always found in this form, but in the empty space between the stars, each hydrogen molecule would be separated into two hydrogen atoms by the small amount of ultraviolet light from distant stars. According to the rules of chemistry, hydrogen molecules could not form there. When two separate hydrogen atoms collided, they both went their separate ways afterwards rather than binding together: the two atoms had too much energy to bind together and no way to dispose of this

[1] 'Dark clouds' are unrelated to so-called 'dark matter'. In dark clouds, all background light is blocked by small grains of sand or soot. 'Dark matter' (to be introduced in Chap. 8) comprises elementary particles that neither emit nor absorb light or any other electro-magnetic radiation.

H. J. Habing, *The Birth of Modern Astronomy*, Historical & Cultural Astronomy,
https://doi.org/10.1007/978-3-319-99082-8_4

energy. For a molecule to form, a third body had to be present at the encounter to absorb the excess energy. In interstellar space, the space between the stars, there were more than enough ultraviolet photons from distant stars to dissociate any existing hydrogen molecule. But inside dark clouds, dust particles absorbed these ultraviolet photons and as the particles had to be very cold a hydrogen atom would stick to any particle with which it collided. Now, it was also known that atoms that stuck to a dust particle were able to move over its surface, so if a second atom also collided with the same particle and stuck to it, there was a good chance that the two would meet one another and then form an H_2 molecule. Calculations showed that this process was highly efficient (Gould and Salpeter 1963). The conclusion was: if hydrogen was present in a dark cloud, it was in molecular form. Outside a dark cloud, it was in the form of an atom.

Was it possible to detect the presence of these molecules? The symmetrical shape of the H_2 molecule meant that it only emitted very faint radiation, as dictated by the rules of quantum mechanics. For the time being there was little hope of detecting hydrogen inside dark clouds, so the attention shifted to the possibility of finding other gases. The first evidence came in 1968 by the detection of an interstellar radio-line of the molecule OH at a wavelength of 18 cm (see Chap. 3), the second radio line after the 21 cm line (Box 4.1).

Box 4.1 Natural Masers

The four lines of OH at wavelengths close to 18 cm (frequencies: 1612, 1665, 1667, 1720 MHz) are expected to have prescribed ratios in average conditions (thermal equilibrium); the lines were first detected in 1963 (Weinreb et al. 1963). In 1965 radio astronomers from Berkeley measured these lines in a galactic source (Weaver et al. 1965) and found to their surprise that some of the lines were strong and others were totally absent. The lines must come from a source in which the conditions were far from average. After some discussion it was generally accepted that the source of the radiation had to be a natural 'maser'. Masers were known as instruments that, like lasers, amplify very strongly the radiation in one spectral line. This is achieved by adding energy to certain atoms (or molecules) so that these atoms are on the verge of emitting a photon in a spectral line (this part of the process is called 'pumping the atoms'). If another photon of the same line passes the charged atom, it will induce the charged atom to emit its photon in the same direction ('stimulated emission'): then there are two photons in that direction. This is the beginning of a snowball effect that can lead to a very intense and narrow beam of photons. When the photons have the wavelengths of radio waves (microwaves), the instrument is referred to as a 'maser'; when the photons have a wavelength of visible light, the instrument is a 'laser'. The words maser and laser are themselves acronyms: 'microwave (light) amplification by stimulated emission'.

American physicist, Charles Townes, mentioned elsewhere in this chapter, received the Nobel Prize in 1964 for his contribution to the discovery of masers and lasers.

Many celestial maser sources have been detected, not only OH-line masers but also a water line maser (H_2O) and masers in lines of silicon oxide (SiO) and methanol (CH_3OH). In each case, a somewhat different type of pumping is proposed, usually infrared radiation.

Maser lines are strong and narrow; doppler shifts can be very accurately measured, and the measurements often give detailed information of the structure of the source. Unfortunately it is impossible to obtain information on temperature and density in the source of the maser line. A major achievement of maser astronomy was in 1995 when Japanese radio astronomers used maser line measurement to prove solidly that 'supermassive black holes' truly existed—see later.

Carl Heiles, a radio astronomer, who had just become staff member at the astronomy department in Berkeley, found there a radiotelescope and an 18 cm receiver and decided to see whether radiolines of OH could be seen the direction of dark clouds. He chose a few candidate clouds and directed the radio telescope accordingly. He was rewarded; after many hours staring at these clouds the lines were detected (Heiles 1968). The signals were faint but real. It was a breakthrough. Finally, gas had been discovered in dark clouds and the clouds were open to further research (Fig. 4.1).

Unfortunately, OH was far from communicative: you could measure the Doppler shift and that was about it. There was now one molecular billy goat over the bridge and there were a lot more to follow. The search for more radio lines was given a powerful impetus by Charles Townes—see Box 4.2. Thanks to him, ammonia (NH_3) (Cheung et al. 1968) and water (H_2O) (Cheung et al. 1969) were discovered. A spectral line of formaldehyde (H_2CO) was found in 1969 (Snyder et al. 1969). These four radio lines had cm-wavelengths, thus photons of very low energy produced in so-called 'hyper-fine transitions'. The discoveries sparked a race in the US to find as many other molecules as possible. Chemists knew that many molecular lines existed at shorter, millimeter wavelengths, lines that were produced in so-called rotational transitions. The race produced a large number of lines from newly detected molecules, all at radio wavelengths of just a few millimeters. That required new receivers and at this short wavelength (and thus high frequencies) the receivers were a technical challenge for the radio engineers. There was also a special radio telescope built in Arizona at a high and dry location especially for the detection of these molecular lines. By 1974 there was already a list of 25 detected molecules, some of them small, such as SO and SiO, some large, such as acetaldehyde, CH_3CHO.

Although molecular hydrogen could not be detected by the radio telescopes, another molecule was found that was a good proxy: the carbon monoxide molecule (CO). It was discovered in 1970 (Wilson et al. 1970) a wavelength of 2.6 mm and occurred in all dark clouds. The molecule produced many lines but all at wavelengths below 2.6 mm and those were often difficult to observe from the surface of the Earth. CO gas was present in practically all dark clouds. The dark clouds went from being opaque to transparent and they were no longer called 'dark clouds' but 'molecular clouds'. A new specialization—astrochemistry—was born (Fig. 4.2).

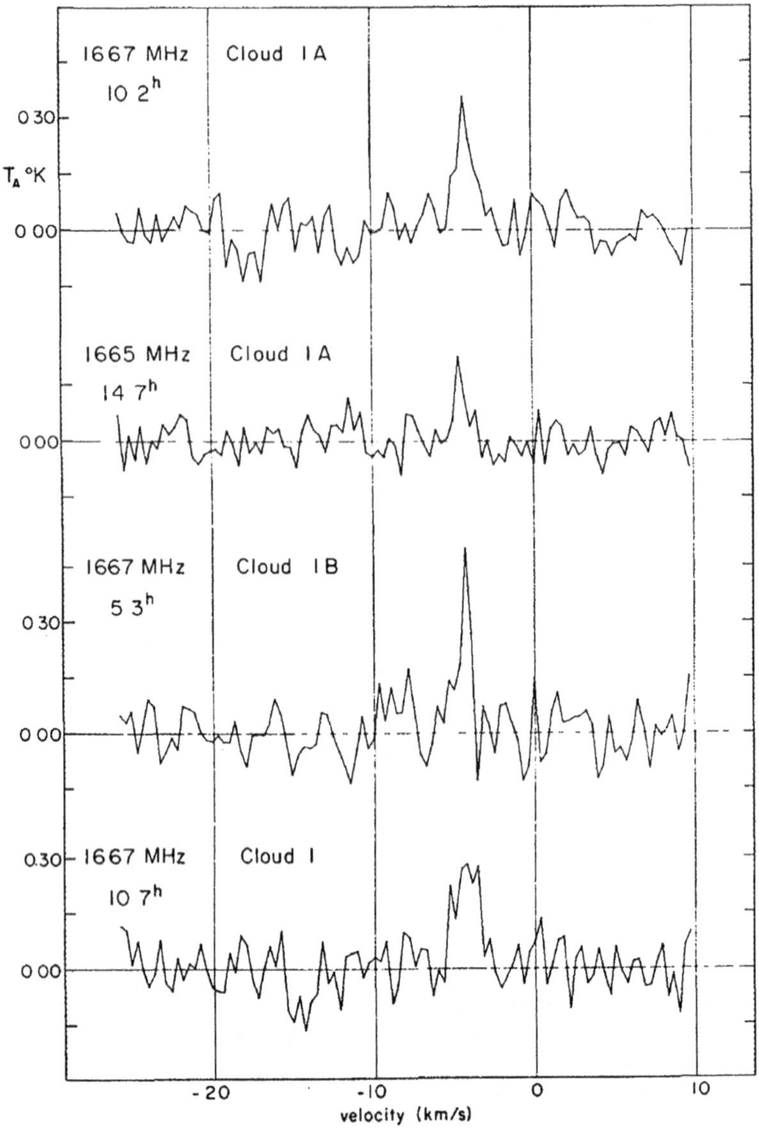

Fig. 4.1 The measurements of two spectral lines of OH in two areas ('Cloud 1A, Cloud 1B respectively) in one dark cloud. Doppler shift in the horizontal direction, signal strength vertically. The upper spectral line of each pair was at a wavelength of 17,998 cm (1667 MHz), and the lower one at 18,018 cm (1665 MHz). In all four measurements there is a convincing peak at around −5 km. In both areas the top signal is slightly stronger than the bottom one, which is in line with theoretical expectations (Heiles 1968)

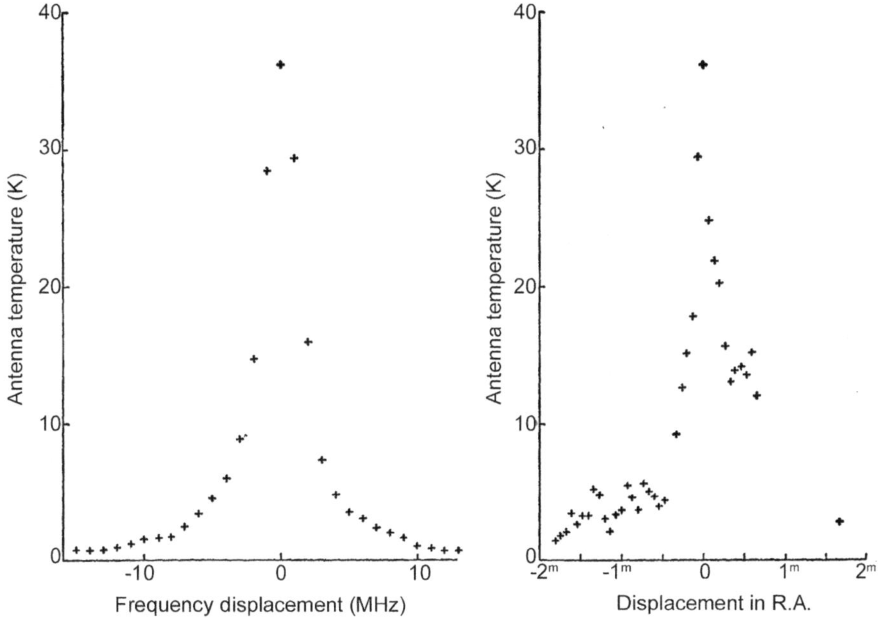

Fig. 4.2 First detection of CO mm lines in Orion. The frequency is indicated horizontally, and the energy that is captured is shown vertically. The frequency can be converted into the velocity of approaching or of going further away via the Doppler effect (Wilson et al. 1970)

Box 4.2 Intuition and Logic

In 1968–1969 I worked as a postdoc with George Field, a famous theoretical astrophysicist at the University of Berkeley (California). A small group of us always had lunch together in the university's restaurant or somewhere nearby. We exchanged scientific news—and there was a lot of news in this period of momentous discoveries. One day a middle-aged man joined us, who was treated with great respect. His name was Charles Townes, and he had won the Nobel Prize for physics because he had contributed greatly to the invention of the laser. He had always worked on the East Coast but had now moved to the American West Coast to avoid the social obligations that a Nobel Prize winner apparently has to deal with. Townes impressed everyone by asking simple and disarming questions about many of the new scientific discoveries we discussed enthusiastically over lunch.

One day George Field told me that Townes had recently gone to see him with a particular question: 'George, I want to discover interstellar molecules with more than two atoms, like water and ammonia. You're a theoretical astronomer and I'm just a simple physicist. What do you think?' Field's answer was very negative. 'Let me explain to you, Charlie, how empty it is in interstellar space. The chance that two atoms will meet and form a molecule is very, very small. The chance that three atoms will collide simultaneously is zero. Forget it, Charlie, it won't work.' Townes replied that he had listened and that Field had made some important points. He left and went a few doors further along to the office of Jack Welch, the Director of the radio-astronomical lab, and said: 'Jack, if I give you 50,000 dollars will you build me a receiver to detect interstellar water and interstellar ammonia?' Welch didn't hesitate; the receiver was constructed and the interstellar lines were found. The lesson to be learned is that scientific advances sometimes come from intuition rather than from rational arguments.

A particular detail that was underestimated when it was discovered in 1972 can be seen in Fig. 4.3. This figure shows a profile of a line of H_2S (hydrogen sulphide), newly discovered and measured in six different molecular clouds. The authors noted that the profile in the upper left has a broad base, which was unusual. But they left it at that. It was not until several years later that the same detail was noticed again, sparking off a detective story that had some exciting consequences (see Chap. 7).

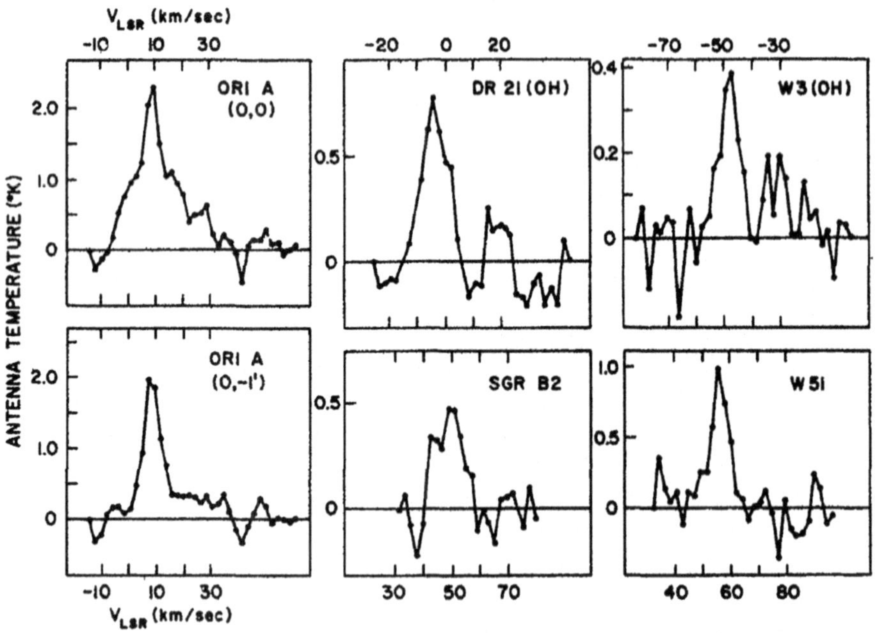

Fig. 4.3 The discovery of a spectral line of H_2S gas in different directions. The measurement in the upper left was taken in the direction of the Orion nebula. The discoverers commented that the line profile in this figure seems to have a broader basis than in the other measurements (Thaddeus et al. 1972)

Finally. Using radio telescopes, many different molecules were found but not the molecule that should have been the most prevalent: H_2. As has already been said, this molecule emitted very faint radiation. In such a case, the best approach was (and still is) to find a light source (star) behind the cloud and to look for dark (absorption) lines in the stellar spectrum, caused by the intervening molecules. For molecular hydrogen the strongest lines were those of the so-called Lyman-series and these were all in the far ultraviolet, around 110 nm, outside the reach of Earth-bound telescopes. In 1970, American astronomer Carruthers managed to make a measurement with a rocket of the spectrum of the star ξ Persei (Carruthers 1970). To his delight, he saw the strong absorption lines of H_2 in the spectrum. A few years later his findings were confirmed by the American Copernicus satellite, which was used to detect many more stars with the same absorption lines of the Lyman sequence. These were all stars that were strongly obscured by dark clouds. The Copernicus satellite was also used to look at stars that were not obscured, and indeed there was no molecular hydrogen to be found in these spectra. Figure 4.4 shows absorption lines in the same star that Carruthers had observed.

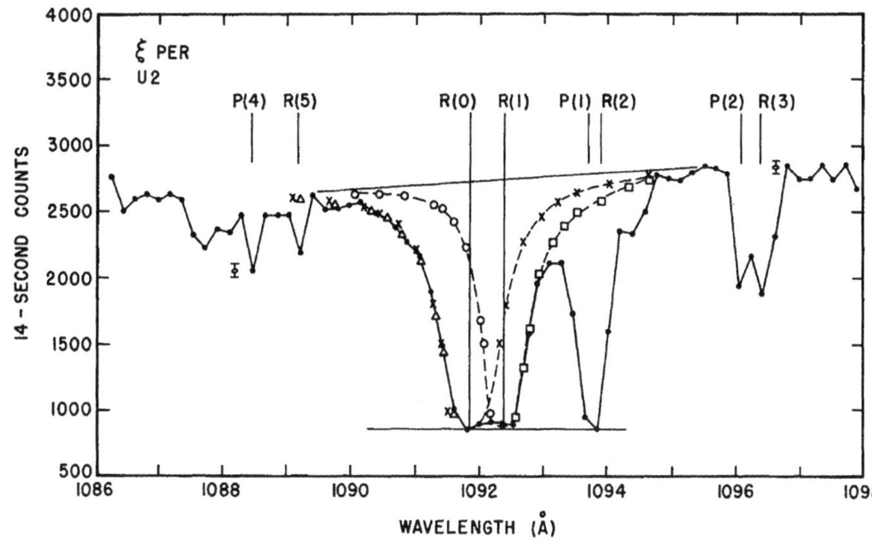

Fig. 4.4 Part of the spectrum of the star ξ Persei. The wavelength in Ångstrom (1 Å = 0.1 nm = 0.1 billionth of a meter) is shown horizontally. The strength of the radiation measured is shown vertically; absorption lines of H_2 cause the deep valleys. The short vertical lines above the spectrum give the spectroscopic names of a sequence of lines of H_2. The two strongest lines, named R(0) and R(1) overlap. The crosses indicate the absorption of light by the line R(0), the circles the absorption by R(1) (Spitzer et al. 1973)

The Orion Nebula as Model

When a new radio receiver was built for the spectral line of a new interstellar molecule, the likelihood of successful detection was greatest if you first aimed at the Milky Way center or the Orion nebula. Look at Fig. 4.5 that shows the heart of the Orion nebula. The four stars in the center are together called the 'Trapezium'. They are very bright and emit lots of very hard UV light, which destroys any molecule or dust particle in their surrounding area. The many molecules detected in the same direction must be located in a cloud behind the nebula. The nebula filled a cavity of very hot gas that the Trapezium stars had dug into the much larger

Fig. 4.5 Frontal view of the heart of the Orion nebula and the 'Trapezium' of four stars that ionize the gas. Photo taken with the Hubble space telescope. Source: Astronomy Picture Of the Day, 6 October 2012, Hubble Legacy Archive, Robert Gendler

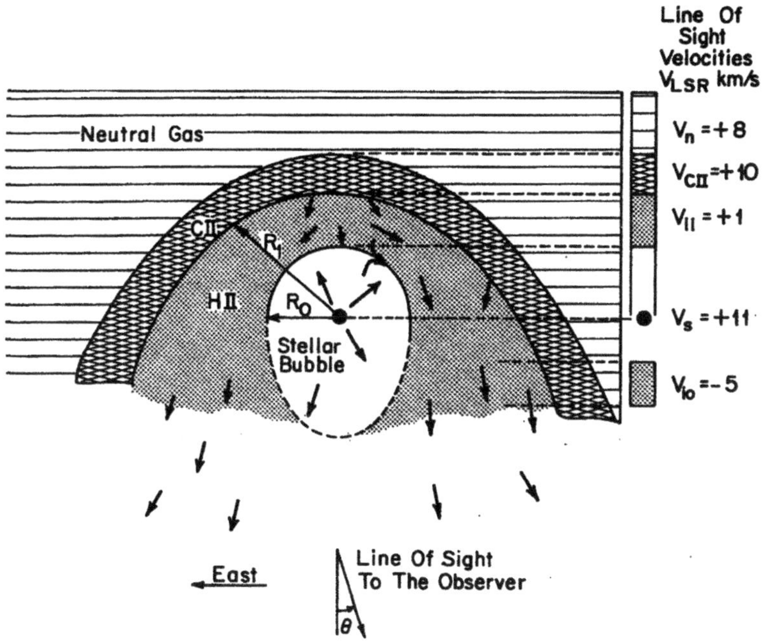

Fig. 4.6 A three-dimensional model of the blister caused by hot stars at the edge of a molecular cloud. The blister is known as the Orion nebula. The thick dot in the middle is the group of four stars called the Trapezium. Light pressure from these four stars 'etches' a cavity in the dark cloud and fills the cavity with ionized hydrogen (indicated by the spectroscopic name 'HII') (Pankonin et al. 1979)

molecular cloud. The hot gas even bulges out slightly. Frank Israel named such cavities on the skin of a molecular cloud 'blisters', a name that stuck.

Different researchers have tried to make a three-dimensional model of the Orion nebula. The main aspects correlate but the details of the models differ from one another. Figure 4.6 shows such a model.

In 1970, American astronomers Aumann and Low measured the infrared radiation of the Orion nebula and other objects at wavelengths of between 5 and 1500 μm (Low and Aumann 1970). This radiation did not penetrate through the atmosphere and the researchers used a small jet aircraft from NASA with which they could reach an altitude of 15 km, high enough for successful measurements. The Orion nebula proved to be an exceptionally strong infrared source, much stronger than one would expect on the grounds of the visible light or of the radio radiation. Aumann and Low suggested that the infrared radiation came from the dark cloud behind the Orion nebula. Later observations confirmed this. See Fig. 4.7.

Fig. 4.7 A photo of the Orion region. To the upper right is the well-known Orion nebula that is shown in more detail in Fig. 4.5. The bright spot inside the nebula is the sum of the four Trapezium stars. To the lower left is another nebula, of which the 'horse head' is a well-known detail. Both nebulae are small, light-emitting parts of a large, dark cloud that stretches over the whole photo. This dark cloud shows up very clearly in lines of CO molecules. Source: Astronomy Picture Of the Day, 10 March 2009, Dale J. Martin

T Tau and HAeBe Stars

Chapter 2 already mentioned the work by Merle Walker, who found young, clearly immature stars in a well-known cluster of stars, NGC 2644. T Tau[2] stars were also present in this cluster (see Chap. 2) and Walker drew the conclusion that these were also young, immature stars on their way to the main sequence. The spectra of T Tau stars were special because they contained not only dark absorption lines but also several very bright lines. Similar lines also occurred in the spectrum of the Sun, where they had been radiated by a very hot thin layer of gas, called the 'chromosphere' that is located on top of the photosphere, the most important light-emitting layer of the Sun. T Tau stars apparently also had a chromosphere but theirs was much stronger. In 1960 the American astronomer George Herbig added to the list of very young stars a further 20 stars (Herbig 1960). These had a much hotter

[2]A short note about the naming of variable stars. T Tau stars were variable stars, of which the star T Tau was the prototype. It was the third star in the constellation of Taurus, of which the brightness was variable. The first variable star detected in that constellation was called R Tau; S Tau was next and T Tau the third. This weird naming practice was introduced at the start of the nineteenth century for reasons comprehensible at that time.

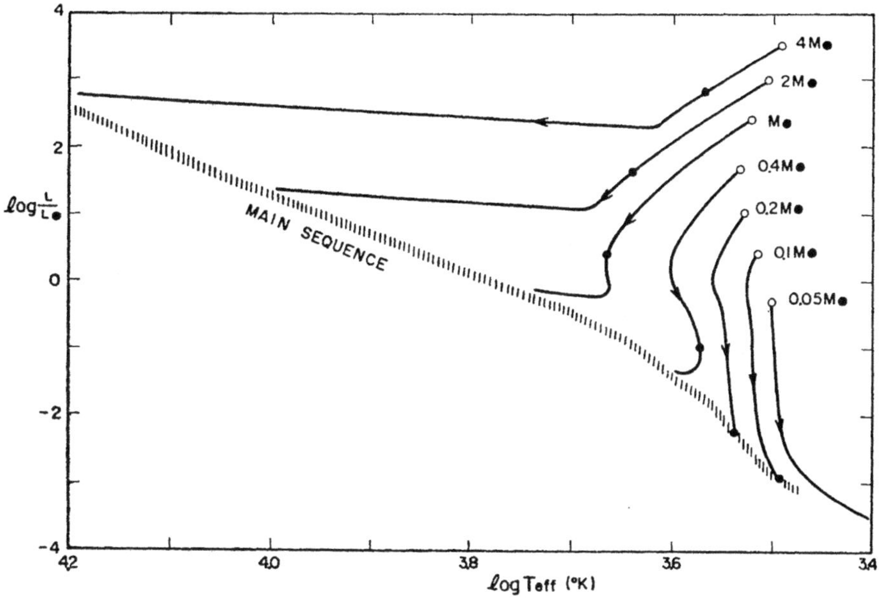

Fig. 4.8 A Hertzsprung-Russell diagram with the tracks along which young, immature stars move to the main sequence. At the start of every track is the mass of the star. The tracks are followed in a short space of time, at least in astronomical terms (Hayashi 1966)

spectrum and they were also located close to dark clouds. Herbig reasoned that these were heavier colleagues of the T Tau stars. The course of history has shown that Herbig was right. The stars were first called 'Herbig's Ae and Be stars' (the capitals A and B indicated the stellar spectral type and the small letter 'e' stood for 'emission') and were later renamed 'HAeBe stars'.

In 1961, Japanese astronomer Hayashi (1961) calculated the path along which stars evolved to the main sequence; see Fig. 4.8. Hayashi demonstrated convincingly that earlier calculations had a major shortcoming in that they had neglected 'convection': heat transfer through flows of hot gas from the centre to the outside and of cold gas from the outside to the centre. Convection is a well-known phenomenon: radiators spread warmth in a room primarily through the flows of warm air.

Associations of Young Stars

High-mass hot and bright stars only exist for a short time and dissolve as a group at or close to their birthplace. Another characteristic is that they seem to exist in clusters and not in isolation. Adriaan Blaauw, his name has already been mentioned, published a paper in 1964 (Blaauw 1964) on three associations of young stars. In the Orion association—see Fig. 4.9—he distinguished four groups of

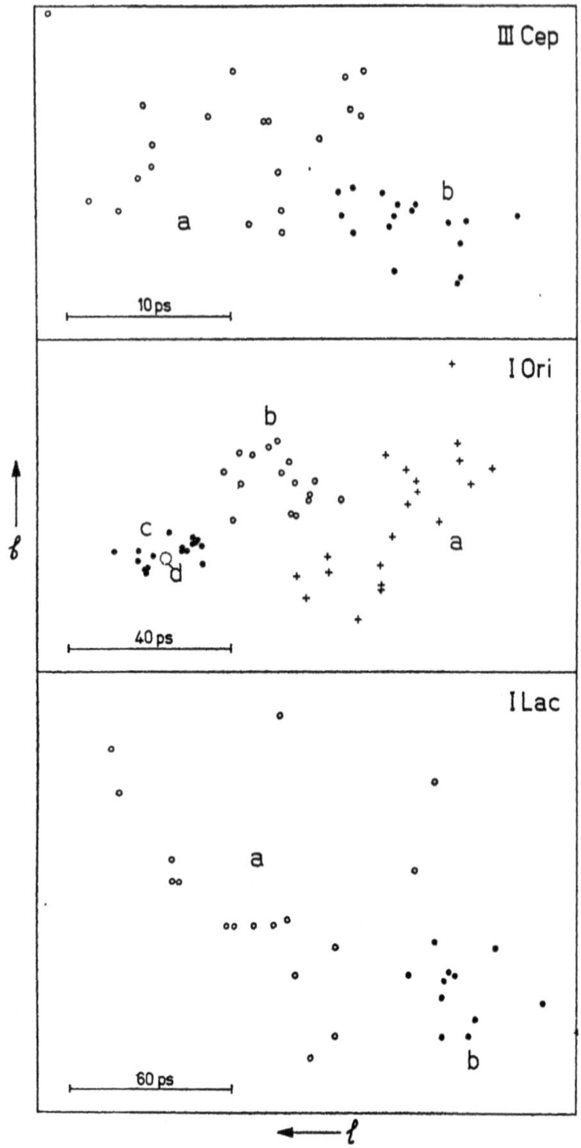

Fig. 4.9 Three different areas of the sky, each belonging to a cluster of young stars. Just consider the Orion cluster in the middle in which four groups are indicated by the letters a, b, c, d. The average age increases from group a to group b to c and to d. The youngest groups are the most compact; older groups have become more dispersed. The conclusion to be drawn from this is that although all these stars are born in small groups the force of gravity is not strong enough to hold the group together, so each group becomes larger as it becomes older (Blaauw 1964)

different ages, marked with the letters a, b, and c. The circle with the letter d encloses the so-called 'Trapezium', a group of four young and heavy stars close to one another. Group d is the youngest and the most compact and group a is the oldest and the most dispersed.

Cocoon Stars: Stars under Construction

In 1967, American astronomers Davidson and Harwit published a short theoretical article on how stars—in particular high-mass stars—were formed (Davidson and Harwit 1967). They reasoned that late in the process of evolution a stage occurred where the star already shone brightly but was so tightly packed in a cocoon of gas and dust particles that all the light from the star was converted into infrared radiation. This publication appeared at around the same time as the discovery by Becklin and Neugebauer, pioneers of infrared astronomy, of a strong infrared source in a corner of the Orion nebula (Becklin and Neugebauer 1967); later it was named the 'BN object'. This source was probably a cocoon star. Shortly after this, a strong maser source of OH 18 cm line radiation was discovered precisely at the site of the BN object. The existence of such a strong 'maser source' indicated the presence of a compressed layer of matter that was heated by infrared light. This could be a piece of such a cocoon. Apart from this, there was nothing to see at visible wavelengths nor at other radio wavelengths.

Further discoveries followed. Gareth Wynn-Williams, a student of Martin Ryle's, had mapped the radiation from a number of bright nebulae in the Milky Way using the Cambridge synthesis telescope and had discovered small, strong and rather isolated radio sources that were not visible on a photo. He went as a postdoc to Pasadena in 1971 and together with Becklin and Neugebauer he observed strong infrared radiation from the same small radio nebulae that he had discovered in Cambridge. Most of what had been detected by the Cambridge radio telescope was present in the infrared map, but there was also an interesting difference. Maser radiation from lines of H_2O came from a very small area in a dark cloud without other radio radiation, but with infrared radiation. This was certainly a star within a cocoon. The three authors calculated that this infrared source emitted 30,000 times more light than the Sun; the amount of light radiated by a main-sequence star five times as massive as the Sun (Wynn-Williams et al. 1972) (Fig. 4.10).

Masers could therefore lead you to very young objects. In the seventies the Westerbork telescope was used to look for radio sources at the site where masers were encountered. Several 'ultra-compact' radio sources were detected and it was concluded that they were just slightly older than cocoon stars: the star had become so hot that the gas closest to the star had already been ionized. Maser radiation is a phenomenon that occurs shortly before heavy stars break out of their cocoon.

4.2 The Spectrum and the Life Course of Stars

Schwarzschild was the first person to calculate correctly when and how a star left the main sequence to become a red giant. He used pen and paper, a simple arithmetic calculator and some brilliant simplifications. All this work was taken over by electronic computers, and as computers became more readily available and ever

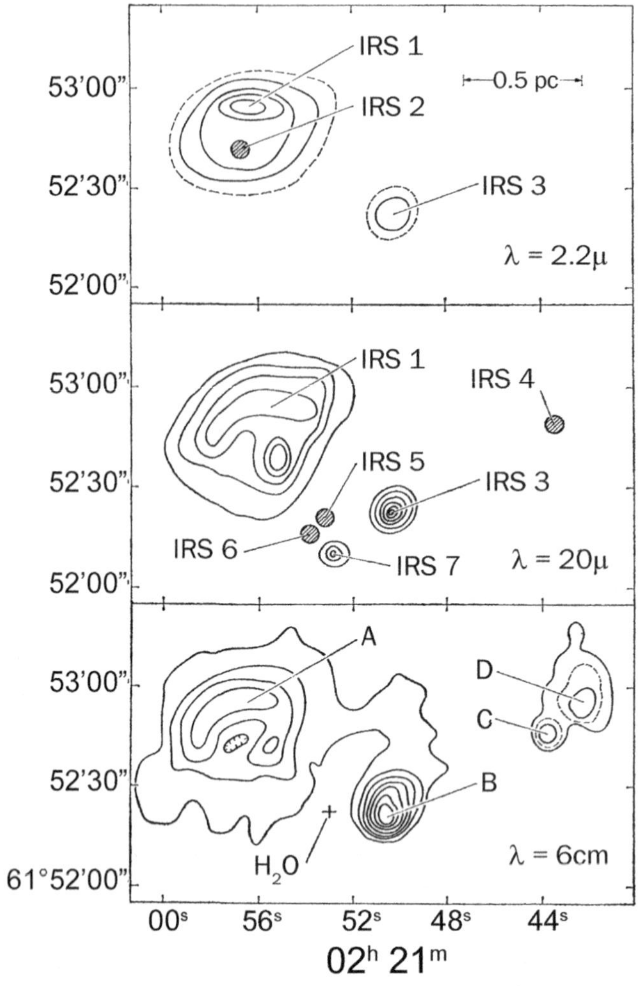

Fig. 4.10 Three detailed maps of the same area. From top to bottom: a map obtained at an infrared wavelength of 2.2 μm, then an infrared map at 20 μm and at the bottom a map of the radio radiation at 6 cm. There are some major similarities between the maps but there are also some significant differences. The IRS 1 source returns as source A in the radio map and the IRS 3 source returns as radio source B. This is remarkable because the radiation mechanism is completely different in the lowest map from in the two upper ones. Even more interesting are the infrared sources IRS 4, 5, 6 and 7. They do not feature in the radio map and they must be stars that do not ionize the surrounding gas, either because they are too faint to do so or because they are too young and not yet hot enough. The infrared source IRS 5 and the cross in the radio map coincide with the word 'H_2O': this is the location of a 'natural maser' of water at a wavelength of 1.25 cm (Wynn-Williams et al. 1972)

more powerful, the army of researchers constructing models increased and the insights that Schwarzschild had achieved with so much effort were confirmed with ease and in greater detail. These results will be discussed shortly, but first let us consider the question of the spectrum of stars. How does the light—the energy that a star produces internally—find its way to the outside? To obtain the answer, models were needed of the star's atmosphere and calculations of the spectrum of the light being emitted.

Models for the Atmosphere of Stars

Stellar models like those devised by Schwarzschild and his successors predicted how much energy emerged per square meter of the surface of the star. How would the atmosphere of the star react to such a flow of energy? What color would the outgoing light have and what spectral lines would be seen? The theory for this calculation was understood fully in the thirties. In 1938 German astronomer Albrecht Unsöld (1905–1995) wrote the standard reference work 'Physik der Sternatmosphären', that in essence contained the complete theory in its first edition. The significance of the book is apparent from the fact that, although it was never translated into English, it was nonetheless widely cited by Anglo-Saxon scientists.

Atmosphere models were calculated by starting at the surface of the star and moving outwards in small steps. At each step, a calculation was made for each wavelength of how transparent the atmosphere was and how much radiation was able to pass through the atmosphere at that wavelength. There were so many steps to be taken and each step demanded calculations that were so time-consuming that until the sixties drastic simplifications had to be made. This resulted in model spectra that were good enough to indicate the observations in general terms, but not in detail. This was unfortunate because the observations of the stellar spectra increased in quantity and quality, and they contained information that could not be deduced from the fairly rudimentary atmosphere models.

In the period from 1960 to 1974, attention was focused on calculating better atmosphere models using computers. It was a silent revolution: gradually all the simplifications were removed, giving rise to higher quality models. The insights they provided were better, but no spectacular discoveries were made. In 1975 and 1978 'atlases' were published with large numbers of model atmospheres (1,100 and 3,500 respectively). Even in 2015, atlases of model calculations were in frequent use. Figure 4.11 gives two examples of a model spectrum.

Just think about this development. For an observer it was no longer necessary to calculate a complex model atmosphere himself in order to arrive at an in-depth understanding of his observations. This was something new. If there were enough off-the-peg suits available, why go to the expense of made-to-measure?

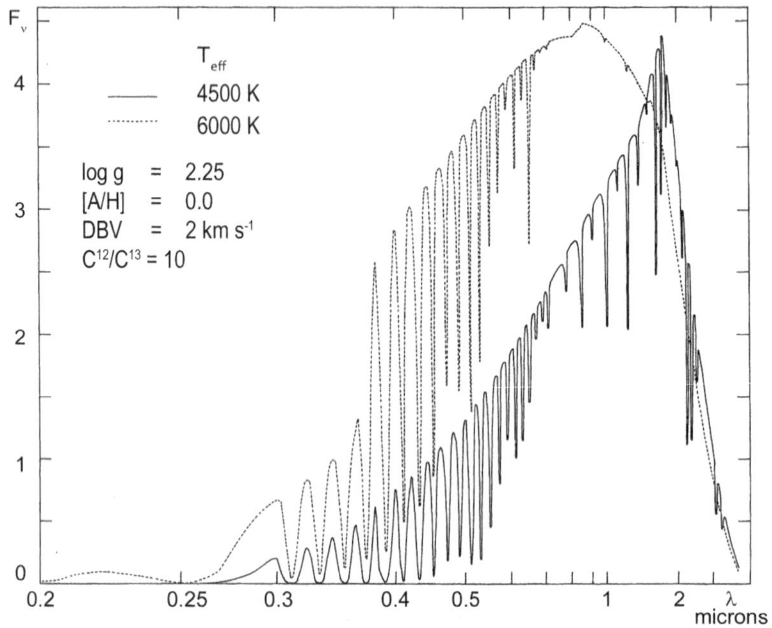

Fig. 4.11 Model spectra of two stars, one with a surface temperature of 4,500 K (slightly below the temperature of the Sun) and the other 6,000 K (slightly hotter than the Sun). The amount of energy radiated is shown vertically and the horizontal line shows the wavelength. The horizontal scale is logarithmic, but the wavelengths are indicated by their value in microns (1 μm = 1 millionth of a meter) (Gustafsson et al. 1975)

Tracks of Evolution in the HR Diagram

Calculations made by Schwarzschild had already revealed the life history of stars after they had left the main sequence. He had also demonstrated that the ageing process was different for a light star than for a heavy star. From the late fifties electronic computers became increasingly faster and a new generation of researchers was busily programming and calculating. This generated a huge amount of information that in many cases was confirmed by observations.

Nonetheless, all these enthusiastic calculations did not reveal that the one-dimensional models had shortcomings. It had been assumed that stars were perfect spheres that did not rotate on their axis and that had no magnetic field. If researchers wanted better models without these restrictions, models should be calculated in three dimensions, something that computers were not yet capable of. These shortcomings did not take away the feeling that the available 1-dimensional models contained the essence of stellar evolution.

Figure 4.12 gives the results of model calculations by Polish astronomer Paczynski (1970) of the development of stars with a mass between 0.8 and 15 M_{sun} after they have left the main sequence. An initial conclusion was that light stars became red as a result of ageing and were then 1,000 times brighter than on the

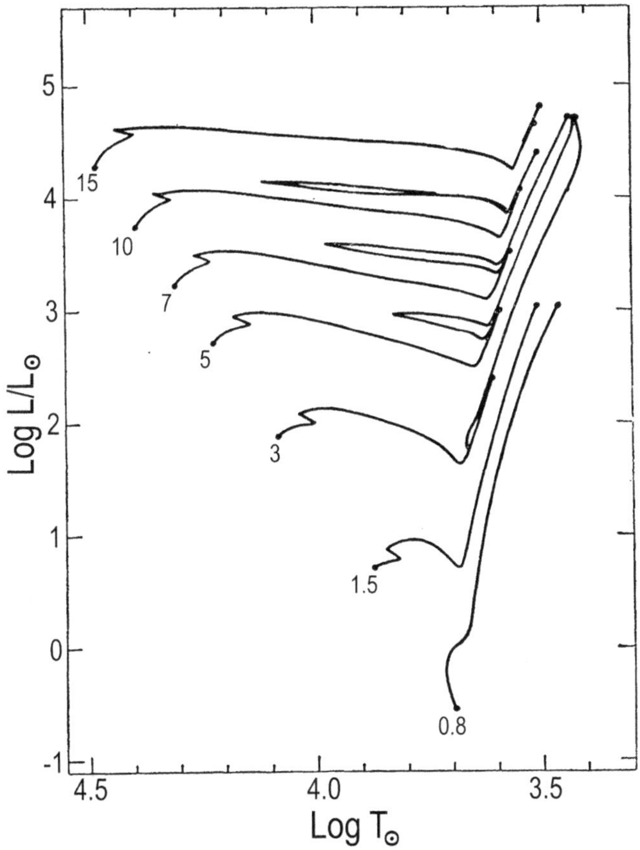

Fig. 4.12 An HR diagram with the calculated 'evolutionary tracks' of stars with different masses. The temperature at the surface of the star is given horizontally. This temperature, expressed logarithmically, determines the color. To the far left the temperature is 32,000 K and the star is blue; to the far right the temperature is 2,000 K and the star is deep red. The total amounts of energy radiated by a star, its luminosity, is given vertically, also logarithmically. The bottom line with '−1' indicates a luminosity of one-tenth that of the Sun; the line with the figure '5' at the top indicates a luminosity of 100,000 times that of the Sun. Each curve is an 'evolutionary track' that indicates how the luminosity of a star and its surface temperature change as the star ages. The mass of the star is given at the start of each track. A star stays at the start of its track for most of its lifetime, and only leaves this position when it is nearing the end of its life. This means that the track is traversed in a very short space of time. Most stars that we observe are therefore at the start of their track; these starting points constitute the main sequence. A star of 0.8 times the mass of the Sun starts as a faint red star and develops towards the end of its life to become a red star that is 1,000 times stronger than the Sun. A star of 10 solar masses starts as a blue star that emits more than 10,000 times as much light as the Sun; it continues to emit this amount of light, but becomes more and more red. What happens after a star has reached the end of its track was largely unknown in 1970 (Paczyński 1970)

main sequence, while heavy stars also became red but not much brighter; see the track that they follow in Fig. 4.12.

In 1974, at the end of a symposium on stellar evolution, Schwarzschild summed up the discussions (Schwarzschild 1974). He concluded that the lifecycle of a star depends critically on its mass. The featherweights, stars with a mass below 0.5 M_{zon}, were too light to ever engage in nuclear fusion. They had to exist but in 1974 they had not yet been discovered. Stars of the lightweight class, between 0.5 and 3 M_{zon}, eventually became red giants and ended as white dwarfs after a brief phase as planetary nebula. The Sun is a youthful example of this group. Then there was the middleweight class with a star mass between 3.5 and 10 M_{zon}. As they aged, they were transformed into red supergiants and ended as supernovae. Schwarzschild was unsure about precisely how that occurred. Finally, there were the heavyweights, heavier than 10 M_{zon}, which were later referred to as hypergiants. Very few of this class were known and there was little to be said about them. They would probably end as supernovae. The work that Schwarzschild started in the early fifties had now—20 years later—reached provisional completion.

I will now discuss briefly the subsequent insights about lightweight, middleweight and heavyweight stars; the featherweight class will only be discussed once the first examples have been observed, decades later (Fig. 4.13).

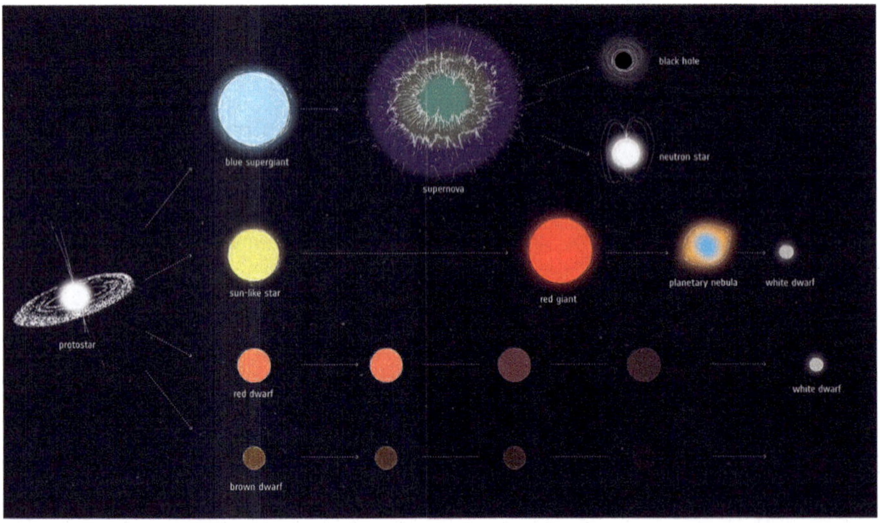

Fig. 4.13 The time runs from left to right. On the left, a protostar is visible in white. Four horizontal series of stars can then be seen. The topmost series is of massive large blue stars that ultimately explode as supernova, leaving a black hole or a neutron star. The second series proposes the evolution of stars comparable to the Sun. They develop to become red giant stars and then peacefully disintegrate to become a so-called planetary nebula and eventually white dwarfs. Stars lighter than the Sun gradually become smaller and end as white dwarfs. The lower series is of brown dwarfs, which never produce energy themselves and simply cool down, a process that can take as long as the Universe exists. Brown dwarfs have only been found since the 1990s and they are described in a later chapter. Source: ESA

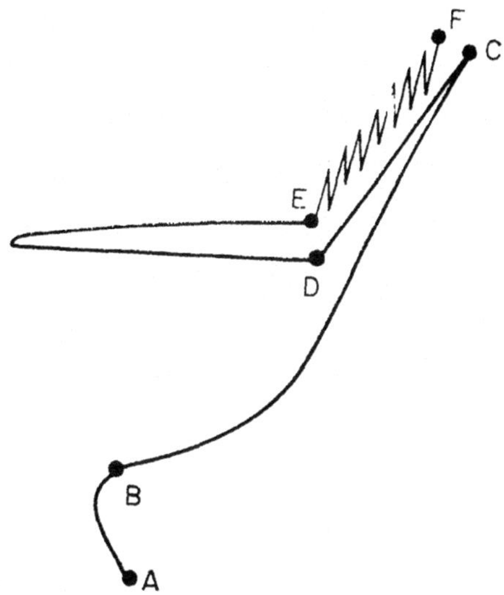

Fig. 4.14 A sketch of the path followed by a light star through the HR diagram. See text (Schwarzschild 1970)

The Lifecycle of Light Stars such as the Sun

Chapter 2 discussed how light stars such as the Sun develop into red giants. In 1960 Schwarzschild was awarded the 'Gold Medal' of the English Royal Astronomical Society for his contribution to the new insights. At the ceremony he gave a succinct yet almost complete description of the evolution of light stars and used a simple diagram to illustrate his findings (see Fig. 4.14) (Schwarzschild 1970). This kind of simple figure that compactly summarized all the data was Schwarzschild's trademark; he was an excellent teacher. A light star started its life in A, on the main sequence and developed, as discussed in Chap. 2, much later to become a red giant (point C). At that point the burning of hydrogen became inefficient and the star shrank, falling back to point D. This is when the burning of helium started. A further period of peaceful development along the so-called horizontal branch followed until the star arrived in E. And then some unresolved difficulties were encountered: the calculations showed that the star was somewhat unstable. The star probably moved via the zigzag line to point F. To calculate that final part of the evolutionary track required many more calculations than computers were capable of in 1970. In any event, it seemed that once the star arrived at point F it again became a red giant. This new, second red giant phase was different from the first, but nonetheless had similarities. This second giant 'branch' was named the Asymptotic Giant Branch, known by the acronym AGB. The word 'asymptotic' is a mathematical term: an asymptotic line comes closer and closer to a straight line, but never reaches it.[3]

[3] Maths: if you calculate y = 1/x for an increasing value of x, you will see that y slowly but surely approaches the x axis, but never reaches it.

The zigzag line came about because there were two sources of energy that were active alternately: at some points in time helium burned to carbon and there were other periods when hydrogen was converted into helium. The star rose in fits and starts to point F.

What happened when the star arrived at point F? It was vaguely assumed that the star would move from F horizontally to the left and in doing so would change form and become a 'planetary nebula'. (This kind of nebula has a historical and generally accepted name. It consists of a small hot star surrounded by luminous gas. A 'planetary nebula' has nothing to do with a planet and the name is very unfortunate.) At this time (1970) AGB stars and planetary nebulae were widely observed objects, but no stars in between, those that were no longer AGB stars and not yet planetary nebulae, had been found. Research in the infrared and at radio wavelengths would remedy this a few years later.

The Formation of 'Dust Particles' and the Emission of Gas by AGB Stars

In Chap. 3 the first infrared astronomical survey of a large part of the sky was mentioned. It produced a catalogue with about 5,600 entries (Neugebauer and Leighton 1969). It contained mainly red stars that were already known and the infrared observations added hardly anything new. But, and that was the bonus, a small number of stars were found that emitted a lot more infrared radiation than could be expected. Worthy of mention are NML Cygni, IK Tau and IRC+10216: all three embedded in a then unusually thick, dusty envelope. NML Cygni is also one of the most luminous stars known: its luminosity is 270,000 times that of the Sun, or 14 magnitudes brighter.

Later infrared measurements at longer wavelengths showed the presence of envelopes that were much thinner. The red giant star χ Cyg was an example from 1968; see Fig. 4.15. The black dots at wavelengths shorter than 6 μm fell precisely on a curve that was an approximation of the stellar spectrum. More radiation than expected was received at wavelengths longer than 8 μm, This was not emission by the stellar photosphere but by dust particles in the outflowing gas.

The discovery of dusty envelopes was important for two very different reasons. The first was the confirmation that red giants could be surrounded by thick shells of gas and dust, that were indications of the material expelled on a large scale by these old stars. If this ejection of matter continued long enough, ultimately the hard, helium core of the red giant would become bare and a white dwarf had been born. The ejection of matter was the last link in the explanation of the existence of white dwarfs. The second reason that dusty envelopes were important is that they proved that decaying red giants produced such small particles. The particles themselves had been encountered at least 30 years previously (see the Prologue), but nobody knew where and how they were made. The answer had now been found: some very cold, and very old stars emitted gas; as the gas flowed away, it cooled down, creating circumstances that were favorable for the condensation of particles of matter.

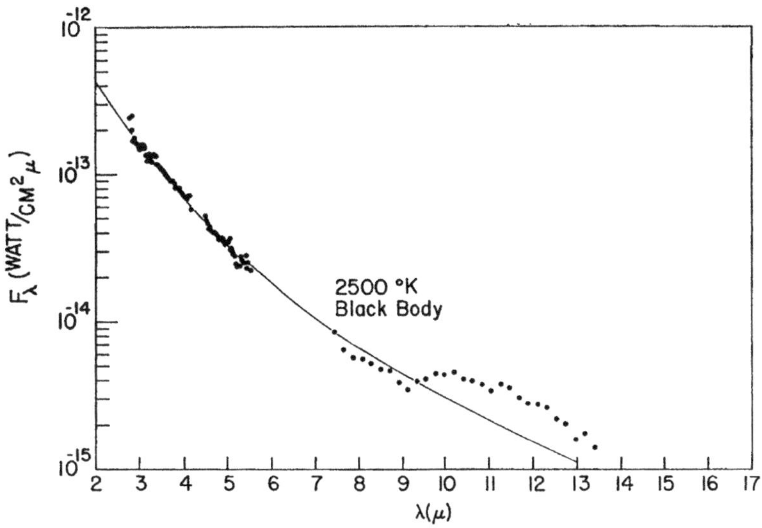

Fig. 4.15 The radiation received from the star χ Cygni between the wavelengths of 2 and 17 μm. The line drawn, indicated as '2500 K Black Body', was an estimate of what the atmosphere was expected to radiate—at longer wavelengths there is clearly more radiation than expected. On the horizontal scale the wavelength is expressed in μm, one millionth of a meter (Gillett et al. 1968)

Further infrared observations showed that there were two kinds of evaporating stars: in some, the infrared particles had to be silicates, that is they are related to very small grains of sand, and in other stars, the particles were more carbonaceous, like very tiny soot particles. Surprisingly quickly, a young American astronomer, Robert Gilman, found an explanation for these two different kinds of dust particles (Gilman 1969). It had already been known for just under a century that there were two kinds of red giants: most had more oxygen in their atmosphere than carbon, and a minority had more carbon than oxygen. Simple thermodynamic reasoning showed that silicate type particles would grow in the shell of the first kind of stars and carbonaceous particles in the second kind. This meant that Hoyle and Wickramasinghe, who as early as 1962 had contended that the so-called carbon stars produce soot-like particles and added those to the interstellar dust, were proven right. There was an even wider perspective: Earth is a rock that was formed from dust particles that had been produced in such stars as in Fig. 4.15. Precisely how a planet like Earth was created was not clear, but it was now clear where the building blocks were made.

Chromosphere, Corona and Magnetic Activity

It was already mentioned in Chap. 2 that there were a number of interesting phenomena to be observed on the surface of the Sun that were too faint to be seen in other stars. In 1963, O.C. Wilson from the Mount Wilson and Palomar observatories noticed a subtle phenomenon in spectra of stars like the Sun (Wilson 1963). These spectra contain very dark lines produced by calcium ions. But in the middle of these lines there was a small reversal, a small hill where the valley was at its deepest. This hill, Wilson said, was produced by the chromosphere of the star. He showed that the hill was higher in a young cluster of stars than in old clusters: the height of the hill showed the age of the star. Chromospheres were strong when a star was young; TTau stars, young, pre-main-sequence stars, had strong chromospheres. Several years later, Wilson reported a connection with the rotation of the star: older stars rotated more slowly. Stars of similar mass as the Sun had a lot more to show than had been expected.

Solar Oscillations or Helioseismology

In 1962, American astronomers Leighton, Noyes and Simon (1962) announced that the Sun is oscillating: the photosphere of the Sun, where all the light that we receive is formed, moves up and down at a velocity of less than half a kilometer per second. Soon it was shown that several oscillations took place simultaneously. It was not a sonata for violin solo, but a symphony for a full orchestra. A few years later Ulrich (1970) proved that under the surface of the Sun oscillations ran up and down, being reflected once deep inside the Sun and the second time close to the surface. These oscillations created standing waves and the effects of these waves caused the oscillations that had been observed.

Ulrich's explanation made it possible to get empirical information on pressure and temperature deep inside the Sun, signaling the beginning of a subdiscipline called 'helioseismology'. This is a combination of the Greek word for Sun (helios) and 'seismos', the Greek word for earthquake. The classical Greeks, inventors of science, were scientifically not sufficiently advanced to need a word for 'sunquakes'. In the following decades helioseismology would give birth to yet another discipline, 'asteroseismology'.

The Lifecycle of the Mid-Weight and the Heavy Stars

Back to Fig. 4.12. The evolutionary tracks of the heaviest stars (10× and 15× the mass of the Sun) ran almost horizontally. The stars were much brighter and were referred to as supergiants. In the main sequence they were to the left of the diagram and were blue; the star Rigel in the constellation of Orion was one example. As they aged, they became red supergiants, Betelgeuze, also in Orion, was such a red supergiant. The evolution from blue supergiant to red progressed rapidly and it was generally expected that supergiants would be mainly blue or red. Naturally, evolution did not stop with the red supergiants and there was the general expectation that red supergiants would eventually explode as type II supernovae (see later in this chapter).

But how this explosion came about was unknown. There was even an indication that a red supergiant changed again into a blue supergiant before exploding. Wolf-Rayet stars were examples of these, as will be discussed later. The understanding of the evolution of heavy stars was still far from complete.

The calculations for the heaviest stars had their limitations. These stars often rotated very fast on their axis and were then no longer completely spherical, as the calculations hypothesized. Another complication was that heavy stars often occurred as binary stars with a slightly less heavy star as companion. If the distance between the two stars became too small, the development of one star influenced that of the other—a phenomenon that will be discussed later in more detail.

While calculations such as those by Paczynski occupied the theoreticians, the observers managed to record increasingly better and more detailed spectra of these hot and heavy stars. In time, it would be possible to pluck the fruits of these gradual improvements. An important new result warrants a separate discussion: the discovery that hot blue supergiants ejected matter.

Loss of Mass of Heavy Main Sequence Stars

Heavy main sequence stars were so hot that they radiated mainly ultraviolet light. This light was absorbed by the Earth's atmosphere and never reached its surface. This had been a serious limitation in the research, as appeared from rocket observations made by Don Morton from Princeton in the 1960s. He was granted three flights with an Aerobee rocket to record the ultraviolet spectrum of these kinds of stars. The observations were made with a camera that was switched on once the orientation control mechanism had located the stars. All three flights failed. With the first two, at first everything went well until the control mechanism was supposed to direct the camera, which it failed to do. The third flight went according to plan until during the planned fall from a height of 180 km the parachute opened too early, the lines to the camera broke and it descended in a freefall. Searches in the desert located the remains of the camera, but the film cassette had broken open and the film was overexposed. Against better judgement, Morton examined the film some weeks later again and surprisingly found a very small part with readable information. The result was spectacular. He was able to identify the stars δ and π Scorpii, hot and without doubt heavy stars whose spectrum exhibited some spectacular details: there were dark lines of ionized carbon and ionized silicon (C III and Si IV)[4] and the width of the lines suggested that the material seen here had left the star at a high velocity. This was something new (Fig. 4.16).

Morton was now granted a further three flights, all of which were successful. Again, but this time better documented, he saw a flow of material emerge from the star at velocities of between 1500 and 2000 km/s. The stream was so strong that it meant that the star lost around half a solar mass in 10,000 years; such a great loss

---

[4] C III stands for the twice-ionized carbon atom: C^{2+}. Similarly, Si IV stands for Si^{3+}.

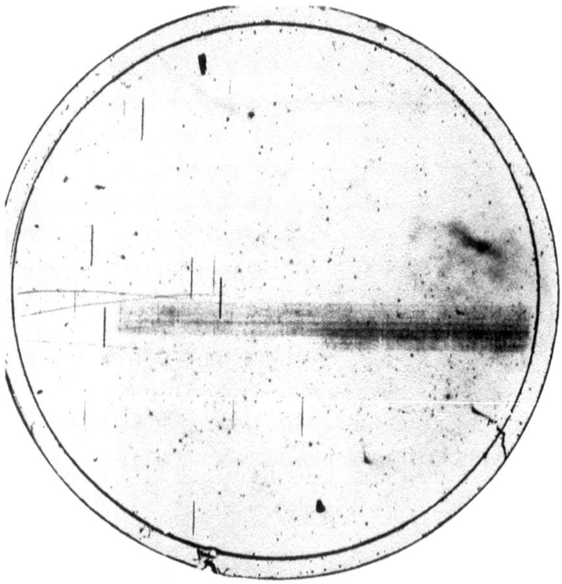

Fig. 4.16 A small part of the visible sky photographed during a short rocket flight. The broad horizontal bands consist of spectra between 126 and 170 nm of the stars δ and π Scorpii (Morton and Spitzer 1966)

must have an enormous influence on the star's development. Morton's conclusion was confirmed a year later by highly detailed spectra of several supergiants by J.B. Hutchings in Canada (Hutchings 1968) and again several years later by more precise observations with NASA's Copernicus satellite.

Radiation pressure was considered as a possible explanation for this efflux. A light particle (photon) has a particular momentum, and if an atom or ion absorbs the light particle, it not only gains energy but also changes velocity. It can be compared to a toy boat in a pond into which stones are thrown from the bank: the boat will sail away. Because, in the atmosphere of the star, the photons come from the inside outwards, the atoms will also be forced outwards. Initially there was some doubt whether the pressure from the photons was great enough. If the stones are replaced by grains of sand that are thrown one by one into the boat, a great many would be needed to have any effect. Initial calculations indicated that the radiation pressure would be much too low, but in 1975 Americans Castor, Abbott and Klein (Castor et al. 1975) demonstrated that if you take all the absorption lines into consideration, the radiation pressure was high enough. The atmosphere of these hot bright stars could not be static: light pressure ensured that gas always had to flow away.

Wolf-Rayet Stars, an Evolutionary Enigma

It has been mentioned earlier that the Wolf-Rayet stars (WR stars) did not fit the simple model in which red supergiants exploded as supernovae. WR stars were hot and had much broader and brighter lines of emission of helium. They had been

Fig. 4.17 The light-emitting nebula NGC 6888 with the Wolf-Rayet star HD 192163. The red color of the nebula comes from the strongest spectral line of the Balmer series of hydrogen. Source: Astronomy Picture of the Day, 11 November 2007, Franck Bugnet

discovered early in spectroscopy, in 1867, by visual inspection of the spectrum well before the introduction of photographic plates. Hydrogen lines were missing from the spectrum, indicating that hydrogen did not occur on the stars. This suggested old age as all the hydrogen was either expelled or burnt, but the blue colour was an indication of young age. In 1973 a young Australian astronomer, Lindsey Smith (1973), proposed with good arguments that WR stars remained after hot and heavy O stars had expelled a lot of material. In some cases, this expelled matter could still be seen as a bright gas cloud around a WR star, for example the nebula NGC6888 that can be found around the WR star HD 192163; see Fig. 4.17.

Wolf-Rayet stars were blue supergiants; but they had lost much matter and that suggested that they had previously been red supergiants. Did these supergiant traverse the HR diagram from blue to red more than once? It was puzzling.

Very Heavy Stars, the Hypergiants

In the course of time, several extremely bright stars were found that exhibited puzzling behaviour and it was generally assumed that these were very heavy stars. Historically two objects were known. The star P Cygni exploded around 1603; the data were scant and little was known, except the fact that an explosion had taken place (de Groot 1969). The other star was η Carinae, a bright star on the southern hemisphere (30 degrees from the south pole) that was embedded in a light-emitting

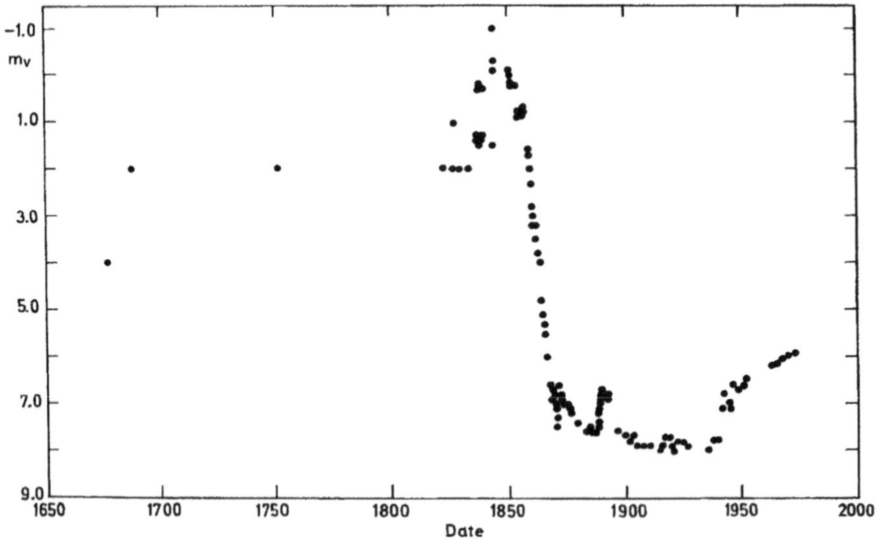

Fig. 4.18 The brightness of the star η Carinae between 1650 and 1974. Time in the horizontal direction and magnitudes vertical (Feinstein and Marraco 1974)

nebula. The brightness of the star had changed considerably over the previous 200 years, particularly in the 1840s, when by chance John Herschel spent several years at the Cape of Good Hope. John was the famous son of his even more famous father William. In around 1840 the star achieved its brightest intensity and then reduced its light output by 9 magnitudes, thus emitting 4,000 times less light: see Fig. 4.18.

The star belonged to an open cluster of stars, of which the distance could be measured. The star itself and its environment were rewarding objects for visual observations. There was a lot of color in the figure and there were numerous details that stimulated the imagination. In 1950 the Argentinian Gaviola (1950) showed contourlines derived from a series of photos each taken after an increasingly long exposure. From this series he deduced the structure (see Fig. 4.19) and commented that it looked like a 'homunculus' (or little man).

Much more recently, some beautiful figures have been recorded using the ESO VLT telescope, in which it can be seen that the central star (η Carinae) has formed two large balls of gas on both sides: see Fig. 4.20.

In the 1960s, η Carinae caused a surprise because at wavelengths between 10 and 20 μm it appeared to be the strongest infrared source in the sky. The infrared radiation was even so strong that the total amount of energy emitted was now the same as in 1840. The star had not changed in light intensity! (Westphal and Neugebauer 1969). It seemed that in around 1840 a cloud shifted in front of the star that absorbed the visible light but that then emitted this again in infrared.

The star η Carinae continued to be an enigma: a star that emitted three million times more radiation than the Sun was—and is—very unusual.

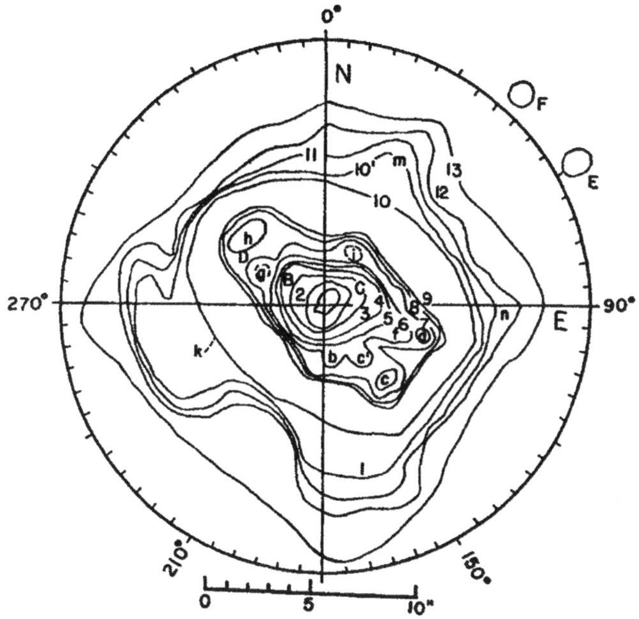

Fig. 4.19 Isophotes (lines of equal brightness) in the area around η Carinae measured on a series of photographic figures, each with a different exposure time for the same photographic plate. The central object looks like the figure of a gingerbread man, and is referred to as a 'homunculus' (Gaviola 1950)

Close Binary Stars

The calculation of models of single stars was now to some degree complete. It was known that as stars age they acquire a very dense core of helium surrounded by a mantle of hydrogen. At a later stage and under higher temperature and pressure, helium would burn to form carbon and oxygen and at an even later stage, first carbon would burn to form neon and magnesium and possibly ultimately oxygen would burn to form silicon and sulphur. What was uncertain was how this last phase in the development to a white dwarf or supernova would proceed. What was certain was that at the start of a new incineration the core became smaller, but the mantle larger.

The enormous expansion of an ageing star could have important consequences if the star formed a binary star with another star. As long as the two stars were distant enough from one another, each could follow its own life course, but that stopped if one of the two expanded so far that matter started to overflow to the other. And then anything could happen. Chapter 2 described how a nova, the unexpected and explosive lighting up of a star, was the consequence of an explosion on a white dwarf after it had received a lot of matter from its partner.

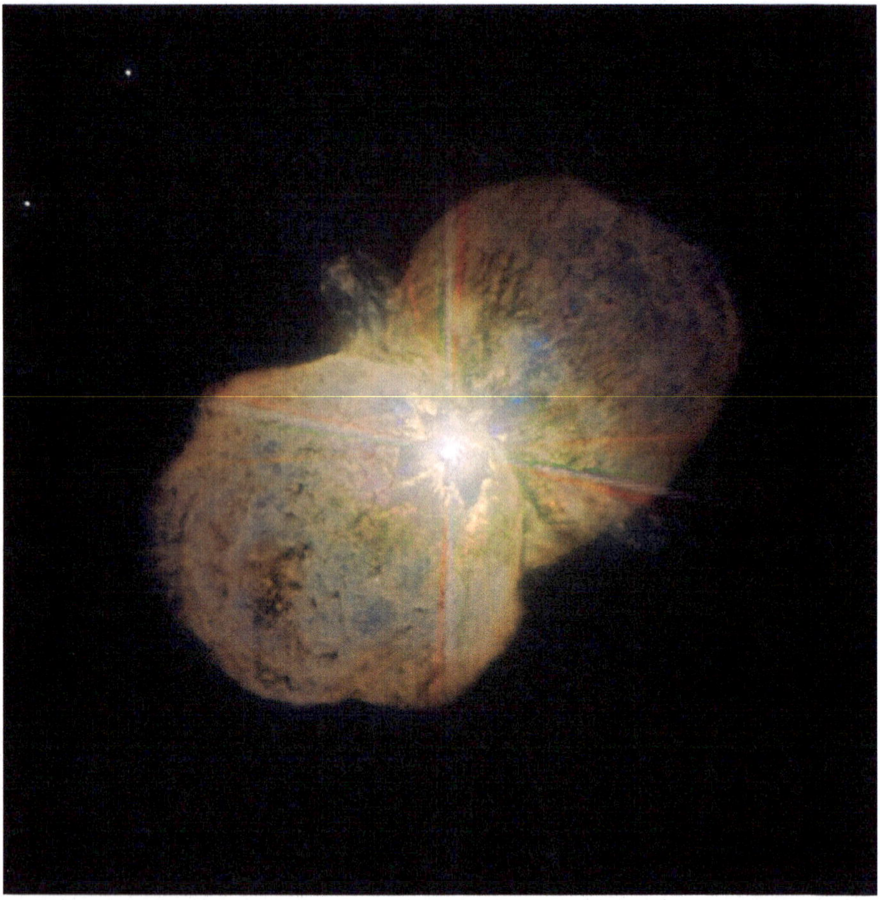

Fig. 4.20 The star at the centre is η Carinae. The two spherical objects were produced by the star, probably by two streams of radiation in opposite directions. Parts can also be seen of a flat surface through the star and perpendicular to the two flows. A thin, colored cross can be seen centred on the star; this is caused artificially by wires in the telescope. Source: ESO

In general, it was impossible to calculate how a constellation of two stars would look (the shape that both stars would adopt) and how this constellation would change during evolution. The problem had to be simplified through a number of assumptions. Moreover, the mass of both stars was an important fact, as was the distance between them and an infinity of different models could be calculated. There were simply too many possibilities to calculate them all. I will not go deeper into all the possibilities, and only warn the reader that the evolution of close double stars is an ocean of possibilities to take in.

4.3 Dying and Extinct Stars

In around 1930 Chandrasekhar had explained the structure of white dwarfs. The general assumption was that white dwarfs were the final stage of a star; a white dwarf was the still warm remnant of a star. This suspicion was reinforced when in the fifties it became clear that a white dwarf could be formed from the core of a red giant. In white dwarfs the material is made up of atomic nuclei and electrons. The scientific imagination did not, however, end with white dwarfs. In 1932 the neutron was discovered, a particle as massive as a proton, but electrically neutral, and in 1934 Baade and Zwicky (see the Prologue) suggested that beside the white dwarfs, stars, too, could exist that were made up of neutrons. Neutron stars would be much smaller than white dwarfs. It was a matter of wild speculation; there was no indication whatever that they were actually present in space. Until 1968, that is (see later).

White Dwarfs and Planetary Nebulae

White dwarfs were a phenomenon where the theory was more advanced than the observations. They were such faint light sources that it was difficult to measure their properties accurately. The brightest white dwarf Sirius B, was of the 9th magnitude but was so close to the bright star Sirius A that every spectrum of Sirius B contained a considerable amount of light from Sirius A. Almost all other white dwarfs were fainter than the 11th magnitude and the recording of a detailed spectrum required a large telescope.

Theory predicted that the higher the mass, the smaller the white dwarf. This went against every intuition and it was therefore interesting to confirm with observations. If, as seemed very probable, white dwarfs were formed from the nucleus of a red giant, then the composition of the dwarf depended on how far the giant had evolved. A white dwarf could in principle comprise helium, but could it equally comprise a mixture of carbon and oxygen, or a mixture of neon and magnesium? A nucleus of silicon and sulphur was a theoretical possibility. If you could determine the mass and radius of a white dwarf accurately, then you knew whether you had a light dwarf with helium or a heavy variant with silicon and sulphur.

Stars on the main sequence or on the giant branch generated the energy that they radiated through nuclear fusion, but nuclear fusion did not occur in white dwarfs. They emitted light and thus energy, which caused them to cool down, just as a hot water bottle slowly cools down while warming the bed. This cooling down of white dwarfs took place very slowly because of their very small surface area; calculations indicated that it could take billions of years before they were really cold. Was it possible to find cold white dwarfs? It was an intriguing idea that the complete cooling took longer than the Milky Way was old and that no dwarfs existed that had cooled to zero degrees. So could white dwarfs be used to calculate the age of our Milky Way? And finally, there was the most important question: how were white dwarfs formed?

The impasse in which the observers found themselves was exemplified by the best photographs of the spectra of white dwarfs, photographs taken with the largest telescope, the 5-meter one at Palomar. These photographs were still not good enough to deliver important news. Then there was the issue that hot stars emitted most radiation in the far ultra violet and this radiation was intercepted by the Earth's atmosphere. And the white dwarfs were too faint for the satellites that could be sent above the Earth's atmosphere.

Nonetheless, a discovery was made that caused a minor shock: a white dwarf was discovered whose light was 'circularly polarized', as it was described, which meant that there was a very strong magnetic field (Kemp and Swedlund 1970). The observation was very delicate and the result was a triumph for the observers. But the pleasure at the discovery brought some concerns with it. A magnetic field considerably complicated the theoretical study of the structure of a star, on the one hand because it added a force with complex effects, and on the other hand because the observations revealed very little about the structure of the magnetic field.

Then there was the question of numbers of white dwarfs. Naturally, astronomers wanted to find more white dwarfs than the few that had been discovered more or less by chance. A major study was carried out from the forties by Willem Luyten, an American astronomer with Dutch roots. He used photographic plates that were taken at points in time as far as possible from one another and looked for faint stars that had changed their location between these times. Relatively large displacements were expected since white dwarfs were always nearby: at large distances they were too faint to detect. Luyten published books with long lists of such candidate white dwarfs; there was some discussion about the quality of the lists, but they did generate a number of bona fide white dwarfs. In 1968, German astronomer Weidemann estimated the number of known white dwarfs at around 1,500 (Weidemann 1968).

The spectrum of a white dwarf was formed in an atmosphere that was just a few hundred meters thick and in which the pressure was around 10,000 times greater than, for example, on the Sun. (Incidentally: a white dwarf has a mass comparable to the Sun, but its radius is 100 times smaller, and the gravitational acceleration on the surface is 100^2 times greater). Only very broad lines of hydrogen and helium were found in the spectra; the width was caused by the enormous high pressure on the stellar surface: a well-known phenomenon in physics.

It was difficult to analyse such broad and shallow spectral lines, which represented a problem en route to the 'holy grail': measuring the 'Einstein shift'. In his general theory of relativity, Einstein had predicted in 1915 that light particles that left the star surface on their way to the outside had to overcome gravity and in so doing lost some of their energy and became slightly redder. For the light leaving the Sun, this Einstein shift was very small, but it was present, as demonstrated in 1962 by J. Brault. The Einstein effect was much greater on the surface of a white dwarf. From the measurement of the shift, one could calculate the parameter M/R^2 and because the radius of the star, R, could be calculated from the observations, the mass M of such a white dwarf could be deduced from the Einstein shift.

Unfortunately, there was another mechanism that shifted the spectral line: the Doppler effect. How could this be distinguished from the Einstein effect? This was possible if the white dwarf was the partner of another star: then both would have the same Doppler shift and you could eliminate the Doppler shift from the spectrum. In 1967, Greenstein from the California Institute of Technology and his graduate student Virginia Trimble measured the spectral lines in several white dwarfs that were members of a binary system (Greenstein and Trimble 1967). Apparently they considered the individual measurement per star not to be precise enough so they averaged their results and obtained an Einstein shift of 51 km/s. After determining the radius R, they calculated that the average mass of the dwarfs was 0.86 times that of the Sun—well below the Chandrasekhar limit of 1.4 times. This was quite a success but at the same time the measurement of the mass was not precise enough to determine the composition of the dwarf: was it helium or a combination of carbon and oxygen?

White dwarfs were the cores of red giants after these had decayed totally. It was generally assumed that on their way to becoming white dwarfs, the stars would pass through an intermediate stage, that of the 'planetary nebula'. These are small nebulae, with very diverse structures and in beautiful colors. At the centre is a faint, white star that is the core of an AGB star (discussed above) and the nebula is what is left over of the AGB star. The nebula expands and will soon disappear (astronomically speaking: within a few tens of thousands of years). The central star is still brighter than a white dwarf, but it is cooling down and will soon become one.

Supernovae

Supernovae were stars that suddenly—within a few hours—became up to 100,000 times brighter. Walter Baade and Fritz Zwicky argued for the existence of supernovae in 1934—see the prologue. How the study of this subject developed up to 1974 can be divided into three parts. What are we seeing? What happens in the inside of the star? What is left over?

What We See, or the Exterior of a Supernova

From 1945, new supernovae were continuously being found and Zwicky's programme that had already started before 1940, contributed considerably to this. The distinction that Baade and Minkowski had made in around 1940 between type I and type II was confirmed time and again and also the difference in location of the two types: type I SNe were found in all types of galaxies, while type II supernovae occurred in the arms of spiral galaxies, where hot and heavy stars were also found. The conclusion seemed justified: 'Type I supernovae are produced by old, light stars' and 'type II only by heavy stars'. Further, it appeared that the light curves (how the brightness of the supernova changed over time) of all type I supernovae were strongly similar to one another. The type II supernova, on the other hand, differed considerably from one another. This was reason for Zwicky to divide type II novae in a number of separate types that are not relevant for this book.

Other characteristics of the two types of supernovae also differed and the contrast was so strong that in 1960 Shklovskii (Shklovskii 1960) concluded, 'We need entertain no doubt that we are observing completely different phenomena in type I and type II supernovae.' Type I supernovae reached the magnitude -19.04 ± 0.23 and type II -17.70 ± 0.34, hundreds of thousands of times brighter than the brightest star in our Milky Way galaxy. In addition, little matter was expelled from type I supernovae (1/10 of a solar mass) but for type II the expelled mass was much higher (ten times the mass of the Sun). Finally, there was a difference in spectra, as Minkowski had already seen. Spectra of type I supernovae comprised broad peaks and troughs and observers did not understand what they were seeing. Spectra of type II were made up of very broad hydrogen and helium lines.

In 1965, Halton 'Chip' Arp, an astronomer at Mount Wilson and Mount Palomar measured in each of three color bands how much light was received from a type II supernova (Arp 1965). He tracked the supernova for 3 months while the brightness decreased and the color changed. Arp saw that the color changes were the same as those one could observe when running through a sequence of increasingly cooler supergiants. It was as if a thick shell of a hot star became larger and cooler. This insight was the basis for diverse models that describe the light curves of supernovae, both type I and type II, with increasing success in the following decade.

In 1969, Colgate and McKee (1969) studied the energetics of such an expanding shell and discovered that the gas in the expanding envelope radiates more energy than it received during the explosion. There should be a continuous source of energy in the expanding envelope. Colgate and McKee calculated that the explosion produced so much radioactive nickel (^{56}Ni) that the decay of this material to cobalt (^{56}Co) and then to ^{56}Fe (iron) could be this source of energy. This proposal was confirmed when it appeared that the light of the supernova diminished at the same rate as the decay of the radioactive nickel and cobalt atoms.

Let us turn to the spectra of supernovae. Spectra of type II supernovae contain well-known lines, especially of H and of He, but the spectra of type I had remained a complete mystery since the first measurements by Minkowski in around 1940. In 1969, Soviet astronomer Y. Pskovskii (1969) put forward a new interpretation of the spectra of type I supernovae. He made the assumption that the spectra are a collection of broad valleys in a high plateau and not, as had previously been supposed, a series of hills on a low plateau. Pskovskii's proposal was important because it meant that researchers should no longer try to identify the wavelength of the high points in the spectrum with those of known spectral lines, but the deep points. Using this idea, Pskovskii made several convincing identifications. His Soviet colleague Mustel derived high expansion velocities that he estimated at about 3% of the velocity of light (9000 km/s) (Mustel 1971). In later years, Kirshner, Oke, Penston and Searle (1973) obtained spectra of superior quality and these confirmed Pskovskii's analysis. The nut had finally been cracked. The mystery of the spectra of type I supernovae had been solved.

The absorption lines were all broad and asymmetrical, but the shape corresponded with the assumptions. The spectra of type I supernovae contained no lines of hydrogen and helium, but apart from this they generally resembled the spectra of type II. Exploding white dwarfs had been thought to be the origin of supernovae of type I and, indeed, these objects contain very little to no hydrogen in contrast to the heavy stars that were thought to generate the type II supernovae.

What Happens Inside a Star before and while it Explodes as a Supernova?

One could predict in very general terms what happened in the interior of a star of high mass. In a new star, hydrogen is converted into helium and this delivers the energy that the star radiates. When hydrogen is exhausted, the star will transform helium into carbon; and this process can go on for some time: carbon burns to oxygen, oxygen to silicon and finally silicon to iron. But then the process stops: to transform iron into a heavier atom you do not get energy, but you have to feed energy to the process. Chemists distinguish between 'exothermic processes', those in which some energy is released, and 'endothermic' processes that require energy to be provided from another source. When atoms lighter than iron fuse, the process is exothermic (it delivers energy), for atoms heavier than iron the process is endothermic (it needs energy).

Let's look what this means in the history of stars. A new star would consist of hydrogen. First the star would burn hydrogen into helium in its center, which would produce a stellar core of helium with a very high density. At some point the core grew too large and the burning of hydrogen became inefficient. The core cooled and contracted but this contraction stopped when helium began to be burned into carbon. Thus a core consisting of carbon would be formed inside the helium core. Then an oxygen core would develop inside the carbon core. This would continue until finally a core of iron was at the stellar centre. The star now looks like an onion: an iron core surrounded by shells of lighter nuclei (see onion skin Fig. 4.21). The star has now reached a catastrophic condition, because there is no way in which it can produce the energy it radiates: the star becomes unstable; the iron core cools, loses pressure and collapses. This is the beginning of a 'core-collapse' supernova.

Only stars with more mass than nine times that of the Sun develop an iron core; in lighter stars the developments come to a halt before the iron core is formed. At some point in their development the core becomes rigid because of the quantum mechanical pressure discussed earlier. The core is then a white dwarf in statu nascendi. More massive stars have an interior that is too hot for the quantum mechanical effect and so these heavier stars ultimately explode as supernovae of type II.

On very general grounds it has thus been argued that the core of high-mass stars would ultimately collapse. In one way or another—how was not known—the collapse was followed by an expansion deep in the star and when the effects of the

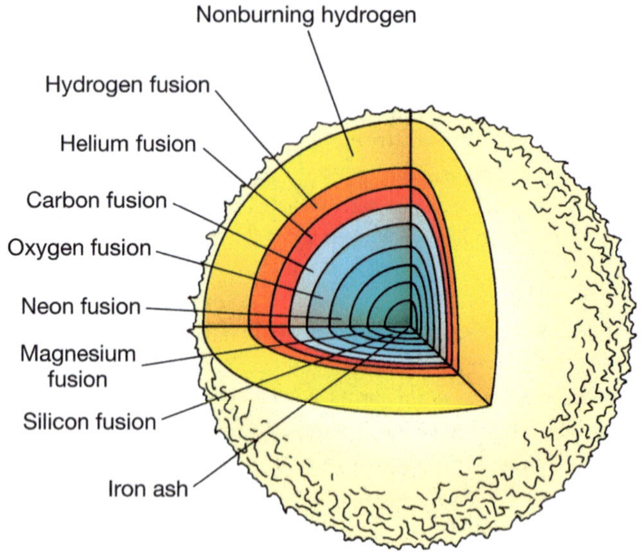

Fig. 4.21 Figure of the onion skin model. Source: Penn State Department of Astronomy & Astrophysics

explosion reached the surface of the star, a type II supernova would be visible. The transition from collapse into expansion could not be seen directly, but could perhaps be detected indirectly by the new chemical elements that would be produced during these events. The production of new atoms became the basis for an important paper on the origin of supernovae by Fowler and Hoyle in 1960 (Hoyle and Fowler 1960). They discussed what mixtures of atomic nuclei and electrons would be explosive if put under high pressure and at temperatures as high as a billion degrees. They also discussed which chemical elements would result; they were thinking particularly of the production of atoms of iron and related atomic nuclei. The considerations were complex but they brought the authors to a clear conclusion. Based on the most probable mixtures, there were two potential types of supernovae that could result: explosions of white dwarfs and heavy stars. This fitted in with the existence of two types of supernovae.

As a student I wrestled with this paper, mainly because it was so totally different from the textbook science that I had to study. My experience then may be of more general interest. The article by Fowler and Hoyle was full of equations with the symbol: '~'. This meant that whatever was to the left of the symbol should behave more or less the same as what is to the right of it. A concrete example would be the statement: '$\rho \sim T^2$', which meant that if the T was twice as high then the density ρ would be approximately four times (2^2) as great. This was not exact science, but an approximate science. The remark is not intended negatively. This kind of 'approximate reasoning' was and still is part of astronomical research and it has generated a great many insights. The necessary condition is that a true specialist is at work, who can estimate the dangers. Hoyle and Fowler were among this category of specialists.

It is not surprising that the 'approximate' article inspired young astrophysicists with a good understanding of nuclear processes to calculate precisely what was happening using ever-faster computers. An example is an article by Americans Colgate and White who came to the conclusion in 1966 (Colgate and White 1966) that there were fundamental reasons why the explosion mentioned in the article by Fowler and Hoyle could not possibly take place. They favored an alternative explanation based on the large number of neutrinos that would arise in the imploding nucleus and would heat the exterior areas to such a degree that it would cause an explosion. But the results by Colgate and White were also disputed. It was basically very difficult (there are those who would say 'impossible') to calculate just what was happening within the supernova at the critical moments that perhaps lasted only seconds: there were too many different factors operating at the same time. Even with the most powerful computers, it was still necessary to simplify the calculations, which, of course, made the results somewhat uncertain. In 1969, one of the better known model builders, American astrophysicist David Arnett, wrote an article setting out such a model (Arnett 1969), but he ended his argument with a critical discussion about the question of why there was still no model that everyone could agree on. How a supernova started and how it evolved was still debated.

One major question to which theoretical research should give an answer was whether, after the explosion, a white dwarf remained, or nothing at all. In the years prior to 1968 there were a few brave souls who suggested that a neutron star would remain, as Zwicky speculated.

In spite of the criticism, considerable progress was being made. The important role of radioactivity as a source of heat in a supernova has already been mentioned. Another prediction arose from the presence of radioactive nickel nuclei: in the supernova, high-energy radiation should emit spectral lines of gamma radiation, that is gamma radiation with a sharply defined energy. It was a prediction that was confirmed 40 years later. A further prediction concerned a great many neutrinos that had to be released during the implosion/explosion. This prediction was confirmed when a nearby supernova was detected in 1987 (see Chap. 14).

In 1973, Americans Whelan and Iben (Whelan and Iben 1973) put forward a proposal that explained the explosion of white dwarfs and thus type I supernova. They based their assumption on a binary star where one of the two stars was heavier than the other. The heavier star aged faster and would ultimately become a red giant with a helium nucleus. The star would lose the matter outside the nucleus, so that ultimately a white dwarf remained. It was presumed that this white dwarf had a mass just below the Chandrasekhar limit. Inevitably, the other member of the binary star would also evolve to become a red giant and would then start to emit matter. Part of the matter emitted matter landed on the white dwarf and if this process continued long enough the mass of the white dwarf would at some point in time exceed the Chandrasekhar limit. The object would become unbalanced and explode: a type I supernova. Type I supernovae were exploding white dwarfs. This proposal quickly acquired the status of a paradigm.

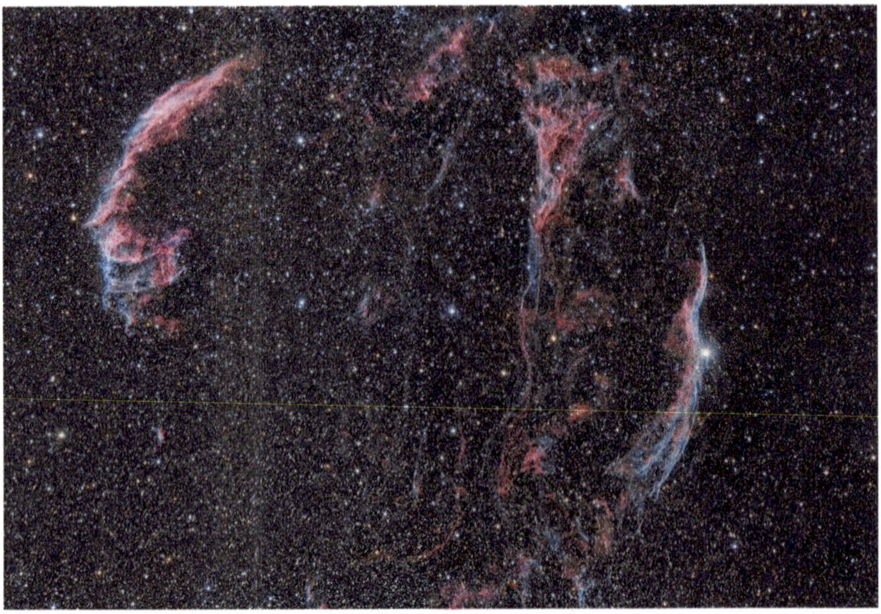

Fig. 4.22 Optical image of the remnant of a supernova that exploded some 10,000 years ago. The nebula is also known as the 'Veil nebula'. Source: Astronomy Picture of the Day, 26 November 2012, Joaquin Ferreiros

Supernova Remnants

Tau A was the first radio source outside the Sun that was identified with an optical object. The object was the Crab nebula and that, as was well known, was what remained after a supernova had exploded 900 years previously. The strong radio source Cas A subsequently also proved to be a remnant of a supernova dating from 1700[5] and comparable radio sources were found on the positions of the supernovae discovered by Tycho Brahe in 1572 and Johannes Kepler in 1604. In 1959, Australian radio astronomer B.Y. Mills made a summary of the 38 radio sources in the Milky Way that were very probably all the remnant of a supernova (Mills 1959). A large, and well-known nebula, called the Veil nebula or the 'Cygnus loop', also turned out to be the remnant of a supernova that exploded 10,000 years or more ago. See Figs. 4.22 and 4.23. And this was only the beginning. Many more remnants were to be discovered in the following years.

The Crab nebula proved to be significantly different from the other supernova remnants. It is filled with radiation, synchrotron radiation at radio and at optical wavelengths. All other remnants consisted of an expanding, thin shell, like a balloon

[5] It is remarkable that there were no known observations of this supernova, even though, at the time of the explosion a large number of astronomers were using telescopes to search the skies for something unusual.

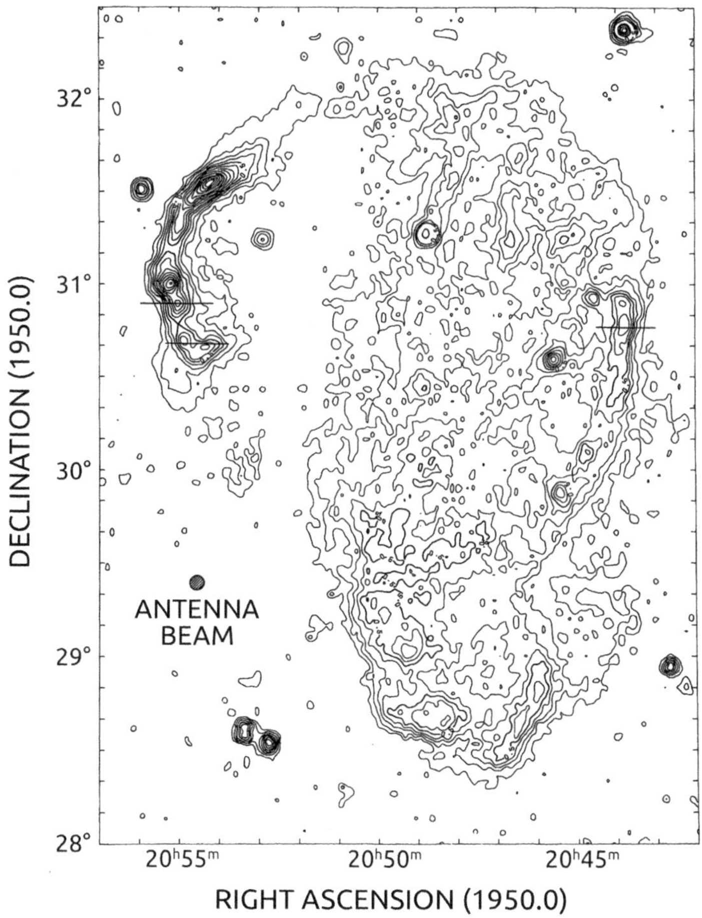

Fig. 4.23 A map of the radio radiation from the Cygnus loop obtained using the German 100 m telescope (Keen and Wilson 1973)

being inflated. After a pulsar had been discovered inside the Crab, it explained why the Crab differed from other supernova remnants.

Remnants were worth studying because they provided an estimate of the energy that was released during the supernova explosion by simultaneously measuring the amount of matter and the expansion velocity. The rate of expansion would slow down because the expanding skin 'swept up' interstellar gas just as a snowplough will sweep up more and more snow. From elaborate models of supernova remnants Chevalier concluded in 1974 (Chevalier 1974) that type I supernovae eject about half a solar mass at a velocity of 10,000 km/s, and the (kinetic) energy of this is at least ten times the energy radiated by the supernova: most of the energy that is liberated during the explosion is not in radiation but in kinetic energy of the expanding gas. The corresponding numbers for the type II supernovae are six times more mass ejected at half the velocity.

In 1964, an X-ray source was detected that coincided with the Crab nebula (see Chap. 3). Later, many more supernova remnants were found to emit X-radiation. Theoreticians even predicted a spectral line emitted by highly ionized iron atoms (Fe^{25+} and Fe^{26+}) with an energy of 7 keV. The prediction was convincing but the instruments were not yet adequately developed to record the line.

Neutron Stars and Black Holes

What Did We Know in 1968?

In 1926, R. Fowler from Cambridge (UK) showed that quantum physics was needed to explain the density of matter inside a white dwarf. There was too much matter (1 solar mass) inside such a small volume (the size of the Earth) to be explained by classical physics. The explanation came through quantum physics, which was something of a surprise. Quantum physics applied to very small things like electrons, atomic nuclei and molecules, but apparently it also applied to large-scale objects like white dwarfs. Even more surprising was the speculation in 1934 by Baade and Zwicky that neutron stars existed with the same amount of matter as in white dwarfs but in an object 100 times smaller: all the atoms and electrons had recombined and the matter consisted of neutrons, particles with the mass of a hydrogen nucleus, but without its electrical charge. Theoreticians were surprised by this suggestion, to say the least: Could such a star exist at all? Was it stable? In the 1930s Robert Oppenheimer, who later became ‘the Father of the atomic bomb’, carried out some fundamental, theoretical studies on neutron stars, but stopped after a few years. When the War was over, nobody continued the work. These hypothetical neutron stars were forgotten.

Neutron stars were not the smallest imaginable stars. Famous astrophysicist Karl Schwarzschild, father of the famous Martin Schwarzschild, received from Einstein his paper on General Relativity (Allgemeine Relativitätstheorie [ART]). It was 1916 and the First World War raged on. Schwarzschild served as a commanding artillery officer on the German side at the Russian front. In spite of the ongoing war he studied Einstein’s equations and showed that the theory implied that all spherical objects had a critical radius. If the object was smaller than this radius, matter or radiation from inside the object could never leave it. A star like the Sun, compressed to a diameter of 6 km, would become such an object. The object remained totally dark: it could not produce any light. A less familiar name is that of Johannes Droste, a student of Lorentz in the Netherlands, who found the same solution independently only a few months later than Karl Schwarzschild; Lorentz published the results in 1917—in Dutch: to underline that the Netherlands were a neutral party during the Great War (Droste 1917). More than 50 years later, American physicist Wheeler discussed such small objects and was the first to call them ‘black holes’. A helpful image is to think of a spherical membrane with a diameter as calculated by Schwarzschild; matter and radiation can pass through the membrane from outside to inside, but never the other way. Another thought is that you cannot recognize a black hole in a totally dark sky.

The analysis by Schwarzschild was a 'Spielerei' (to use an appropriate German word) for mathematically gifted physicists, with no practical meaning. After the first excitement about the General Theory of Relativity had dissipated in the 1920s, scarcely any research was carried out on this theoretical concept. But in the 1960s the interest of theoretical physicists was again aroused by the questions raised by this subject with as its high point a discovery in 1963 when New Zealand mathematician Roy Kerr found a solution of the ART equations for a spinning black hole. Such objects had a differently shaped membrane (somewhat ellipsoidal) and close to such a spinning black hole phenomena occurred that did not exist in Schwarzschild's solution.

Neutron stars and black holes were theoretical concepts and they had never been detected. General relativity had received much attention immediately after World War I but later the interest 'waned' and the theory went into deep sleep, probably because it was not needed anywhere. But then, 15 years after 1945 theoretical physicists renewed their interest. John Wheeler is usually credited for this revival, although there were other scientists involved as well. The discovery of the first neutron star in 1965 and the discovery of the first black hole in 1972 made general relativity a timely object for research again. And it was not only astronomers and physicists who became inspired; many science fiction writers and film-makers also became interested! This may be a good moment to remind the reader that even stranger objects have been proposed but until now they have not been detected: quark-stars or strange-stars, for example. You can look them up in Wikipedia.

The Discovery of Neutron Stars

Chapter 1 introduced the idea that there had to be a 'hellish machine' in the Crab nebula that constantly spat out electrons at almost the velocity of light. There were two rather white stars in this nebula, and Minkowski had indicated that one of them was special, without understanding why. In 1965 Hewish and Okoye in Cambridge (England) discovered within the Crab nebula an unusual radio source at a (long) wavelength that fluctuated in strength (Hewish and Okoye 1965). In 1967, Bell and Hewish (Bell and Hewish 1967) concluded that at a frequency of 38 MHz (wavelength 7.9 meters) the source is smaller than 1 arcsec if—and this was very likely—the fluctuations were ascribed to local irregularities in the solar wind.

Hewish decided to look further for more of such small sources at long wavelengths. He constructed a new radio telescope and in November 1967 he and his PhD student Jocelyn Bell started their search. They looked week after week at the same sections of the sky trying to find strongly fluctuating sources. Very quickly Bell noticed something unexpected—see Fig. 4.24: a source that emitted short-duration signals. The signals could be strong or faint but each subsequent signal came exactly 1.337 seconds after the previous one (Hewish et al. 1968).

At first the discoverers thought it was a man-made signal, but very soon it became apparent that the source was fixed firmly on the celestial globe and had to be outside our solar system. Was it an extra-terrestrial signal from intelligent life? Bell and Hewish toyed with the thought for a short while and christened the source the 'Little Green Man #1', referring to the science-fiction cartoon. But when similar sources

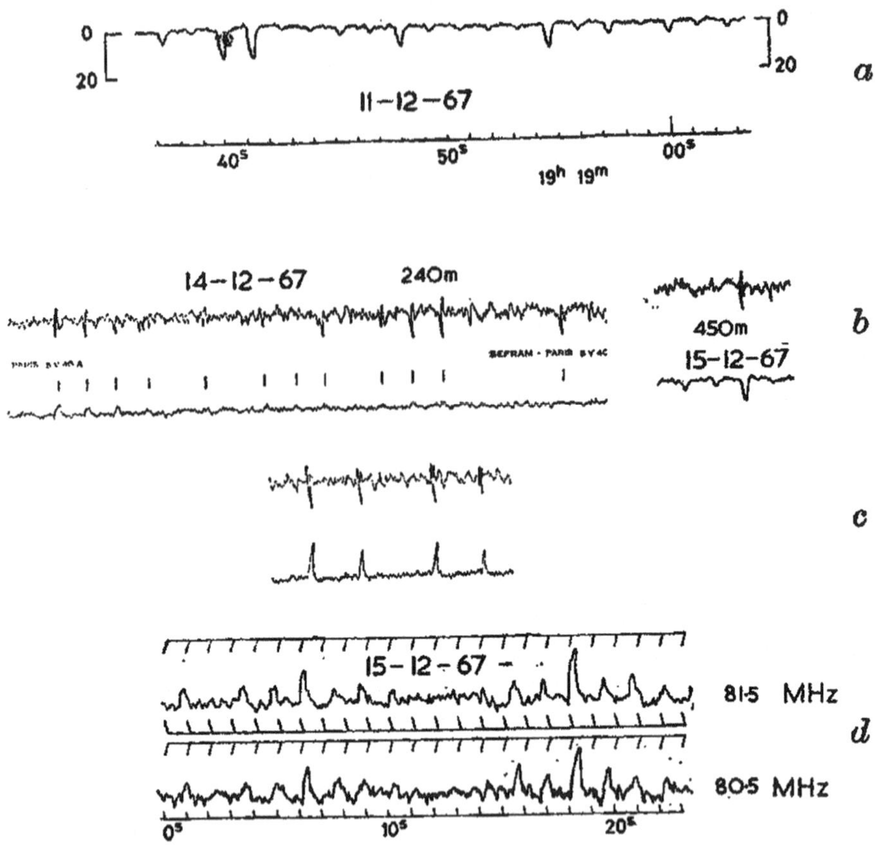

Fig. 4.24 The recordings of the first pulsar to be discovered. The frequency of the radio radiation was 81.5 MHz (3.7 m wavelength). Although the separate signals varied in strength, each appeared at precisely 1.337 seconds after the previous one. The date of the observation is indicated for each recording (Hewish et al. 1968)

were found in other directions, a physics explanation seemed more appropriate than a biological one. On 24 February 1968 the result was published in Nature under the title *Observation of a rapidly pulsating radio source*. The pulse kept repeating after 1.3372795 ± 0.000002 seconds. The many decimals were for once right and showed that the pulse period was very stable. The authors concluded cautiously that this could be a vibrating white dwarf or (!) a neutron star.

The discovery came as a bombshell. Neutron stars had been predicted, but had never previously been seen. Neutron stars could have the mass of the Sun but the size of the city of Amsterdam. Moreover, shortly before this discovery, a young astrophysicist, Franco Pacini, from Florence, had predicted the existence of a rotating, strongly magnetic neutron star (Pacini 1967).

Pulsars were 'hot news'. In the British journal *Nature* in the 14 months after February 1968 more than 20 articles on neutron stars appeared, each with a new

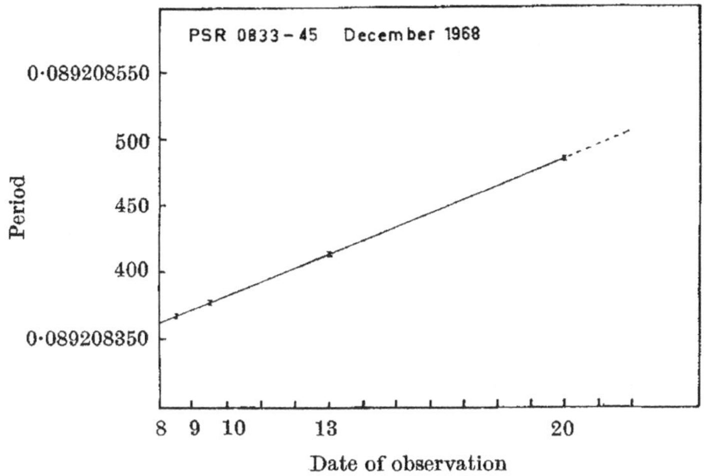

Fig. 4.25 The slow increase in the period of the pulsar and thus the slowing down of the velocity of rotation (Radhakrishnan and Cooke 1969)

viewpoint. The star prize in this race went to Thomas Gold[6] from Cornell University in the US (Gold 1968). He proposed that a pulsar was a rapidly rotating neutron star with a strong magnetic field: the kind of object that Franco Pacini had predicted. The axis of the magnetic field was directed at a small angle with the axis of rotation, geometrically similar to that of the Earth, where the North Pole and the magnetic pole do not coincide. Electrons were thought to be captured in the magnetic field and then to emit radiation in the direction of the axis of the magnet. During the rotation of the star a pulse of radiation was received every time that the magnetic pole was directed towards us. Later this was referred to as a 'lighthouse effect': when the next flash appeared, the star had completed a full rotation. Neutron stars rotated very rapidly on their axis—the first pulsar in just 1.337 seconds. Exotic, but in terms of physics perfectly possible!

An important point in favor of Gold's model was the exceptionally high stability of the rotation period. This also gave the first verifiable prediction. Such a rotating magnet produces much more radio radiation of very low frequency than the (high-frequency) radiation that was captured with a standard radio telescope. The energy for this low frequency radiation had to come from somewhere and the rotation energy seemed to be the most likely—and possibly the only—source. The gradual loss of this energy would cause the radiation period of the pulsar to lengthen. And hey presto: within a few months it was shown that all pulsars appeared to slow down: it was a small effect, but easily measurable (Radhakrishnan and Cooke 1969). See Fig. 4.25. This slowing down made it possible to estimate the number of years that a pulsar had left, by dividing the pulse period by the change in the period.

[6]Together with Bondi and Hoyle, in 1946 Gold had come up with the idea of 'steady state' cosmology. See Chap. 3.

Admittedly, Baade and Zwicky had already advocated the existence of neutron stars in the 1930s. After some 30 years they were proven right: neutron stars did exist!

In October 1968, a pulsar was discovered in Australia in Vela X, a large radio source, probably the remnant of supernova explosion (Large et al. 1968). That certainly made itself heard: a neutron star as the nucleus of the star that exploded as a supernova. And the discovery led to a following question: does the Crab nebula have a pulsar? There is something in the history of pulsars that I do not understand. Hewish and Bell had started their research after Hewish had found a small source of long-wave radio radiation in the Crab nebula, that within the measurement precision coincided with one or the other of the two small stars in the centre of the nebula. Remarkably, it was not them, but other researchers who looked at the Crab. In October 1968 two variable radio sources (NP0529 and NP0532) were detected near the Crab nebula, (Staelin and Reifenstein 1968) and in November observations with the huge Arecibo radio telescope showed that one of the two, NP0532, coincided with the center of the Crab nebula to within the measurement errors (Comella et al. 1969). These observers also measured the pulse rate: the period was 33.09114 ± 0.00001 milliseconds and the object thus made 30 revolutions per second—faster than all pulsars known at that time. There was another surprise to come. In January 1969 strong light flashes were discovered near to the radio pulsar (Cocke et al. 1969) and later the same month the source of the flashes was found to be an optical pulsar, with the same pulse rate as the radiopulsar NP0532 (Nather et al. 1969). The same authors also showed that the optical pulsar coincided with the star that Minkowski in 1942 had identified tentatively as the core of the nebula; Minkowski had no convincing evidence. The best illustration of the presence of the optical pulsar is given in Fig. 4.26.

The Crab pulsar also emitted X-ray and gamma pulses. Figure 4.27 shows that the shape of the pulse is in effect the same at all wavelengths.

The Magnetosphere of a Pulsar

There are no direct observations of the pulsar itself; all that we see are the radio pulses (except in the Crab). American astronomers Jim Gunn and Jeremy Ostriker devised a simple model in 1969 (Ostriker and Gunn 1969): a magnetic dipole that is diagonal to the axis of rotation and that rotates with the neutron star. Gunn and Ostriker demonstrated that the characteristics of the 27 known pulsars, their number, their lifespan and their division over the Milky Way, were in line with the supposition that they had been created in supernova explosions.

Within a month of the discovery of the Crab pulsar, it was noticed that its pulse period was becoming longer, meaning that the pulsar was starting to rotate more slowly. Taking account of the change in rotation period, it proved possible to calculate how much rotational energy the pulsar contained. And that amount was equivalent to the total amount of radiation energy that relativistic electrons in the Crab nebula emitted per second (the diffuse blue light of the nebula). In the early 1950s Shklovskii had proposed that the radio emission from the Crab was in fact

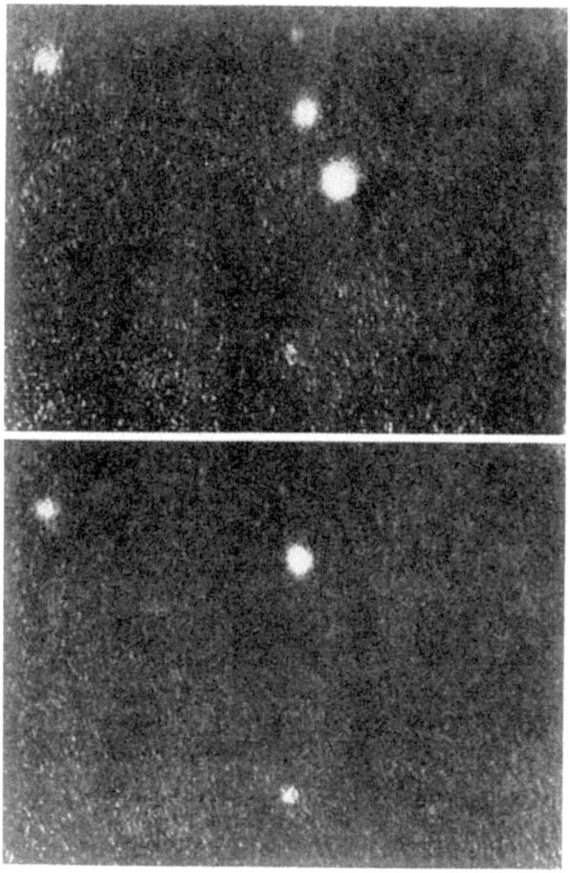

Fig. 4.26 Two recordings of starlight in a small area in the heart of the Crab nebula. The upper image gives the sky in the centre of the Crab nebula. The two brightest stars had been described first by Minkowski, who suspected that the brighter of the two was the nucleus of the exploded star—but he had no decisive evidence. Below is an image obtained with a camera that is shut off every 33 milliseconds, coinciding in rhythm with the periodic disappearance of the pulsar (Miller and Wampler 1969)

synchrotron radiation. If that was the case, there had to be a continuous source of energy in the Crab nebula. This source of energy therefore had to be the pulsar, this small rotating star, the 'monster of the Crab'.

Gunn and Ostriker offered no explanation for how radio radiation was generated. It is a question with which astronomers had struggled for more than 40 years and that has not yet been fully answered. The radiation must be generated in the atmosphere of a neutron star. But the properties of the atmosphere were not determined by gas laws, temperature and gravity, as with all normal stars, but by electromagnetic forces. These were determined precisely in Maxwell's famous laws, but mathematically the laws were much more complex than the laws of gravity: you had to take into account both the electrical axis and the magnetic force, and the two exerted a complex influence on one another. This magnetosphere has never been seen. Neither is there a model that explains all the aspects of such a magnetosphere.

An artist's vision of a pulsar is shown below (Fig. 4.28).

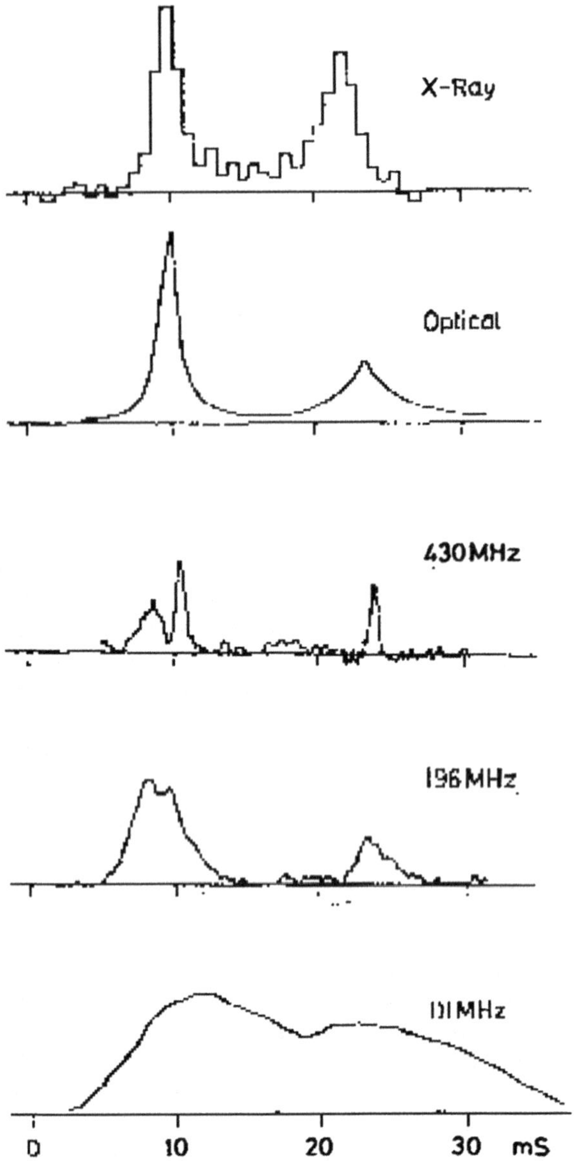

Fig. 4.27 The shape of the pulse of the Crab pulsar at different wavelengths. The lower two images were obtained at radio wavelengths of 1.5 and 2.7 m. The intrinsic pulse shape is the same in all cases; the differences in the lower two profiles were caused by radiation transfer effects between the source and us (Hewish 1970)

The Structure of a Neutron Star

In 1972 a model appeared for the structure of a neutron star—see Fig. 4.29. The model shows the essential points: the small sizes, the kinds of matter at different depths and the thin atmosphere (10 meters). More models appeared in subsequent years but did not provide major new perspectives.

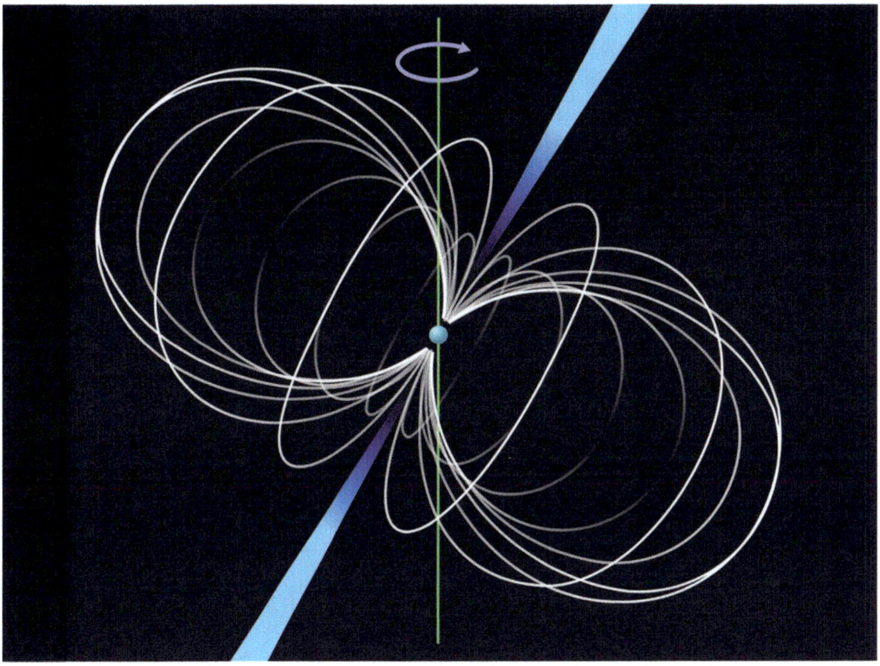

Fig. 4.28 An artist's view of a pulsar. The neutron star is the blue object in the centre. Magnetic field lines connect one magnetic pole with the other—similar to the magnetic field lines of the Earth. In green is the rotation axis of the neutron star. Blue is the beam of radio radiation emerging from the magnetic poles. The observer receives radio radiation (and light in case of the Crab pulsar) each time the beam of radio radiation hits the Earth. Source: Wikipedia, Creative Commons

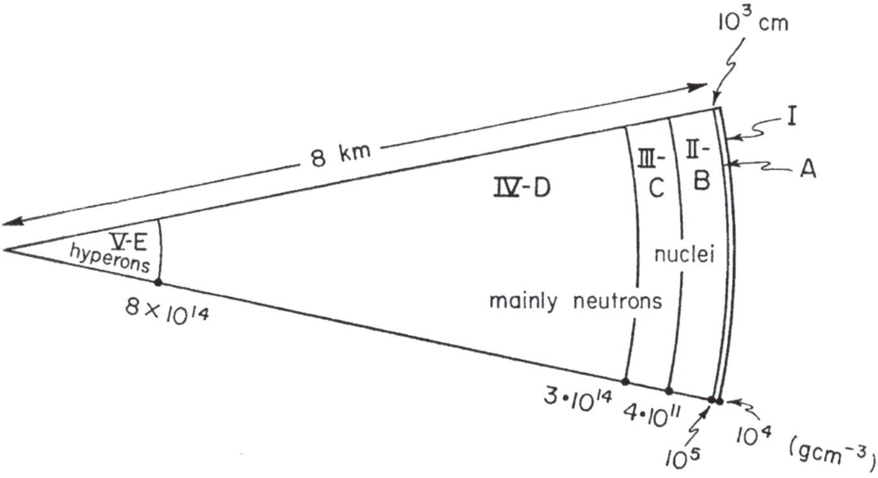

Fig. 4.29 The structure of a neutron star of average mass. The outer layer is probably a hard crust. Notice the dimensions (Ruderman 1972)

The rotation period of pulsars, was, as has been said, very stable. More information could be gleaned from this conclusion. If, for example, a pulsar formed a binary star with another star, and rotated around the point of gravity of both stars, then the Doppler effect would ensure that the period of the pulsar varied and if the orbit was a circle, the period changed to become a familiar mathematical curve, the sine. Figure 4.34 later in this chapter shows such a sine. This same change in another binary star of two pulsars was to result in a second Nobel Prize for pulsar research as will be shown farther on (Box 4.3).

Box 4.3 Jocelyn Bell

The discovery of pulsars meant that a completely new kind of star had been found with properties that were even more extreme than those of the white dwarfs. Large-scale objects made up of neutrons had never been seen in natural objects or in the laboratory. The discovery of the first neutron star was important enough to warrant a Nobel Prize.

In 1974 Ryle and Hewish both won the Nobel Prize for physics. Ryle was awarded the prize for the development of the already mentioned Fourier synthesis; his achievements were undisputed. Hewish received the prize 'for his decisive role in the discovery of pulsars' and that did generate some debate: it was not Hewish himself, but his student who had made the discovery. The world of professional astronomers resounded with indignation. Here was a young researcher, and a woman, who had done the work and her boss, a man of course, walked away with the honours! There were phone conversations and indignant heads were put together at conferences. The general conclusion was that if Hewish were an honest guy, he would refuse the prize. Hoyle, too, who was still at war with the radio astronomers in Cambridge, wrote a venomous piece in a newspaper.

Jocelyn Bell, on the other hand, remained calm; she did not join in the criticism. She said, not just once but many times, that Hewish had had the vision of embarking on this research, that he had found the funding and had taken responsibility for building the telescope. He had selected and appointed her to do the actual work. OK, she had found the first pulsars, but it was a harvest that Hewish had sown, not her. In an after-dinner speech in 1976 she stated clearly: 'I am not upset about it myself'. And there is no evidence to indicate that she had been 'upset'. She was a convinced Quaker and it could be that her faith had protected her.

Jocelyn Bell received a large number of awards—honorary doctorates and other prizes—in the subsequent years. She was made a member of the Royal Society and in 2007 she even became 'Dame Commander of the Order of the British Empire' (DBE).

Close Binary Stars with a Compact Object

We will put the pulsars to one side and go to X-ray astronomy, where we will at last encounter black holes.

X-Ray Binary Stars and Accretion Discs

Sco X-1 was discovered in 1962; it was the first X-ray source outside the solar system and its detection was convincing. But the location of Sco X-1 was so uncertain that it was impossible to identify with an optical object. Gradually, with more balloon and satellite flights, the position of Sco X-1 was measured more precisely and in 1966 Sco X-1 was identified with a star of the 13th magnitude; see Fig. 4.30.

It was immediately apparent that the star was brighter in blue than in red. And it was quickly discovered that the intensity of the light varied somewhat in an irregular fashion: see Fig. 4.31. All this was rare but it had been seen before, particularly in the remnants of novae. The spectroscopic peculiarities could indicate an unknown nova.

I have a personal memory of this identification of Sco X-1. In the summer of 1966 I took part as a novice in a symposium in Noordwijk on radio radiation in the Milky Way. Of course, it was primarily scientific news but it was also an international

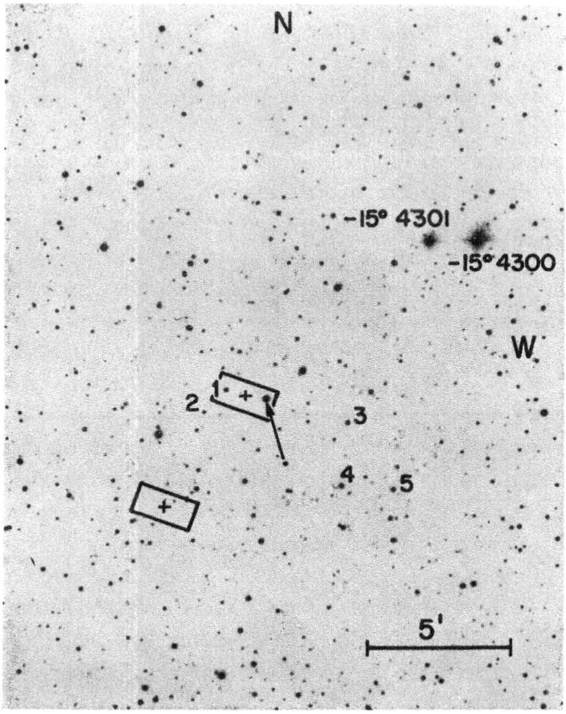

Fig. 4.30 The two crosses indicate the two equally probably positions of X-ray source Sco X-1. The squares around them indicate the uncertainty of this position measurement. The arrow indicates the position of the star that (see text) must be the correct identification of the X-ray (Sandage et al. 1966)

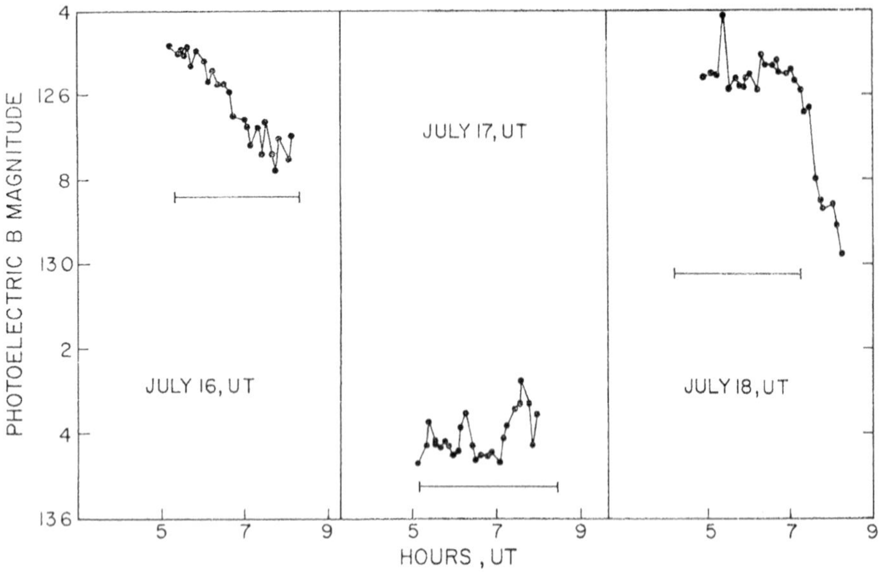

Fig. 4.31 Measurements of the magnitude in blue light of the star associated with X-ray source Sco X-1. The source varies in strength from day to day, but also from minute to minute (Sandage et al. 1966)

celebration of the 65th birthday of Jan Oort. Among the participants were many eminent colleagues from the US, the former Soviet Union and various West European countries. One of the participants was Bruno Rossi, who was mentioned in Chap. 3 as the 'godfather' of the AS&E group. He reported the recent identification of Sco X-1 with a blue, unobtrusive star that put one in mind of a nova at rest. Participant American astronomer George Herbig, with his extensive knowledge of variable stars, recalled that according to the newest insights novae were close binary stars with a red star that filled its Roche loop and created a stream of gas to a white dwarf; the blue-white color of the object was the light of the 'accretion disk' of a gas around the white dwarf. Rossi's announcement surprised those present and caused a group of prominent astronomers (G. Burbidge, Rossi and Woltjer from the US, Ginzburg and Shlovskii from the USSR) to withdraw and during the symposium to consider how such an unobtrusive object could be the brightest X-ray source in the Universe! After a few days, Geoffrey Burbidge reported on behalf of the group. He calculated that in such a binary star the overflowing matter would arrive with such force on the white dwarf that it would generate enough energy to explain the X-radiation of Sco X-1. This message generated a lively discussion that can be read in the report of the conference published by Hugo van Woerden (1967).

Shklovskii pointed out two possibilities. One of these was: imagine that the white dwarf was replaced by a neutron star, the radius of which was around 60 times smaller, then the impetus of the imploding matter would be 60 times more powerful. Shklovskii's suggestion was unusual, because the Noordwijk conference took place more than a year before the first neutron star was discovered: neutron stars were still purely hypothetical objects in which only Zwicky (and Shklovskii?) believed.

In the very preliminary model in Noordwijk, the imploding matter landed on a disk that rotated around the white dwarf. This disk was the object in which almost all X-radiation was generated. Chapter 2 already mentioned that a disk-shaped source of radiation was a completely new element in astronomy and it was not surprising that the theoreticians immediately went into action.

Two articles had a strong influence on further research, one article from Moscow by Shakura and Sunyaev (1973), and another from Cambridge (UK) by Pringle and Rees (1972). These articles were written apparently independently from one another but agreed in their conclusions. Figure 4.32 shows two different configurations. In the upper one, a low-mass star (slightly lighter than the Sun) fills the left-hand loop and in the right-hand loop—sure enough—there is a black hole. Because the donor star here is a star of low mass, this configuration is often referred to as an LMXB: 'low mass X-ray binary'. In the lower figure, the left-hand star, the donor,

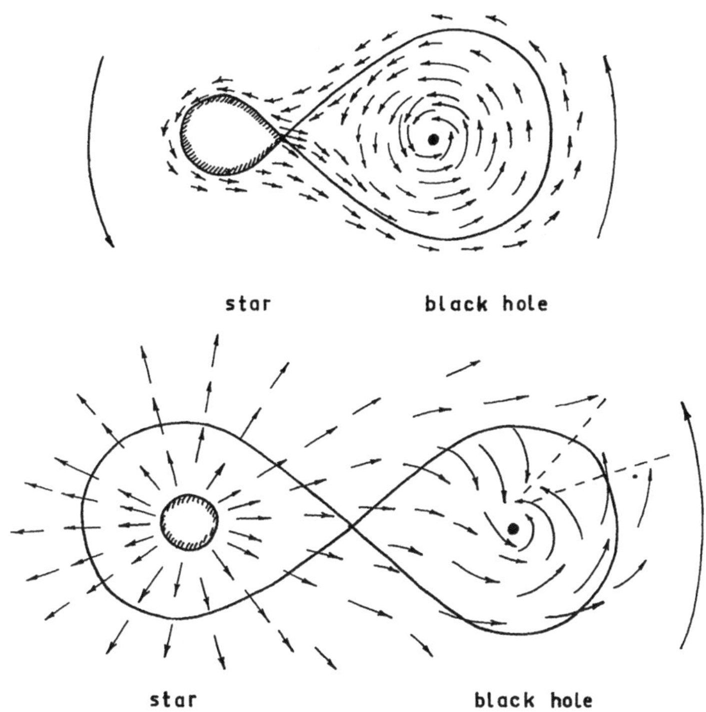

Fig. 4.32 We are looking from above at two stars that rotate around one another at a short distance apart. In both configurations matter flows from the left-hand star (the donor) to the right-hand and smaller star (the recipient). This smaller star can be a neutron star or a black hole. In the upper figure the donor is a light star, that is old and that has grown in size to such an extent that the Roche loop is filled and matter is flowing through point L1 to the recipient. Matter can also move freely over the potential surface. In the lower diagram, the donor is a heavy star that loses mass in all directions. Part of the outflowing matter can land on the recipient (Shakura and Sunyaev 1973)

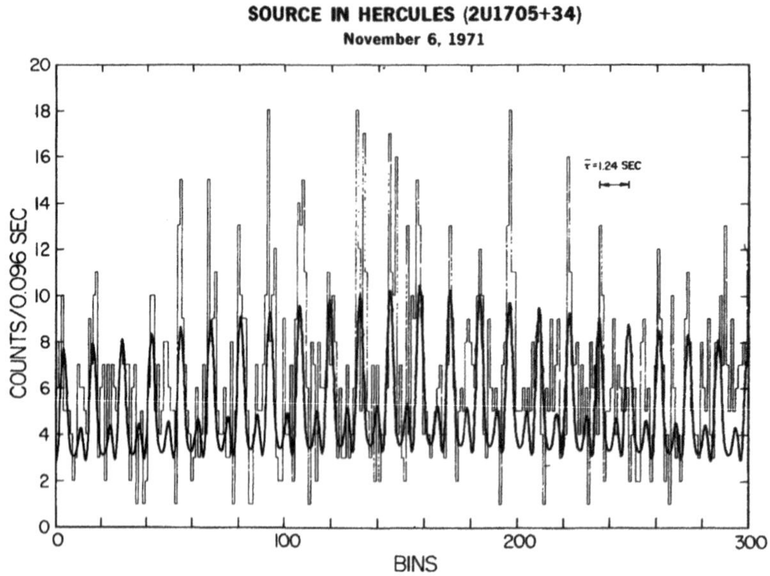

Fig. 4.33 The pulsation of the X-ray source X-1 (Tananbaum et al. 1972)

is a heavy star that sheds a lot of material: an HMXB ('High-mass X-ray Binary'). It is a useful distinction because the two types are different in all kinds of respects although the two configurations do have some characteristics in common.

X-Ray Pulsars

The first X-ray pulsar was found in 1972 by the AS&E group: Hercules X-1—see Fig. 4.33.

Following the discovery of the regular pulsations, the X-radiation suddenly disappeared, only to reappear a short while later with the same strength as originally. This must be an eclipse: the X-ray star disappeared behind the donor star. The double star character was confirmed when it appeared that the period of the pulsar was not constant, but varied between 1.70017 ± 0.00004 day (see Fig. 4.34).

Although with Her X-1 the X-ray source is so dominantly present, in terms of mass and size the object is very small in comparison with the donor star. Figure 4.35 shows the relative relations between donor star pulsar/recipient.

Cygnus X-1: An X-Ray Binary Star with a Black Hole

In 1971 it was reported that an X-ray source Cygnus X-1 was likely to be the first X-ray pulsar. This was later proven not to be the case, but because of the raised interest, the position of the source was good enough for a rapid observation by George

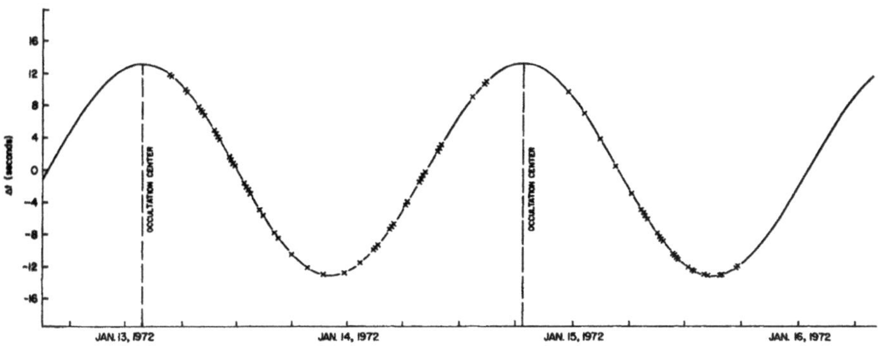

Fig. 4.34 The variation in the length of the period of the pulsar in Her X-1 measured over several months. The curve closely follows a sine, which can be explained by a circular orbit of the pulsar around another star (Tananbaum et al. 1972)

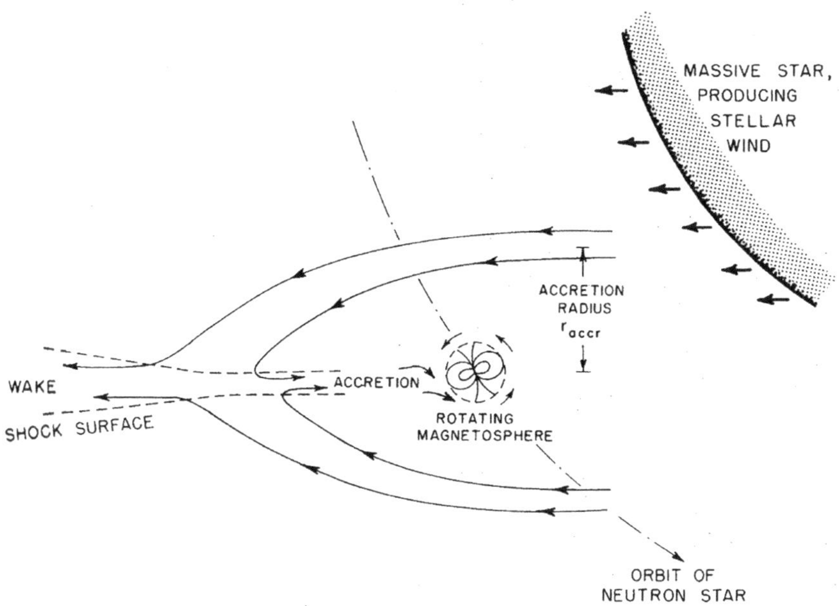

Fig. 4.35 Model of a neutron star receiving a jet of gas from a nearby, old star. Notice the small size of the neutron star in comparison with the donor star (Davidson and Ostriker 1973)

Miley and Luc Braes (Leiden) using the Westerbork telescope (Braes and Miley 1971). They found a small radio source whose position could be determined to within a few arc seconds. American radio astronomers C. Wade and Bob Hjellming confirmed this result a short while later (Hjellming and Wade 1971). This prompted two English researchers, Paul Murdin and Louise Webster (1972), to look whether a known star was located at this position. And they found one! It was a rather bright star that occurred in a catalog dating from around 1900: HD 226866. Using the Isaac

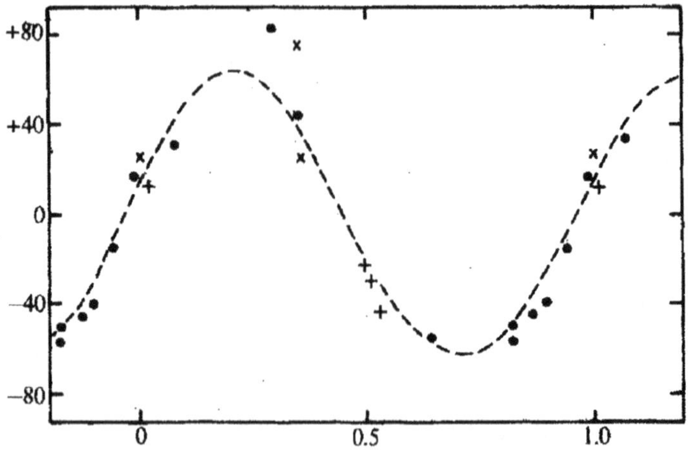

Fig. 4.36 The doppler shift of the lines in the spectrum of HD 226868. Time in the horizontal direction; the value 1.0 equals one period of 5.6 days. The vertical axis contains the Doppler shift expressed in km/s. The sine-shaped variation shows that the star has a partner that orbits in a circle around their shared centre of mass. This partner was assumed to be the X-ray source Cyg X-1. Each spectrum showed lines from the giant star but none from the X-ray source and its encircling disk (Webster and Murdin 1972)

Newton telescope, which was still located in England (see Chap. 3), the two researchers repeatedly recorded a spectrum over a period of a few months and saw that the spectral lines changed position periodically: they attributed the varying shift to the Doppler effect and concluded the star was in fact a double star. See Fig. 4.36.

The main star of the binary looked like a normal main-sequence star and Webster and Mudrin could find its mass in the existing literature. The size of the Doppler shift then could be used to calculate the mass of the partner star. To their surprise, this star was too heavy to be a neutron star. Webster and Murdin happily drew the conclusion that Cygnus X-1 coincided with a black hole. This was a sensational piece of news and the result impressed many people, but was the reasoning sound? The weak point appeared to be the estimate of the mass of the main star: if that was a regular star, then the mass was known; but was it a regular star? As a compromise Cyg X-1 was thereafter referred to as a 'candidate black hole'. For years people continued to use this word 'candidate' when considering a black hole until eventually, in around 1990, a number of courageous researchers started to omit the word 'candidate'.

4.4 Mysterious Sources of Gamma Radiation

Photons of gamma radiation have far more energy than X-ray photons. This makes them easier to detect, but the difficulty lay in determining from which direction they came, as has been discussed in Chap. 3. As an observer once commented, a gamma telescope is more like a nuclear accelerator laboratory than an optical telescope.

Researchers looked for gamma radiation from, for example, the Crab, but the first results of directed observations in 1972 were negative or highly uncertain. However, in the late sixties, a persistent rumour did the rounds that American reconnaissance ('spy') satellites, the Velas, had detected explosions of gamma radiation (see Chap. 3). Ultimately in 1973 in the (publicly accessible) Astrophysical Journal an article appeared by authors residing in the 'secret weapons lab' in Los Alamos in the US (Klebesadel et al. 1973). The authors reported on 16 explosions of gamma photons of very high energy, between 200 and 1500 keV. The explosions lasted between 0.1 and 30 seconds. Where the bursts came from was not mentioned, but in any event they did not come from the solar system. 'Gamma-ray bursts' were to remain a mystery for 30 years—it was conceivable that these explosions came from far off in the Cosmos (in which case they were sensational), but nobody knew for certain.

References

Arnett, W.D.: Astrophys. Space Sci. **5**, 180 (1969)
Arp, H.: Astrophys. J. **142**, 402 (1965)
Becklin, E.E., Neugebauer, G.: Astrophys. J. **147**, 799 (1967)
Bell, S.J., Hewish, A.: Nature. **213**, 121 (1967)
Blaauw, A.: Annu. Rev. Astron. Astrophys. **2**, 213 (1964)
Braes, L.L.E., Miley, G.K.: Nature. **232**, 246 (1971)
Carruthers, G.R.: Astrophys. J. Lett. **161**, 181 (1970)
Castor, J.L., Abbott, D.C., Klein, R.I.: Astrophys. J. **195**, 157 (1975)
Cheung, A.C., Rank, D.M., et al.: Phys. Rev. Lett. **21**, 1701 (1968)
Cheung, A.C., Rank, D.M., et al.: Nature. **221**, 626 (1969)
Chevalier, R.A.: Astrophys. J. **188**, 501 (1974)
Cocke, W.J., Disney, M.J., Taylor, D.J.: Nature. **221**, 525 (1969)
Colgate, S.A., McKee, C.: Astrophys. J. **157**, 623 (1969)
Colgate, S.A., White, R.H.: Astrophys. J. **143**, 626 (1966)
Comella, J.M., Craft, H.V., et al.: Nature. **221**, 453 (1969)
Davidson, K., Harwit, M.: Astrophys. J. **148**, 443 (1967)
Davidson, K., Ostriker, J.P.: Astrophys. J. **179**, 585 (1973)
de Groot, M.: Bull. Astr. Inst. Neth. **20**, 225 (1969)
Droste, J.: Koninklijke Nederlandsche Akademie van Wetenschappen Proc. **19**, 197 (1917)
Feinstein, A., Marraco, H.G.: Astron. Astrophys. **30**, 271 (1974)
Gaviola, E.: Astrophys. J. **111**, 408 (1950)
Gillett, F.C., Low, F.J., Stein, W.A.: Astrophys. J. **154**, 677 (1968)
Gilman, R.C.: Astrophys. J. Lett. **155**, L185 (1969)
Gold, T.: Nature. **218**, 731 (1968)
Gould, R.J., Salpeter, E.E.: Astrophys. J. **138**, 393 (1963)
Greenstein, J.L., Trimble, V.L.: Astrophys. J. **149**, 283 (1967)
Gustafsson, B., Bell, R.A., et al.: Astron. Astrophys. **42**, 407 (1975)
Hayashi, C.: Publ. Astr. Soc. Japan. **13**, 442 (1961)
Hayashi, C.: Annu. Rev. Astron. Astrophys. **4**, 171 (1966)
Heiles, C.: Astrophys. J. **151**, 919 (1968)
Herbig, George, H.: Astrophys. J. Suppl. **4**, 337 (1960)
Hewish, A.: Annu. Rev. Astron. Astrophys. **8**, 265 (1970)

Hewish, A., Okoye, S.E.: Nature. **203**, 171 (1965)
Hewish, A., Bell, S.J., et al.: Nature. **217**, 709 (1968)
Hjellming, R.M., Wade, C.M.: Astrophys. J. **168**, L21 (1971)
Hoyle, F., Fowler, W.A.: Astrophys. J. **132**, 565 (1960)
Hutchings, J.B.: Mon. Not. R. Astron. Soc. **141**, 219 (1968)
Keen, N.J., Wilson, W.E.: Astron. Astrophys. **28**, 197 (1973)
Kemp, J.C., Swedlund, J.B.: Astrophys. J. **161**, L77 (1970)
Kirshner, R.P., Oke, J.B., et al.: Astrophys. J. **185**, 303 (1973)
Klebesadel, R.W., Strong, I.B., Olson, R.A.: Astrophys. J. **182**, L85 (1973)
Large, M.I., Vaughan, A.E., Mills, B.Y.: Nature. **220**, 340 (1968)
Leighton, R.B., Noyes, R.W., Simon, G.W.: Astrophys. J. **135**, 474 (1962)
Low, F.J., Aumann, H.H.: Astrophys. J. Lett. **162**, 79 (1970)
Mills, B.Y.: Publ. Astron. Soc. Pac. **71**, 267 (1959)
Miller, J.S., Wampler, E.J.: Nature. **221**, 1037 (1969)
Morton, D.C., Spitzer Jr., L.: Astrophys. J. **144**, 1 (1966)
Mustel, É.R.: Sov. Astron. **15**, 1 (1971)
Nather, R.E., Warner, B., Macfarlane, M.: Nature. **221**, 527 (1969)
Neugebauer, G., Leighton, R.B.: NASA Publication, Washington (1969)
Ostriker, J.P., Gunn, J.E.: Astrophys. J. **157**, 1395 (1969)
Pacini, F.: Nature. **216**, 567 (1967)
Paczyński, B.: Acta Astron. **20**, 47 (1970)
Pankonin, V., Walmsley, C.M., Harwit, M.: Astron. Astrophys. **75**, 34 (1979)
Pringle, J.E., Rees, M.J.: Astron. Astrophys. **21**, 1 (1972)
Pskovskii, Y.P.: Sov. Astron. **12**, 750 (1969)
Radhakrishnan, V., Cooke, D.J.: Astrophys. Lett. **3**, 225 (1969)
Ruderman, M.: Annu. Rev. Astron. Astrophys. **10**, 427 (1972)
Sandage, A., Osmer, P., et al.: Astrophys. J. **146**, 316 (1966)
Schwarzschild, M.: Quart. J. Roy. Astron. Soc. **11**, 12 (1970)
Schwarzschild, M.: Int. Astron. Union Symp. **66**, 256 (1974)
Shakura, N.I., Sunyaev, R.A.: Astron. Astrophys. **24**, 337 (1973)
Shklovskii, I.S.: Sov. Astron. **4**, 243 (1960)
Smith, L.F.: Int. Astron. Union Symp. **49**, 15 (1973)
Snyder, L.E., Buhl, D., et al.: Phys. Rev. Lett. **22**, 679 (1969)
Spitzer, L., Drake, J.F., et al.: Astrophys. J. Lett. **181**, L116 (1973)
Staelin, D.H., Reifenstein III, E.C.: Science. **162**, 1481 (1968)
Tananbaum, H., Gursky, H., et al.: Astrophys. J. **174**, L143 (1972)
Thaddeus, P., Kutner, M.L., et al.: Astrophys. J. Lett. **176**, L73 (1972)
Ulrich, B.: Astrophys. J. **162**, 993 (1970)
van Woerden, H.: Proc. IAU Symp. **31**, 279 (1967)
Weaver, H., Williams, D.R.W., et al.: Nature. **208**, 29 (1965)
Webster, B.L., Murdin, P.: Nature. **235**, 37 (1972)
Weidemann, V.: Annu. Rev. Astron. Astrophys. **6**, 351 (1968)
Weinreb, S., Barrett, A.H., et al.: Nature. **200**, 829 (1963)
Westphal, J., Neugebauer, G.: Astrophys. J. Lett. **156**, L45 (1969)
Whelan, J., Iben Jr., I.: Astrophys. J. **186**, 1007 (1973)
Wilson, O.C.: Astrophys. J. **138**, 832 (1963)
Wilson, R.W., Jefferts, K.B., Penzias, A.A.: Astrophys. J. **161**, L43 (1970)
Wynn-Williams, C.G., Becklin, E.E., Neugebauer, G.: Mon. Not. R. Astron. Soc. **160**, 1 (1972)

Chapter 5
1960–1974: Galaxies

This chapter is about galaxies and their study in the period from 1969 to 1974. Galaxies that are strong radio emitters—a phenomenon that seemed to originate from the heart of the system—play an important role here. These galaxies will be discussed in Sect. 5.2 as 'AGN galaxies': galaxies with an active nucleus. The most spectacular examples of these galaxies are quasars. First, in Sect. 5.1, galaxies will be discussed that are not particularly strong radio emitters. Sect. 5.3 will look at clusters of galaxies, along with the sum total of all galaxies: the Universe.

5.1 Galaxies without Strong Radio Radiation

In the 1930s, Hubble had classified galaxies according to their external appearance—their morphology. It was an elegant system that earned him a lot of prestige, as is apparent from the many textbooks in which the subject 'galaxies' starts with an exposé of 'Hubble's tuning fork'. But it was a system based on external characteristics, which always raised some doubts: was the true nature of galaxies not hidden below the surface? Gérard de Vaucouleurs (1918–1995) was a driven astronomer who was not averse to attacking some sacred cows if they were not to his liking. He published an extensive catalogue of galaxies, categorising them according to his own system, which was basically an extension of Hubble's classification. It involved a complex naming system that added very few new insights and consequently soon fell into disuse. In 1966, Halton Arp, an astronomer at Mount Palomar, published an 'Atlas of Peculiar Galaxies' that contained good photos of galaxies that did not fit into Hubble's tuning fork, a collection based in part on publications by Zwicky and Russian Vorontsov-Velyaminov. In the introduction to the atlas Arp wrote: 'It is difficult to resist an oversimplified impression of what a galaxy is because the Hubble classification divides the galaxies into the well-known categories of smooth, amorphous ellipticals, and flattened spirals with star-studded arms. But by no means all galaxies fit the Hubble sequence of types of nebulae. In fact, if looked at closely

H. J. Habing, *The Birth of Modern Astronomy*, Historical & Cultural Astronomy,
https://doi.org/10.1007/978-3-319-99082-8_5

enough, every galaxy is different. It is important to have an appreciation of their peculiarities in order to build a realistic picture of what galaxies are really like.' Fifty years later it is clear just how right Arp was, even if the publication of his atlas was an act of defiance. Arp's book signalled some real problems but without offering any solutions. Hubble's tuning fork will feature frequently in the rest of this book, particularly in the final chapters.

How are Galaxies formed?

In 1957 a conference on stellar populations took place at the Vatican, where many of the leading astronomers of the time came together (see Chap. 2). The conference created a general confidence that a new, much more physical view of the Milky Way had emerged that encompassed all its elements (gas, stars, globular clusters); this description could also be used for other galaxies. Looking back, it is not surprising that a short time later (1962) a model would be generated of the history of our Galaxy, a model that explained the formation of the stellar populations. The high quality of this paper and its authoritativeness were impressive. The paper was written by the British/American trio of Eggen, Lynden-Bell and Sandage (1963), and is usually referred to by the acronym ELS. The starting point of the paper is the evidence that stars with lower metallicity (a lower iron content) move along very elongated orbits that take them from the innermost parts of our Galaxy to the outer regions. These stars are the oldest that we know. The conclusion followed that our Galaxy was formed out of one large, contracting cloud of gas that had few if any atoms other than hydrogen and helium. The globular clusters and the first stars were formed during this contraction. Most of the gas settled into a disk, where new stars have formed continuously up to the present day. Stars of different generations were present in the disk that formed an assemblage that we now see in the small but accessible part of the Milky Way near the Sun. The contraction started 10 billion years ago (10 Gyear) and was completed quickly, within a few times 0.1 Gyr. The impression that the ELS paper made can be judged by the fact that it has been quoted over 1,800 times in the scientific literature.

The next step was taken 8 years later in two papers, one by a young Australian astronomer, Ken Freeman (1970), the other by Freeman together with Alan Sandage, visiting from California, and N.R. Stokes (1970). For more information, return to the Prologue and look at Fig. 9 and the 11 possible shapes that galaxies could take. Hubble asked in vain why galaxies could have only these shapes. In the 1930s he ventured a preliminary answer: could it be possible that all galaxies evolve over the course of time from right to left, beginning as a spiral nebula and ending as an elliptical galaxy, from ordered motion of the stars in a disk to random velocities in an elliptical galaxy? (Did Hubble see an analogon to the fundamental theorem in statistical mechanics of the continuous increase of entropy, the increase of chaos?)

Let us forget Hubble and look at the Australian papers. First, there are two distinct types of families of galaxies: disk galaxies and elliptical galaxies. The distinction is easy. The disk galaxies have a flat and thin disk in which stars and clouds of gas move in well-ordered orbits around the center. There is almost always a spheroid of old stars around the center that looks similar to an elliptical galaxy. Second, the elliptical galaxies lack a disk; they appear as slightly flattened spheroids and have an even distribution of stars that is adequately described by well-known mathematical equations. The flattening of an elliptical galaxy is never very strong; the minimal value is 1 in 3. Because elliptical galaxies are always less flat than disk galaxies, the conclusion is that elliptical galaxies cannot develop into disk galaxies or vice versa: the two families constitute distinct specimens. Hubble's preliminary answer (see above) is to be rejected.

The colours and spectra of the elliptical galaxies showed convincingly that their stars were predominantly old and so was their host, the elliptical galaxy: at least 10 billion years. But disk galaxies: could they be young? No, they also had to be old - even those with loose spiral arms that were still making new stars. In a careful analysis of the colors of such galaxies Searle, Sargent and Bagnuolo concluded in 1973 that even these star-forming galaxies were as old as the ellipticals and the rate at which they had been forming stars had been declining from the beginning (Searle et al. 1973).

Evolution of Galaxies or the Relation between the Age of a Star and the Chemical Elements it Contains

It all began when a huge cloud of gas collapsed under the pressure of its own gravity. It broke up into pieces and from these pieces the first generation of stars was formed. The stars contained only the atoms that had been produced in the early stages of the Big Bang: 90% hydrogen atoms, 10% helium atoms and a sprinkling of very light elements like deuterium, lithium and boron. During their lifetime these stars produced, deep in their interiors, some atoms of elements heavier than boron. When the stars decayed and died, they returned most of their gas to interstellar space and added these heavier atoms as 'enrichment'. A new generation of stars was born out of the 'enriched' interstellar matter that contained a larger proportion of elements heavier than boron. The process of death and birth was repeated for several generations and this led in the 1950s to the idea that the history of our Milky Way can be read from the abundance of the various atoms in the atmosphere of the stars, or, rather, from the abundance of the different populations of stars. A paper by Salpeter (1955) was probably the first to attempt to reconstruct this history. The information that he had at his disposal was still minimal but, nonetheless, he wrote an impressive paper. Many studies followed that were based on the increase in knowledge especially in the domain of nuclear reactions. In spite of these efforts, the progress was very limited. Cameron and Truran, two of the astronomers involved, stated in 1971 that, '(...) it might be considered very premature to attempt a grand synthesis (....) to account for the evolution of the Galaxy as a whole' (Truran and Cameron 1971).

At first, a simulation of this history looked like a simple bookkeeping issue. Start with an initial population of stars with no or very few heavy atoms; add new stars and subtract those that had died. When a star dies, it distributes its interior over its whole volume, and new stars are formed containing more heavy atoms than the previous generation. So far so good? Not really. One uncertainty lies in the question of whether all the heavier atoms inside the star are distributed after the star dies or whether a considerable residue will remain inside a permanently dead object like a white dwarf? And how much fresh gas and how many stars will fall on the galaxy from the outside? Immigration and emigration, as we know, plague the demographics of any country. In the case of galactic demography, none of the simple models could pass a certain test and no one knew why this was so. The test is called the 'problem of the missing G-dwarfs' (Schmidt 1963a). Simulations produced too many stars comparable to the Sun (astrophysically: a G-dwarf) mass, that formed early in the lifetime of the Milky Way, i.e. stars that lacked atoms such as C, N, O and Fe. Where were the old, 'low-metallicity', low-mass stars? There was no ready answer.

Gas in the Disk of our Galaxy

The 21 cm line measurements had shown that the disk of our Galaxy is surprisingly flat and thin; its thickness is between 1% and 2% of its radius. Only on the periphery is this layer curved upwards in the northern sky and downwards in the southern sky—possibly because of gravitation from the Andromeda galaxy. A remarkable feature is the low-density void inside a circle of 3 kiloparsec (some 10,000 light-years) around the galactic center: the void contains stars but no gas, which is a mystery. A similar void in the gas layer is also found in the Andromeda galaxy. Closer in to the galactic center, inside 300 pc (1000 light-years), the disk of gas reappears and contains high-mass molecular clouds that produce new stars.

Interstellar gas had been discovered in around 1900 with the detection of thin absorption lines in spectra with Doppler shifts that differed from those of stars. By far the strongest and best observed absorption lines were those of calcium and sodium. Often the lines were multiple, which led to the conclusion that the interstellar matter was concentrated in discrete clouds and not spread out to an evenly distributed gas. This was puzzling. If the clouds existed in a vacuum, they would quickly expand and disappear. Their continued existence needed to be explained. The only viable answer was to assume the existence of an intercloud medium with the same pressure but hotter and with a much lower density than the clouds. Boyle's law prescribes that the pressure is the product of density and temperature. The intercloud medium would be in balance with the clouds if it were much hotter than the gas inside the clouds.

The intercloud medium had thus been predicted but it had not been detected. The analysis of the 21 cm line measurements did not distinguish clearly between clouds and the intercloud medium. A theoretical argument, in favor of the existence of an intercloud medium, appeared in 1969, based on calculations of the heating and cooling of interstellar gas. The atoms in the clouds frequently collided with other atoms and occasionally a photon was produced that escaped from the cloud. As a result, the cloud lost energy permanently, a mechanism that was well known. The heating of the clouds had remained a mystery until in around 1968 a new heat source was found: collisions between energetic electrons of the cosmic rays and atoms of the interstellar gas. In a paper in 1969, Field, Goldsmith and Habing showed that cosmic ray heating led to the existence of two stable phases for the interstellar gas. These two phases could be in pressure equilibrium: a cool, and dense phase and a warm and rarefied phase, cloud and intercloud medium. This two-phase model corresponded with all existing observations.

The best-studied interstellar lines were those of calcium and sodium (Na), but there was every reason to suppose that all the other atoms (N, O, Fe, etc.) were equally present but their lines were all at inaccessible ultraviolet wavelengths. This was the reason Lyman Spitzer from Princeton had argued for many years for an observatory in space. Spitzer was an influential man and the OAO 3 satellite launched in 1973 was NASA's response. The satellite was named Copernicus because it was launched precisely 500 years after Copernicus was born. The satellite provided what Spitzer and co-workers had hoped for: measurements of interstellar absorption lines of atoms of many different chemical elements—see the left part of Fig. 5.1. It shows

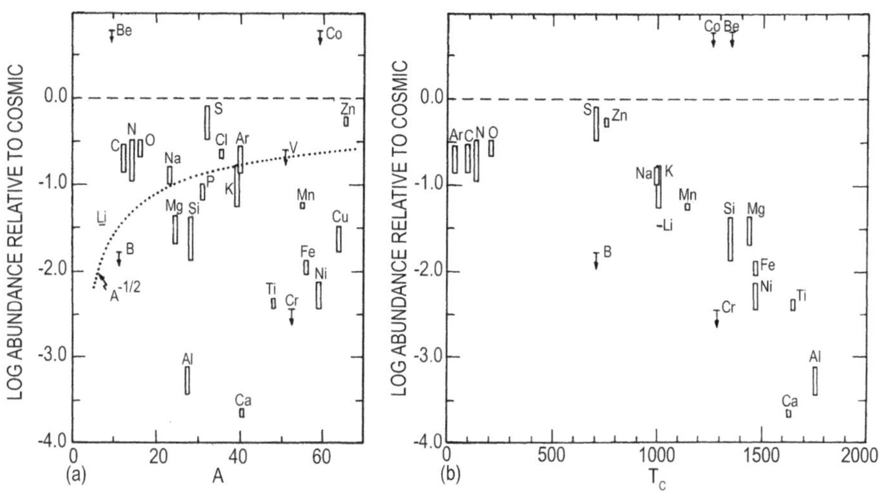

Fig. 5.1 The left-hand diagram shows the abundance of interstellar atoms, that is, the number of atoms in a cloud as compared to what is expected from cosmic abundances (see Chap. 2). Horizontally, the atoms are ordered by their atomic number. The dashed line indicates how many atoms of each element are expected. It is obvious that all atoms are 'underabundant'. The worst case is calcium that is found in interstellar space more than 3000× less than in stellar atmospheres. The right-hand diagram provides a large part of the explanation for these underabundances. Assuming that the interstellar dust particles condense from the gases expelled by red giants, one can assign a condensation temperature T_c to each element (Field 1974). The right-hand diagram shows the underabundances arranged according to T_c. Atoms that have a high condensation temperature will condense close to the star where the density is higher, and the density more efficient. Atoms with a low condensation temperature will condense out far away from the star where the density is low and the condensation less complete (Spitzer and Jenkins 1975)

the presence of interstellar atoms of many more elements. But it also highlights a fundamental problem that had been known for many decades: the lines of the interstellar atoms were much too faint, and there were much fewer atoms than expected. For example, the calcium lines were much too faint compared with the sodium lines. The question was, why was this so?

The explanation was found by considering the formation of the interstellar dust particles. These particles, either silicate-like or carbon-like, had been discovered in the 1930s and their formation had been explained in the 1960s: they condensed in warm gas that was expelled by red giants. In the outflow the temperature of the gases decreased gradually with increasing distance from the stellar surface. Assuming that the condensation was most complete near the star, then those atoms that condensed at the highest temperatures would be most deficient. The right-hand-side of Fig. 5.1 shows clearly that the shortage of atoms is highest for atoms with a high condensation temperature; these atoms were the first to condense on the particle-in-formation.

The Copernicus satellite delivered a real surprise in 1974 when interstellar absorption lines were detected from O VI ions, five times ionized oxygen atoms or O^{5+}. The Princetonian discoverers argued that these lines were formed within large, more or less spherical cavities in the interstellar medium and that these cavities were filled with a hot rarified gas. To strip five electrons from an oxygen atom is a rather violent act and it was concluded that the gas must be very hot, a temperature of about a million degrees (Jenkins and Meloy 1974).

At the same time it was noted that the frequency of supernovae in our Galaxy (once every 50 years) was so high that the bubble blown by a previous supernova was still present when the next supernova exploded. One should expect the bubbles to form 'tunnels' in the interstellar gas. A number of different observers searched for such tunnels and found them. The results were interesting but never convincing. Shortly after these papers were published, it was proved that stars of high mass during their lifetime might produce equally large bubbles as supernovae.

These one-million-degree bubbles formed a third phase in the interstellar medium, adding a further phase to the two mentioned previously. McKee and Ostriker (1977) therefore proposed a new theory of interstellar space full of hot, supernova-blown spheres filled with a very hot and rarefied gas and with clouds consisting of a hot outside layer and cool gas inside. Add to this the molecular clouds with emission nebulae on their outside, and the model thus expanded became a paradigm for interstellar matter in the Galaxy (McKee and Ostriker 1977).

Center of our Galaxy and of the Andromeda Nebula

Radio astronomy began in the 1930s with the discovery of radio waves from the Milky Way and, in particular from the center of our galaxy. Later studies showed that the radiation was stronger at longer wavelengths: one measured synchrotron radiation and there should be a magnetic field and a source of relativistic energies:

a 'monster' of unknown nature. Radio measurements showed that the radio source in the galactic center, referred to as Sagittarius A (or Sgr A), had a complex and extended structure. It should therefore be mapped by aperture synthesis. The center, however, is prominent in the southern skies and synthesis radio telescopes were all in the north. In 1974 a very small source was detected inside the complex with a diameter of less than 0.1 arcsec, or 0.04 pc and it was proposed that this small source was the real center of the Milky Way (Balick and Brown 1974) and our own 'monster', a modest relative of the monsters in such radio galaxies as Virgo A or Cygnus A. What was the nature of this monster? Nobody knew. For the moment the word 'monster' meant only that everyone was impressed by the small size and the large output. Some courageous people whispered: 'of course it's a black hole', but nobody could substantiate the statement.

How far away is the galactic center from the Sun? This simple, elementary question has been very hard to answer. The center cannot be seen optically; from infrared observations (see below) we can estimate that light from the galactic center is weakened 60 million times by dark clouds en route to the Sun. The center is clearly present in radio observations but these do not give any information on distance. To obtain any estimate, astronomers had to rely on counting stars in a few, somewhat transparent fields ('windows') in the general direction of the galactic center. In 1963 Baade discussed the distribution of RR Lyrae variables in one of these 'windows' and decided that the star counts reached a maximum value at a distance to the galactic center of 8.16 kpc. This second decimal after the decimal point suggests a much greater accuracy than should be believed. In a different—equally large—study in 1974 the distance was estimated to be between 7.7 and 10.2 kpc. When Baade's analysis was repeated with better data, the result was a distance between 8.1 and 9.3 kpc. The uncertainties in the measurements reflected a general problem in astronomy: it was still exceedingly difficult to determine such large distances inside the Milky Way. It would take decades before new radio techniques finally led to accurate and reliable distance estimates.

A positive, new and potentially very important observational result was the detection of infrared radiation from the galactic center at a wavelength of 2.2 μm by Eric Becklin and Gerry Neugebauer (Becklin and Neugebauer 1968) from Caltech (see Fig. 5.2). They recognized a cluster of stars through clouds so dense that the optical light had been weakened by 27 magnitudes (60 million times).

In 1969 F. Low and co-workers published (Low et al. 1969) measurements of the infrared radiation from the galactic center over the full range of infrared wavelengths, from 1.6 to 1200 μm (0.0016 to 1.2 mm). They used an airplane to raise their telescope above 15 km and thus above the Earth's atmosphere. The center and its surroundings were a strong infrared source at all wavelengths with a peak at 100 μm. The map of the infrared radiation was complex and too difficult to disentangle with existing observations. The same peaked distribution of brightness was found in the center of a number of other galaxies.

Let us look now at optical research on the Andromeda nebula (M31). This galaxy has one big advantage over research on our own Milky Way since we are looking at it from outside. But the details are much smaller because M31 is around a 100 times

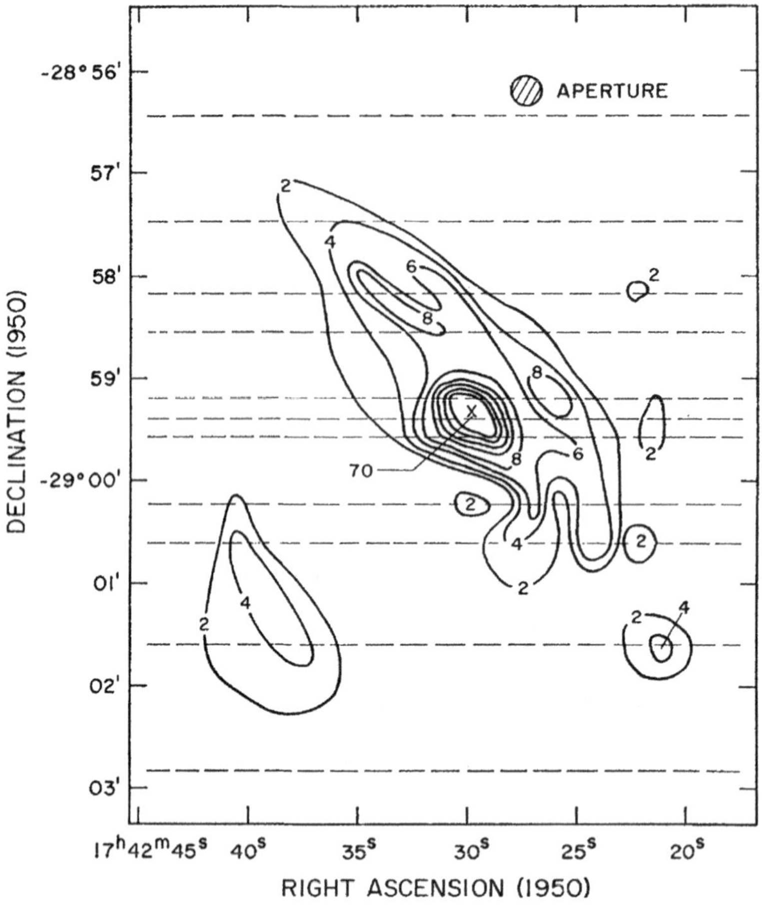

Fig. 5.2 Map in the direction of the center of our Milky Way at a wavelength of 2.2 μm. The strongest source is indicated by a cross and it coincides with the radio source Sagittarius A. The source is elongated along the galactic plane that runs roughly from upper right to lower left (Becklin and Neugebauer 1968)

further away than the center of our Milky Way. The core of M31 is even so small that details have been made invisible by variable irregularities in the Earth's atmosphere, an effect called 'seeing'. Astronomers at Princeton headed by Martin Schwarzschild built a telescope with a 91 cm mirror that they raised by balloon to take photos at a height of 25 km. The critical point of the whole undertaking was to make sure that the telescope remained focused on its target and did not start to waver. It worked: the telescope was held in the right direction to within 0.2″, a great achievement. An image of the telescope and an impression of how big this contraption was can be found in Fig. 5.3. Once the telescope was safely on the ground again (bringing it back, although risky, was successful), the film proved to have delivered results, as can be seen from Fig. 5.4.

The authors drew the conclusion that two different structures coincided at the center of M31 : the central part of the bulge of M31 and embedded in this a separate

Fig. 5.3 The 'Stratoscope II' telescope used by Light, Danielson and Schwarzschild to observe the central area of M31. Comparison with the chair at the bottom of the image gives an impression of the size of this contraption that was suspended under a balloon (Danielson et al. 1974)

Fig. 5.4 The central area of M31 measured with the Stratoscope II telescope at a resolution of 0.2 arcsec or 0.75 parsec. The core is around 0.5 arc seconds, which corresponds with a length of some 1.5 parsec or 5 light-years (Light et al. 1974)

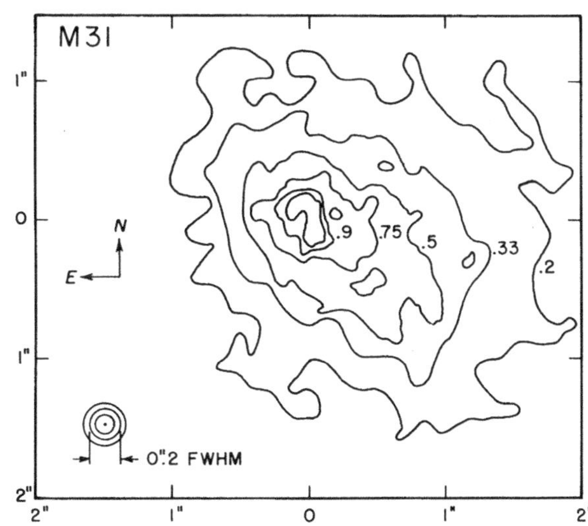

core with a radius of 2 arc seconds—around 3 pc. The density of stars in this core was several million times the density of those in the environment of the Sun and around a 1,000 times more than was found in the spherical star clusters. The result underlined the notion that centers of galaxies are very special places indeed. Of course the question was asked: 'Is there a black hole at the center of M31?' but at this time the question met a standard answer: 'There might be, but there is no convincing evidence'.

Stability of the Disk of our Galaxy and a Dark Matter Halo

You strike a gong and then hear a noise. What happens in terms of physics? The gong starts to vibrate: using the right photographic technique you see on the surface a quickly varying pattern of hills and valleys where the copper becomes slightly raised or lowered. This raising and lowering of the surface causes the air to vibrate; the vibrations move at the velocity of sound and reach our ears. The vibration of objects like a gong is a well-known effect that can be described in mathematical terms.

Similar vibrations could take place in a disk of a galaxy, an object that was rather loosely held together. In 1964 Alar Toomre (his name will recur later) showed that vibrations (similar to those of a gong) will destroy the disk unless the motions of the stars are much more chaotic than was observed (Toomre 1964). In 1973 Ostriker and Peebles (Princeton) used a computer to simulate a rotating disk of stars. To their surprise they saw that the disk lost its shape and the stars began to move chaotically in a short space of time: rotating disks of stars are unstable (Ostriker and Peebles 1973). Because the galaxy had existed much longer, its disk had been stabilized by another force. This force, the authors discovered, could be the gravity from a halo that surrounded the galaxies, provided that this halo had a mass much greater than was known. This led to the suggestion that the halo contained a great deal of matter that somehow was not seen. More about this unexpected conclusion will follow shortly.

How are the Spiral Arms Maintained?

There are many spiral galaxies. How did the spiral arms form and how are they maintained? One naive idea is that the arms are formed because the outer parts of a disk rotate more slowly around the center than the inner parts. This idea is completely wrong: taking the rotation velocities as we know them, it is easy to see that rotation would draw out the spiral arms in a short time, causing the arms to disappear. A more permanent explanation was needed. Two Chinese-American theoretical physicists from MIT, C.C. Lin and F. Shu, provided the explanation in 1964 (Lin and Shu 1964). In line with the discussion on the stability of the galactic disk (see above), they demonstrated that theoretically the disk could have a small increase in matter in the form of a spiral; they called the increase a 'density wave'. This density wave would be maintained for a long time, while it rotated very slowly around the galactic center, much more slowly than the stars that orbit around the center. Stars would encounter the density wave twice in each orbit. When the stars entered the spiral, they slowed down slightly, which in turn slightly increased its density, but then they left the arm and moved on with the same velocity as before. The very modest effect of this first slowing down and then speeding up would barely be noticeable in the distribution of the stars.

The increase in density would, however, have a large effect on molecular clouds: the clouds would be compressed, which would force them to form new stars. The spiral arms are thus marked by a higher concentration of interstellar gas and by young and bright stars. Because these stars have only a short lifespan, they would only be seen near the place where they were born: in the spiral arms. The mass increase in a spiral arm is therefore small, but it has conspicuous effects.

Support for the Lin-Shu theory came in 1972 from measurements with the then new Westerbork telescope. Mathewson, Van de Kruit and Brouw measured the synchrotron radiation in the spiral galaxy M51 (Mathewson et al. 1972). The map (see Fig. 5.5) showed clearly that the synchrotron radiation is stronger at the beginning of a spiral arm. That is what the Lin-Shu theory predicted: there the magnetic field is compressed (slightly) and thus the synchrotron radiation is also compressed. I remember the excitement when Don Mathewson, a visitor from Australia, and PhD candidate Piet van der Kruit showed their map for the first time in the coffee room at the Leiden Observatory. People jostled to catch a glimpse of it. None of us was in any doubt that this was an important discovery.

Intruders in our Galaxy

In the late 1950s two astronomers, Americans Münch and Zirin, used the 2.5 and 5.0 m telescopes in California to take spectra of very distant and hot stars far above the plane of the Galaxy (Münch and Zirin 1961). The spectra showed absorption lines of interstellar calcium and sodium atoms with large Doppler shifts. The two concluded that it was very likely that not only the stars but also the interstellar clouds were very distant from the galactic plane. The discovery inspired Oort to begin a search for such clouds at 21 cm with the Dwingeloo telescope. It took him quite some time before he had convinced the programme committee for the telescope of the importance of this search (as a young student I was a witness to what happened). The sky was large, the expected signal was very faint and the clouds could appear at all possible velocities. Who was going to do all the work to try to catch one single fish in such a vast ocean? But Oort insisted and finally one cloud was discovered and the programme became urgent. Figure 5.6 shows the detection of one of the first 'high-velocity clouds'. Oort immediately hired a new PhD student, Aad Hulsbosch, to continue the search, so there was suddenly more manpower and less observation time. To let the things move forward Oort also gave six very different explanations of the phenomenon of high-velocity clouds (Oort 1966).

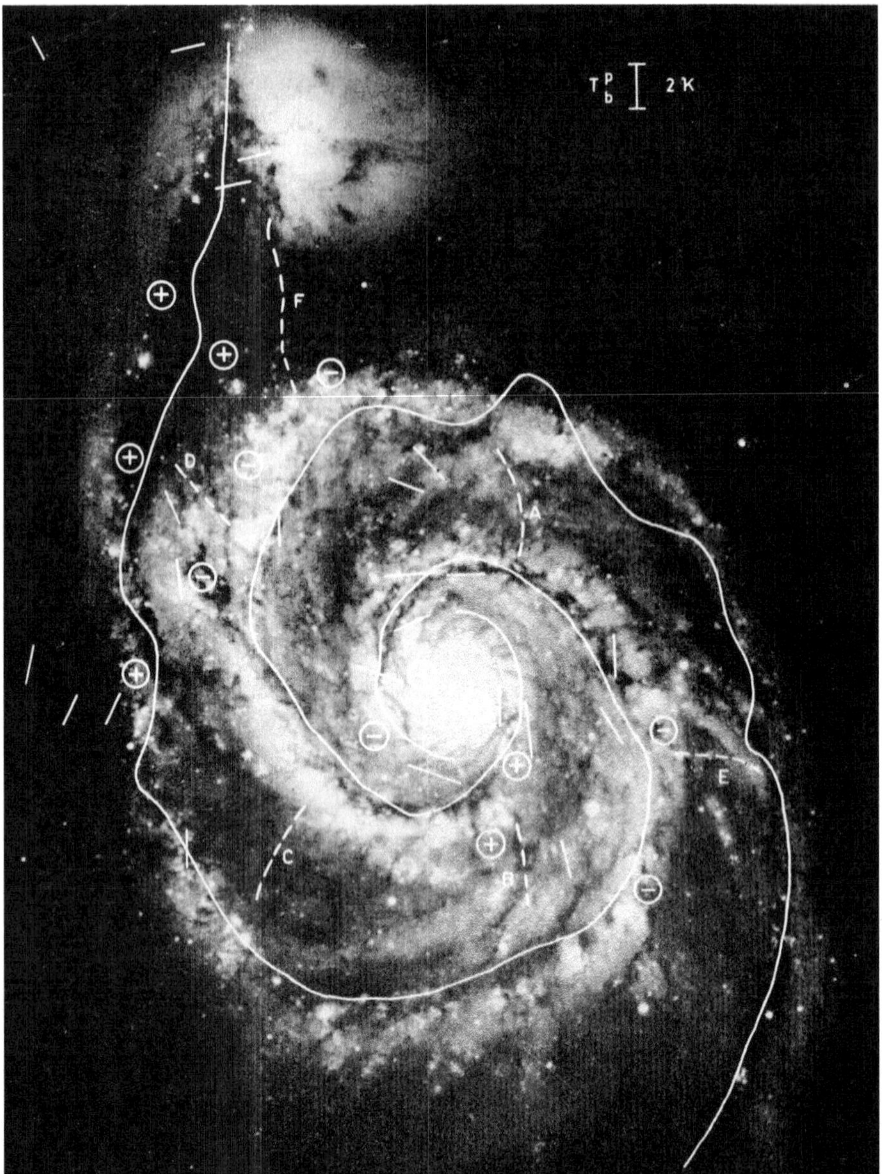

Fig. 5.5 A photo of the spiral galaxy M51 with white curves that indicate where the synchrotron radiation is stronger. The curves coincide with the places where gas and stars pass a shockwave and are slightly compressed. This compression causes star formation and this is also the place where the heavy, short-living stars are found—it is their light that we see as spiral arms (Mathewson et al. 1972)

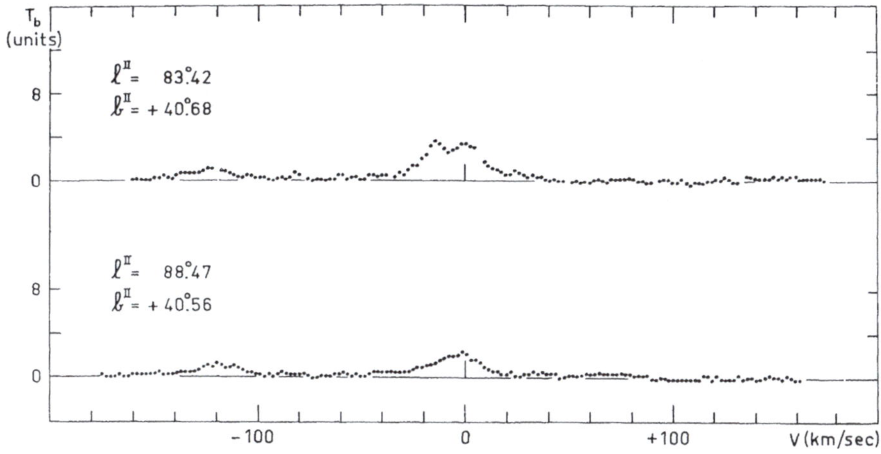

Fig. 5.6 Two profiles of the 21 cm line in the direction of a high-velocity cloud. The Doppler velocity is shown horizontally and the intensity of the signal vertically (Hulsbosch and Raimond 1966)

Fig. 5.7 A montage of the Milky Way, the Magellan Clouds and the stream of gas that links the clouds with our Milky Way. The clouds themselves can be seen in the image as two white spots in a red coloured river (Astronomy Picture Of the Day, 25 January 2010)

The HVCs ('High Velocity Clouds') were the start of a series of discoveries on gas that enters a large galaxy from outside. One particularly important discovery was that of the Magellanic stream. In 1974 two Australian radio astronomers, Mathewson and Cleary, found a very long wisp of hydrogen gas that starts in the Magellanic clouds and stretches out to our Milky Way. Figure 5.7 shows in somewhat theatrical colors where in space the stream can be seen (Mathewson et al. 1974). Mathewson and Cleary may not have been the first. Not everyone will have heard of him, but a Leiden student, Joop van Kuilenberg, had already remarked that the HVCs could possibly be a link between the Large Magellanic Cloud and our Milky Way. But the idea was deemed too speculative and was not published. Bad luck for the student.

Collisions and Mergers between Galaxies

For many years Zwicky had already pointed out unusually shaped galaxies that did not fit in Hubble's tuning fork. These were, for example, galaxies with large projections. In 1966 Arp published an atlas full of peculiar galaxies. It was impressive. However, further investigation showed there was nothing exceptional going on: galaxies are relatively loose structures and if there are two too close together, gravity will cause all kinds of deformities, as shown by computer simulations made by Alar and Juri Toomre, two American brothers from Estonia. They showed that Newton's laws were able to explain a number of Zwicky's very remarkable galaxies. Figure 5.8 shows the two spiral galaxies that were named 'the Antennae' and Fig. 5.9 shows the computer simulation.

The Toomre brothers produced other successful examples of complex galaxies as a result of two merging galaxies. The 'antennae galaxies' will be discussed again later on account of their interesting infrared characteristics.

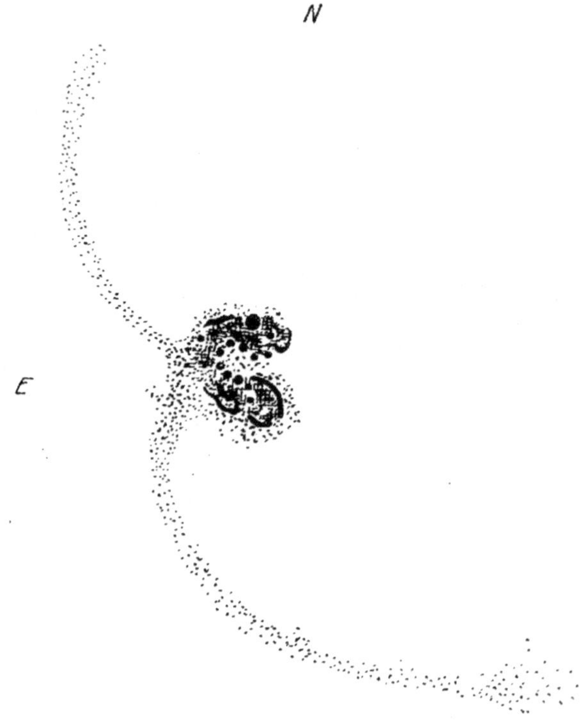

Fig. 5.8 Two spiral galaxies in a close encounter (NGC 4038 and 4039, also referred to as 'the antennae') (Zwicky 1956)

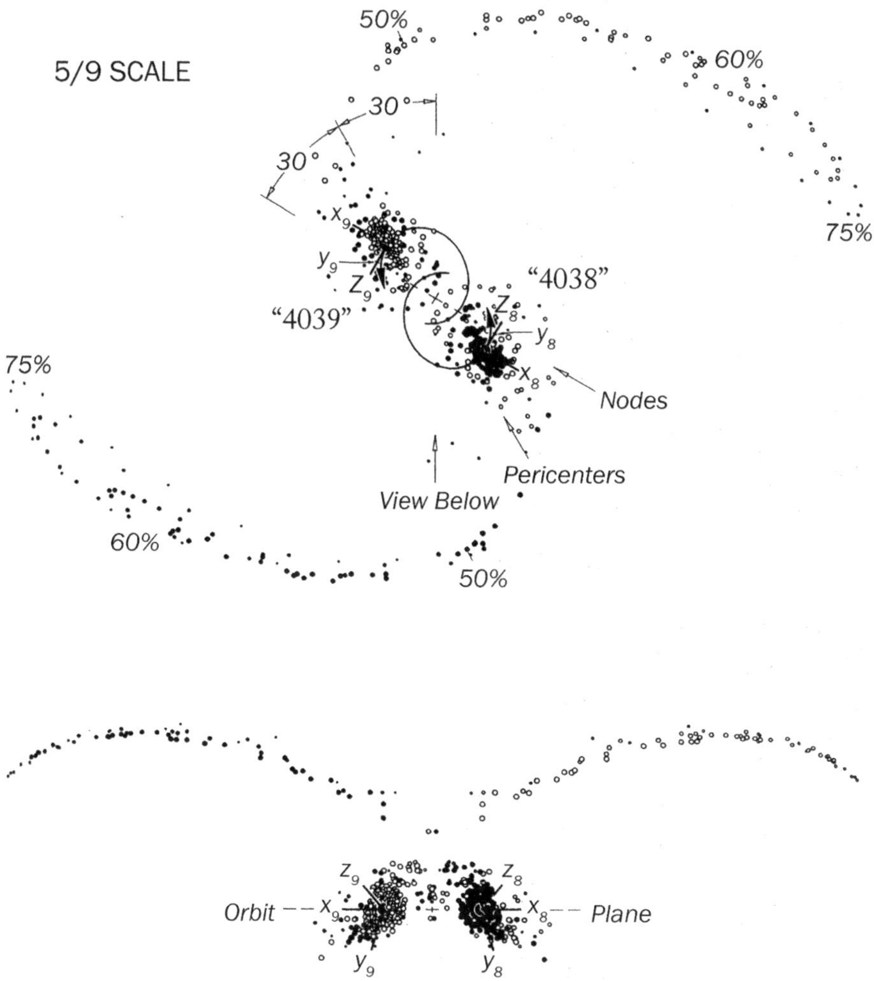

Fig. 5.9 The 'antennae' in a simulation by A. and J. Toomre. We can see the two galaxies at a certain point in time during the collision from viewpoints that are perpendicular to one another (Toomre and Toomre 1972)

M82: Massive Star Formation?

M82 is an irregular galaxy in a group that is dominated by the larger and better-known M81 galaxy. In 1963, Americans Lynds and Sandage noted that a system of filaments could be seen on the plates of the Palomar Sky Survey (Lynds and Sandage 1963). They made photos using all kinds of narrow color filters and saw that M82 was filled with fibres. The fibres were explained as having been caused by an enormous explosion that had taken place several millions of years previously and that released as

Fig. 5.10 A Hubble photo of galaxies M81 (left) and M82 (right). The thin red curves around M82 are indications of violent activity going on (Astronomy Picture Of the Day, 1 June 2004)

much energy as 10,000 supernovae! A tsunami of new stars traversed M82 from the center. M82 also proved to be a strong, fairly large infrared source and a modest radio galaxy. No active nucleus was seen; the many supernovae that arose from the young stars were able to generate enough radio radiation. M82 was therefore unlikely to be a radio galaxy; it underwent a 'starburst'—a new word that will recur (Fig. 5.10).

A map of the 21 cm line radiation from 1972 had already shown that the two galaxies together with a third (outside the figure) were embedded in a single cloud of hydrogen (Roberts 1972). The tsunami in M82 could well have been caused by a recent encounter of M82 with the far larger M81—a prediction that would be confirmed some 20 years later with a better quality and more detailed map of the 21 cm radiation.

The infrared measurements also caused some surprise. In 1970 a very strong but small infrared source was found in M82 at a wavelength of 10 μm (Kleinmann and Low 1970). This was new and unexpected. The source of this infrared radiation could be a black hole in an active nucleus, but there was also another possibility: there could be a cluster of very hot, short-lived stars surrounded by an impenetrable cloud of matter and gas, something we recognize from our own Milky Way. In this case, the dust particles could absorb light from the stars and retransmit it at infrared wavelengths. M82 thus became the first example of galaxies in which a large number of new stars were formed in a very short space of time—indeed: a starburst!

Starburst Galaxies

Fritz Zwicky had searched on plates from the POSS for ‘compact galaxies’: objects with an image just slightly bigger than a star. He found quite a number of them but did not publish the data. Instead he kept the information in a number of lists. He sometimes passed this list onto astronomers from the Mount Wilson/Palomar community. One of the chosen recipients was English-American astronomer Wal Sargent. Together with his colleague Leonard Searle he recorded spectra of promising objects and came across IZw18, number 18 on Zwicky’s first list. The object was of the 16th magnitude (Fig. 5.11a–c).

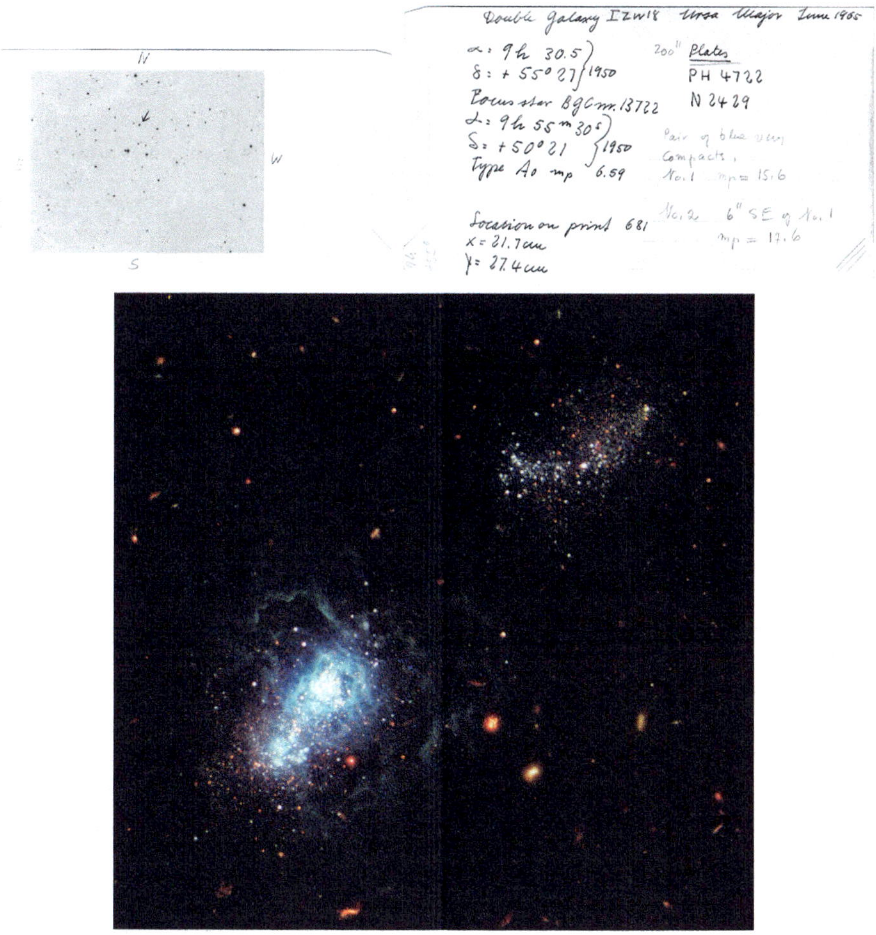

Fig. 5.11 Image of the plate on which Zwicky noted the ‘compact IZw18’. Zwicky’s hand-written note on the discovery of IZw18. HST recording of IZw18. It is a double galaxy and in the galaxy on the left a starburst is occurring, as is shown by all the bluer stars. (These three figures are taken from the ‘NASA/IPAC Extragalactic Database NED’)

The object was ‘compact’ and it radiated a high amount of ultraviolet light. Sargent and Searle were able to deduce the distance from the redshift, and concluded that it was a dwarf galaxy. The surprise was that IZw18, and another Zwicky galaxy, IIZw 40, formed new stars at such a rate that they would exhaust all the gas present in an astronomically short time. This was a brief ‘intense burst of star formation’. The authors Searle and Sargent (1972) did not find the cause of this burst; perhaps star formation could have been triggered during a close encounter with another galaxy, but there was no obvious candidate for such an encounter. IZw18 and M82 were examples of a new class of galaxies because they had no active nucleus, but they nonetheless did emit a huge amount of infrared radiation.

Do we know all the Types of Galaxies?

Before we go any further, let us first ask a question: do we know all the kinds of galaxies? Probably not. The discovery of quasars (see below) was an example: galaxies as small as a star had never been seen before. But the opposite is also true: there are also galaxies that are so dispersed that their light does not exceed the low light of the night sky. Figure 5.12 illustrates this.

1974: A New Paradigm—the Existence of Dark Matter

It is 1974 and something happens. Almost 15 years later, the event was described by Scott Tremaine, a Canadian astronomer (Tremaine 1987), and Janiste and J. Einasto from Estonia (Jaaniste and Einasto 1987), using the words of Thomas Kuhn’s book on ‘The Structure of Scientific Revolution’: ‘1974 was a year when an important paradigm was questioned and replaced by another’. A paradigm is ‘a body of intertwined theoretical and methodological beliefs’, a coherent set of generally shared ideas, that may not always be explicitly formulated. The paradigm under dispute was this: all matter in the Universe is in stars plus a small fraction that is in gas clouds. (Planets can safely be ignored.)

The doubts about this paradigm had three roots, two of which were planted before World War II. First there was Zwicky who had discovered in 1933 that the Coma cluster of galaxies contained hundreds of times more matter than could be accounted for by counting the galaxies in the cluster—see the Prologue. Zwicky’s conclusions were not questioned, but until 1974 they were rarely discussed. In the first 40 years after the article was published, it was cited only 15 times in other scientific papers, and in the subsequent 25 years it was cited 134 times. Yet, even in these 40 years of negligence, Zwicky’s paper had not been forgotten.

In 1954 Martin Schwarzschild (Schwarzschild 1954) reviewed the evidence that all matter in the Universe emits light. He concluded that for the time being ‘mass followed light’; there is no mass without light. Zwicky’s contrary conclusion,

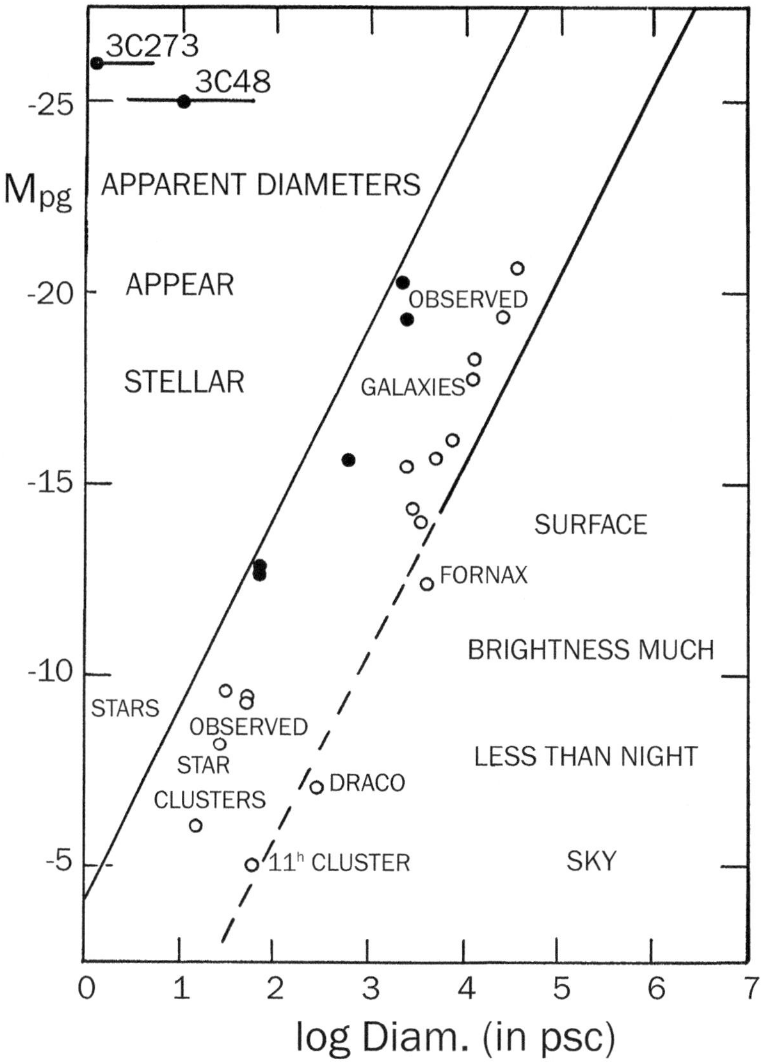

Fig. 5.12 Diagram in which galaxies are indicated by their size horizontally and their magnitude vertically. Two straight lines define the area with the galaxies that were known before 1963. Above the upper line are the quasars, the first of which was discovered in 1963. Below the bottom line are galaxies that barely show up against the vague light of the night sky and so they are overlooked. Two such galaxies ('Fornax' and 'Draco') were discovered because they were at a very short distance from our Milky Way (Arp 1965)

Schwarzschild said, was based on very uncertain data. I have a high regard for Schwarzschild, but here he was wrong: each time Zwicky's analysis had been repeated with newer, better data, his result was confirmed, The Coma cluster contained 200–300 times more mass than corresponded to the amount of light that the galaxies produced.

The second root of discoveries that led to the crisis in 1974 began with the thesis by the American Babcock in 1939 (Babcock 1939). He took spectra of nebulae in the disk of the galaxy M31 and deduced velocities from the doppler shift. He showed that the disk rotates around a central axis and was then able to derive the distribution of mass in the disk. Babcock concluded that there was a lot of mass but no light on the periphery of M31. In 1957, as discussed already in Chap. 2, Babcock's conclusion was confirmed from data obtained with the Dwingeloo radio telescope at a wavelength of 21 cm. In 1970, much improved optical measurements allowed Rubin and Ford (1970) to confirm Babcock's result, and a few years later new 21 cm line data showed a 'significant amount of matter' outside the photographic image (Roberts and Rots 1973; Roberts and Whitehurst 1975).

There was a third independent indication that the Andromeda galaxy (M31) and our galaxy were embedded in much more matter than could be seen. The Andromeda galaxy is approaching us at a velocity of 300 km/s and Kahn and Woltjer (1959) argued that this probably meant that our galaxy and M31 form a double system. The age of the Universe, so they said, was the longest period for the orbits of both around a common point of gravity. These arguments led to the conclusion that the total matter in this double system is about 1.5×10^{12} M_{sun}. That was definitely much more than the sum of the two galaxies. The authors assumed that this extra matter was distributed as an ionized gas in intergalactic space, but it had not been detected: no radio radiation or X-radiation had been recorded. In retrospect, the missing matter might as well be dark matter in the haloes of our Milky Way and of M31.

The impossibility of explaining the problem of the unseen matter finally reached crisis point in 1974. To solve it, a change in paradigm was needed and the courage to state this clearly can be read in two different papers from the two sides of the Iron Curtain. The first paper by Jaan Einasto, Ants Kaasik and Enn Saar from Tallin in Estonia appeared on 26 July 1974 in *Nature* (Einasto et al. 1974). Its title was: 'Dynamic evidence of massive coronas of galaxies'. The second paper was published on 1 October 1974 in the *Astrophysical Journal*. It was written by Princetonians Jerry Ostriker and Jim Peebles and Israeli A. Yahil and bore the title 'The size and mass of galaxies and the mass of the Universe' (Ostriker et al. 1974).

Einasto et al. took as their starting point the conclusion of the recent analysis of the Coma cluster mentioned above: there has to be a lot of matter inside this cluster. Recently X-ray telescopes had measured the presence of very hot gas inside clusters (see farther down) but Einasto et al. concluded that the emission was too faint to correspond to all missing matter. The missing matter had not yet been seen. The authors added the argument that unseen matter was also present in spiral galaxies such as M31. The conclusion: galaxies are embedded in more unseen matter than in luminous matter: light traces matter only partially.

Ostriker, Peebles and Yahil also considered the rotation curves of disk galaxies and drew the same conclusion as the Estonians. They added their expectation that this unseen matter would be enough to 'close the universe' and contradicted earlier conclusions, namely that the Universe is 'almost empty'. No conclusions could be drawn about the nature of the unseen matter. Let us speculate. If the simplest answer is provided by Occam's rule, an observer will assume that the dark mass consisted of very light stars, say stars with a mass of 0.1 M_{sun}, stars that had not yet been discovered but that were generally assumed to exist. A theoretical physicist, also using Occam's rule, will suppose that dark matter consists of elementary particles of an as yet unknown kind: particles with mass (because they cause gravity) but without electrical charge (because they do not emit light or any related radiation).

Zwicky was proved right for the second time; in the 1930s he had suggested that invisible matter was present in clusters of galaxies (see Chap. 1). In the same period, he had also indicated the existence of neutron stars. It now appeared that both these fantasies were real; the cosmos was stranger than ever (Box 5.1).

Box 5.1 Fritz Zwicky 1898–1974

Zwicky was born in Bulgaria, the son of a Swiss businessman and a Bulgarian mother. He studied at the 'Eidgenössische Technische Hochschule' in Zurich, possibly the most prestigious technical university in Europe, and obtained his doctorate there in 1922. In 1928 he left for Pasadena, California, to work with Robert Millikan, a professor at 'CalTech', where he was appointed in 1942 as the first professor of astrophysics. In Pasadena he worked with the equally young German astronomer Walter Baade, who was a staff astronomer at the Mount Wilson observatory. In 1934, the two researchers reached the conclusion that there must be 'supernovae' in addition to standard novae. The supernovae become far, far brighter, but they are very rare events. In a second article in the same year, the two suggested that supernovae might be formed when a normal star changes to become a neutron star. A neutron star was an almost completely new concept and quite speculative: the object was maintained by the same quantum mechanism as in a white dwarf but now the gas was not made up of free electrons and hydrogen and helium atoms: all free electrons and protons had been compressed to form neutrons, an elementary particle whose existence had been shown only a few years before. Baade and Zwicky devoted no more than a few lines to this idea and the suggestion was met with disbelief. Theoretical physicist Oppenheimer with some of his PhD students was the only person to take the idea further, examining the theoretical implications of the proposal.

In 1933 Zwicky published an analysis of radial velocities of galaxies in the Coma cluster. He used the virial theorem (see Chap. 1) and concluded that for an average galaxy the ratio between mass and the light it produces (referred to as M/L) was 500 times larger than in the Sun. In other words: all matter in this

Fig. 5.13 Fritz Zwicky in a good mood

cluster taken together produced 500 times less light than the Sun. This small amount is almost incomprehensible. It was not until the 1970s that Zwicky's idea was generally accepted: clusters of galaxies contain a lot of matter that is invisible to us (Fig. 5.13).

Zwicky was an eccentric man who became embroiled in conflicts with almost everyone who worked with him or who did work in which he was interested. If he got into an argument, he might well call the person concerned 'spherical bastard'. Why spherical? Something that always looks the same, regardless of the angle from which you view it, has to be a sphere, was the mathematician's explanation.

Zwicky had some indisputably important insights long before they were confirmed and adopted or endorsed, But even now, more than 30 years after his death, his achievements are at times still belittled; he incited a lot of anger, and that has not yet been completely forgotten.

5.2 Radio Galaxies and other Unusual Galaxies

In 1960 radio astronomy had left behind its teething troubles and the radar experts realised that they had become astronomers. The information in the third Cambridge catalogue (the '3C') appeared to be very reliable, which restored the confidence in Ryle and his group after the debacle of their 2C (see Chap. 2). Practice confirmed that interferometry was needed if people wanted to compare radio sources with optical images. Identification of radio sources with visible objects progressed rapidly and as we will see shortly, people found radio galaxies with ever bigger redshift values, z: in 1974 $z = 3.53$ had already been achieved (see Box 5.2).

Box 5.2 Redshift Parameter z

If we receive light from an object that is moving away from us, all the lines in the spectrum have a longer wavelength: a redshift. If the object is approaching us, we see that the spectral lines have a shorter wavelength: a blue shift. Blue shifts occur just as often for stars as redshifts, but the spectra of galaxies always show a redshift with a very few exceptions. This shift is interpreted as the doppler effect, after the Austrian physicist Johannes Doppler who predicted it in 1842.

To set a value to the redshift, a 'redshift parameter' is introduced, indicated by the letter 'z': If the wavelength of a spectral line measured on Earth has a value of λ_{lab} and in the spectrum of the galaxy λ_{obs}, then the 'redshift parameter' z is indicated by $\lambda_{obs} = (1 + z)\,\lambda_{lab}$. If $z = 0$, there is no redshift. For small values of z, the relation between z and the velocity v is simple: $v = z \times c$. But if z is greater than 1, this simple multiplication predicts that v is greater than c, the velocity of light. This is unphysical and wrong. The correct formula can be deduced from the special theory of relativity. If an object is flying away from us at a velocity v the special theory of relativity states that $1 + z = \sqrt{\{(1 + \beta)/(1 - \beta)\}}$ where $\beta = v/c$; v is the velocity that we are looking for. Assume that for a particular galaxy $\lambda_{gem} = 2\,\lambda_0$. Then $1 + z = 2$ and so $(1 + \beta)/(1 - \beta) = 4$ and $\beta = 0.60$. The galaxy is moving away from us at a velocity of 180,000 km/s. If $\lambda_{gem} = 3\,\lambda_0$, then $\beta = 0.8$ and so the velocity is $v = 240{,}000$ km/s.

The further a galaxy is away, the larger its redshift. When z is small, the distance d can be calculated by the 'Hubble relation' ($z \times c = H \times d$; where H is the Hubble constant). However, when z is larger than a few times 0.1 the relation between z and distance is no longer simple. One has to use a more complex formula that depends on the cosmological model, the model that one assumes for the evolution of the Universe. This will be discussed in Chap. 16. Reader, if you do not want to wait that long: you'll find a useful table on the NASA site http://apod.nasa.gov/apod/astropix.html of 8 April 2013. The site http://www.atlasoftheuniverse.com/redshift.html also has some useful information.

In 1963 a discovery was made that had a lasting influence on all further research on galaxies and the Universe: the discovery of 'quasars', objects that after some discussion were regarded as super-bright cores of distant galaxies. Quasar research led to two further fields of research: studies of gas between the quasar and us and studies of the distant or early Universe (thanks to the finiteness of the velocity of light, 'far' means 'a long time ago').

Slightly less spectacular, but nonetheless of major importance, was the discovery of other types of galaxies related to quasars: so-called Seyferts, BL Lacs and D- and N-type galaxies.

Radio Galaxies

In 1953 it was discovered that the very strong radio radiation from Cyg A originated from two sources located symmetrically on both sides of a faint galaxy (see Fig. 1.13 in Chap. 1). Cyg A was no exception: there were many more radio galaxies with a comparable structure; see Fig. 5.14. Five of the six sources have a double structure, the sixth, Vir A (M87), is described as a ‘core with halo’. Later observations with a somewhat sharper resolution in Cambridge (England) revealed a small but oblong source in the center. The image is not as sharp as one might like but it is good enough to able to conclude that it comprised three parts (see Fig. 5.15). The central source is the core of M87, the right-hand one is the famous ‘jet’, already discussed in Chap. 2. The left-hand source is considered as the ‘contra-jet’ that one expects but that cannot be seen optically—probably obscured by dark clouds in

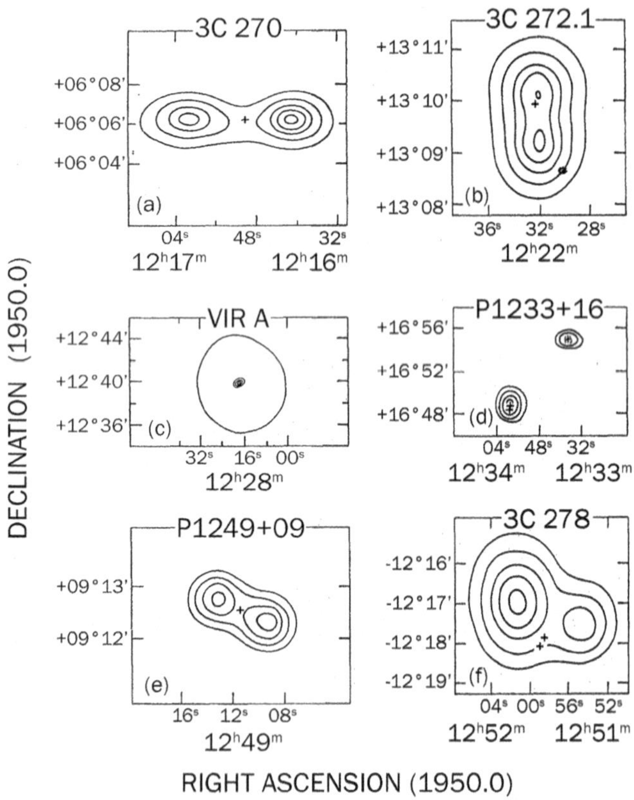

Fig. 5.14 Contours of the radio radiation at 21 cm (continuum) of several radio galaxies, including Vir A = M87 (Fomalont 1971)

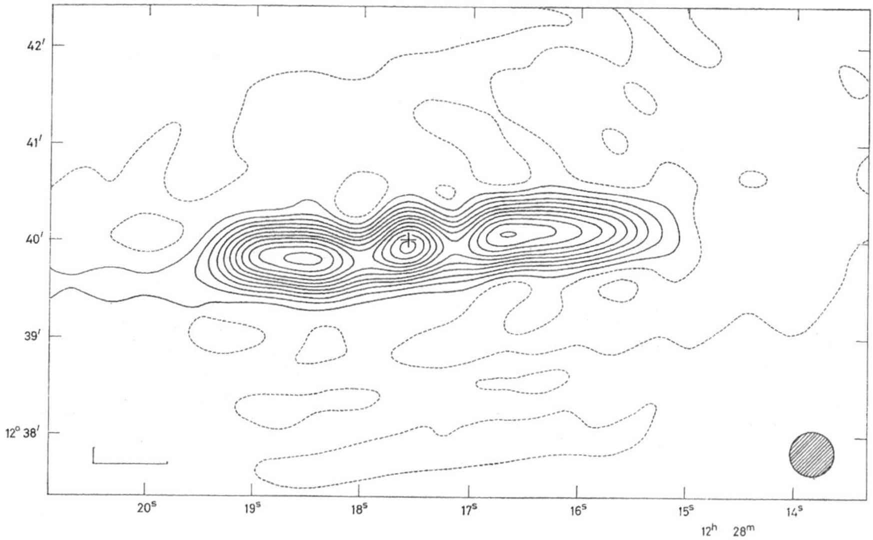

Fig. 5.15 The contours of the radio radiation of the central part of Virgo A = M87 at a wavelength of 6 cm, with the nucleus at the center. The source on the right is the well-known jet. The source on the left-hand side is not visible and is taken to be the counter-jet (Graham 1970)

front of it. In 1969 VLBI measurements (see Chap. 3) made with two telescopes, one in California and the other in Australia, showed that there is a radio source in the center of M87, approximately 3 light months, a quarter of a light year. Could this be a black hole?

Seyfert Galaxies

In their pioneering article in 1952 (see Chap. 2) Baade and Minkowski referred to an article from 1943 by Carl Seyfert (1943), in which he described six galaxies that share some peculiarities. In the spectrum very bright lines could be seen that were known from studies of light-emitting nebulae although this was not unusual. What was peculiar was the huge width of the lines. This was explained by the doppler effect and meant that gas was present at the core of these galaxies with big, random movements at velocities of up 8500 km/s—a very high value. There was obviously something going on, although it was not clear what that ‘something’ might be. Two of the galaxies, Baade and Minkowski said, were already known as radio source NGC 1275 (3C84 = Perseus A) and NGC 1068 (3C71), while the spectrum of Cygnus A was similar to that of a Seyfert galaxy. Seyfert galaxies were reintroduced by the Burbidges and Sandage in 1963 (Burbidge et al. 1963). From then on they played an important role in extragalactic research.

Quasars

The Discovery

At a meeting of the American Astronomical Society on New Year's Eve 1960 the announcement was finally made that a radio star had been discovered; this was radio source 3C48 (number 48 in the third catalogue from Cambridge). Its position was determined precisely to just a few arc seconds using a new interferometer in California, and researchers at Jodrell Bank determined that the radio source was smaller than 1 arc second. The radio source coincided with a star of the 16th magnitude (Fig. 5.16). Radio stars did exist! After several decades Ryle and others were proven right. But was it a star? The object had unusual colours. The ultraviolet was relatively strong and the star also varied irregularly in brightness. The colours reminded the authors of a nova or a white dwarf, 'but it couldn't possibly be that'. In spite of the doubts, it was too much like a star to think it could be anything else. A spectrum was recorded that displayed bright, broad lines but it was not possible to say from which elements they came. And then the research came to a standstill and the results were for the time being not published. An announcement was made in the popular scientific magazine 'Sky and Telescope' and that is the oldest reference.

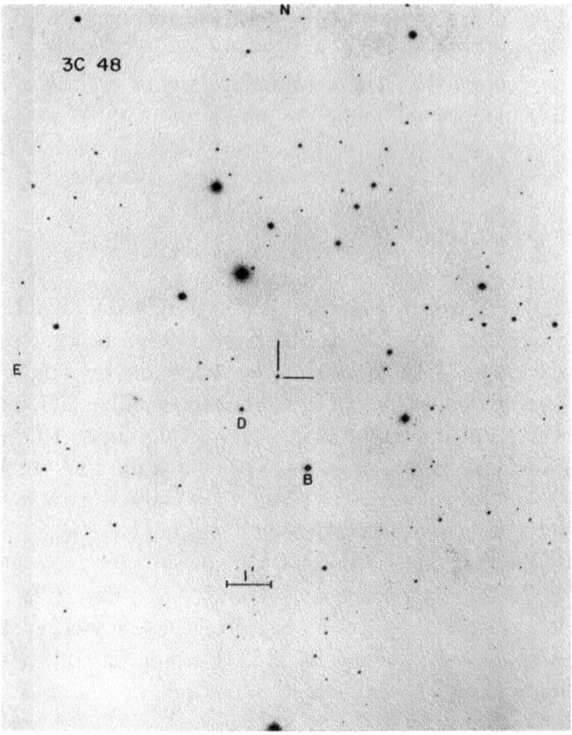

Fig. 5.16 Photo of a field around 3C48 taken with the 200 inch telescope on Mount Palomar. The radio source coincides with the star (Matthews and Sandage 1963)

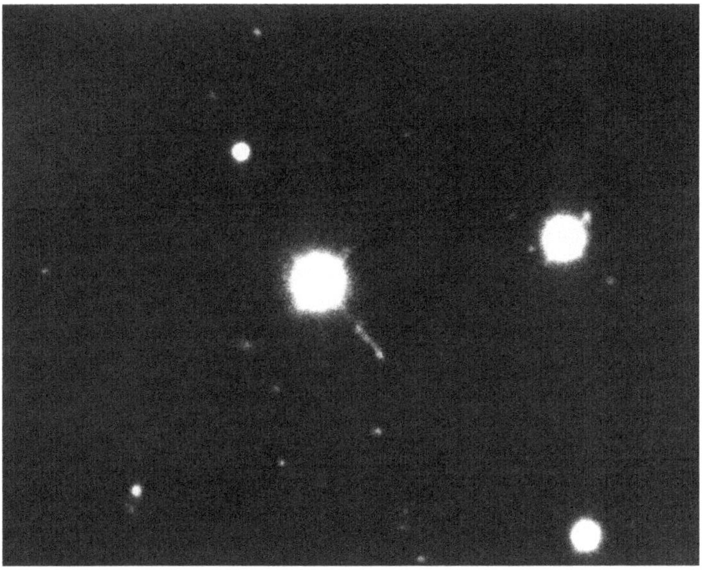

Fig. 5.17 Photo of 3C273, star and jet. Source: NOAO/AURA/NSF

The breakthrough and the sensation came in 1963. It started when the Australian Parkes radio telescope tracked the radio source 3C273 as it entered and left occultation by the moon. The beginning and end of the occultation were timed very sharply. The analysis left no room for doubt. 3C 273 was made up of two parts, A and B, that were 20 arc seconds apart. The position of each was determined with an error of less than half an arc second. Optical identification was 'a piece of cake'! The B part coincided with a star and part A with a faint 'jet' that reminded one of the 'jet' in M87 (see Fig. 5.17).

Maarten Schmidt, a professor at CalTech, originally from the Netherlands, recorded a spectrum of the star on Mount Palomar. This spectrum, too, was baffling: there were a number of bright lines that he was unable to identify. After a lot of fruitless puzzling, one afternoon he worked out the ratios between the measured wavelengths. Surprise! He recognised these ratios: they were the same as the Balmer lines, a well-known series of lines of hydrogen. And then Schmidt understood what he saw: the unidentified lines had to be Balmer lines with a redshift of $z = 0.16$. This was not a star! It was a galaxy with a very big redshift! See Box 5.3. Schmidt told his colleague Greenstein about it and he looked at the spectrum of 3C 48 again and saw that the lines could all be reconciled with a redshift of $z = 0.36$. This was not a star, either. When Schmidt returned home that evening after this exciting afternoon he told his wife, 'I think I've discovered something unusual!'

Box 5.3 The Wavelengths Measured and the Redshift of 3C273
Schmidt recorded a spectrum of the star that coincided with the radio source 3C273. He found emission lines at wavelengths of 563.2, 503.2, 475.3 and 459.5 nm. He (and everyone else that he consulted) was puzzled. He knew of no element with lines at these wavelengths. Then, on an impulse, he calculated the relation between the wavelengths: 563.2/503.2 = 1.12, 503.2/475.3 = 1.06 and 475.3/459.5 = 1.03. He recognized these ratios as coming from a known series, the Balmer lines, of the hydrogen atom: Hβ (486.1 nm), Hγ (434.0 nm), Hδ (410.2 nm) and Hε (397.0 nm). Only the brightest line (Hα) was missing. When Schmidt multiplied the wavelengths of the Balmer lines by 1.158 he arrived at the wavelengths in the spectrum of 3C 273. The strongest Balmer line Hα at a wavelength of 656.2 nm; multiplied by 1.158, the wavelength is 759.0 nm; this was outside the measured spectrum. Schmidt's colleague Oke then looked at this wavelength with a different instrument and found indeed a strong line. Conclusion: all spectral lines in the spectrum of 3C273 were 1.158 times longer than in the laboratory and that could only be explained with a doppler shift. The object 3C273 moved away from us at a velocity of 22,000 km/s. Such major redshifts had been encountered one or two times in normal galaxies, but they were rare.

Unsurprisingly, the first thought was that these radio sources were the nuclei of galaxies with a very large redshift that were thus a very long distance away. This raised a number of questions. Why was the surrounding galaxy not visible? And was there some other way of explaining the redshift? The first question remained unanswered for the time being. Schmidt (1963b) discussed the second question and had little difficulty in excluding the possibility that this was a redshift caused by gravity and therefore the object could not be a true star. Schmidt stood by his first impression, namely that i3C273 was very distant radio galaxy and that the large redshift was due to the expansion of the Universe. The discovery promised that in the future similar objects could be traced deeper in the Universe. The objects were called 'quasi-stellar radio sources', later contracted into the new word 'quasar'. The acronym QSO is another possibility.

This discovery made Schmidt world famous and in 2008 it brought him the Kavli prize of half a million dollars. I remember the broadcast of a current affairs programme on Dutch television in 1966. During the programme there was not only telephone contact with Maarten Schmidt himself (his Dutch was still faultless) but there was also a camera team at his father's house in Drenthe, where his father proudly took out the American weekly 'Time', featuring his son on the cover, a distinction few scientists can claim (Fig. 5.18).

Fig. 5.18 Portrait of Maarten Schmidt on the cover of Time. Source: Time, 11 March 1966, cover by Robert Vickrey

Are They Really So Far Away? The Controversy

Schmidt's claim that quasars were galaxies was not universally accepted. Doubts were soon raised. Why was no galaxy to be seen around the quasars? The resistance took on the nature of a rebellion against a 'politically correct' conclusion. The rebels united in the supposition that quasars were fireballs that were shot out at high velocity from the core of a normal galaxy; e.g. (Burbidge 1968). The rebel camp was not big, but it was made up of notable researchers. Two outspoken opponents were Geoffrey Burbidge and Halton 'Chip' Arp. The dispute escalated so far that Arp resigned as staff astronomer at Mount Palomar out of anger at the opposition he believed he was encountering when he submitted proposals for observations to support his viewpoint. The director decided that Arp wanted more of the same and that the proposed observations would not resolve the conflict. Mount Palomar was rocking on its foundations.

Arp never admitted he was wrong but the heat gradually went out of the discussion. No new rebels appeared and the missing galaxies were actually found around the quasars (Kristian 1973). At the end of his life Geoffrey Burbidge once again stated that rebellion is healthy—and even necessary—for research. I endorse this view, but I have my doubts whether rebellion always leads to better insights. The assumption that a quasar was a galaxy was slowly but surely confirmed by all kinds of new observations.

More and More Quasars with and Without Radio Radiation

The discovery of a completely new kind of galaxy led to a hunt for more. Whenever the position of a radio source was measured precisely, the spectrum was quickly recorded. But was it necessary to wait for that? The first quasars to be discovered emitted much more ultraviolet light than normal stars. So people started to look at existing photographic plates for objects that emitted much more blue than red light. And, as a pleasant surprise, it worked; more quasars were found and some without radio radiation! It was even estimated that the number of radio-silent quasars was many times greater than quasars with measurable radio radiation.

In 1965 Sandage remembered that Dutch American Willem Luyten and Mexican Guillermo Haro had been searching for years for faint, blue-colored stars on the assumption that this would lead them to more white dwarfs. They had hundreds of objects on their lists. But, Sandage thought, these are not white dwarfs at all, but quasars! He wrote an emotionally charged article entitled: 'The existence of a major new constituent of the Universe: the quasi-stellar galaxies' (Sandage 1965) and used all his authority in the struggle for as rapid a publication as possible. Chandrasekhar, the formidable editor-in-chief of the Astrophysical Journal, was persuaded to place the article as an enormous exception in the list of articles to be published and not at the bottom of the pile. As Manuel Peimbert discovered years later, the date of receipt indicated on the publication was pre-dated—a highly unethical act. Unfortunately, Sandage's dream was short-lived. Yes, there were quasars in the lists of Luyten and Haro, but they made up no more than a couple of percent of the total. If you looked at much fainter objects, the number of quasars increased, but Luyten and Haro only found very few of them. Sadly, this signalled the demise of a wonderful idea. The few new quasars were 'red herrings'. Years later Luyten's and Haro's work was continued by Green and Schmidt, but this time using photos with a longer exposure that went deeper. This hunt for distant quasars was successful.

In 1965 Schmidt published a list of nine quasars, including one (3C9) with a redshift equalling $z = 2.012$. The strongest line of hydrogen, 'Lyman-α' had a wavelength in the lab of 121.6 nm and thus in the spectrum of 3C9 a wavelength of

(1 + 2.012) × 121.6 nm = 366 nm. Thus this line had been shifted to a wavelength that could be observed from the surface of the Earth. This was astounding. In 1956 (9 years earlier) the redshifts had been published that had been measured in the previous 30 years. The highest value in this collection was z = 0.13. In 1960 Minkowski proudly reported that he had found a radio galaxy with a redshift of 0.2 and a velocity of recession of 60,000 km/s, or 20% of the velocity of light. Now, 5 years later 2012 was the highest value of z. Using the relativistic equation given in a box somewhat higher up one calculates that z = 2.012 corresponds to a velocity of recession 240,000 km/s, 80% of the velocity of light. The fact that the Lyman-α line could be detected from the ground led to a new direction for research: the study of clouds of gas between the quasar and us. See the next section.

In 1968 Schmidt wrote an influential article based on the data from 33 quasars. He showed, that there had previously been more many quasars than now. Quasars are the dinosaurs of the Universe; they used to be numerous but now they are almost extinct. This article was one of the first studies of the history of the 'distant Universe', the Universe that we see from a great distance and of which the image is from long ago: looking at the far distance means we are seeing events from the distant past. The conclusion that the Universe is now different from previously was of course deadly to 'steady state' theory, in which the Universe always stays the same. But in 1968 'steady state' theory had already been overturned by the discovery of background radiation in 1965 (see later).

Absorption Lines and Intergalactic Matter

The first quasar spectra showed only bright lines and no thought was given to the dark absorption lines. But in a quasar with a redshift of z = 2 the strongest line of hydrogen had a wavelength of around 364.8 nm, putting it within the spectra taken from the surface of the Earth! This line was named the Lyman α line (abbreviated to Lyα) and in the lab has a wavelength of λ = 121.6 nm.

This realization led to an immediate response from CalTech astronomers Gunn and Peterson (1965): imagine that there was a small cloud of gas of hydrogen atoms between us and the quasar, then that cloud would absorb Ly-α photons and the absorption line would be visible in spectra recorded from Earth. If there were more clouds, then more lines would be found. And if this hydrogen gas was not collected in clouds but filled the whole space, it would consume all the light of the quasar at wavelengths shorter than $(1 + z)$ × 121.6 nm. This would lead to the existence of what was later described as the 'Gunn-Peterson trough'. Their prediction was confirmed in 1971: Lynds (1971) found a large number of absorption lines in a quasar spectrum all at a smaller redshift than that of the quasar and concluded that the line of sight to the quasar passed through many hydrogen clouds. When the same was found in spectra of several other quasars these lines clouds obtained the name of 'Lyman-alpha forest'.

A difficult question arose: had this gas been there from the earliest times or had it been expelled by the quasar? Expulsion was quickly rejected as an explanation: some absorption lines had redshifts that differed so much from those of their quasar that expulsion was out of the order. Figure 5.19 shows a bar diagram of the

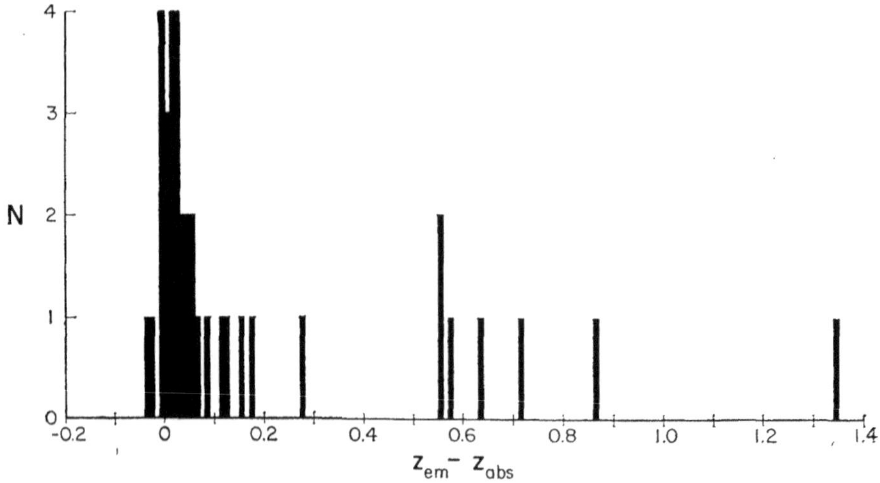

Fig. 5.19 Histogram of the values of $|z_{em} - z_{abs}|$. Here, z_{em} is the redshift of the bright lines in the quasar spectrum that originate from the quasar itself. But z_{abs} is the redshift of absorption lines in the quasar spectra and these are almost all smaller than z_{em}. Where there is a big difference the absorption line must have come from 'something' between the quasar and us (Burbidge 1971)

difference between the redshifts of the quasars and that of the absorption line. Several redshift differences are too large to suppose that intergalactic gas and quasar were somehow related.

BL Lacs or 'Blazars': Not Quasars But Closely Related

Quasars were a completely new type of galaxy but they proved not to be unique. Similar objects were found, such as the so-called 'BL Lac objects'. In 1972 it was shown that a variable star, BL Lac coincided with a radio source (Schmitt 1968). This variable star had been discovered in Germany in 1929 but was barely studied any further. The spectrum of the star showed no bright or dark lines and at first the object could not be explained. Other similar objects were found and some coincided with some smudge of diffuse light. In 1974 a spectrum was obtained of the smudge around BL Lac and that proved this to be an elliptical galaxy with a redshift of $z = 0.07$ (Oke and Gunn 1974). BL Lac and the other similar sources were extragalactic and clearly related to quasars.

What Is a Quasar? Do Dead or Dormant Quasars Exist?

There is a widely cited article from 1964 by Salpeter (1964) in which he suggests that the energy source of quasars could be a black hole with a mass of a million times the mass of the Sun. The black hole would become active if it was 'fed' by a sufficient amount of matter falling into the hole. This suggested the possibility that black holes were sometimes not fed and then remained undetected. A few years later the suggestion came (Lynden-Bell 1969) that at the center of very many galaxies there was a dead or dormant quasar! It was a great idea, but there was no evidence, unless the radio sources SgrA, in the center of our Milky Way, could be shown to be a dormant heavy black hole. Thinking about dormant black holes gives the same feeling as when you visit old volcanic craters in a natural park and then imagine that once they produced huge eruptions. Or remember when, in your imagination as a child, you had to tiptoe around a sleeping dragon.

In Summary: Galaxies with Active Nuclei ('AGNs')

A careful discussion of the types of galaxies that could be concealed by a strong radio source led in 1964 to the Fig. 5.20. Standard disk galaxies (symbol 'x' in the image) were relatively faint radio sources. Below there was also the M82 system, a prototype

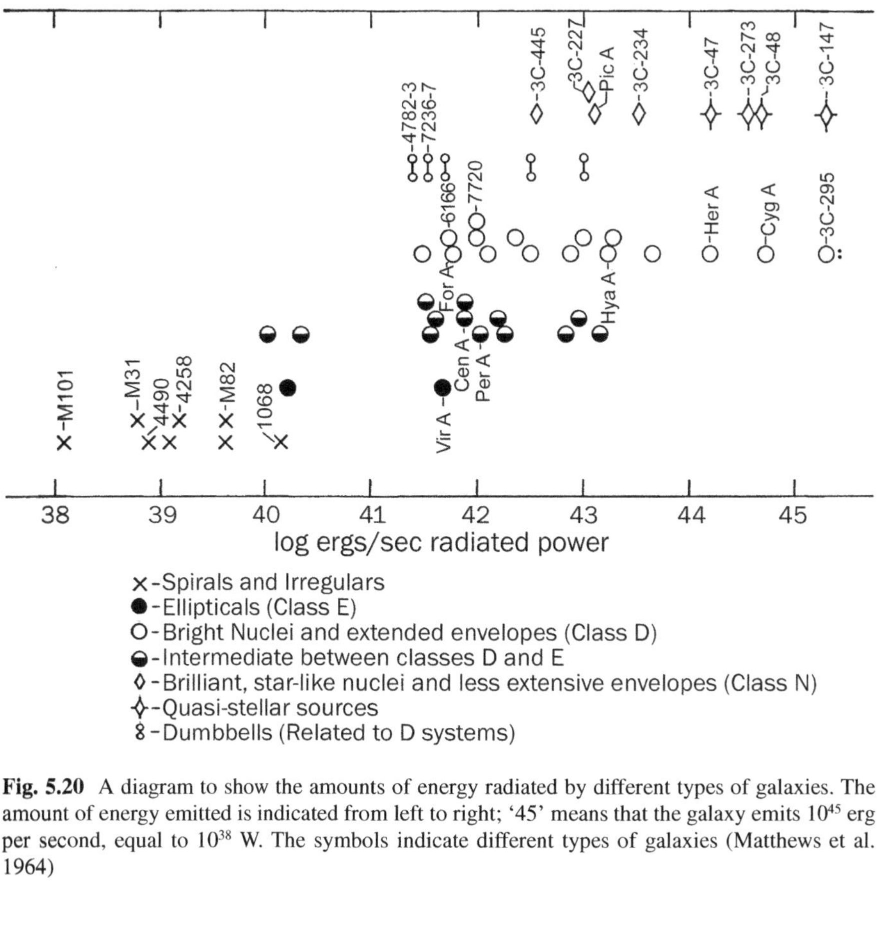

Fig. 5.20 A diagram to show the amounts of energy radiated by different types of galaxies. The amount of energy emitted is indicated from left to right; '45' means that the galaxy emits 10^{45} erg per second, equal to 10^{38} W. The symbols indicate different types of galaxies (Matthews et al. 1964)

of a starburst galaxy. There were two elliptical galaxies ('E') both of which were brighter as a radio source than the disk galaxies and the irregular galaxies. But the really big monsters were the D galaxies (open circles) and the D/E sources (half black and half white circles). The D systems constituted a new type of galaxy that did not occur in Hubble's tuning fork. They were like elliptical galaxies but they were much heavier. They were found mainly in the center of a cluster of galaxies and that led to the suspicion that they were so heavy as a result of swallowing many smaller galaxies—an activity that was referred to as 'cannibalism'. Of the radio galaxies discussed in Chap. 2, NGC 5128 (Centaurus A) is a D/E galaxy and M87 (Virgo A) an E galaxy. In spite of the existence of these huge E and D-systems the ultimate conclusion was that the cores of these systems, the quasars, emitted the most light.

All these unusual galaxies have in common that at their center there is a small core from which jets spurt left and right. The jets consist of relativistic but the jet will probably also contain positively charged protons. Galaxies with this kind of core were subsequently referred to as 'AGN galaxies' (Active Galactic Nucleus) and I will do likewise.

The characteristics of the core of AGNs—the 'hellish machine'—became an important question that for the time being could not be answered. It has already been mentioned that black holes might be the 'monster', holes as heavy as millions of times the mass of the Sun. In 1974 Blandford and Reed in Cambridge (England) (1974) brought some clarity to the image. Figure 5.21 shows a very small source of energy A in a hollow C that expelled relativistic gas in two opposite directions. Eventually the jet collided far from the source with 'intergalactic' gas that had been swept up. The magnetic field was disturbed and the electrons started to emit synchrotron radiation. This tied in with the observation that in many (but not all) sources synchrotron radiation is strong at the end of a jet stream (see Fig. 5.22).

5.3 Large and Largest Scale

I will start with clusters and superclusters, probably the biggest separate and coherent structures in our Universe. Thereafter will be the Universe.

Clusters and Superclusters

Clusters of galaxies were introduced in the prologue. They are large objects that appear to be held together by gravitational attraction. In spite of their being very far way, they are still large objects in the sky found on large-scale photographic plates. In the 1950s the Mount Palomar Sky Survey (see Chap. 1) offered a good opportunity: they cover fields of 4 deg × 4 deg and show objects as deep as the 19th magnitude. In 1958, American George Abell (1958) published a catalogue of more than 2700 'rich' clusters of galaxies, where 'rich' meant 'a lot of galaxies'. After 60 years, Abell's catalogue is still in use, a clear sign of its quality (Fig. 5.23).

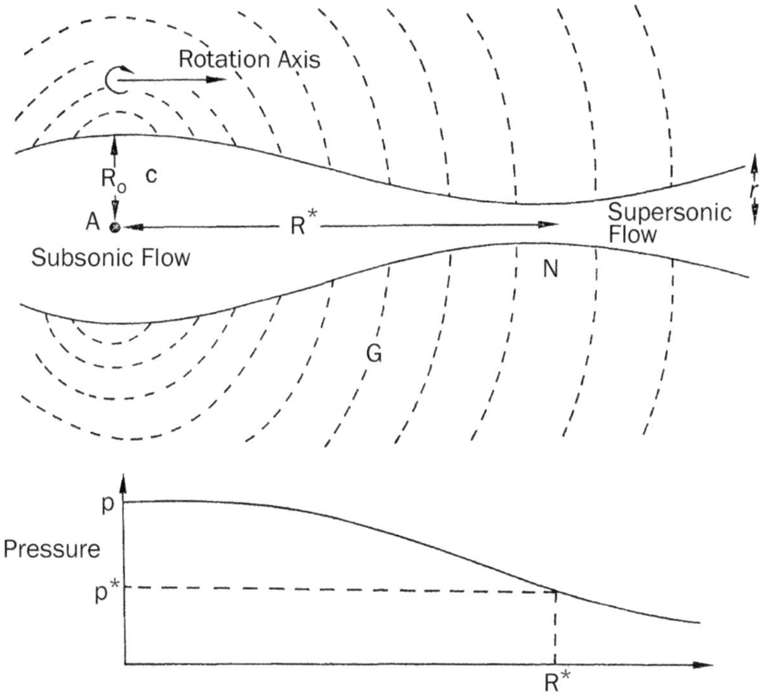

Fig. 5.21 Model of the center of a radio source. The source in A is at center of a cavity in a huge ball of matter. Relativistic gas flows in two narrow jets in opposite directions, to the left and the right. Only the start of the jet to the right is drawn. The velocity of sound in the relativistic gas is approximately $c/\sqrt{3}$ (=170,000 km/s), where c is the velocity of light. At the start the jet is subsonic but when the jet thins out the velocity is supersonic (Blandford and Rees 1974)

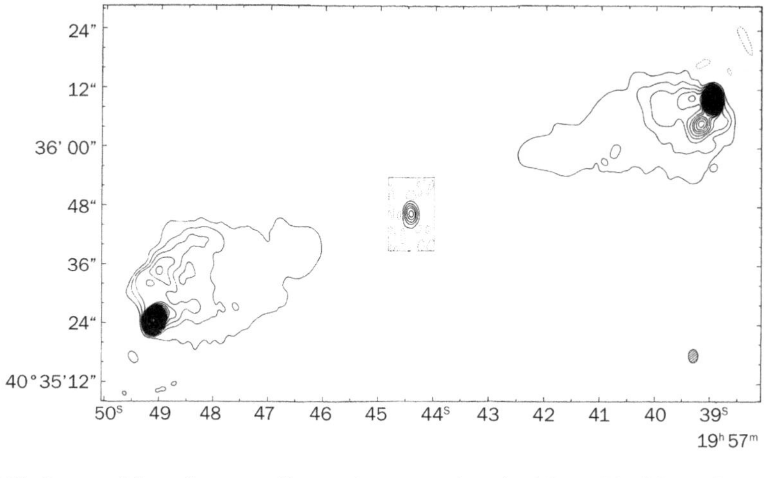

Fig. 5.22 Image of the radio source Cygnus A at a wavelength of 5 cm. The faint radio source in the center coincides with the galaxy described for the first time by Baade and Minkowski. A strong radio source can be seen at the end of each jet where the relativistic gas of the jet is stopped by swept up matter (Hargrave and Ryle 1974)

Fig. 5.23 Abell2151, also known as the Hercules cluster. Most of the objects in this picture are galaxies that belong to A2151, but a few are in front of the cluster and many are behind it. There are also a few stars present that belong to the Milky Way. To determine the true members of the cluster, the radial velocity of each object is needed. This requires the measurement of the spectrum which is a time-consuming procedure. Source: Astronomy Picture of the Day, 16 July 2009, Tony Hallas

The catalogue lists the galaxies, their morphology (elliptical or disk-like), how much light they emit and their precise location, but unfortunately not one other essential piece of information: the velocity of the galaxy inside the cluster, as derived from the doppler effect. There is a good chance that a galaxy seen inside the boundaries of the cluster is actually a foreground or background object. But all galaxies in a cluster have about the same velocity and therefore the measurement of the redshift will prove or disprove that this galaxy belongs to the cluster. The redshift is derived from a spectrum and measuring a spectrum took quite some time on a large telescope, time that was difficult to obtain. In retrospect it is clear that the study of clusters of galaxies has been slowed down for decades because of the lack of redshift measurements.

In the prologue Zwicky was mentioned who in 1933 determined the total matter content of the Coma cluster of galaxies by two different methods, with the surprising outcome that one method predicted some 400 times more matter than the other. Zwicky could not find any error in either method or in the data he used and he eventually concluded that the Coma cluster was full of 'dark' or 'invisible' matter that had not been included in one of his two methods. In later years, Zwicky's study was repeated a few times, but always confirming his result. The problem was there and did not go away by looking away. It was an 'inconvenient truth'.

Then, in the 1960s, X-ray astronomy took off. Rockets were fired with X-ray detectors on board and directed at known objects that might emit X-rays. Radio galaxies (Virgo A, Perseus A and others) were among the first candidates that brought success. Then, in 1970, the UHURU-Xray satellite was launched. It brought new detections and soon it became clear that there were two types of X-ray sources among the radiogalaxies. In one type the X-radiation was confined to the galaxy; it was an isolated source. The other type consisted of sources much larger than the radio galaxy and the actual X-ray source was not the radio galaxy, but the cluster of galaxies to which the radio galaxy belonged.

The discovery that clusters of galaxies were a source of X-radiation caused quite a sensation. Different proposals were made but it was quite quickly agreed that the clusters contained a lot of hot gas with a temperature of a few millions of degrees. Initially there was hope that this gas held the cluster together and that it was the 'dark matter' that Zwicky had postulated. But no, calculations soon proved that the mass of the hot gas was at most a few percent of the missing matter. Within a few years it appeared that a cluster with more matter also had stronger X-radiation (and thus more cluster gas). The intracluster gas was from now on an important ingredient of a cluster.

The existence of the intracluster gas received independent confirmation from radio astronomy. Many radio galaxies had two jets in opposite directions from the central galaxy. But in 1968 a very different radio galaxy turned up. The two jets have been turned around because the galaxy is 'ploughing' through the gas inside the cluster—see Fig. 5.24.

A different aspect of clusters arose in 1970 when American cosmologist Peebles produced an analysis of how a cluster such as Coma could have developed from a small irregularity in the distribution of matter in the Universe. This will be discussed in more detail later. Peeble's article inspired Americans Gott and Gunn to think about what happened with matter just outside the gas that formed the cluster. In what proved to be a very influential paper in 1972 they suggested that even after the cluster was formed a lot of gas still flowed into it (Gunn and Gott 1972). This gas could have a range of different effects; it could in particular find its way to the very heavy galaxies in the center, causing a change the characteristics of these galaxies. This could explain how the heavy galaxies (cD galaxies; see earlier in this chapter) in the center of some clusters came about. The research reinforced a general feeling, namely that some—or even most—clusters are still under construction! In other words, the Universe is still evolving.

An important contribution came from the Soviet Union in 1972 by Rashid Alevich Sunyaev and Yakov Borisovich Zel'dovich (1972). Wherever there was a cluster, they argued, the background radiation would be slightly fainter. The gas in a cluster would be able to absorb part of the cosmic background radiation (to be introduced later) and the level of absorption would reveal the amount of gas present. For the time being such measurements were not possible technically, but decades later the search for the Sunyaev-Zeldovich effect would become a prominent way of detecting very distant and very young clusters.

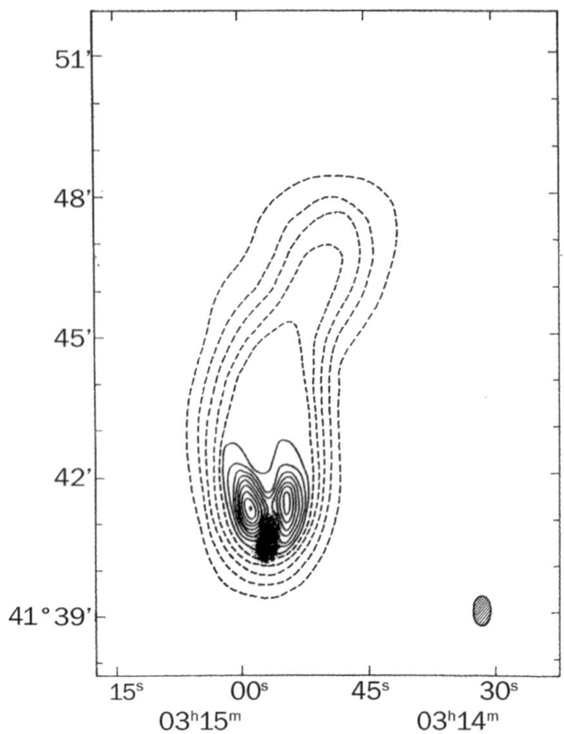

Fig. 5.24 NGC 1265, a radiogalaxy of unusual shape inside the Perseus cluster of galaxies. The black rectangle is the optical galaxy. The contour lines represent the intensity of the radio emission. The two maxima in the radio contours are the two jets from the center of the galaxy, but the jets have been bend around because the galaxy is moving through the hot X-ray gas in the cluster (Ryle and Windram 1968)

Cosmology: the Outlook in 1961

It has been said in the prologue and in Chap. 3 that the study of the origin of the cosmos had come to a halt in the 1930s and that progress could be expected once observational data of much higher quality had been obtained. Theory had progressed as far as it could and it was up to the observers to provide the missing data. Unfortunately, in the 1940s and 1950s, the required accuracy of the measurements was far beyond the technical possibilities.

True, the steady-state theory had brought some diversion. The proposers were highly competent researchers whose ideas could not be ignored but it is my guess that few astronomers ever 'believed' in the theory. Other cosmological research of great importance by George Gamow and collaborators had been forgotten, unjustifiably.

In 1961 Alan Sandage, Hubble's successor, wrote a widely cited article entitled 'The possibility of the 200-inch telescope to distinguish between selected Universe models'. These 'selected' models were the three mentioned in the Prologue. Sandage proposed concentrating on clusters of galaxies and then using the third brightest galaxy in the assumption that that galaxy always has the same absolute magnitude—which is questionable. Next, he showed a diagram of the relation between the distance of the cluster and its redshift. He then indicated his intention to obtain redshifts of much higher accuracy of even more distant clusters. Even to the master-student that I was in 1961 it was clear that Sandage's plane could never be successful. The project appeared doomed—as many thought. Admittedly, Sandage was a formidable researcher and very driven: so who knows? He himself had his doubts: at around this time he remarked during a conference that in every field there are optimists and pessimists, except in cosmology, because in that field only optimists survive. The comment by Russian Zeldovich is also very telling: 'Cosmologists are often in error but never in doubt' (Fig. 5.25).

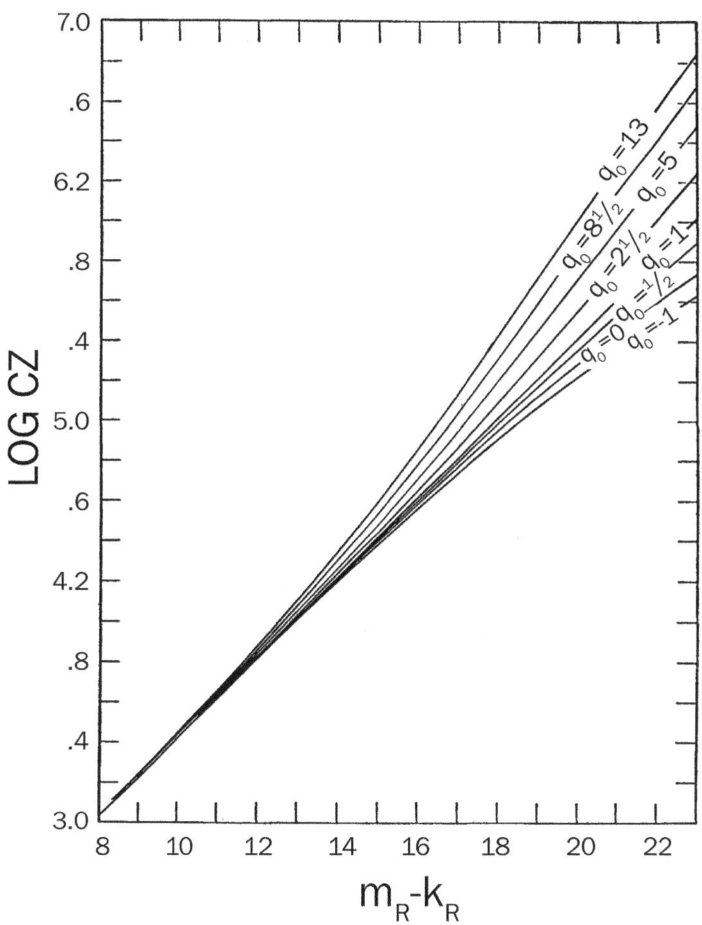

Fig. 5.25 A diagram of the distance of stellar galaxies against radial speed. The different lines correspond with different models for the Universe; the distances and speeds are indicated logarithmically. By measuring the curve precisely you learn automatically which curve applies to our Universe. It is a project that was doomed to failure (Sandage 1961)

Fig. 5.26 Bob Wilson (left) and Arno Penzias (right) in front of the telescope with which they in 1965 discovered 3 K background radiation, not knowing the treasure that they had in their hands. But they still received the Nobel Prize in 1978 for their discovery. Source: ZUMA Press, Inc./ Alamy Stock Photo

1965: Background Radiation: A New Foundation for Cosmology

Sandage's plan from 1961 was never successful. The real breakthrough in cosmology came 4 years later, completely unexpectedly. Theoreticians and observers made the same discovery at the same time.

Let us start with the theory. The research by Gamow in the late 1940s was described in Chap. 2. He was interested in the formation of the atoms shortly after the Big Bang. Together with his student Alpher, he concluded that the Universe should still contain a homogeneous 'ocean' of photons at a temperature of 7 K. Gamow's work was apparently forgotten but at least 10 years later researchers in both in Princeton and Moscow, independently and separately from one another, came to the same conclusion as Gamow (and subsequently discovered that Gamow had thought of all before them). And then in 1965, suddenly and unexpectedly, an observation was made that raised this theory to a new foundation for cosmology (Penzias and Wilson 1965) (Fig. 5.26).

The observers in question, Americans Arno Penzias and Bob Wilson, worked for Bell Labs, the US telecommunications giant.[1] In the early 1960s they decided to

[1] In the 1930s another Bell engineer, Karl Jansky, discovered that radio radiation came from the Universe—see Chap. 1.

examine whether all the contributions to astronomical radio radiation were known. They had an existing, somewhat unusual radio telescope that was suitable for their purpose and they opted for measurements at a wavelength of 7 cm. They worked very thoroughly and were eventually able to trace all the contributions to radiation. One component remained that they did not understand: radiation that came from everywhere and that had a temperature of −270 °C or 3 K when compared to some emitting surface. Penzias and Wilson were certain of their result, but had no explanation for it. Then they heard that a group of physicists at Princeton who were analysing the events immediately following the Big Bang had come to the conclusion that the Big Bang had left some remnant radiation that filled the Universe. What Penzias and Wilson had measured could well be that radiation. The contact was made and two articles were written that were printed in succession in the Astrophysical Journal. In one article Penzias and Wilson (1965) described their finding and the immediately preceding article contained the meaning ascribed to this finding by the Princetonians, Dicke, Peebles, Roll and Wilkinson (1965). The phenomenon was christened 'Cosmic Background Radiation', abbreviated to CBR. Later, it was discovered in the West that Soviet astrophysicists Doroshkevich and Novikov had already predicted this background radiation in 1964. But their article was written in Russian and was not picked up on the other side of the Iron Curtain. Penzias and Wilson received the Nobel Prize for Physics for their findings in 1978, but none of the honours—not even the slightest part—went to the theoreticians.

Penzias and Wilson published their findings in 1965 under the very prosaic title 'A measurement of excess antenna temperature at 4080 Mc/s'. You really had to be one of the insiders to understand that this boring title concealed such explosive content. Directly before this article, one from Princeton appeared containing a reconstruction of the history of the Universe from approximately one second after the Big Bang. Figure 5.27 shows this reconstruction of the past. It can be seen as an expanded, up-to-date version of Fig. 2.19 in Chap. 2. Just consider this figure for a moment and realise what a masterpiece it was: a single measurement with a single telescope at a single wavelength explained the complete history of the Universe, from shortly after the Big Bang to the present day. This history still holds today. And yet, there was still no Nobel Prize for the theoreticians?

It is worth taking the time to explain this figure. The text along the lower border indicates how, as a result of the expansion, the distances in the Universe increase; the 'radius of the Universe' is the same as the previously mentioned scale factor and this is the value 1 measured in the 'now'. Along the upper border is the time measured from the Big Bang. Under the horizontal bar 'Now' is approximately 10 billion years after the Big Bang, so after 3×10^{17} seconds—see the upper border.

Read the diagram from left to right, and also from top to bottom. In the earliest stage, within 1 second after the Big Bang, the energy density was so high that photons continually 'materialised' in elementary particles, which transformed again into photons: radiation and matter were one. The relation between photon energy and the mass of the particles that originated from it is indicated by Einstein's most famous law: $E = mc^2$. After around 1 second, the temperature and density had

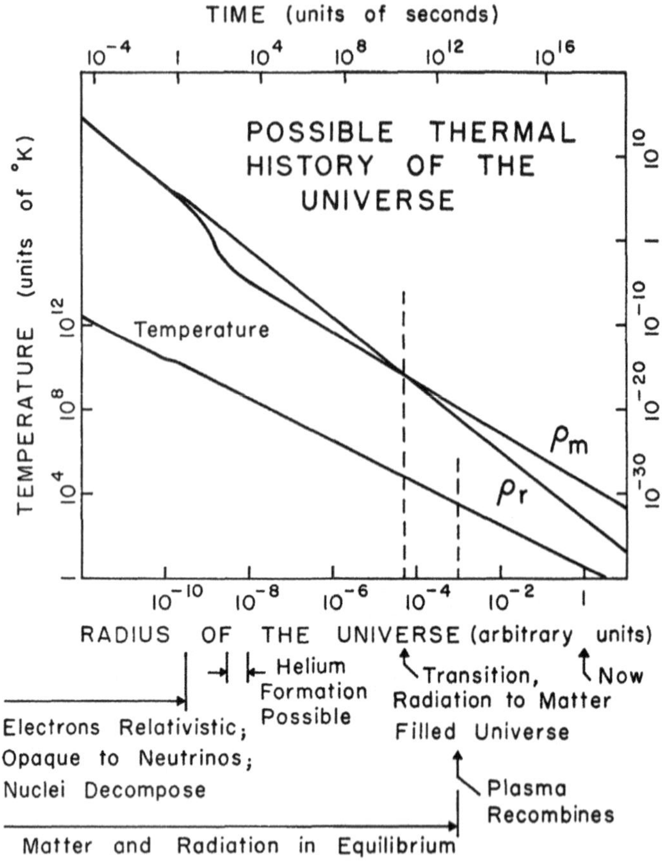

Fig. 5.27 The history of the Universe from (almost) the Big Bang to the present day and later. Compare this with Fig. 2.19 in Chap. 2; the scales of both figures are the same. See the text (Dicke et al. 1965)

decreased so far that matter particles no longer converted into photons (and, vice versa, photons into particles); from that instant the density of the gas of photons (indicated by ρ_r) and the gas from matter (ρ_m) evolved in different ways, which is indicated in the figure by two different curves. The matter comprised mainly protons (H^+) and electrons (e^-), although there was a brief period in which helium was formed (He^{2+}, also referred to as the α-particle). The gas of protons and electrons remained in close contact with the photons: the electrons continually absorbed and re-emitted radiation (photons); electrons continually collided with protons so that they maintained the same temperature. Initially there was more radiation than matter, but after some time this was reversed—at the point in time indicated in the figure by 'Transition, Radiation to Matter Filled Universe'. Later still, the temperature dropped to 4000 K and then protons and electrons combined permanently to form atoms (along the horizontal axis this is indicated as 'Plasma Recombines'). At this point in time the Universe was around 300,000 years old.

The 'recombination of plasma' had significant consequences. Hydrogen and helium atoms were no longer able to emit or absorb the photons when these had cooled below 3000 k; this inability is the accidental consequence of the structure of these atoms. The period following recombination is often called the 'dark ages', which should be understood as the period in which the matter did not produce photons. The cosmic background of photons was still around. It cooled down further as a result of the expansion of the Universe, giving it a present-day temperature of 3 K. This cooled photon gas forms 'Cosmic Background Radiation' (CBR).

The Princetonian theory ends more or less at the moment that the first well-known objects appear, galaxies, stars, clusters of galaxies. Three Soviet astrophysicists, Doroshkevich, Zeldovich en Novikov, reasoned what the sequence of events could be from that moment on; they published their paper in 1967 (Doroshkevich et al. 1967). The authors state explicitly that theirs is a speculation; it may and will be improved upon in many details, but it gives further research a direction. 'DZN' started their story about 300,000 years after the Big Bang, when the recombination has been completed, the short time interval in which protons and electrons united and formed hydrogen atoms. Under the then ruling circumstances the authors predict the formation of stars with a mass one million times that of the Sun. These live even shorter than the most massive stars that we know (such as η Carinae, discussed in Chap. 4). The explosion of such superheavy stars is that of a super-supernova and it fills space with very hot gaseous debris. Then a second generation of objects contract comparable in mass to a smaller galaxy. These break into parts and then the third generation appears: galaxies, as we know them, stars and clusters of galaxies. These three-generation events happen within about 1 billion years after the Big Bang. Remember that 'the present' is a time 12 billion years later.

During the Dark Ages, the hydrogen gas emitted 21 cm of radiation. Detecting this radiation has become an important aim for radio astronomers throughout the world as will be discussed in Chap. 17.

From Gamow's work and that of Dicke and colleagues it followed that the strength of the CBR at other wavelengths could be predicted precisely using the equation that had been developed in 1900 by the German physicist Max Planck—see Box 5.4 below. The radiation at every wavelength in every direction has to be of equal strength, 'isotropic'. Numerous radio astronomers applied themselves to detecting this radiation. The Planck formula predicted that the radiation reached its peak at a wavelength of 1 mm, and it was therefore important to see whether the radiation reduced in strength at shorter wavelengths. Unfortunately, such short-wave radiation did not penetrate—or barely penetrated—the atmosphere. In Fig. 5.28 there is a summary of what had been achieved in 1972. The curve through these measurement points was in line with Planck's formula at a temperature of 2.8 K. Measurements of the CBR at wavelengths shorter than 1 mm were clearly desirable.

A change can be seen here in the approach to cosmology. Sandage's cosmology, expressed in his 1961 paper, was about a mathematical, geometric description of the Universe. His focus was on the future of the Universe. The articles by Gamow and Dicke and their colleagues were much more based on physics: they discussed the temperature of the background radiation and of the matter. Research was no longer fixated on the future, but was mainly concerned with what happened in the past.

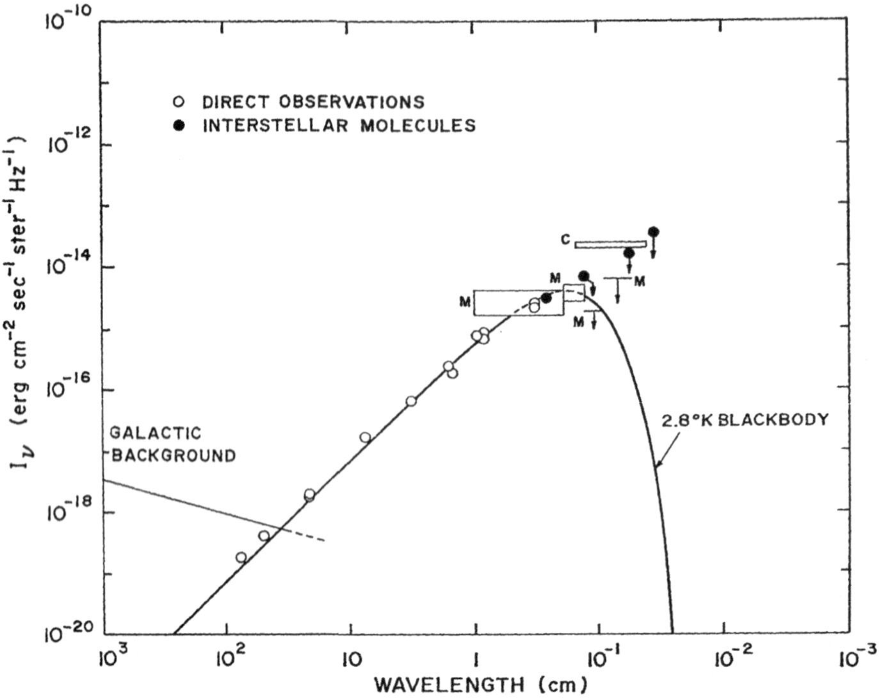

Fig. 5.28 Results of measurements at different wavelengths of the background radiation displaying the Planck curve at a temperature of 2.8 K. The open circles come from measurements from the Earth's surface. The measurement indicated with the letter C comes from a rocket flight by astronomers from Cornell and the letter M relates to measurements with a balloon by astronomers from MIT. The black circles come from an analysis of the strength of interstellar lines (Thaddeus 1972). Chapter 16 contains a more recent version of this same curve

Box 5.4 Planck Formula, Background Radiation

In the final decades of the nineteenth century, a great deal of physics research focused on the link between matter and electromagnetic radiation. An important theme was to calculate how much radiation there was in a cavity of which the walls had a uniform temperature. Eventually in 1900 Max Planck found the correct formula for the strength of this radiation. In German this radiation is referred to as 'Hohlstrahlung' (English 'cavity radiation'), which is a better term than the English-derived 'black body radiation'.

I will try to explain this by way of a personal observation. In 1947 when I was a child, I stayed a few days with a baker. His oven was a hollow cavity with thick walls that was accessible via a small metal door. The baker burned wood in the oven until the walls were hot enough. He then removed the ash and slid plates with dough shaped into loaves into the oven, later taking the baked bread out again. I learned that the loaves were baked by the infrared radiation that was emitted by the walls of the oven. I was impressed and the memory of that puzzle of the baking explains why for me 'cavity radiation' is a more appropriate word than 'black body radiation'.

It seems unbelievable, but this Planck equation for 'cavity radiation' describes precisely not only all the existing measurements of the CBR in 1972, but also the many measurements that followed. Planck developed his equation without thinking of an application in cosmology. Nonetheless, the formula describes very precisely the strength of the radiation in the Universe at all wavelengths; the precision will later be 1 in 100,000.

An Independent Argument about the History of the Early Universe

Chapter 3 explained that all elements heavier than hydrogen are produced in the inside of stars. Helium was the only exception: stars do not produce enough helium to explain the amount observed. The only possibility for helium atoms to form lay in the earliest stages of the Universe—and that happened as Gamow had proven. Figure 5.29 shows the results of calculations on the formation of atomic nuclei in the first 10,000 seconds after the Big Bang.

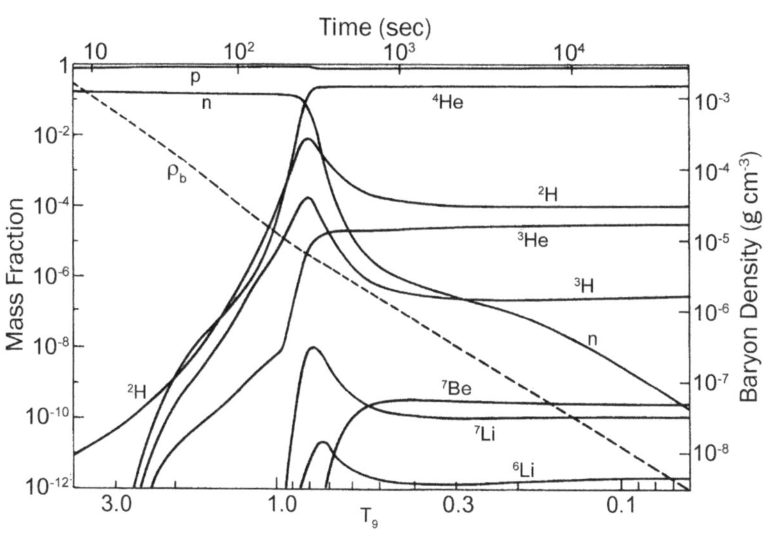

Fig. 5.29 The composition of the original gas during the first 100,000 seconds after the Big Bang. On the horizontal axis at the top, time is given and the vertical axis shows the fraction of certain isotopes present at a given moment. The formation of the light elements was completed 15 minutes (1000 seconds) after the Big Bang (Wagoner 1973)

The Big Bang could produce all the helium atoms found inside stars, but stars could not have produced that much. This was an important argument against the 'steady state theory' that was discussed in Chap. 3.

Focusing Attention on the Past: How were the Clusters of Galaxies formed?

Background radiation gave a detailed picture of the Universe when it was just 300,000 years old and there were not yet any galaxies or stars. And this, the cosmologists understood, ultimately offered some possibility of a theory on the formation of galaxies. In 1974 Martin Rees, summing up at a symposium on the Universe, commented: 'The microwave background has brought the physics of the very early Universe within the framework of serious scientific discussion….' And this expressed probably the general opinion. Cosmology was from now on based on three well-proven principles: the expansion of the Universe recognized in the velocities of distant galaxies (Hubble's discovery), the formation of the nuclei of hydrogen and helium atoms during the first minutes of the expansion after the Big Bang (Gamov's work) and now background radiation.

Based on these three principles, a spectacular development took place over the following decades. The principle was initially that the Universe was extremely homogenous and equal; the sky is equally bright in whatever direction one looks. But today this is no longer the case. Looking up at the night sky makes it clear that the Universe is highly structured, in stars, in gas clouds and in galaxies. This structure must have arisen from this (seemingly) homogeneous start and therefore there must have been small irregularities in the Cosmic Background Radiation (CBR). How these small, minimal wrinkles could develop to become large structures was a conundrum facing many cosmologists and it took until almost the year 2000 before the issue was resolved.

There were earlier attempts but we are now looking at the developments after the discovery of CBR. In 1967 American physicists Sachs and Wolfe (1967) argued that there had to be irregularities in the background radiation as a result of incipient contractions and that irregularities of around 1% might be expected. Later observations contradicted this conclusion but the article remained a source of inspiration for many people. Another example is 26-year-old British astronomer Joe Silk, who, while working at Harvard, discovered a simple mechanism that could halt the start of small contractions, while large, extensive irregularities had the chance to contract. He, too, predicted irregularities in the background radiation (Silk 1968).

It is now time to mention two articles that presented a somewhat complete model for the birth of (clusters of) galaxies. The first appeared in Western publications at a conference in 1974 of physicists from Moscow: A.G. Doroshkevich, R.A. Sunyaev and Y.N. Zel'dovich (1974) and it is a further development of the paper published in 1967—see above. The main new aspect was the prediction that the third contraction would occur in one dimension, creating large two-dimensional objects called 'pancakes'. The pancakes ultimately broke up in clusters of galaxies and of individual galaxies. 'Pancakes' was a word that Zel'dovich used for a long time to refer to this theory with a heartwarming, almost childish pleasure. This Russian view of the formation of galaxies and stars happened is sometimes referred to as top-down: start big and let the smaller parts follow. The ELS theory (see higher up in this chapter) is an example of such an approach.

The second highly influential article was by Americans W.H. Press and P. Schechter dating from 1974 (Press and Schechter 1974), in which they explained that galaxies arose from the clumping together of smaller nebulae; in fact, they contended the opposite of what Doroshkevich et al. claimed. Press and Schechter did some algebraic calculations and verified it with numerical simulations. Their conclusion was that if the Universe started to expand as described by the Friedmann equations, and if small compressions started at random points, this would lead to collisions and the merging of heavier units. The longer this process lasted, the more matter would clump together. But some smaller units always remained; the distribution of these masses, the 'mass spectrum', no longer changed. The conclusion reached by Press and Schechter made quite an impression because it proved that larger bodies in the Universe could indeed have been formed from smaller bodies. It was a formation of structures from small to large, a way of thinking that would later be referred to as 'hierarchical'. This approach was thus to build up larger structures from smaller entities. Just as ELS was an illustration of top-down galaxy formation, there is another observational example for the hierarchical structure that was quite influential. In 1978 American astronomers Searle and Zinn (1978) proposed that globular clusters, separate units in the Milky Way consisting of millions of stars, actually originated outside the Milky Way and were attracted to it at a later stage.

The uncertainty about how galaxies and clusters of such galaxies were created (top-down or hierarchical) is apparent in a review article by Peebles at a symposium on 'the formation and dynamics of galaxies' in 1974. His discussion is very general and unspecified. Peebles was one of the leading cosmologists of the time, but in this article he is pussyfooting around in his approach to the subject. He uses an interesting comparison with geology: in the eighteenth and nineteenth centuries there were important yet conflicting ideas about the formation of the Earth ('vulcanism' or 'neptunism'). Most progress was achieved not by people who defended supreme explanations but by those, like Hutton, who looked for answers to less far-reaching questions, such as how long does it take for a river to carve out a valley? Peeble's moral, it seems to me, was: don't start with the ultimate questions, but first solve the small questions, and the big answers will follow later. Whether in the end he was right or wrong is less important than the expression of the fundamental uncertainty that then reigned in the minds of the cosmologists.

References

Abell, G.O.: Astrophys. J. Suppl. **3**, 211 (1958)
Arp, H.: Astrophys. J. **142**, 402 (1965)
Babcock, H.W.: Publ. Lick Obs. **19**, 41 (1939)
Balick, B., Brown, R.L.: Astrophys. J. **194**, 265 (1974)
Becklin, E.E., Neugebauer, G.: Astrophys. J. **151**, 145 (1968)
Blandford, R.D., Rees, M.J.: Mon. Not. R. Astron. Soc. **169**, 395 (1974)
Burbidge, G.R.: Astrophys. J. Lett. **154**, L41 (1968)
Burbidge, E.M.; Study Week on Nuclei of Galaxies p121 (1971)
Burbidge, G., Burbidge, E., Sandage, A.: Mon. Not. R. Astron. Soc. **169**, 395 (1963)
Danielson, R.E., Tomasko, M.G., Savage, B.D.: Astrophys. J. **178**, 887 (1974)
Dicke, R.H., Peebles, P.J.E., et al.: Astrophys. J. **142**, 414 (1965)
Doroshkevich, A.G., Zel'dovich, Y.B., Novikov, I.D.: Astronomicheskii Zhurnal, vol. 44, p. 295 (1967). (SvA v11 p233)
Doroshkevich, A.G., Sunyaev, R.A., Zel'dovich, Y.N.: Int. Astron. Union Symp. **63**, 213 (1974)
Eggen, O.J., Lynden-Bell, D., Sandage, A.R.: Astrophys. J. **136**, 748 (1962)
Einasto, J., Kaasik, A., Saar, E.: Nature. **250**, 309 (1974)
Field, G.B.: Highlights of Astronomy, vol. 3, p. 37 (1974)
Field, G.B., Goldsmith, D.W., Habing, H.J.: Astrophys. J. **155**, L149 (1969)
Fomalont, E.B.: Astron. J. **76**, 513 (1971)
Freeman, K.C.: Astrophys. J. **160**, 811 (1970)
Graham, I.: Mon. Not. R. Astron. Soc. **149**, 319 (1970)
Gunn, J.E., Gott III, J.R.: Astrophys. J. **176**, 1 (1972)
Gunn, J.E., Peterson, B.A.: Astrophys. J. **142**, 1633 (1965)
Hargrave, P.J., Ryle, M.: Mon. Not. R. Astron. Soc. **166**, 305 (1974)
Hulsbosch, A.N.M., Raimond, E.: BAN, vol. 18, p. 413 (1966)
Jaaniste, J., Einasto, J.: ESOC. **27**, 3 (1987)
Jenkins, E.B., Meloy, D.A.: Astrophys. J. **193**, L121 (1974)
Kahn, F.D., Woltjer, L.: Astrophys. J. **130**, 705 (1959)
Kleinmann, D.E., Low, F.J.: Astrophys. J. **161**, L203 (1970)
Kristian, J.: Astrophys. J. **179**, L61 (1973)
Light, E.S., Danielson, R.E., Schwarzschild, M.: Astrophys. J. **194**, 257 (1974)
Lin, C.C., Shu, F.H.: Astrophys. J. **140**, 646 (1964)
Low, F.J., Kleinmann, D.E., Forbes, F.F., Aumann, H.H.: Astrophys. J. **157**, L97 (1969)
Lynden-Bell, D.: Nature. **223**, 690 (1969)
Lynds, C.R., Sandage, A.R.: Astrophys. J. **137**, 1005 (1963)
Lynds, R.: Astrophys. J. **164**, L73 (1971)
Matthews, T.A., Sandage, A.R.: Astrophys. J. **138**, 30 (1963)
Matthews, T.A., Morgan, W.W., Schmidt, M.: Astrophys. J. **140**, 35 (1964)
Mathewson, D.S., van der Kruit, P.C., Brouw, W.N.: Astron. Astrophys. **17**, 468 (1972)
Mathewson, D.S., Cleary, M.N., Murray, J.D.: Astrophys. J. **190**, 291 (1974)
McKee, C.F., Ostriker, J.P.: Astrophys. J. **218**, 148 (1977)
Münch, G., Zirin, H.: Astrophys. J. **133**, 11 (1961)
Oort, J.H.: BAN, vol. 18, p. 421 (1966)
Oke, J.B., Gunn, J.E.: Astrophys. J. Lett. **189**, L5 (1974)
Ostriker, J.P., Peebles, P.J.E.: Astrophys. J. **186**, 467 (1973)
Ostriker, J.P., Peebles, P.J.E., Yahil, A.: Astrophys. J. Lett. **193**, L1 (1974)
Penzias, A.A., Wilson, R.W.: Astrophys. J. **142**, 419 (1965)
Press, W.H., Schechter, P.: Astrophys. J. **187**, 425 (1974)
Roberts, M.S.: International Astronomical Union (IAU) Symposium, vol. 44, p. 12 (1972)
Roberts, M.S., Rots, A.H.: Astron. Astrophys. **26**, 483 (1973)
Roberts, R.S., Whitehurst, M.N.: Astrophys. J. **201**, 327 (1975)

Rubin, V.C., Ford, W.K.: Astrophys. J. **159**, 37 (1970)
Ryle, M., Windram, M.D.: Mon. Not. R. Astron. Soc. **138**, 1 (1968)
Sachs, R.K., Wolfe, A.M.: Astrophys. J. **147**, 73 (1967)
Salpeter, E.: Astrophys. J. **121**, 161 (1955)
Salpeter, E.E.: Astrophys. J. **140**, 796 (1964)
Sandage, A.: Astrophys. J. **133**, 355 (1961)
Sandage, A.: Astrophys. J. **141**, 1560 (1965)
Sandage, A., Freeman, K.C., Stokes, N.R.: Astrophys. J. **160**, 831 (1970)
Searle, L., Sargent, W.L.W.: Astrophys. J. **173**, 25 (1972)
Schmidt, M.: Astrophys. J. **137**, 758 (1963a)
Schmidt, M.: Nature. **197**, 1040 (1963b)
Schmitt, J.: Nature. **218**, 663 (1968)
Schwarzschild, M.: Astron. J. **74**, 757 (1954)
Searle, L., Zinn, R.: Astrophys. J. **225**, 357 (1978)
Searle, L., Sargent, W.L.W., Bagnuolo, W.G.: Astrophys. J. **179**, 427 (1973)
Seyfert, C.: Astrophys. J. **97**, 28 (1943)
Silk, J.: Astrophys. J. **151**, 459 (1968)
Spitzer Jr., L., Jenkins, E.B.: Annu. Rev. Astron. Astrophys. **13**, 133 (1975)
Sunyaev, R.A., Zeldovich, Y.B.: Astron. Astrophys. **20**, 189 (1972)
Thaddeus, P.: Annu. Rev. Astron. Astrophys. **10**, 305 (1972)
Toomre, A.: Astrophys. J. **139**, 1271 (1964)
Toomre, A., Toomre, J.: Astrophys. J. **178**, 623 (1972)
Tremaine, S.: Int. Astron. Union. Symp. **117**, 547 (1987)
Truran, J.W., Cameron, A.G.W.: Astrophys. Space. Sci. **14**, 179 (1971)
Wagoner, R.V.: Astrophys. J. **179**, 343 (1973)
Zwicky, F.: Ergebnisse der exakten Naturwissenschaften, vol. 29, p. 344 (1956)

Chapter 6
1975–1984: Developments in Instruments

6.1 Planning for the Future

In 1982 the Field Report was published, a study in the US that aimed to look 10 years ahead to predict the most important developments and to advise on the major scientific equipment that would be needed.

The plan foresaw the need for some significant investments. First, it mentioned the need to be able to make observations at all wavelengths, from gamma rays to radio waves, an activity known as 'all-wavelength data gathering'. In order to make these observations, telescopes would be needed in space. The report stipulated that the most state-of-the-art computers would increasingly be used at all stages of research, from theory through observation to interpretation, and therefore had to be readily available. Then there was dark matter, the existence of which had been discovered because it exercised more gravity than visible matter. Two- and three-dimensional computer simulations of contracting and rotating clouds of gas would be of crucial importance for research on the formation of stars and planets. The discovery of explosions of X-ray and gamma radiation also warranted some priority.

The committee further recommended that a large X-ray satellite should be constructed (15 years later this was to become the Chandra satellite); its members argued in favour of the 'Very Long Baseline Array', a system of radio telescopes with telescopes further away from one another than the VLA telescopes but not as distant as their VLBI cousins. The recommendations also included a 15-meter telescope for optical research from the surface of the Earth and a 10-meter telescope in space that was eventually to become the 2.4-meter Hubble Space Telescope. The 15-meter telescope on the Earth's surface was to be constructed using new computer technology that was to be able to hold a very thin mirror in the right shape.

The conclusions from the Field report were taken seriously, not only in the US but in other countries too, and by organizations like ESO and ESA. Both European organizations developed their own plans for the future, with ESA constructing a

H. J. Habing, *The Birth of Modern Astronomy*, Historical & Cultural Astronomy,
https://doi.org/10.1007/978-3-319-99082-8_6

very rigorous programme for the next 20 years, 'Horizon 2000', which was adopted in 1980. Amazingly for a plan that was looking 20 years ahead, it was actually implemented! Competing communities of astronomers had faith in ESA, with the X-ray astronomers allowing the infrared astronomers to take precedence, fully confident that they would get what they wanted later. As indeed they did.

6.2 Research from Space

In the period from 1975 to 1984, 14 space telescopes were put into operation. Five telescopes were launched by NASA alone, five by NASA in collaboration with ESA, the UK and the Netherlands, two by ESA, one by the USSR and one by Japan. Three satellites were dedicated to gamma-ray astronomy, eight to X-ray astronomy, three to UV astronomy and one to infrared. This summing up shows that NASA was the most important player in the field, with the other countries following behind. The large number of X-ray telescopes was due to the successes of the rocket flights and of the UHURU satellite up to 1974. These successes had caused something of a gold-rush: X-rays were where every telescope could generate some rewarding surprises. The X-ray telescopes focused on the detection of more sources, but also on obtaining good maps, for example of supernova remnants, and on the variability of the sources. Important results were achieved at all wavelengths (gamma-ray, X-ray, UV, infrared) as the following chapters will show.

An exceptional survey was carried out by small infrared telescopes, that were mounted as piggybacks on classified rocket flights by the US Department of Defense. The observations yielded genuine astronomical data, although this data reached the general public in a somewhat curious fashion, as will be discussed later.

Spectral-line observations have always been enormously important for astronomy, because the strength and the line profile contain so much information. It is more difficult to detect them than broad-band radiation. In this period lines were detected for the first time at X-ray energies (Truemper et al. 1978) and at gamma-ray energies (Mazets and Golenetskii 1981). Figure 6.1 shows the detection of gamma-ray photons with an energy of 1808 keV, produced by radioactive aluminum, the isotope ^{26}Al. The isotope is produced during a supernova explosion and then ejected into interstellar space. The isotope lives long enough to be detected afterwards along the Milky Way.

In 1975, ELDO, the European organization that built launchers (or rockets), and ESRO that conducted scientific research, were merged to form the European Space Agency (ESA), which led via a complex route to Arianespace, a separate European organization that built and launched the successful Ariane 5 rockets. ESA had a broad programme of space activities (including meteorology and telecommunications) and in addition an obligatory, relatively small but financially stable programme of scientific research in and from space. An important side-effect of the European space efforts has been that it brought together scientists from all over Europe for specific tasks, such as the preparation of plans for new satellites and the

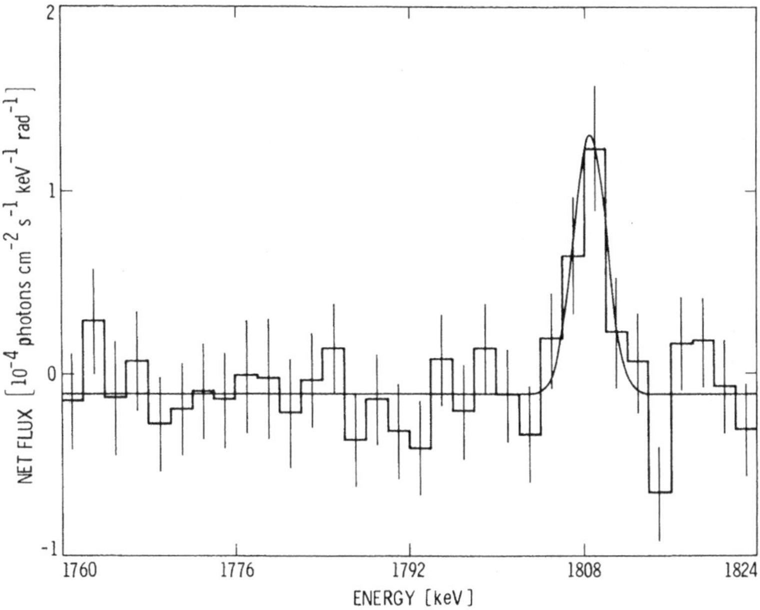

Fig. 6.1 Emission line emitted by radioactive atom ^{26}Al at 1808 keV. The image is made by adding up numerous observations along the galactic plane with NASA's HEAO-3 X- and gamma-ray telescope (Mahoney et al. 1984)

scientific supervision of activities during the long years of construction. ESA strongly enhanced inter-European collaborations as did ESO. The Soviet Union and Japan had their own programmes within which they conducted similar space research. This gave rise to competition but also to collaborations between all the players; those involving ESA and NASA were the most intensive.

6.3 The Growing Significance of Computers: Automation

Computers were becoming increasingly important. Besides improving and speeding up work that was already being done by computer, new and at times completely unexpected applications also emerged. Universities and observatories were homes to the 'mainframes', the huge machines where each component was the size of a large cupboard. Data were stored on magnetic tapes on 60-cm diameter reels. 'Portable' computers also became available. Through my direct involvement in the IRAS project (see later), I had the use of an 'Osborne' as early as 1982, a device in a briefcase as big as a sewing machine and just as heavy. I used the ARPAD network, a predecessor of internet, to contact California and I surprised friends and family with my stories of the wonders of the time, but in all honesty, the Osborne was very

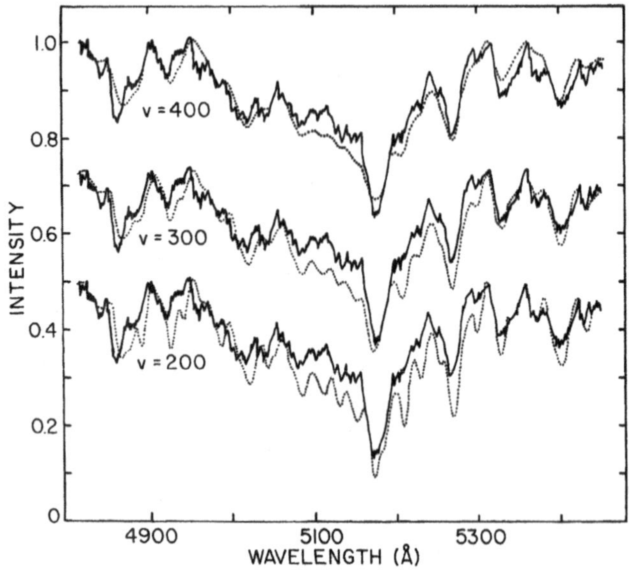

Fig. 6.2 A small part of a spectrum of the elliptical galaxy NGC 4472 is compared with three model spectra to see which gives the best result. The dark line is the actual measurement; the dotted line is from a model. This model is the spectrum of a red giant star (type K3III) but broadened according to a Gaussian profile, assuming that the velocities of the stars in the galaxy follow a Gaussian distribution. For different dispersions (400 km/s, 300 km/s, 200 km/s), the spectrum of all stars together appears as the dotted line. It is clear that a velocity dispersion of 400 km/s gives the best match between model and observations. This result is much more reliable than before, when comparisons had to be judged by eye (Faber and Jackson 1976)

limited. It had very little memory and was not advanced enough for sending and receiving images, manuscripts and reports.

The 'mainframes' were used for numerical simulations of astronomical processes and for processing the results of observations. Researchers began to work with images, resulting in early forms of 'photoshopping'. Their observations were subjected to complex processing, which had not been possible previously due to the length of time it would have taken. The findings of such a complex analysis were much more reliable than the results that could be obtained using a slide rule. Figure 6.2 contains an example of a computer analysis from 1976; it is one example of many.

6.4 Innovations in Optical Astronomy

For all the excitement that the new astronomies caused, additional optical observations frequently seemed to be of more decisive importance. Often many of the objects discovered with a radio or X-ray telescope could be interpreted only after successful optical observations had been made.

Telescopes

Telescopes that were under development in the previous period were now coming into operation. In 1976 the Southern European Observatory (ESO) 3.6-meter telescope finally came into use, the instrument for which Oort and Baade were the first to roll up their sleeves in 1957. In 1974, the 3.9-meter Anglo-Australian telescope became operational, and also in 1974 a 4-meter telescope on Cerro Tololo in Chile for AURA, a North American consortium of universities (see Chap. 4). In Arizona, a telescope was commissioned in 1979—the MMT, or 'Multi-Mirror Telescope'—comprising six mirrors, equivalent to a 4.5-meter telescope. A French-Canadian-Hawaiian consortium constructed a 3.6-meter telescope on Hawaii and the German Max Planck Company built a 3.5-meter telescope in Spain. With the newer detection methods and improved spectrographs, it was now possible for Americans, Europeans and Australians to conduct competitive research in all areas of astronomy.

ESO now had its 3.6-meter telescope, the main objective of the European collaboration, but in all honesty there were a lot of astronomers who regarded the design and implementation of this showpiece as strongly outdated. They had placed their hopes in ESO's plans for constructing a bigger 'New Technology Telescope'; in 1980 work began on a trial telescope with a mirror 3.6 meters in diameter. Chapter 1 mentioned that a telescope with a mirror larger than 5 meters could never be built because of the enormous weight of the glass. This objection fell by the wayside when it was discovered that the distortion of a lighter mirror could be compensated using magnetic rods under the mirror that were controlled by computer. The 'New Technology Telescope' was constructed as a prototype. It became a great success.

A mirror with its shape controlled by computer offered even more benefits. The air above a telescope is always in turbulent motion and, as has been said earlier, this means that an image of a star is never the actual star picture, but is always a small 'speck' that lasts less than a second, after which another speck appears. During normal observations the recorded image is the sum of all the specks. If, however, one continuously monitored the transparency of the atmosphere one might be able to distort the mirror and compensate for the irregularities in the path of the rays. This was the goal of 'active optics': the shape of the mirror was continuously adjusted to the atmospheric circumstances so that the 'reception' provided a sharp (or rather, a much sharper) image. The principle is explained in Fig. 6.3.

In 1977 a major conference took place at ESO on large telescopes, at which numerous new telescope plans were discussed. In 2013, the report of the conference seemed to have disappeared from several libraries, and this absence can be seen as an indication of the importance of the conference. The participants felt that optical astronomy was entering a golden age with telescopes 10 meters in diameter. It became apparent that Europe was playing its own independent role in this innovation, and that the initiative in telescope building was no longer exclusively in the hands of the US. ESO's 3.5-meter 'New Technology Telescope' was first used in 1989 and from the very start it met all the specifications, including that it should be considerably less expensive than the outdated 3.6-meter telescope had been and be constructed in much less time.

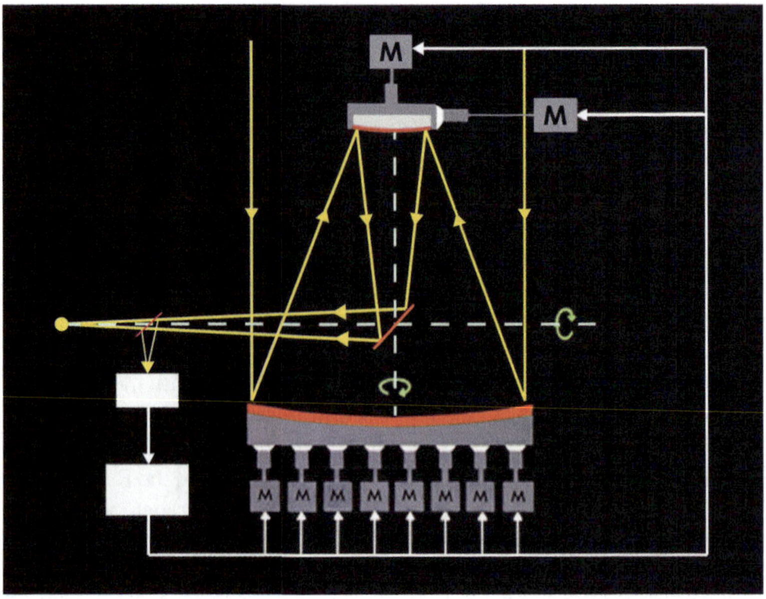

Fig. 6.3 Active optics: the light that falls on the mirror is ejected sideways via a second mirror and is then divided into two bundles: most light goes to the detector, but a small amount goes to a sensor. The sensor determines how the Earth's atmosphere has distorted the image and passes this information to a computer that works out how the mirror has to be manipulated by magnetic rods (labeled M) to compensate for the atmospheric error. Source: ESO

Detectors

Chapters 2 and 4 have already explained that the use of photographic plates had reached its limits. 'Photo-electric' techniques clearly held the future, but had one obvious shortcoming: you could observe only one 'pixel' of one object at a time. The big challenge was to measure a whole image in one go photo-electrically. In other words, a well-functioning TV camera was needed that could produce images in circumstances where there was only very weak light. Trials were made with different instruments, and with some success, as, for instance, in the case of Lallemand in France and Boksenberg in England.[1] Their experimental instruments worked well, at least as long as the scientist building the telescope was himself involved in the observations. Boksenberg's device attached to the 200-inch telescope on Mount Palomar as a visiting instrument made it possible for astronomers at Mount Palomar to make observations that were way ahead of their time, such as, for example, spectra of absorption lines in white dwarfs, lines that are very broad and weak. These measurements set a new standard for the quality of optical measurements.

[1] Boksenberg's instrument later became part of the European contribution to the first generation of instruments for the Hubble Space Telescope.

The actual replacement of the photographic plate came in around 1975 as a result of the invention of the CCD (charged couple device) plate. This is the same plate that has today replaced photographic film in all cameras. The detector plate is split into thousands of picture elements (pixels) and when a photon hits such a pixel it releases an electron. The electrons are regularly counted in an elegant and effective way. The CCD is much more sensitive and more precise than the photographic plate and makes it possible to see details that in the past were lost. The inventors of the CCD, George Smith and Willard Boyle, were awarded the Nobel Prize for Physics in 2009 for their discovery of this technique, an award that met with wide approval from astronomers.

More Precise Spectroscopy

In 1967, English astronomer R. Griffin devised a new instrument for measuring Doppler shifts. With hindsight, the idea was very simple, as good ideas often are. Take a star of which you want to measure the Doppler shift, then choose a star with a spectrum that is very similar to the spectrum you want to study. Place transparent photos of both spectra on top of one another and slide one of the two until you have the best overlap. How far you have to slide the photo indicates the difference in redshift. If you know the redshift of one of the two spectra, you also know the redshift of the other one. Griffin determined the position of the optimal shift by transmitting light through both spectra and establishing when most light was allowed to pass through. Using an instrument he built himself, he achieved a precision of ±1 km/s, a considerable improvement on what had been possible previously. His instrument naturally attracted worldwide attention. Jim Gunn at Mount Palomar invited Griffin to come to Pasadena so that they could together construct an equivalent instrument for the 200-inch telescope. The instrument was built and was even better. Doppler shifts were measured at a precision that was previously unimagined (Gunn and Griffin 1979). With further improvements in the technique and more usage of computers, 15 years later it was possible to detect the first exoplanet (see Chap. 14).

Multifiber Spectroscopy

When recording a spectrum, a narrow slit was placed in the light path and the light was then separated into wavelengths by, for example, a diffraction grating. If there was more than one object on the agenda of the programme, the objects had to be observed one after the other—a very time-consuming process. In the 1980s glass fibers became available: thin, flexible threads that directed light where you wanted it. This made it possible to direct the light from a number of objects (stars, galaxies) via fiber and thus to make images of spectra of different objects next to one another.

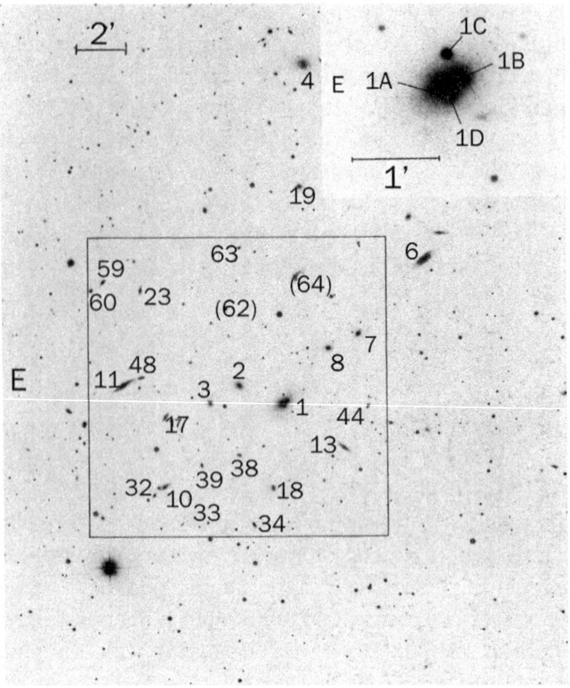

Fig. 6.4 A photographic plate of a cluster of galaxies taken with the AAT, the Anglo-Australian Telescope. The galaxies are indicated with numbers. The light from each numbered object was transported by fiber to a spectrograph, where all spectra were recorded separately and simultaneously (Ellis and Gray 1984)

It was now possible in one go to measure the spectra of many different objects. Following some predictable teething troubles, multifiber spectrographs became the standard at most observatories. Figure 6.4 shows a field/region with galaxies where the spectra of the numbered objects have been recorded in one go. The biggest advantage was the saving in time because the spectra obtained were not of better quality. This new technique made it possible in the 1990s to measure radial velocities of almost 400,000 galaxies with the Anglo-Australian telescope and it was also used in the even bigger 'Sloan Digital Sky Survey' in around 2000.

6.5 Radio Astronomy

Radio astronomical research developed largely in two distinct directions, one of which was measuring the continuous radiation of radio systems, supernova remnants and pulsars. The other was the measurement of radio spectral lines that gave insights into the interstellar matter in other galaxies and the formation of new stars from molecular clouds. As discussed in Chap. 4, aperture synthesis was the method adopted for obtaining sharp images in both these lines of research.

This first line of research benefited from the launch in 1980 of the 'Very Large Array', an aperture synthesis telescope, America's answer to the Ryles telescope in Cambridge and to the Dutch telescope in Westerbork. The VLA outperformed the Westerbork telescope in terms of its sensitivity (27 telescopes rather than 12) and its

Fig. 6.5 The 'Very Large Arrray' (VLA) viewed in an easterly direction. The separate telescopes have a diameter of 25 m. Source: NRAO/AUI, Dave Finley

ease of use, just as in 1970 Westerbork had outdone the Cambridge telescope in both these aspects. The first director of the VLA was Australian Ron Ekers, a professor in Groningen, who had for several years been one of the most skillful users of the Westerbork telescope. Radio astronomy and optical astronomy were now neck and neck (Fig. 6.5).

All these developments together made it possible in 1984 to detect radio sources that were 10,000 times weaker than sources in the so famous third Cambridge catalog (3C) from 1962.

A successful, and apparently financially modest development was the construction of MERLIN, a network of largely already existing radio antennae in the UK. MERLIN came into use in December 1980 and was able to make much sharper images of many interesting objects than had previously been possible.

The second research theme started in 1951 with the discovery of the 21-cm line of hydrogen, and between 1980 and 1984 expanded to include the study of molecules in dark clouds in our Milky Way and (gradually) into other galaxies.

Figure 6.6 shows where clouds with atomic hydrogen are found in the M101 galaxy. This image can be compared directly with a recording of the blue light of M101 in Fig. 6.7. The blue light comes from young and hot stars that were created a short time before in the hydrogen clouds.

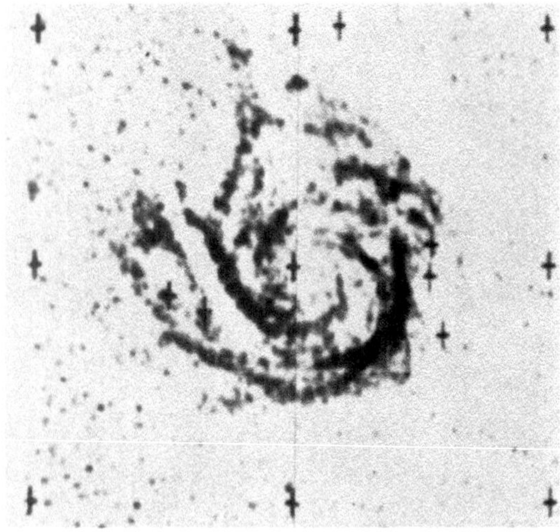

Fig. 6.6 The distribution of atomic hydrogen gas in the M101 spiral galaxies measured using the Westerbork telescope (Allen et al. 1973)

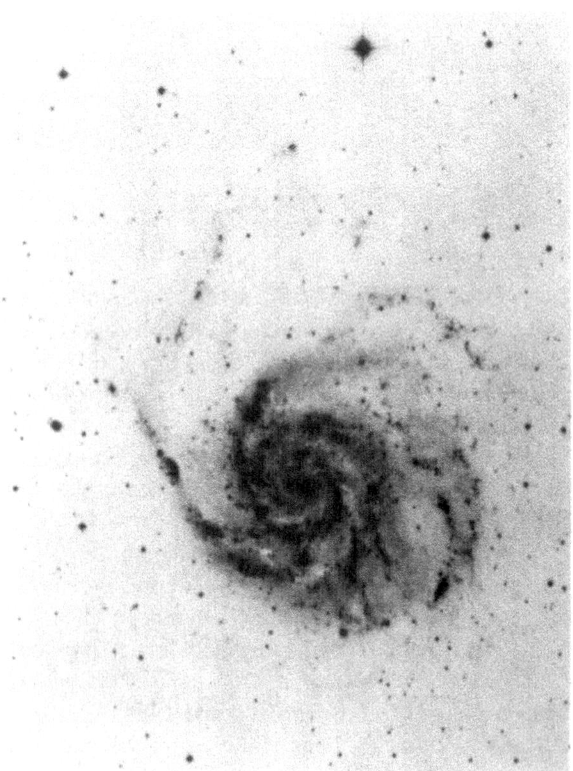

Fig. 6.7 Photo in blue light of the M101 spiral galaxies, from the Arps 'Atlas of Peculiar galaxies' (Allen et al. 1973)

Radio receivers could measure spectral lines with an accuracy that far exceeded that of optical spectroscopes. However, their tuning range was very limited and in most cases it was impossible to tune them to another spectral line: a new receiver had to be built. In addition, it was cutting-edge work to build receivers that worked at a wavelength of a few millimeters or less. A second problem was the reflection of the radio radiation to the focal point. The surface of the Dwingeloo telescope, for example, was made up of a coarse mesh with openings of several centimeters. This made the telescope less vulnerable to strong winds but also allowed radio waves of millimeter wavelengths to pass through the mesh without reflecting it. Millimeter spectroscopy called for new telescopes, and these gradually became available. I will mention a few of them. In 1973 a 20-meter telescope became operational at the radio observatory at Onsala near to Gothenburg (Sweden). In 1984, a German-French 30-meter telescope was commissioned in the mountains around Granada in Spain, and a 14-meter mirror was built for a collaborative effort by five colleges in Massachusetts.

Personally, I would award the first prize to Pat Thaddeus for a small mirror with a diameter of 1.2 meters on the roof of the physics lab of Columbia University in the heart of New York. Following the highly successful Leiden-Sydney chart of our Milky Way, he wanted to make a chart of the clouds of molecular hydrogen, H_2. As the H_2-molecule is such a poor radiator, he opted for CO as a 'proxy', a molecule with a suitable line at 3.5 mm. Thadeus reasoned that if a 7.5-meter radar antenna was good enough for that Leiden research at 21 cm, then a 1.2-meter telescope would be good enough at a wavelength of 3.5 mm. He then recalled that it freezes so hard in New York in the winter that the humidity of the air is practically zero—an ideal clear sky for a millimeter wave telescope. The research was so splendid that Thaddeus later left for Chile with a similar telescope to chart the southern hemisphere (Fig. 6.8).

Fig. 6.8 Thaddeus' 1.2 m radio telescope on a roof at Columbia University in New York. The skyline of New York with its famous buildings, including the Empire State Building is in the background. Source: Harvard CfA Millimeter-Wave Group

6.6 Infrared Astronomy

The quality of the infrared detectors improved steadily, although most of this progress was achieved in the laboratories paid for by the US Department of Defense (DoD). There was therefore a lot of secretiveness in contacts between American and non-American infrared astronomers. As far as the observations are concerned, optical telescopes were becoming better equipped for infrared observations, including such modifications as making the secondary mirror more mobile. In 1979, two telescopes were constructed for infrared research at an altitude of 4,200 meters on the extinct volcano Mauna Kea in Hawaii: a 3-meter telescope (IRTF) was built by NASA and a 3.8-meter (UKIRT) one by the United Kingdom. In 1977 some unexpected funding was made available in Wyoming to construct an infrared telescope with a diameter of 2.3 meters at 2,900 meters altitude on a high mountain top.

Aircraft, Balloons and Satellites

Figure 1 in the Prologue showed that the extra-terrestrial infrared radiation only penetrated through to the surface of the Earth via a limited number of 'windows'. Between 13 μm and 1 mm (=1000 μm) there were no more windows and the atmosphere closed off the heavens to us. To make observations at these wavelengths, the telescope had to be as high as possible above the surface of the Earth. Three types of 'carriers' were used: aircraft (12–14 km altitude), balloons (40 km) or satellites (200 km).

A telescope suspended from a balloon was an obvious option. The operations were costly but the investments were more modest. Observations were made during carefully planned campaigns from launch stations in isolated locations where a crash landing by a balloon would not cause any accidents. At first sight, the campaign seemed to be a great adventure, but the excitement soon began to wane. It was absolutely no fun staying for weeks on end in Palestine, a village in a boring part of Texas. You had to unload the telescope in parts from the boxes, assemble it and then conjure up the patience to wait for good weather and a good wind. You always came across balloon astronomers from another country but all you could do to alleviate the shared boredom was to indulge in secret drinking in what was a no-drinking region. When it was your turn, it was far from certain that the flight would be a scientific success. The camera might not open at the vital moment, or the basket would make a hard landing to put it euphemistically.

And then there was the option of using aircraft. Experiments with small planes showed that at 12–14 km altitude the atmospheric conditions were excellent. In May 1974 NASA started to use aircraft for these infrared observations that could only be made at a high altitude. Eventually a C-141A transport plane was converted into an observatory: a 36-inch (90 cm) telescope could look outwards through a hole in the fuselage. The aircraft was named the 'Kuiper Airborne Observatory' (KAO) after astronomer Gerard Kuiper, who had lobbied very hard for this airplane. In line with

Fig. 6.9 The Kuiper Airborne Observatory. To the left of the main wing a small rectangle can be seen: this was the opening through which the telescope looked outwards. Source: NASA

good NASA tradition, proposals for observations with the KAO were also accepted from outside the US (Fig. 6.9).

And finally, there were rockets. They achieved high altitudes but each flight had only a few minutes available for observations. When these flights were successful, they delivered a spectacular measurement but usually this provided little new knowledge. An exception was the AFGL catalog to be discussed a few lines down.

If you are wondering how much new scientific knowledge the satellites, balloons and aircraft generated, the answer today has to be 'very little lasting knowledge'. The observations raised intriguing questions but there were no possibilities for finding the answers: there were too few observations and they lasted for too short a time. In 1983 the IRAS satellite proved what all infrared astronomers already knew: astronomical research could only be rescued from a dead end if there was enough observation time. And this was only possible with satellites.

The Curious AFCRL Catalog

The above conclusion was confirmed by the curious history of the AFCRL/AFGL catalog. From around 1975, European astronomers became aware that there was a list, probably a catalog, on the desks of some of their American colleagues. This was quickly hidden away when a visitor came into the room or was explained (although never shown) to the visitor with some embarrassment. The catalog contained infrared sources measured during military rocket flights in several wavelength bandwidths between 5 and 30 μm and the embarrassment of the American astronomers showed the conflict between military secrecy and scientific openness. By selectively distributing the list of sources to non-military American astronomers, the (military) astronomers concerned (Stephen Price and Russ Walker) hoped to encourage further research. But they were disappointed in the reactions. A group from Arizona reported that they could only find 34 of the 281 sources. They wrote: 'The AFCRL catalog contains a large number of spurious sources.' Others were not so mild in their criticism.

Shortly afterwards a revised catalog appeared that was accessible to everyone, the AFGL catalog. The previous measurements had been analysed again, this time more carefully, and the result was more reliable. There were some questions about how complete the catalog was. It was more or less certain that there were large parts of the sky that the instrument had not observed. On the other hand, positive things could be said about the AFGL catalog. It included a number of new sources with unexpected and interesting characteristics. One of these was AFCRL 2688. The (optical) photo of the object showed an image that was not recognised. A broadband spectrum without details showed an energy distribution that had never been seen before. What was it? A new-born star or an almost defunct one? The object was to become known as the 'Egg nebula', and it continued to generate new insights for many years with every new piece of research. Ultimately, the object was proven to be not an infant star but a low-mass dying star, the pre-stage of a planetary nebula.

IRAS, the 'Infrared Astronomical Satellite'

On 25 January 1983, an American-Dutch-British satellite was launched to explore the whole of the sky at four wavelengths: 12, 25, 60, 100 μm. NASA had the telescope built and provided the 62 detectors; in the space laboratory at the University of Groningen a spectrograph was constructed that worked at between 8 and 22 μm. The large telescope, a mirror with a diameter of 57 cm, was made of beryllium. The mirror and detectors were placed in a 'dewar', a kind of thermos flask with helium that escaped slowly and would keep the temperature of the telescope a few degrees above zero over the period of a year. The software for directing the instrument was in Dutch hands. The ground station, the English contribution to the project, was near to Oxford, where the magnetic band was read twice a day with measurements and new commands were sent to the satellite. The heart of the control system was a large room full of electronics. The launch took place from the Vanderbilt base in California using a Thor-Delta rocket. It went without a hitch and the mission was more or less perfect. The data reduction had been demanded by NASA and the amount of manpower was several times greater than what the Netherlands could have contributed. The catalog contained more than 300,000 sources, the majority of which had previously been unknown.

Fig. 6.10 A drawing of the IRAS satellite above the Earth. The solar panels collected the electricity for powering the instruments and enabling the communication with the Earth. They also kept sunlight out. The entrance to the telescope can be seen at the top. Source: NASA/JPL

Right from its launch, IRAS was a success. When, after several days, the instruments were switched on, the team watching the incoming signals on monitors realized that IRAS would give us much more than even our most optimistic moments had foreseen. If this continued, the findings would be fantastic. And it did indeed continue for 10 months, around a month longer than planned. More than 30 years later, IRAS is still famous in the world of astronomy. It has laid the basis for a great deal of follow-up research and will be referred to in coming chapters (Fig. 6.10).

Box 6.1 IRAS, the InfraRed Astronomical Satellite
In the 1960s a working group was set up in Groningen for space research that gradually came to focus on infrared research using balloons. During a balloon flight, instruments were kept cool by putting them in a 'dewar', a kind of thermos flask that was supplied by Ball Brothers in Boulder, Colorado. In the mid-1970s, Ball Brothers pointed out to Van Duinen, head of the Groningen group, the possibility of an infrared satellite: their dewars were 'mission

proof'. No details were given, so Van Duinen guessed the dewars had probably been proven on military rockets. At that time, the Netherlands had successfully completed its first space mission with the Astronomical Netherlands Satellite (ANS). Van Duinen set to work and did some fierce lobbying, eventually gaining the approval of Prime Minister Lubbers. The Dutch mission was led by the Netherlands Institute for Space Research (NIVR). The decision was taken to ask again whether NASA was prepared to launch a Dutch satellite. Having had a positive experience with ANS, NASA found the idea appealing, and responded that the two parties could discuss cooperation but that the situation was more complex than when ANS had been launched. An infrared project was already underway in the US!

In the US, by 1975 a number of groups of infrared researchers had already been active for over 10 years. They had pioneered many new techniques, which had generated some interesting discoveries. The time was now ripe for an infrared satellite. The competition between the different groups was very fierce and there were also strong personality conflicts. When the letter from NIVR arrived at NASA, a scientific team had just been formed there comprising representatives of each of the American groups. NASA's response to NIVR suggesting that they could discuss cooperation was made with the proviso that the scientific supervision had to be in the hands of a team with an equal number of American and Dutch astronomers. On the Dutch side, the matter had also become more complex because the Brits were taking part as a 20% partner of the Netherlands. The UK, too, had plans for its own infrared satellite but had no funds to put the plans into action.

The partnership between the British and Dutch scientists proceeded smoothly. The technical and operational collaboration with NASA, NIVR and the British SERC was never easy but there was mutual respect and in general they did achieve some positive benefits. However, the collaboration of the European and American astronomers was blighted by tensions and annoyances. The American group was dominated by several very strong personalities, all of whom had impressive reputations—successful pioneers who were more than aware of their merits.

The Dutch members of the science team were astronomers with minimal experience in infrared research and even then only with balloon research. Around a year before the launch, Van Duin stood down as leader of the Dutch/British scientists and took up a good job in industry. It was with mixed feelings that the author of this book took over his leadership. Whatever our advantages may have been, the Dutch/British part of the team lacked researchers with the scientific stature of the Americans Frank Low, Gerry Neugebauer, Fred Gillett and Jim Houck. In particular Low, a Texan farmer, never missed an opportunity to draw attention to this difference. From the viewpoint of harmonious cooperation, the science team was a disaster, but it did eventually achieve its objective: it initiated an excellent, highly promising project, directed it competently and made a success of it (Fig. 6.11).

Fig. 6.11 The launch of IRAS was, as usual for NASA events, a major happening in the full gaze of publicity. On the morning of the launch, for 1 or 2 hours no attention was paid to the Science Team. Gerry Neugebauer, Jim Houck, Frank Low and I (HJH) went exploring: could we see the rocket? We could; in fact we came close enough to make a photograph. Five minutes later we were arrested: we had broken all the safety protocols. We were locked up in an office and it was only with great difficulty that we were freed in time to see the launch, this time from a safe distance. Photograph by Frank Low

6.7 The Far Ultraviolet

The FUV spectral region extends from approximately 90 to 330 nm. For the short wavelengths it connects to the area of soft X-rays, and on the long side it connects to the visible or optical part of the electromagnetic spectrum. All FUV radiation is absorbed by the Earth's atmosphere. It provides important additional information, particularly on the spectra of hot stars. In 1972 NASA launched a telescope named 'Copernicus', after the Polish/German astronomer who had been born 500 years earlier. This telescope made high-precision measurements of the profile of dark lines in the spectrum of hot, bright stars. These were lines that had been produced by gas clouds between the stars and us. Such lines had been known since 1900 in the visible part of the spectrum although a lot of information was missing and some measurement results were contradictory. The FUV was thought to be a much richer area for observation, which proved to be true. Among the most important results was the detection of lines of molecular hydrogen H_2. Another important conclusion was that the Sun was located in a bubble that is 30 light years in size and is filled with gas at a temperature of millions of degrees.

In 1974 NASA launched ANS, the Netherlands' first satellite (the Astronomical Netherlands Satellite), which orbited the Earth at an altitude of between 300 and 1,100 km. There were two instruments on board, a Dutch one for the FUV and an American one for X-ray observations. The FUV measurements, although they yielded no surprising results, were very useful. The satellite gained national prominence when a state secretary had the satellite switched off due to cutbacks at his ministry. The resulting protest from the scientific community was strong and effective, and the telescope was turned back on. The American instrument discovered bursts of X-ray radiation from globular clusters that contain only small, light stars. The discovery of these eruptions was unexpected. Two other items of news were the discovery of X-rays from the star Capella, the first 'normal' star outside of our solar system that emits X-rays, and an (X-ray) eruption of a 'flame star'.

American and European astronomers had access to the IUE, the 'International Ultraviolet Explorer', a space telescope for measuring spectra between 115 and 320 nm. The telescope had a diameter of 45 cm and was parked in a geostationary orbit at 30,000 and 40,000 km above the Earth. IUE was a 'real observatory'. An astronomer whose observation proposal was accepted would go to one of the ground stations and could then conduct his observations in 'real time', supervised by an expert 'operator'. The project was set up from England, with ESA joining later. IUE particularly made observations of hot stars, but there were also sightings of AGNs. The telescope was launched in January 1978 and was to operate for a period of 3 years. However, things turned out differently. The demand for observation time was so great and the telescope continued to function so well that it ultimately lasted until 1996 (18 years after the launch!) before being finally switched off, even though it was still working. At that time, I was a member of an advisory committee for ESA and I witnessed at first hand the disappointment of ESA officials who were constantly refused permission to pull the plug on the telescope, even though they so desperately needed the money for other, new projects.

6.8 X-Ray Radiation

X-ray telescopes were mentioned in Sect. 6.1 of this chapter. Two of these—NASA's EINSTEIN satellite (also known as HEAO-2) and ESA's EXOSAT—stand out because of the large number of publications they yielded (1200). Thousands of new X-ray sources were discovered with the Einstein satellite. After the release of the data, many specialists successfully searched the data for specific objects, both stellar objects (cataclysmic variables and white dwarfs, for example) and extragalactic objects (Seyfert systems, radio galaxies and clusters of galaxies). However, there was another important characteristic of X-ray sources, namely their variability. Pulsars were variable X-ray sources, but there were also transient and explosive sources ('bursters'). Variability was clearly an important property of X-ray sources. Usually satellites could not monitor a source long enough to detect its variability. ESA's EXOSAT satellite was therefore launched in a highly elliptical orbit around the Earth. This orbit ensured that the satellite could observe an object for several hours. Probably EXOSAT's most important discovery was made in 1984: oscillations caused by matter near a black hole (see Chap. 13).

Another important technical step forwards was the use of so-called Wolter telescopes that could project the image of an X-ray source on a detector. X-ray photons did not reflect on the surface of a mirror but passed through it. In the Wolter telescopes, the X-ray photons struck the reflective surface at a very small angle ('grazing incidence') and were then directed to the focal point. The telescope was made up of a large number of elongated, rather curved cones that guided the X-ray photons to the focal point. This technique was limited to X-ray photons with energy less than about 15 keV (Fig. 6.12). Photons of higher energy were no longer reflected at the grazing angle. One way or another, an alternative imaging technique had to be found for these photons. This was accomplished eventually, but not during this period. For the time being, researchers were limited to a single opening entrance of the detector and the limited angular resolution that that offered.

To illustrate, an image of the 'veil nebula' in Zwaan, a well-known and much photographed object that is considered to be the remnant of a supernova. Figure 6.13 shows the outline of the X-ray radiation drawn on an (optical) picture. The X-ray radiation reached the detectors in the Einstein satellite via a Wolter telescope.

6.9 Gamma Radiation

The first detection of gamma rays, originating from the vicinity of the center of the Milky Way, was mentioned in Chap. 4. They were measured using detectors attached to balloons. Cosmic gamma photons were rare, but enough were produced by cosmic objects to allow them to be detected.

Because the photons were so slow in coming, satellite measurements were preferred over balloon or rocket observations. Only satellites offered the possibility of collecting statistically enough gamma photons after long periods of staring at a

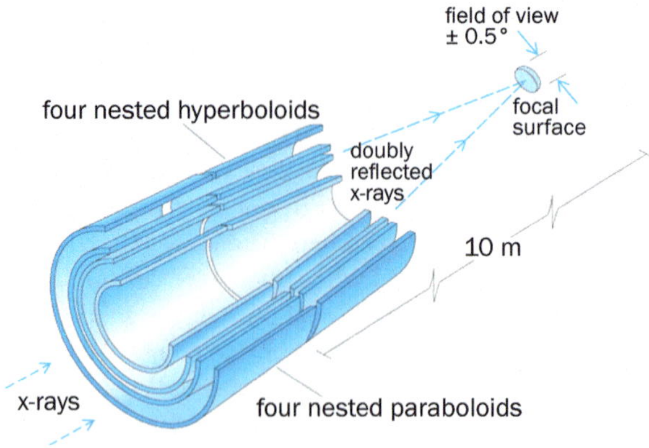

Fig. 6.12 Sketch of a 'Wolter' telescope that images an extended X-ray source in a focal plane. X-ray photons of less than about 15 keV are reflected if they strike the right material at a very small angle ('grazing incidence'). The Wolter telescope was born from this idea. The 'mirror' consists of a large number of layers of parabolic/hyperbolic surfaces that guide the photons from an area of half a degree in the sky to the focal point. The distance from the aperture of the telescope to the focal point is generally several meters. Source: D. Willinggale (2016)

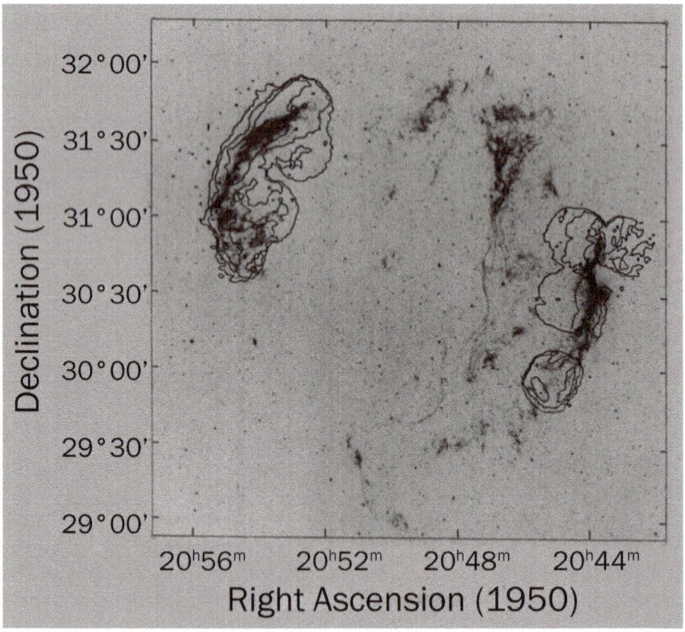

Fig. 6.13 A photographic image of the 'loop' nebula in the constellation Swan ('Cygnus loop') and superposed the contours of X-ray radiation with energy between 0.4 and 4 keV. This 'loop' is the remnant of a supernova and the image is obtained with the Wolter grazing-incidence reflector in the Einstein satellite (Ku et al. 1984)

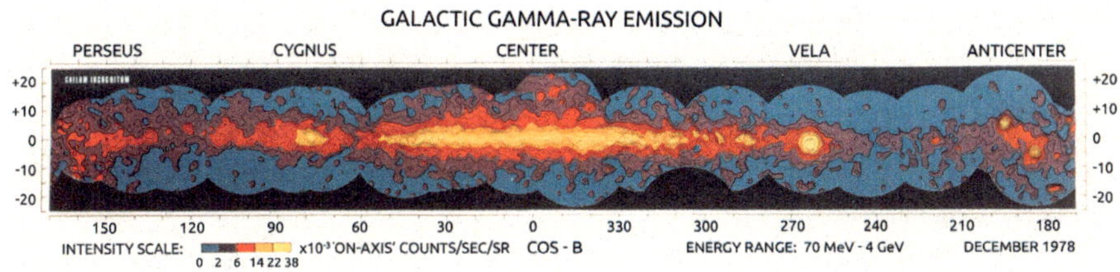

Fig. 6.14 The distribution of gamma radiation along the Milky Way as measured with COS-B. The horizontal scale is galactic longitude and the galactic latitude is shown vertically. The galactic center is in the middle. At a longitude of about 270 the supernova remnant in the constellation of Vela is a prominent feature. There are two sources between longitudes 180 and 200. The right-hand source is the Crab Nebula, the left, named Geminga, remained unidentified for two decades, but was then proven to be a gamma pulsar. Source: ESA

source. In November 1972, NASA launched the SAS-2 satellite for detection of gamma rays. however, it had to be aborted in June due to a technical failure. ESA had more luck. It launched its gamma satellite COS-B in 1975, which remained active far longer than planned, finally being decommissioned in 1982. From the measurements made, a map could be drawn of the Milky Way (see Fig. 6.14). It was a result of much patient waiting while nature took its slow course. The somewhat tongue-in-cheek comment has been made (and never denied by the parties involved) that the COS-B astronomers phoned each other when another gamma photon arrived.

Twenty-five discrete sources can be found in the map, of which the majority could not be identified with an object known at other wavelengths. The pulsar in the Crab nebula was present (see Fig. 6.14, coordinates 185, −5, far right) and, indeed, the gamma radiation appeared to be 'pulsed'. Diagonally and to the left above this was a source (Geminga, coordinates 195 + 4) that for a long time could not be identified. Eventually, in 1996, 20 years after the discovery of the source, it proved to be a pulsar. The quasar 3C273, the closest quasar, was also detected, but it is far from the Milky Way region and does not feature in the image (Fig. 6.15).

6.10 GEV and TEV Photons

Finally, mention should be made of the numerous efforts to detect photons with even higher energy than the photons that have just been discussed. This can be done with a telescope on the Earth's surface. Photons with very high energy are trapped in the upper atmosphere, where they produce many new particles that continue in almost the same direction. Ultimately there is a cascade of photons that reaches the Earth's surface: the effect is known as an 'air shower'. These secondary photons can be captured using conventional telescopes and the direction from which they originate can also be determined. For a long time, there were no suitable instruments for

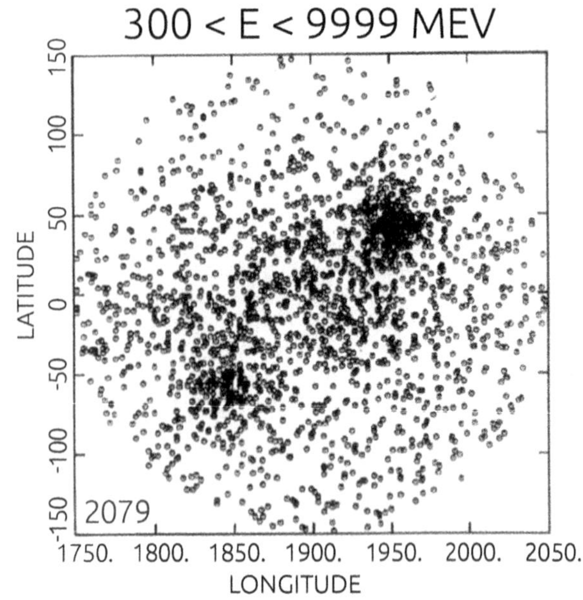

Fig. 6.15 The image shows a patch of sky in which each point is the location where a gamma photon struck the detectors of COS B. The coordinates are galactic longitude and latitude expressed in 0.1°. The photons mark two sources: the lower one is the Crab pulsar and the upper one is Geminga. Geminga was a mysterious source for over 20 years (Bignami and Caraveo 1996)

detecting ‘air showers’. It was in the period from 1975 to 1984 that several such instruments started delivering their first results. In 1983 there was a claim from two German nuclear physicists from Kiel, Samorski and Stamm, that the X-ray source Cyg X-3 emits photons with an energy of about 10^{16} eV. Others later confirmed these results (Samorski and Stamm 1983).

References

Allen, R.J., Goss, W.M., van Woerden, H.: Astron. Astrophys. **29**, 447–451 (1973)
Bignami, G.F., Caraveo, P.A.: Annu. Rev. Astron. Astrophys. **34**, 331 (1996)
Ellis, R.S., Gray, P.M.: Mon. Not. R. Astron. Soc. **206**, 285–292 (1984)
Faber, S.M., Jackson, R.E.: Astrophys. J. **204**, 668–683 (1976)
Gunn, J.E., Griffin, R.F.: Astron. J. **84**, 752 (1979)
Ku, W.H.-M., Kahn, S.M., et al.: Astrophys. J. **278**, 615–618 (1984)
Mahoney, W.A., et al.: Astrophys. J. **286**, 578–585 (1984)
Mazets, E.P., Golenetskii, S.V.: Astrophys. Space Sci. **75**, 47–81 (1981)
Samorski, M.; Stamm, W.: ApJ v268 pL17, 1983
Truemper, J., Pietsch, W., et al.: Astrophys. J. **219**, L105 (1978)
Willingale (2016): https://slideplayer.com/slide/5255371/

Chapter 7
1975–1984: Stars

7.1 Star Formation

Dark Clouds and Compaction

Chapter 4 discussed how molecules were discovered by radio telescopes in dark clouds, clouds that were subsequently referred to as molecular clouds. The molecule that was most prevalent, H_2, molecular hydrogen, was detectable under very special circumstances, such as in a compacted and warm layer between two colliding clouds. However, carbon monoxide, CO, was a very common substitute and has played the role of 'proxy' for H_2, ever since. It occurred in all clouds and even different 'isotopes' were present in measurable quantities. Besides the most common isotope, $^{12}C^{16}O$, the observers also found $^{12}C^{18}O$ or $^{13}C^{16}O$, each at slightly different wavelengths from $^{12}C^{16}O$. Different lines of these molecules were observed and the measurements confirmed what had long been suspected, namely that the temperature in the clouds is very low, up to just a few degrees above absolute zero.

An example of this wealth of molecules was the dark cloud TMC-1 ('Taurus Molecular Cloud #1') in which more than 20 different molecules were identified, from CO to CH_2CHCN. The chemistry of carbon in particular played an important role.

Many dark clouds were examined. Figure 7.1 shows a color photo of the Orion nebula and the surrounding area. Practically all nebulae and the brightest stars are at the same distance from us, roughly 1600 light years. Figure 7.2 shows the CO clouds that were detected; there is a close correspondence with the color photo. Figure 7.3 shows where groups ('associations') of young hot stars ('of spectral type O and B') occurred. The youngest group was in the densest cloud. The correlation between association and clouds decreased as time elapsed.

Research thus showed that clouds had an irregular structure with smaller and larger areas of higher density. Figure 7.4, below, shows a 'globule' (a small, very dense concentration with usually sharp boundaries) in a cloud close to a young star ρ Oph. This globule contains several different molecules.

H. J. Habing, *The Birth of Modern Astronomy*, Historical & Cultural Astronomy,
https://doi.org/10.1007/978-3-319-99082-8_7

Fig. 7.1 Image of the whole Orion region in the same (true) colors as seen by the naked eye. The stars have either an Arabic or a Latin name or carry a Greek letter. This variety shows that several different cultures named the stars. Top left is the red star Betelgeuse (α Orionis, abbreviated to: α Ori). To the right is the blue-white star Bellatrix (γ Ori). The bright blue-white star at the bottom is called: Saiph (κ Ori). In the center are three bright blue-white stars in a row that runs diagonally upwards to the right; from left to right: Alnitak (ζ Ori); Alnilam (ε Ori) and Mintaka (δ Ori). Some distance below Alnilam is the Orion nebula (M42) and the star θ Ori. The well-known horsehead nebula is close to Alnitak. Stellar names are also indicated in the figure below. The broad, red river in the left hand side of the diagram is glowing ionized hydrogen and carries the name 'Barnard's Loop'. Source: Astronomy Picture of the Day, 23 October 2010, Rogelio Bernal Andreo

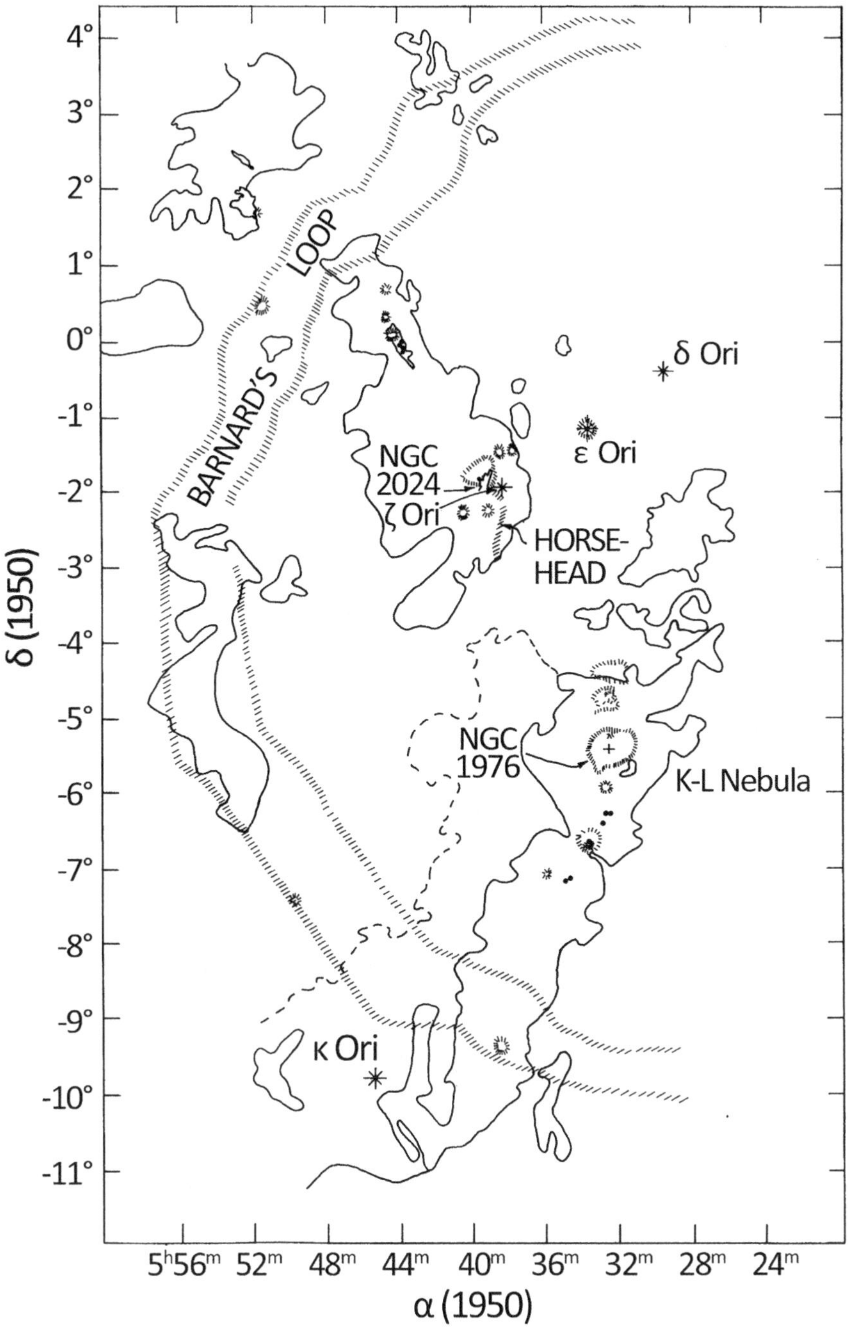

Fig. 7.2 The contours outline clouds of CO in the neighbourhood of the Orion nebula. CO is a tracer of molecules: when CO is present, H_2 and other molecules are present as well. Numbers identify substructures. NGC numbers indicate nebulae that emit hydrogen emission; NGC 1976 is the Orion nebula. Barnard's loop is a very long luminous structure. Objects with an identification starting with 'L' are dark clouds. The image can easily be compared with Fig. 7.1 above (Kutner et al. 1977)

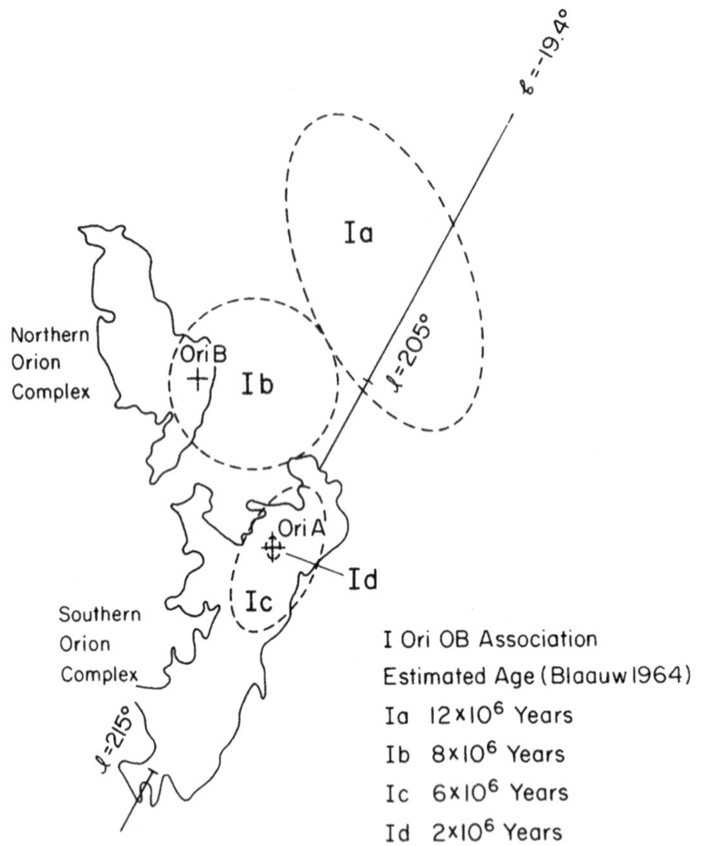

Fig. 7.3 Position of three young 'associations' of high-mass stars in Orion. Association 1a is the oldest and 1d the youngest. The conclusion is justified that the formation of a nest of young, high-mass and therefore bright and hot stars always creates a vacuum around itself. Orion A is another name for the Orion nebula. Orion B is a light emitting nebula near the horsehead. Compare with the two previous figures (Kutner et al. 1977)

Where did all these molecules come from? How were they formed? Chapter 4 already discussed how hydrogen molecules H_2 were formed: on a cold dust particle. But how were other molecules formed? There were two very different mechanisms. One was formation on dust particles, just like the H_2 molecule. The second possibility was that freely moving atoms and molecules in the gas collided and formed a compound. As has been said in Chap. 4, H_2 molecules could not be formed in this way, but other molecules could. This would occur all the more easily if particles from cosmic radiation slightly ionized the gas, which was a possibility.

For the time being it was impossible to choose between these two possible processes. But there was one issue on which there was agreement: deep in a molecular cloud the dust particles had to be surrounded by layers of ice made up of all kinds of different compounds. An image of a dust particle surrounded by ice layers is shown in Fig. 7.5.

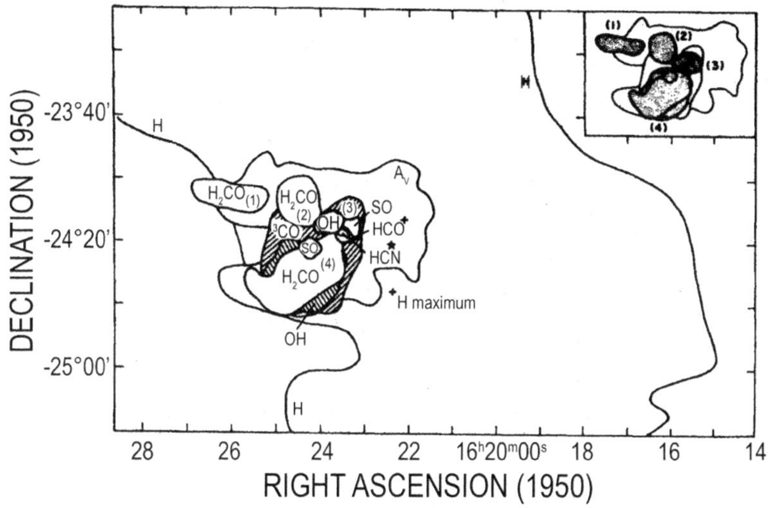

Fig. 7.4 The locations of various molecules in a compact cloud (a 'globule') close to the young star ρ Oph. New stars can be formed in such a globule (Myers et al. 1978)

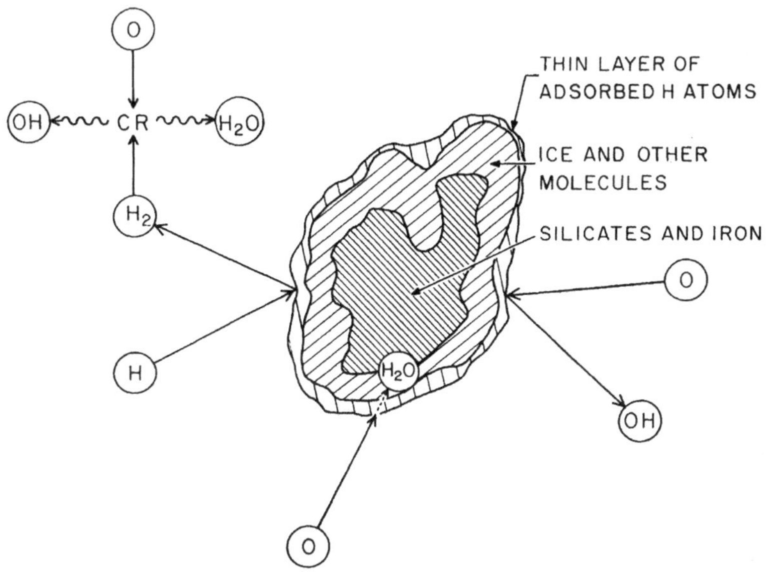

Fig. 7.5 A sketch of an interstellar dust particle inside a molecular cloud; the size of the grain is of the order of the wavelength of ultraviolet light. Silicate atoms (with perhaps some iron atoms) form the core of this particle. Various ice layers grow on the silicate core and there may be a thin layer of H atoms. The arrows indicate various processes that may occur on the surface of the particle; e.g. an O-atom may collide with the particle, stick to the surface, and then combine with an H atom to form an OH molecule. The molecule then leaves the particle. Another possibility is that of an O-atom that meets two H-atoms on the particle and forms an H_2O molecule. Complex reactions may occur when a particle from the cosmic rays ('CR') hits an H_2O molecule or an OH molecule (Field 1974)

In Chap. 5 astrochemistry was mentioned as a new specialism within astronomy and this new specialism too brought its own specialists with it, a repeat of the early history of radio astronomy but with other players. The corps of astronomy researchers was expanding.

Young Stars and Dark Clouds

Just as in Chaps. 2 and 4, stars will be called 'young' as long as they have not yet arrived at the main sequence. Systematic research of dark clouds produced infrared sources that were probably T Tau variables or HAeBe stars and that apparently lay within a cloud.

A study from 1979 by British/American astronomers Cohen and Kuhi had a slightly different approach. They looked at spectra of stars that coincided with a dark cloud and in which bright emission lines indicated the presence of a chromosphere. The chromosphere of the Sun is formed by 'magnetic activity' in the form of sunspots and solar flares. These young stars evidently have a higher degree of magnetic activity. Figure 7.6 shows an HR diagram of a number of young stars extended in two constellations, Taurus and Auriga.

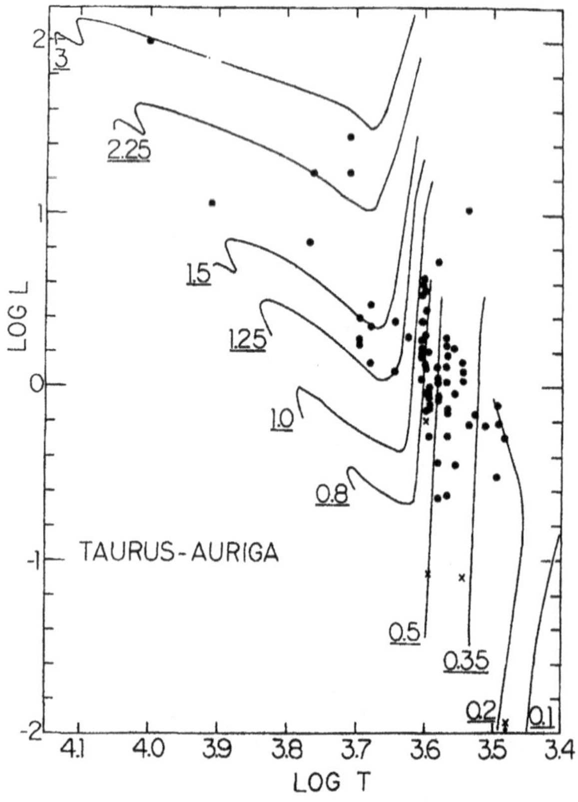

Fig. 7.6 A Hertzsprung-Russell diagram of young stars in the cloud in Taurus and Auriga. Paths are inscribed along which young stars migrate towards the main sequence. They migrate from the top of this diagram downwards and to the left. The paths are labelled with the mass of the star. At the end of its path, a star has reached the main sequence (Cohen and Kuhi 1979)

The magnetic activity of young stars led to the supposition that they can produce X-radiation. Radio astronomers were (and still are?) fascinated by other types of object and did not pay so much attention to young stars. French astronomers Montmerle, Koch-Miramond and Falgarone were looking for something else when they asked for observing time on NASA's Einstein X-ray satellite. With the help of an X-ray expert from Harvard, Grindlay, they discovered that the cloud near to the star ρ Oph was full of rather weak but well detected X-ray stars; there were at least 50 of them. The X-radiation was far from constant but it flared up briefly from time to time. Such behaviour confirmed the presence of a strong magnetic field and a strong chromosphere.

High Speeds, 'Jets' and Herbig-Haro Objects

In Chap. 5 the profile was discussed of a line of H_2S in the direction of the Orion nebula. It had the shape of a Gaussian curve, except that the bottom of the line was much too broad, indicating an excess of high-speed gas. The authors (Thaddeus 1972) mentioned this fact but paid no further attention to it. Interest was aroused in 1976 when a line of CO was measured very precisely in the direction of the Orion nebula and it deviated in the same way from a Gaussian curve: again there was locally a lot of gas at high speeds: this excess is only seen within a small area (radius of less than 1 arc minute) (see Fig. 7.7).

In another celestial object the measurement of the 2.6 mm line of CO showed that the CO gas occurred in two concentrations seemingly moving in opposite directions (see Fig. 7.8). Between the two parts, a cross is drawn that coincides with an infrared star. This faint object proved to be a very bright, hot star of spectral type B, behind a very dark part of the cloud that weakened the light of the star by around a hundred million times (20 magnitudes of extinction).

Figure 7.9 shows how the authors interpreted their measurements: from the central star, tenuous, elongated cavities stretch out that are fed by a stream of gas from the central star. Both figures also show several Herbig-Haro objects. They appear as small fragments in dense parts of the stellar wind. The drawing of Fig. 7.9 was somewhat daring: not all the details were explained, although later research confirmed the accuracy many times over. In 1980, and thus after some 30 years, an answer was found to the question of what these remarkable Herbig-Haro objects actually are: the end point of a jetstream expelled by a young star (Fig. 7.10).

Figure 7.9 brings to mind the radio galaxy Cyg A (see Chap. 6): a central object, a Milky Way galaxy, expels two jets of matter in opposite directions. The similarity in the images is striking but the energy and the size differences are dramatically large and there may well be two very different mechanisms that explain outflows from young stars and outflows from the nucleus of a galaxy.

A detail in Fig. 7.9 is the 'accretion disc' in the center. This rotating disk had not been observed directly; it was assumed to be present in order to explain the two jets that are expelled perpendicular to this disc; a magnetic field can help here: see Fig. 7.11 below. As said before, rotating discs had been a new kind of structure in astronomy but more and more such structures were found in various fields of research.

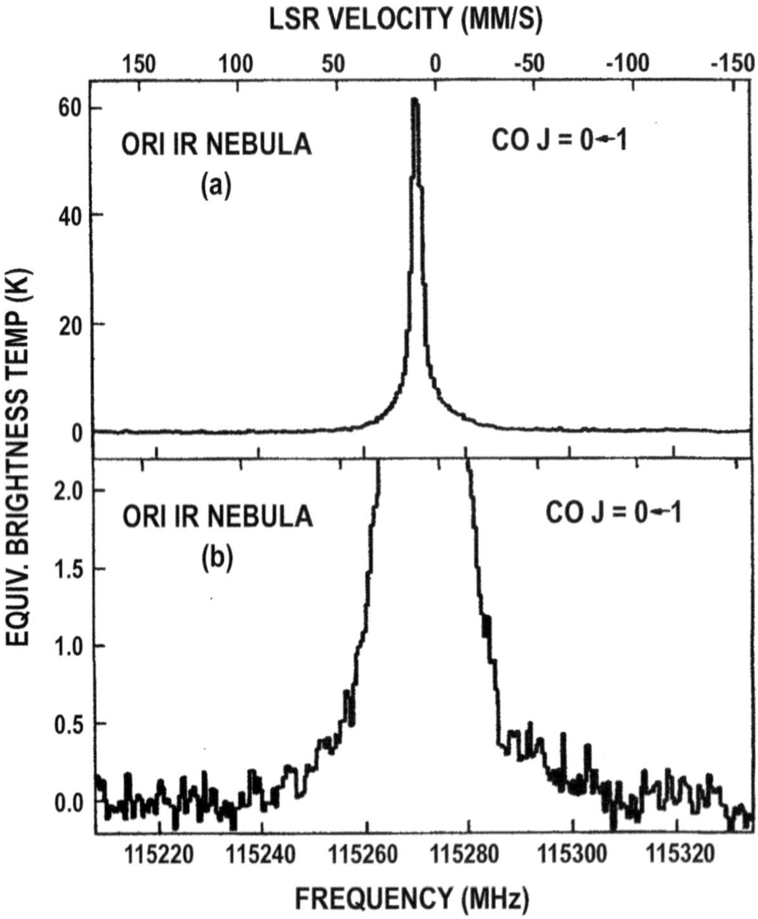

Fig. 7.7 The profile of the spectral line of CO at a wavelength of 2.6 mm in the direction of the Orion nebula. At first sight the profile appears similar to a Gaussian curve, which is what one would expect to find here. However, careful analysis shows that there is too much high-speed gas at the foot of the profile. This indicates an outflow of gas directly from the area of star formation (Zuckerman et al. 1976)

Finally, one more finding relating to Herbig-Haro objects. In 1981, Americans Herbig and Jones published the results of a study lasting many years. It concerned the motion across the sky of two groups of HH objects. The two groups appeared to move away from one another (see Fig. 7.12). In the image the position of a star is shown that had been suggested as the source of these HH objects, but this was rather unlikely: both groups of HH objects migrated at the same speed; one would expect to find the source half-way between both groups. And indeed, in 1985 Pravdo et al. identified a radio source precisely halfway between the two HH-groups (Pravdo et al. 1985). This radio source met all the requirements for being the source of the HH objects.

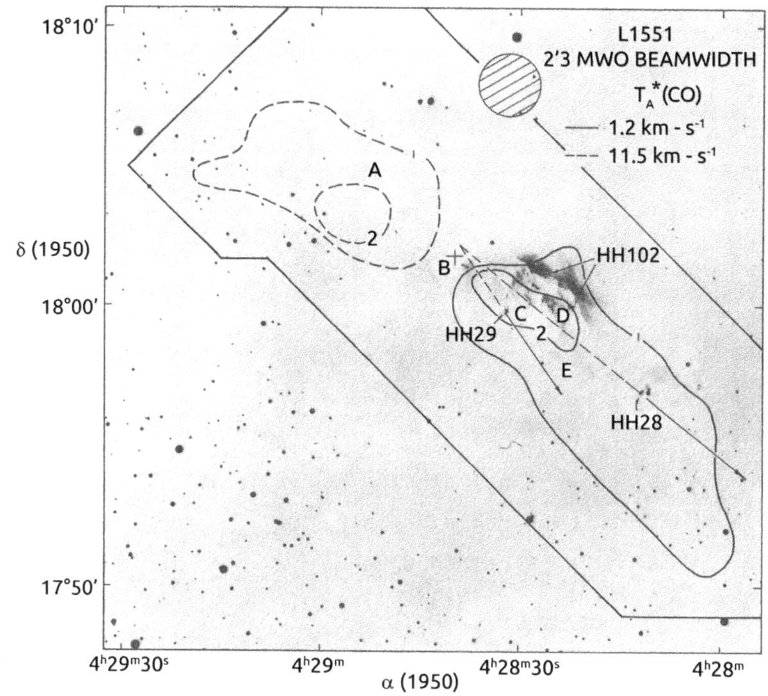

Fig. 7.8 An optical photograph of part of a dark cloud in Taurus. The continuous curve indicates the presence of CO at a velocity of 1.2 km/s and the broken curve the presence of CO at a velocity of 11.5 km/s. Several Herbig-Haro objects are indicated as HH129 etc. The cross with the letter B is the location of an intrinsically very bright star that has been weakened by strong absorption within the cloud (Snell et al. 1980)

Disks of 'Dust' Around Young Stars: Remains of a Planet-Forming Process?

And now, like Monty Python, 'something completely different'. In 1983, the IRAS satellite measured infrared radiation of hundreds of thousand objects. To check the consistency of the results the data on several well-known stars were compared: was the strength of the measured flux of radiation as one should expect? Indeed, for all selected stars the IRAS measurement agreed with what one knew, but there was one exception: IRAS detected a much stronger signal than expected of the star, Vega, α Lyra, the brightest star in the constellation of Lyra and one of the brightest stars in the sky. George Aumann and Fred Gillett (1984), who made this discovery, came to the conclusion that Vega was surrounded by a shell of dust particles. Vega was a relatively young star and the presence of such a shell was interpreted as a remnant from the time that the star was born. Further analysis of the IRAS data showed three more stars with the same feature. One of these was β Pictoris. A few

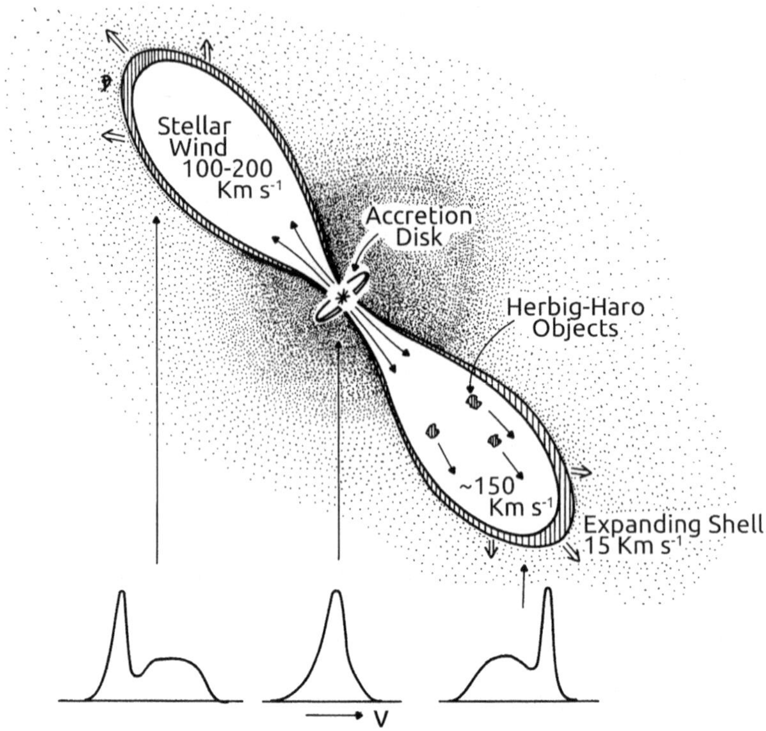

Fig. 7.9 A model that in essence explains the measurements in Fig. 7.8. The model manifests itself in three line profiles of the CO 2.6 mm line at the bottom of the figure (Snell et al. 1980)

months later two American astronomers, Smith and Terrile (1984), took a photograph of β Pictoris, in which they managed to block the light emitted by the star. Figure 7.13 shows the result: the dust is concentrated in a thin disk around the star. It was assumed that the same was the case for the other three stars. One is reminded of a theory, formulated for the first time in 1755 by the famous philosopher Immanuel Kant, who explained why all the planets orbit the Sun in (almost) the same plane: they were born in a thin disk of dust that had formed around the Sun during its formation. Vega and the three other stars were young enough to still have a remnant of such a planet-forming disk! For the first time there was a clear indication that other stars carried planets as well. The discovery of the infrared excess of Vega has become one of the most important discoveries made with IRAS. And it was a totally unexpected result.

Fig. 7.10 A photo of the area around the Herbig-Haro object 102. This is certainly a complex structure. Several red 'bullets' are present. The object that dominates the image is extended in the direction of lower left to upper right and can also be found in Fig. 7.8 where it is shown extending from the top right downwards. Source: Astronomy Picture of the Day, 8 December 2011, Adam Block, Mt. Lemmon SkyCenter, University of Arizona

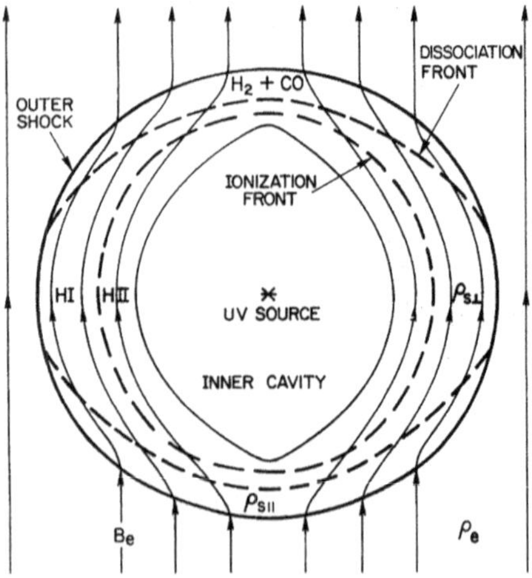

Fig. 7.11 The possible structure at the heart of the central machine in Fig. 7.9. A young star is created in a magnetic field. The star ('UV source') expels matter in all directions but a magnetic field collimates this matter to two jets in opposite directions (Koenigl 1982)

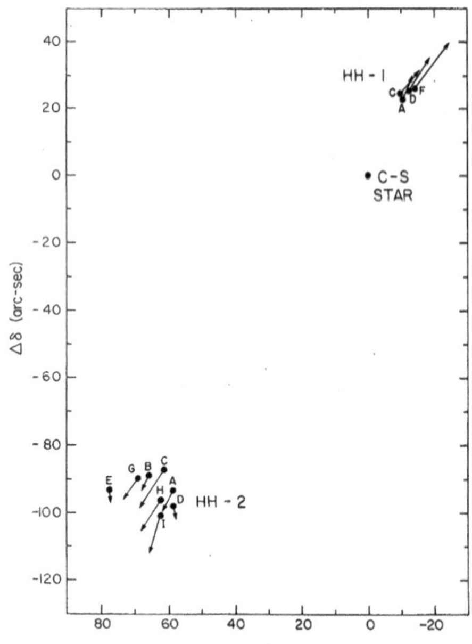

Fig. 7.12 The image shows the displacement of two groups of HH objects over the course of several years. The two groups are moving away from one another. Close to the upper/top group there is a star that was suggested as the source of the HH objects. Later research showed that a radio source between the two groups was the correct identification (Herbig and Jones 1981)

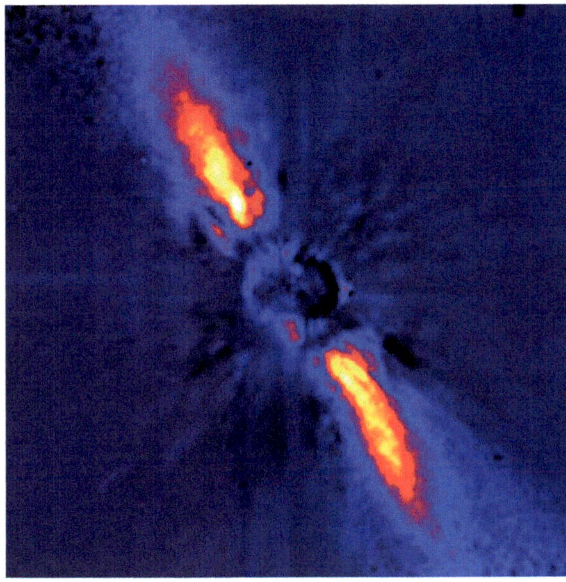

Fig. 7.13 Disk of dust around β Pictoris seen edge-on; a recording with an ESO telescope. The star itself is in the middle of the two parts, but its light has been blocked during the exposure in order to show the very weak emission by the disk. Source: Astronomy Picture of the Day, 28 November 1997, J.-L. Beuzit et al., ESO

7.2 Stars on the Main Sequence

The stars on the main sequence in the HR diagram represent the adult stage of stellar life. They burn hydrogen in or close to their center, and are stable for the longest part of their life. Stars of low mass stay on the main sequence for a time longer that the Universe exists (billions of years), stars with a large mass stay only for a short while (millions of years).

A new discovery was the detection of X-radiation from these main-sequence stars. A few of them that were close to the Sun had already been detected in the years 1974–1976 through rockets and the ANS satellite. A few years later NASA's much more sensitive Einstein satellite produced Fig. 7.14: an HR diagram containing those stars with X-radiation. They covered all the areas in the HR diagram. The authors concluded that, as a general rule, all stars emitted X-radiation, but at the same time it was clear that the energy emitted as X-radiation was always less than 1% of the energy that is emitted as light. This was contrary to the X-ray binary stars that we will come across later, where the X-radiation was much stronger than the energy of the light emitted.

Light Stars on the Main Sequence: Activity

All stars have an atmosphere where the energy coming from below was emitted into space. The part of the atmosphere where practically all the radiation came from was called the 'photosphere'. Observations during solar eclipses showed that the solar

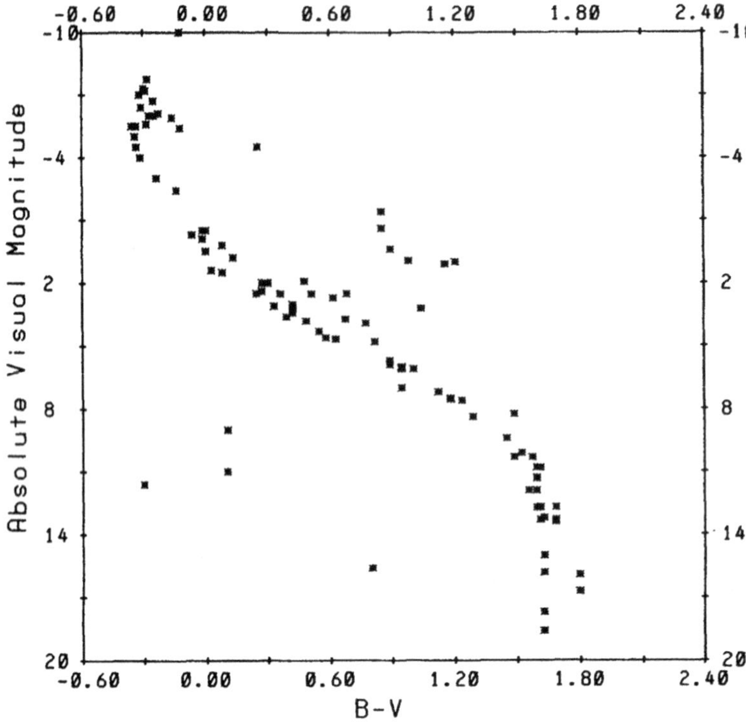

Fig. 7.14 The Hertzsprung-Russell diagram of stars of which X-radiation has been detected by the Einstein satellite. The x-axis runs from blue stars (B-V = –0.6) to red stars (B-V = 1.00). The y-axis gives the 'absolute magnitude', the magnitude a star would have if it were put at a distance of 10 pc. Apparently, all kinds of stars in the HR diagram emit X-radiation: main-sequence stars, red giants and white dwarfs (Vaiana et al. 1981)

atmosphere extended beyond the photosphere. There was a hotter, thinner and more rarified layer, the chromosphere (so called because it was colorful). Also during eclipses a huge 'corona' could be seen, outside of the chromosphere, where the gas was still thinner and hotter. From centuries of counts of the number of sunspots it had been remarked that the Sun had a variable 'activity': the corona was never the same as in the previous eclipse, and there could be periods with more and with fewer solar flares. Most important: there was a cycle of around 11 years with the activity fluctuating up and down.

Other stars were so distant that stellar spots and chromospheres showed themselves through minute details. A chromosphere, for example, was detectable from details in the stellar spectrum. A significant detail was mentioned in Chap. 4: a slight emission line in the deepest point in the two dark (absorption) lines of calcium; hot gas in the chromosphere produced this slight line. The line occurred in the spectra of stars with the mass comparable to that of the Sun (spectral types F, G, K).

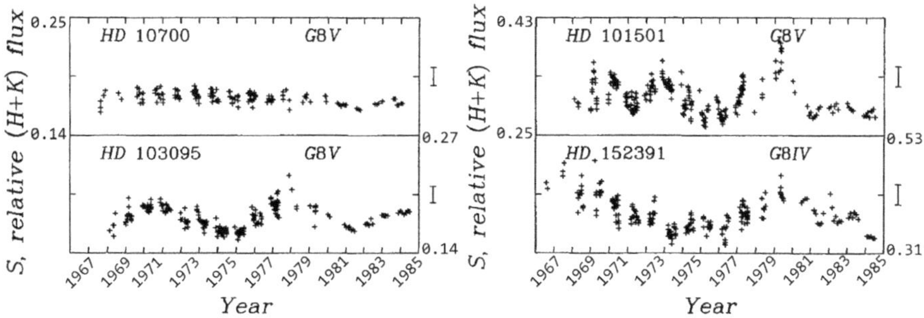

Fig. 7.15 For many years the strength has been measured of an emission line of the chromosphere in stars comparable with the Sun. The variation of the line strength is explained by varying 'activity' of the star (Baliunas and Vaughan 1985)

In this period, 1975–1984, long series of observations of a group of light stars proved that in each star the chromosphere line varied in strength slowly over the course of a few years; see Fig. 7.15. These stars had an activity cycle just like the Sun. Even more surprising was the discovery that the strength of the line (and of the chromosphere) depended on the predicted depth of the 'convection' under the stellar surface. Convection has already been discussed in Chap. 4: in the outermost part of the Sun strong flows of gas were present that transported hot gas outwards from deep inside and simultaneously an equally strong flow of cooled gas moved downwards. According to the theory, stars just as heavy as the Sun or lighter must all have such a convection layer. The variable chromospheric line confirmed this expectation (Noyes et al. 1984). The conclusion supported an already long-existing suspicion: convection and rotation give a light star a magnetic field and it was the magnetic field that produced chromosphere and corona and determined the solar activity.

This small detail proved for the n-th time that the Sun was a very ordinary star. Nonetheless there was one property in which the Sun was still unique: the Sun had planets around it and that had never been seen around another star. A unique property of the Sun until the detection of 'exoplanets' in 1995!

To explain the activity of other stars from one spectroscopic detail was a daring move. Fortunately, support was forthcoming from another source: X-radiation. The corona is the most important source of X-ray radiation from the Sun. Stars that exhibited 'activity' like the Sun could therefore also have a corona. Measurements with the Einstein X-ray satellite also confirmed this suspicion and even showed that the X-radiation was stronger in stars with a rapid speed of rotation. Figure 7.16 shows this.

The Expulsion of Matter by High-Mass Stars

In the previous period cameras built into satellites had shown that even during the quiet period of their existence, the main-sequence stage, high-mass stars expelled a great deal of matter (see Chap. 4). These results were confirmed by measurements

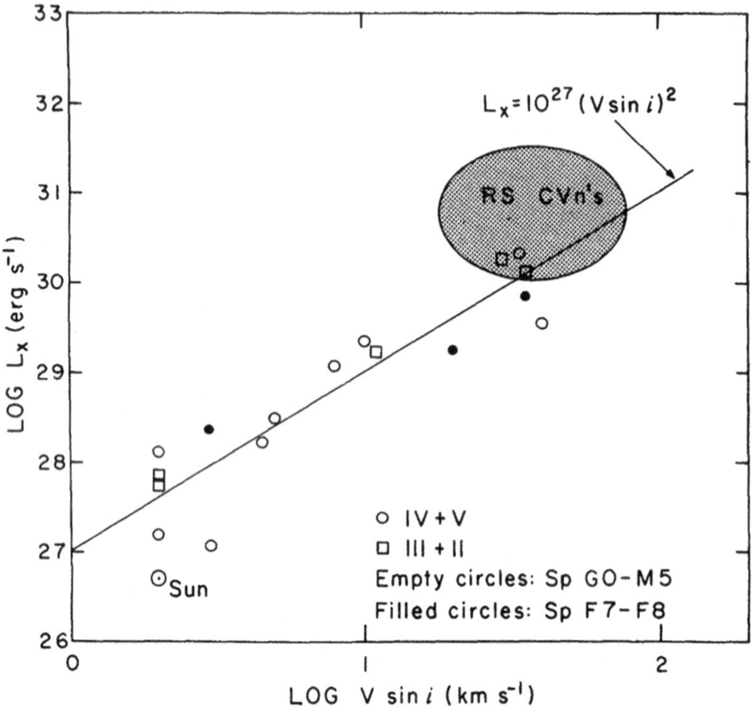

Fig. 7.16 Diagram of the X-radiation of cold stars measured by NASA's Einstein satellite. The horizontal axis indicates the speed of rotation of the stars and the vertical axis the amount of X-radiation, emitted by the star. The ellipsis with 'RS CVn' refers to a special category of binary stars, which we will not address further here (Pallavicini and Golub 1981)

with the Copernicus satellite and with the IUE, both discussed in Chap. 6. The profiles of the dark lines in the spectrum of hot stars also showed that much matter was expelled at high speeds: 1000–4000 km/s. The amount of expelled matter was estimated at one solar mass in a million years. This was such a huge amount that it had to determine the further evolution of the stars. Figure 7.17 shows several of these profiles.

Soon a good quantitative theory appeared that explained the high-speed flow (Castor et al. 1975). The atoms absorbed photons that corresponded with a spectral line and then emitted that photon again later. The photons absorbed all came from the direction of the stellar surface, the photons re-emitted left in all directions. The nett effect was that the atoms experienced light pressure that was directed outwards from the star. The effect of a single spectral line was too weak to explain the flow but there were so many spectral lines that together they produced enough force to explain the high exit velocity.

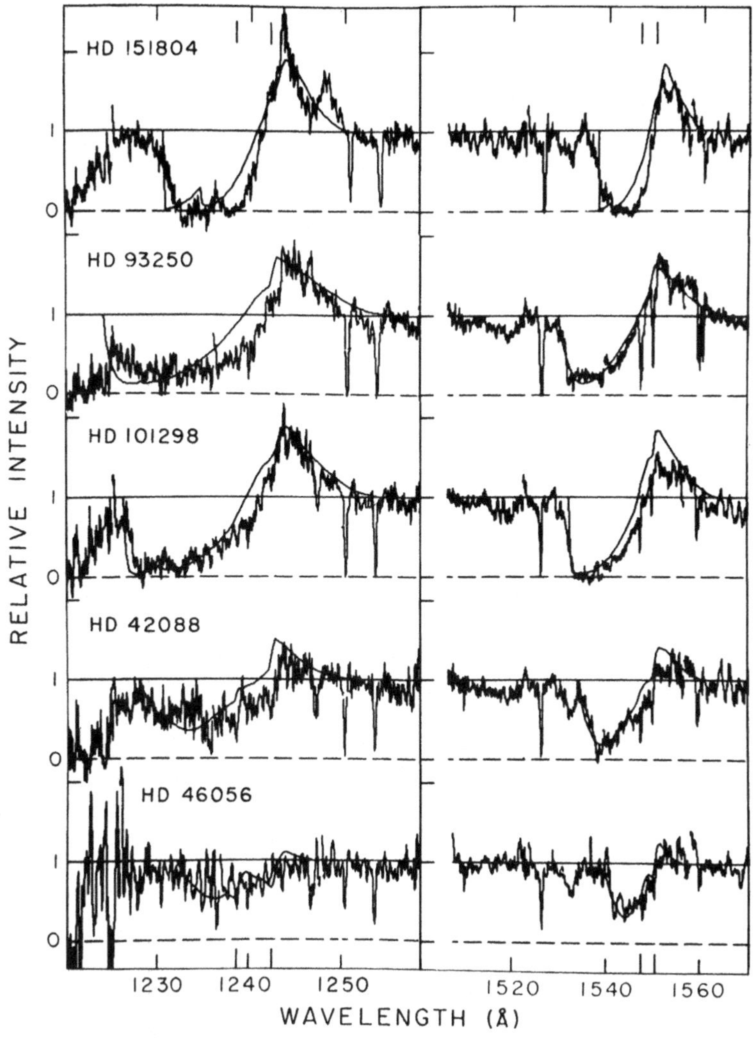

Fig. 7.17 Profiles of spectral lines in five high-mass main-sequence stars measured with the IUE. In each diagram the x-axis contains the wavelength and the strength of the received flux is in the y-direction. In the right hand part the spectral lines were produced by threefold ionised carbon atoms, C^{3+} or CIV. Each profile has a peak to the right and a valley to the left. The peak is caused by atoms closer to the surface of the star, the valley by atoms that are more distant from the star and that move in our direction at great speed. They are farther from the star and colder and therefore they absorb the blue (short-wavelength) side of the line emitted by the hotter atoms. The left hand part of the diagram contains more complex line profiles from quadruple ionised nitrogen atoms, N^{4+} or NV (Garmany et al. 1981)

Blowing Bubbles

The NGC 7635 nebula was a particularly beautiful object that had been known to observers for a long time (see Fig. 7.18). It was difficult to avoid concluding that this was a soap bubble and the eventual explanation went somewhat in that direction.

The bright and high-mass main-sequence stars expel so much matter and at such high speeds that if the matter collided with external matter, this matter was swept along with it outwards. Calculations in 1977 (Weaver et al. 1977) showed that these stars can blow bubbles very comparable to those of a supernova. A supernova injected an enormous amount of energy and impulse in the gas expelled and in a very short space of time, but calculated over the lifetime of a high-mass, hot star, the hot star injected a comparable amount of energy and impulse. An important effect was that in the environment of a group of hot, high-mass stars the distribution of the interstellar matter looked like a piece of Swiss cheese in which holes abound (see Fig. 7.19).

Fig. 7.18 The soap bubble nebula: a thin bubble blown in the surrounding gas by a young, massive star. Source: Astronomy Picture of the Day, 2 September 2010, Dave Jurasevich

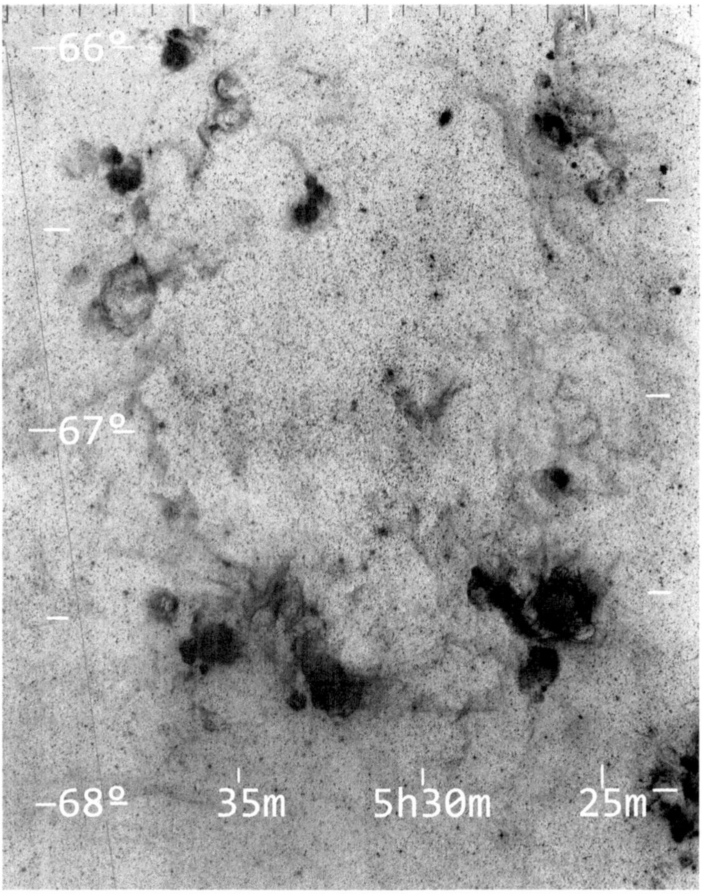

Fig. 7.19 A photo of part of the Great Magellanic Cloud made in the light of the Hα spectral line. Twisting structures can be seen that have been produced by the destructive forces of stellar winds coming from high-mass main-sequence stars (Goudis and Maeburn 1978)

The Most Massive Stars and the Wolf-Rayet Stars

People always are interested in what is the most outstanding, whether it's the number of 'likes' or a sporting record. The same applies to the heaviest and brightest stars. Figure 7.20 shows the Hertzsprung diagram of very bright stars in our Milky Way. Horizontally is the logarithm of the surface temperature of a star. The 'bolometric magnitude', M_{bol}, is shown vertically. This gives the logarithm of all radiation that a star emits in comparison with the radiation of the Sun. In the right-hand half of the diagram the highest value is $M_{bol} = -10$, which means that the star

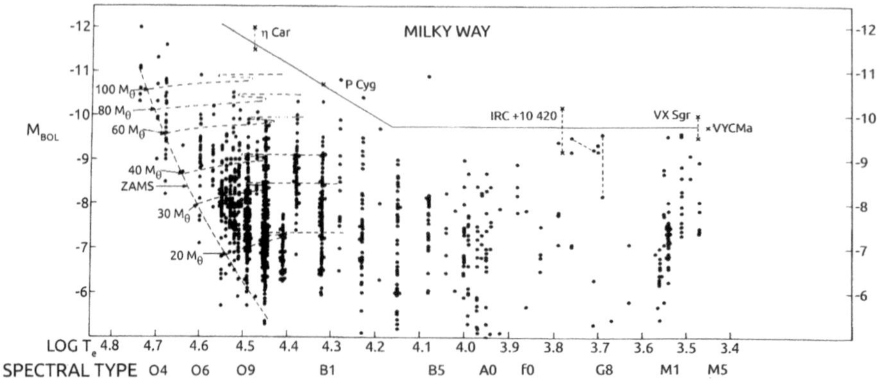

Fig. 7.20 The Hertzsprung-Russell diagram of the (intrinsically) brightest stars in the Milky Way. The dotted line to the left is the main sequence (ZAMS: zero-age, main-sequence); the main-sequence positions of stars that are 20, 30 to 100 times the mass of the Sun are indicated. The two sections of a straight line that meet at the center top apparently form a natural limit for the brightest stars in our Milky Way galaxy. The two stars that project just above this border are famous examples of extremely bright stars. One of these is the star η Carina that emits around five million times more energy than the Sun (Humphreys and Davidson 1979)

emitted a million times as much radiation as the Sun. (For the mathematicians: the bolometric magnitude of the Sun is around +5. A star with $M_{bol} = -10$ is therefore 15 magnitudes brighter; 5 magnitudes difference corresponds with 10^2 times more light, and 15 magnitudes corresponds with to $10^{3x2} = 1$ million). In the left-hand half, the limit goes up to $M_{bol} = -12$, which corresponds with five million times all the light of the Sun. The broken straight line in the diagram is referred to as the Humphreys-Davidson limit. Everyone would like to understand what causes this limit, but no successful explanation has yet been found.

Intrinsically luminous and hot stars have been discussed in Chap. 4, and among them the remarkable object η Carinae. In the course of many years the insight grew of a the existence of a small group of high-mass stars with comparable properties, while each was unique in some aspects. In 1984, at a symposium, Peter Conti (Conti 1984) named the group 'Luminous Blue Variables' or LBVs; it was a name that stuck. In the diagram of Humphreys-Davidson LBVs are close to the two straight line segments that mark the maximum luminosity.

An interesting discussion arose in around 1980 about R136, the brightest star in the 30 Doradus nebula in the 'Large Magellanic Cloud'. Was it an exceptionally bright single star or was it a compact cluster of hot stars? The object was studied with the best available imaging techniques but no satisfying answer was found. There was a proposal by bona-fide astronomers that this was a single star that was 500–1000 times heavier than the Sun and unique in the Milky Way. The discussion lingered on for several years. Finally, in 1995/1996 the nut was cracked; see Chap. 10.

X-Radiation from High-Mass Stars

There was no convection below the surface of a high-mass star, but there was in light stars. Without convection there is no magnetic activity, no corona and thus there should be no X-radiation—see above. But there was X-radiation, see Fig. 7.14, above and Fig. 7.21.

In the period from 1975 to 1984 no good explanation was found for the X-radiation from high-mass stars. Observations showed that the X-radiation became stronger when the star produced more light. A conceivable source of the X-radiation lay in the stellar wind that left the star at high speed and could collide with stationary matter. This would produce X-radiation. But this was all speculation.

7.3 The End and what Remains

In Chap. 4 the hypothesis was discussed that light stars could reach their end calmly and then leave behind a white dwarf of helium or carbon. High-mass stars should explode as supernovae and leave behind a neutron star or a black hole. This reasoning was confirmed and further explained in the period 1975–1984.

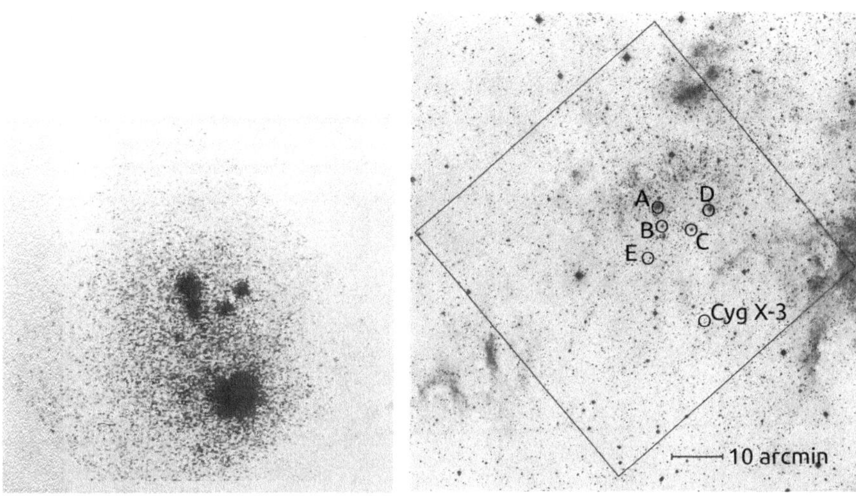

Fig. 7.21 To the right a photo of a section of the sky around the Cyg OB2 association of high-mass stars. A few stars are circled. To the left is an image obtained with X-radiation detected by NASA's Einstein satellite. The four stars ABCD from the right-hand image appear to be X-ray sources. The strongest X-ray source, in the left-hand image below, is a well-known binary X-ray star, Cygnus X-3, comprising a high-mass star and a neutron star (Harnden et al. 1979)

Disintegrating Low-Mass Stars and the Formation of White Dwarfs

A few stars were discussed in Chap. 4 that emitted strong infrared radiation. These were red giants, or old stars, and they expelled a lot of matter; they were probably in the process of declining. This topic now received new support thanks to relatives in 'that curious AFCRL Catalog' in Chap. 6. Some of them produced very photogenic objects. Two examples are given here (Figs. 7.22 and 7.23).

In both cases a star could be seen in the center of the image. The star emitted a lot of gas along an axis through that center. Until shortly before, the objects were AGB stars with thick shells of gas and dust and were only visible in the infrared (see Chap. 4). They were now trying to break out of their shell. Soon the star would turn into a so-called 'planetary nebula' consisting of a faint white star surrounded by a

Fig. 7.22 Image obtained with the Hubble Space Telescope of the object AFGL 5237 that is known as the Calabas nebula, and also, because of the presence of sulphur dioxide, as the 'Rotten Egg'. Source: NASA/ESA

Fig. 7.23 Image obtained with the Hubble Space Telescope of the object AFGL 915, also known as the 'Red Square'. Source: ESA/Hubble and NASA

nebula that would be quickly (10,000 years) dissipated. Then a small, hot star would remain, a white dwarf, that would cool down further and continue to shine—or rather glow—for a long time. These AFGL-objects hadn't reached this stage yet: they still radiated as much energy as an AGB star. They were given the name 'post-AGB objects'. The objects were 'multivisual': they were studied using radio, infrared and optical telescopes and the results of one telescope supplemented the results of the other. They were rewarding objects for observers.

Using a simple ad-hoc assumption about this idea of matter being expelled, the young German theoretician Detlev Schönberner demonstrated convincingly how a computer could simulate the transition from an AGB star to a white dwarf. His results form the basis of the Complete track figure (Fig. 7.24).

In 1978 Canadian astronomers Kwok, Purton and Fitzgerald (Kwok et al. 1978) published a relatively simple model for the intermediate stages a star went through in this post-AGB era. It started with a star that expelled a lot of matter at a low speed and ended when the majority of the matter had disappeared. What remained was a beautiful planetary nebula. This happy transition lasted only a short while because the remnants of gas were expelled at high speed as a result of the high light pressure. Thereafter a white dwarf is all that remains. This quite simple model compared favourably with much more complete and much more complex models. Often it was referred to as 'a model to play with', or a 'toy model'.

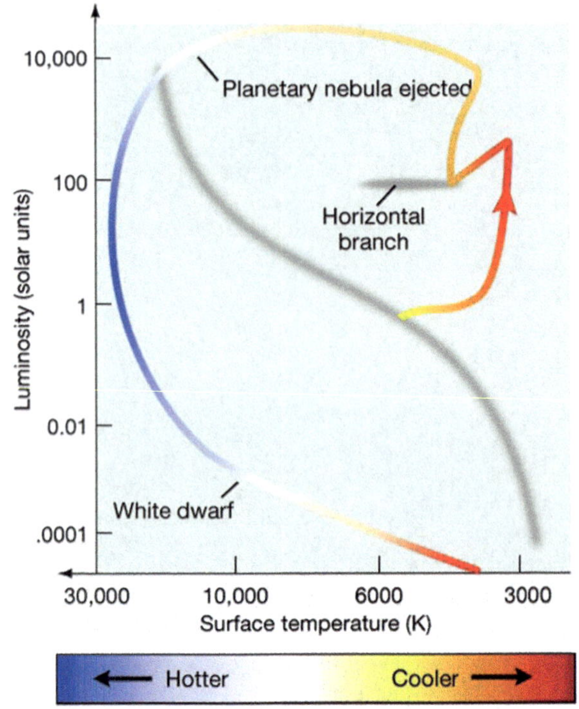

Fig. 7.24 The Hertzsprung-Russell diagram of stars and the evolutionary track a star with the same mass as the Sun will make, when it grows old. The color (temperature) of the stellar surface is indicated horizontally, and the amount of light emitted is indicated vertically. By far the largest number of stars is on the grey curve known as the main sequence. The place of the Sun is approximately at the start of the thick yellow line. An ageing star leaves the main sequence along the first yellow—then later red—line upwards, reaches a maximum as a red giant and then falls back to the grey horizontal line, known as the 'horizontal branch'. After some time the star climbs upwards again and then becomes an 'AGB star'. But this is a short-lived stage. The star moves to the left as a 'post-AGB object'. It goes rapidly on its way: horizontally and to the left. It quickly reaches the point of 'planetary nebula ejected'. After this the star wanders off along the blue line. Its light decreases rapidly but the star keeps its blue color, which means that its surface is and remains hot. Once the object has become a white dwarf, it creeps at a snail's pace further downwards. It takes billions of years before it reaches the end of its life path. Source: Penn State Astronomy & Astrophysics

The formation of a planetary nebula and a white dwarf has been described here as a gradual process. In Chap. 4 it was reasoned that only light stars are capable of this; heavier stars end with an explosion as supernovae. It was not well-known where the boundary was between light stars that developed into a white dwarf and high-mass stars that exploded as a supernova of type II. The best guess was a limit between 5 to 8 times the mass of the Sun.

Supernovae

The existence of supernovae had been discovered in 1934—see the Prologue. Little was known although most astronomers (and the general public) found them intriguing objects. Information was scarce and difficult to come by because a supernova explosion is a rare phenomenon in any galaxy. Supernovae were stars that became 10,000 or more times brighter within a few days. They were studied for several reasons. Firstly, new elements were formed during a supernova explosion, particularly heavy atoms such as iron and these new atoms were injected into the surrounding interstellar space; secondly, such large explosions disturbed the gas clouds in their vicinity on a large scale; thirdly, supernovae are probably the source of cosmic rays. These rays consisted of mysterious particles discovered around 1900 when it was found that photographic plates were always slightly illuminated even when they had been hermetically sealed. The only explanation was the presence of highly energetic atomic nuclei in the daily environment of the laboratory. Where did these particles come from? Tests in balloons that rose to great heights showed that this illumination of sealed photographs became stronger as the balloons rose higher; the illumination particles came from the cosmos: cosmic rays. Their origin was an enigma.

There were two main types of supernovae. The ‘type II supernovae’ were high-mass stars that exploded for the reasons already given. ‘Type I supernovae’ arose from white dwarfs on which matter from outside fell until the white dwarf became heavier than the Chandrasekhar limit. I will first discuss the type II explosions and then type I, and finally, something about the remnants left behind by supernovae.

Type II Supernovae: The Explosive End

The study of exploding high-mass stars gave the same problem as the study of normal stars: you could not look inside the star where all the important processes took place. Reasoning and excluding many of the possibilities was the only way to learn about the construction of the interior and what happened there. This was difficult enough for normal stars and it was even more difficult for supernovae because when the supernova was detected the explosion had already reached an advanced stage. I write this as an excuse for the many, very competent theoreticians who were involved in studying supernovae, and who made only very slow progress. The observers also contributed little: supernovae were rare and were always seen in far distant galaxies.

Let us look again at the developments of a high-mass star. It had a peaceful existence as long as it was able to fuse hydrogen to helium. But eventually this was followed by a brief period in which helium was fused to carbon and a small amount of oxygen. Then, new and ever shorter periods with neon, magnesium and silicon fusion followed in succession until finally the core of the star became iron. And this is where the problem lay: iron could not fuse; it could not 'burn'. It was the end of the road. The innermost core cooled down and the gas pressure dropped, leading to an implosion: a 'core-collapse' supernova, a type II supernova. A neutron star could be formed during the implosion or a black hole—or maybe there was nothing left behind?

The *im*plosion had to be followed by an *ex*plosion, because this was the supernova that we saw time and again. In 1975 it was an open question how this 'bounce' from implosion to explosion came about and in 1985 it was still the same, even though in the meantime there had been an ongoing search for the answer. Numerous scenarios were thought of, and many calculations were made, but there were a lot of factors that had to be taken into account and this called for extensive calculations. I quote one of the experts, Wolfgang Hillenbrandt (1986), who concluded at a meeting of the International Astronomical Union: 'Most of the theoretical models are still quite controversial'.

In spite of these doubts, there were a number of conclusions that seems fairly certain. Stars with a mass smaller than 8–10× that of the Sun could, as was already known, develop peacefully to become a planetary nebula and ultimately a white dwarf. Stars heavier than this limit could become supernovae, apart from the heaviest ones. Stars more than 100 times as heavy as the Sun would probably reach their end stage as Wolf-Rayet stars and never produce a supernova. But this was by no means certain, as can be deduced from the publication of a calculation of a supernova that arose from a star 2000 times as heavy as the Sun. The article was regarded with some skepticism, but it was not rejected (Utrobin 1984). Another uncertain conclusion was that a supernova left a neutron star behind when the exploding star had a mass between 20 and 50 times that of the Sun, but stars heavier than that would produce a black hole.

The light curves of type II supernovae were different from one another, unlike the light curves of type Ia—see Chap. 4. As more observations became available it appeared that two-thirds of the light curves of type II had one common characteristic: the light curve declined briefly after reaching the maximum but very quickly afterwards reached a continuous plateau: there was no reduction in the luminosity during this period. See Fig. 7.25.

Some supernova calculations predicted such a plateau. These calculations were simplified by the fact that the outer parts of the exploding star were hardly affected by the details of what happened deep inside, as Arp had shown in 1961 (see Chap. 4): the light curve of a supernova was like that of an expanding gas layer that was fed by some central energy source. Consequently the light curve gave few or no details of what happened in the center. The spectrum, however, gave such details, because it told how many new elements had been produced.

What the model calculations provided is shown in Fig. 7.26. It is the model of a star in the few seconds before it explodes. The star originally had 15 times the mass of the Sun. It is a complex diagram that needs to be studied carefully. Fortunately, it is not necessary to understand it in order to follow the rest of this chapter!

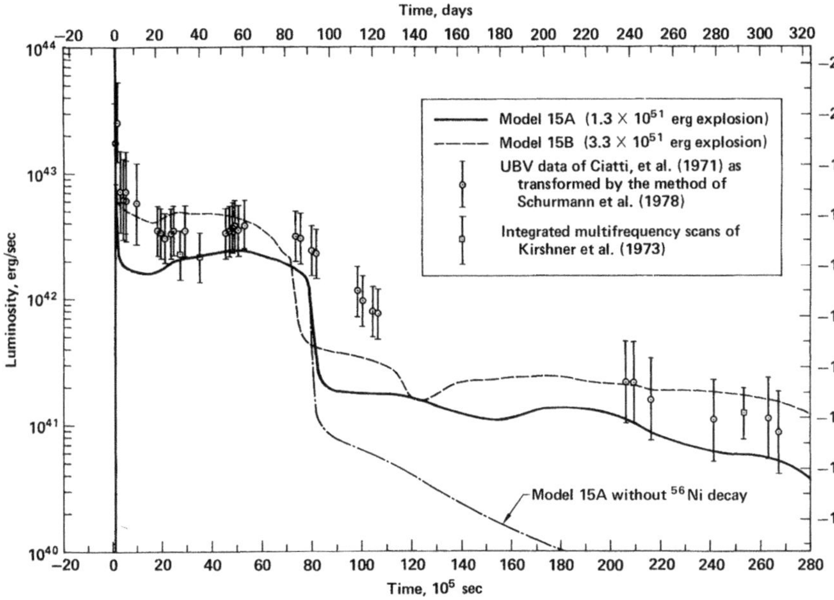

Fig. 7.25 Light curves of supernovae of type II. The points with error marks present observational results and the curves describe model calculations. The essential content is the fact that the light of these supernovae seems to decline in steps with a constant output in between (Woosley and Weaver 1986)

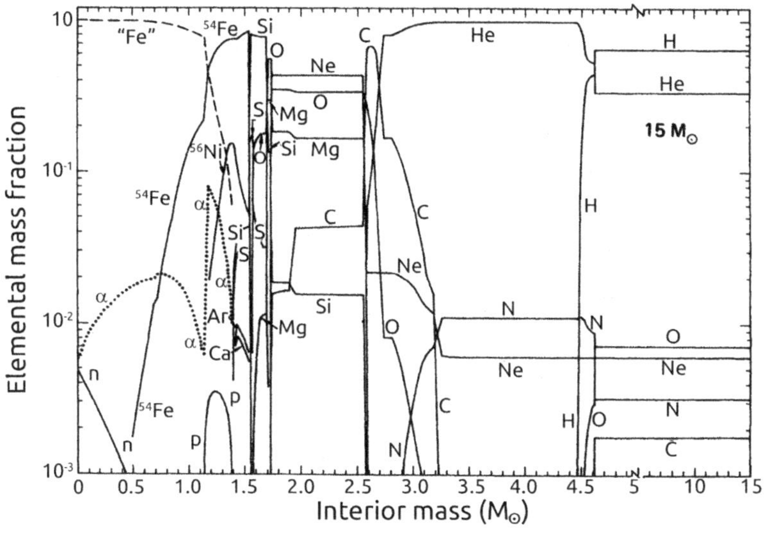

Fig. 7.26 Model calculations of the chemical composition of a star with initially 15 times the mass of the Sun shortly before the star explodes as a supernova. The horizontal axis indicates the amount of matter within the given mass: the figure 2.0 along the horizontal axis indicates that 2.0 times the mass of the Sun lies within this limit. The fraction of a special element in this shell is shown vertically. It is possible to read from this figure that the matter that lies within 1.5 times the solar mass is made up mostly of iron ("Fe" and ^{54}Fe). Around it is a shell with 0.15 times the solar mass of silicon (Si); around this a thin layer of oxygen (O) and around that a somewhat thicker layer of neon, oxygen and magnesium (Ne, O, Mg). The next thin layer is of carbon, then a thick layer of helium and an outermost layer of hydrogen (Weaver et al. 1978)

Type I Supernova: A Binary Star in Difficulty

White dwarfs were very stable objects and could be older than the present age of the Universe. They were—almost—immortal, but they had an Achilles heel: when a white dwarf became heavier than the so-called Chandrasekhar limit (1.4 times the mass of the Sun), the internal pressure could no longer resist gravity and the object was bound to collapse. Type I supernovae were formed when a white dwarf exceeded this limit as a result of the influx of matter from outside. The only realistic situation in which this occurred was when a white dwarf was part of a close binary star: in that case a substantial part of the donor star could settle on the white dwarf and cause a large explosion there.

In the period from 1975 to 1984, the theoreticians were primarily involved with the question of how the white dwarf responded to such an influx of matter. What happened when the mass of the dwarf exceeded this fundamental limit? There were a priori all kinds of possibilities. It was possible that the white dwarf was made up of carbon and oxygen or it could be of helium: this depended on the exact type of red giant from which the dwarf originated. Then, the explosion might start in the center or at a more or less coincidental place elsewhere in the dwarf. Figure 7.27 summarizes the different possibilities that are set out in the subscript. There's no need to read them all; simply be amazed at all that is possible.

There was an important distinction between two frequently used, and apparently similar concepts: 'detonation' and 'deflagration'. With detonation, the combustion progressed at very high speed (physically, many times faster than the speed of

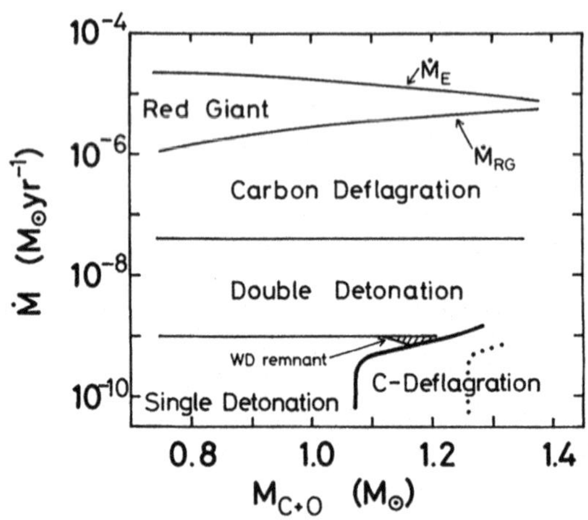

Fig. 7.27 An explanation of the many possibilities for an explosion (supernova) of a white dwarf. A white dwarf with a core of carbon and oxygen (C+O) receives matter from another star. The weight of the star is indicated horizontally. The amount of matter that flows onto the white dwarf is indicated vertically: if the influx is low, in a heavy white dwarf, below right, the carbon fusion will not be explosive but the burning layer will spread slowly ('C-deflagration'). If the star is light, it will explode in one go ('single detonation'). Explosions that start in two places in the star ('double detonation') are a possibility and the carbon can also burn at a higher mass ('carbon deflagration'). Above the uppermost curve incoming matter is expelled by gas pressure and between the two upper curves the white dwarf will regain the characteristics of a red giant (Nomoto et al. 1984)

sound) and with deflagration, the combustion found a way to the outside much more slowly. The consequences of combustion were more difficult to calculate. The real process occurred in three dimensions while the calculations were one-dimensional because of the limited capacity of computers. I quote a calculation termed to be 'a standard model' of a white dwarf initially of 1 solar mass on which a relatively large amount of matter was dumped by a nearby companion. Inside the dwarf, the carbon ignited when the white dwarf became 1.378 times the mass of the Sun. The light curve corresponded well with that of supernovae that had been observed, and the spectrum was also predicted fairly accurately. A solar mass of iron was released and an almost equal amount of silicon. It seemed certain that nothing star-like remained; no new white dwarf, for example.

Even though a 'standard model' had to be named, the researchers were none too sure of themselves. There was another model that remained attractive, based on a narrow binary star made up of two white dwarfs. They emitted gravity waves, causing them to lose energy; consequently they crept close to one another and eventually merged, after which the explosion took place. That the theoretically predicted gravity waves existed only became confirmed in 2016, but they had been demonstrated indirectly in 1974 in a study of a binary pulsar (see later in this chapter).

By now people had learned to calculate a spectrum of the type-I supernova that matched the observations (see Fig. 7.28). This successful technique arose from the development of the pioneering work that was carried out in the previous period (see Chap. 4).

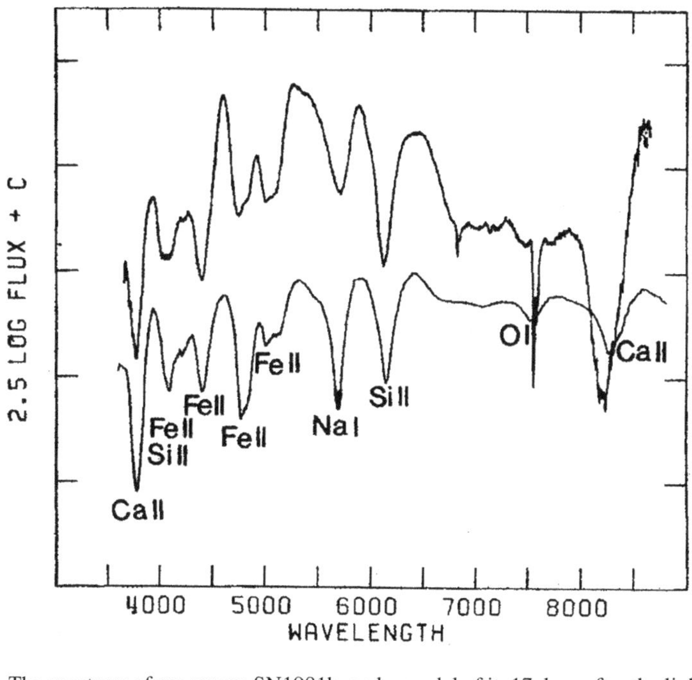

Fig. 7.28 The spectrum of supernova SN1981b, and a model of it, 17 days after the light from the supernova had reached its maximum value (Branch et al. 1983)

Supernova Remnants

In this period, astronomers continued to study the remnants that remained long after the explosion. They hoped to learn how much kinetic energy was present: the models predicted that almost all the energy released from a supernova eventually found its way to the moving matter. This amount of kinetic energy was readily measurable in the spectrum of the fragments. As said before, from these spectra one also gained some insight as to how many heavy elements the supernovae had expelled.

Finally, there was the hope that cosmic rays were produced in these fragments, that is, that atomic nuclei and electrons were speeded up in these fragments to almost the speed of light. When Baade and Zwicky concluded in 1934 that supernovae existed, they at the same time also predicted that the supernovae produced the cosmic rays. This idea continued to hold, but 'how' this took place was unclear. According to the calculations, no particles could be produced at the highest speeds, or rather at the highest energy found in the cosmic rays. The last word had not yet been said on this topic.

The discovery that a pulsar had been found in the center of the Crab nebula led to the search for pulsars in other supernovae remnants. But while more than a hundred supernova remnants were known, only two pulsars were found like those in the Crab nebula. No pulsar was found, for instance, in the well-studied remnant of the supernova radio source Cas A. This was not a dramatic conclusion: the possibility that a supernovae left no remnant at all was held open by several theoreticians.

White dwarfs

White dwarfs were faint objects and difficult to find. The hottest were strongest and as they also were relatively strong in the ultraviolet the first search by Luyten looked for stars stronger on blue- than on yellow-sensitive photographic plates. Luytens samples, however, were polluted with other objects strong in the UV: the quasars, as Sandage had suggested and the so-called hot subdwarfs, a class of presumably narrow double stars. It was highly uncertain how many of each kind were present because the search technique had not been very homogeneous.

To improve the statistics Green and Schmidt carried out a new survey in the late 1970s. They searched for UV-strong stars on plates covering 10% of the sky. They stayed away from the broad band of the Milky Way in order not to be swamped by objects of other kinds. They took photographic plates with the Schmidt telescope that Zwicky had obtained in the 1930s—see the Prologue. In their final analysis (Green et al. 1986) 1715 objects were discussed. Half of the objects were hot subdwarfs, white dwarfs were a quarter and 5% (93 objects) were quasars.

A small group of white dwarfs stood out because they were exceptionally hot. The prototype of this group, PG 1159 (PG stands for 'Palomar-Green') had an estimated surface temperature of 140,000k. Such a hot white dwarf must have emerged very recently from an AGB star and be at the beginning of the 'cooling curve' of white dwarfs in the HR-diagram; compared to other white dwarfs it was very young (see Chap. 4). This suggestion was confirmed after a faint, main-sequence companion star was found for one object in the PG1159-group. The distance to this companion could be estimated reliably and the luminosity of the PG1159 object could be derived. It had an absolute magnitude 6 and was thus about 100 times more luminous than all well-known white dwarfs: it marked the beginning of the cooling path in the HR diagram that white dwarfs will follow.

PG 1159 had one more unusual characteristic: it vibrated (McGraw et al. 1979). This made it a candidate for seismological research—a new research technique that many hoped to become reality. Later, more stars would be found of this new class of 'pre-white dwarfs'. One of these was the central star of a planetary nebula, which underlined the fact that planetary nebulae are temporary disguises of stars on their way to becoming a white dwarf.

Another important conclusion from this period was the absence of strongly cooled white dwarfs, stars that would have to be placed at the bottom of the HR diagram (Liebert et al. 1979). Why were they missing? Maybe the very first white dwarfs were not yet old enough; in other words, our Milky Way galaxy had not yet existed for long enough. This was an intriguing conclusion, but unfortunately there were also other less interesting explanations, such as that people had not searched hard enough. As so often in research, a promise is not a result.

Most white dwarfs were hotter than 30,000 K and some were much hotter—so it was to be expected that they would emit X-radiation. The confirmation came in 1975 when the ANS satellite discovered X-radiation from Sirius. This was a binary star, comprising a hot main-sequence star (Sirius A) and an optically very weak, much hotter star (Sirius B). In around 1900 Sirius B was one of the first white dwarfs to be recognized. The ANS-satellite saw both stars at the same time but it was obvious that the X-radiation must have come from the hotter of the two Sirius stars. Sirius B was the first white dwarf with X-radiation, but there were more to come. In 1985 X-radiation had been found from a dozen or so other white dwarfs, particularly with NASA's Einstein X-ray satellite. The X-radiation could be used as a sensitive thermometer for high temperatures on the surface of the star (30,000–200,000 k). An optical image of Sirius is shown below and also an X-ray image (Figs. 7.29 and 7.30).

In this period a lot of attention was paid to making models of the atmosphere of white dwarfs and to the question of how you could deduce from the spectrum the gravitational acceleration on the surface (Wesemael 1981). An average white dwarf has a mass roughly equal to that of the Sun and a radius 100 times smaller. Thus one expects the gravitational acceleration at the surface of the white dwarf to be 100^2 or 10,000 times larger than on the Sun. For a considerable number of white dwarfs spectra helped to determine the mass of the star from known data. Figure 7.31 comprises a bar diagram of the mass of a number of white dwarfs. On average, the dwarfs had a mass that was 0.6 times that of the Sun.

In the previous period the unexpected discovery of a white dwarf with a strong magnetic field attracted a lot of attention. Further research made it clear that very strong magnetic fields were rare but this did not answer the question of why a strong magnetic field occurred in one star but not in the other. This question was postponed for later (Angel 1978; Angel et al. 1981).

Fig. 7.29 A photo of Sirius A and B taken with the Hubble Space Telescope. The four long lines have been produced by the telescope and are thus artifacts. Sirius B is the minuscule dot to the south-west of Sirius A. Source: NASA, ESA, H. Bond, and M. Barstow

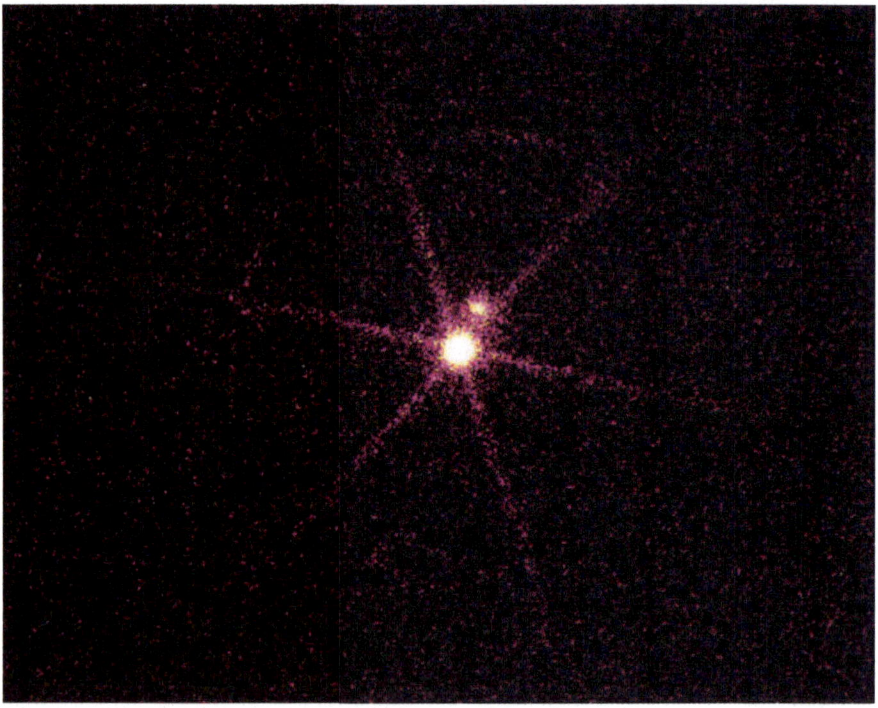

Fig. 7.30 An X-ray photo of Sirius taken with NASA's Chandra satellite. Compare with the previous figure. The roles are reversed: the bright star with the six artificial rays is Sirius B and the weak object is Sirius A. The six lines are a by-product created by the structure of the telescope. Source: NASA, CXC, SAO

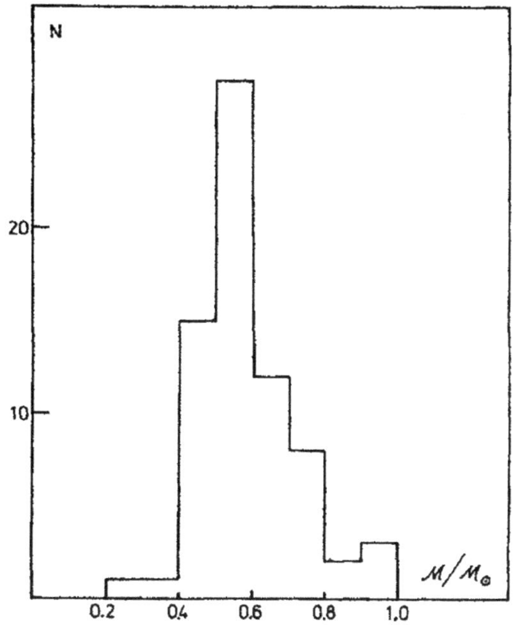

Fig. 7.31 Bar diagram of white dwarf masses that have been determined from the spectra of the stars and the distance (Weidemann and Koester 1984)

White Dwarfs in Close Binary Stars

Completely unexpectedly, it was discovered in the 1950s (see Chap. 2) that stars that had ever exploded as novae always comprised two stars, a binary, at such a small distance from each other that together they fitted easily within the Sun: the whole system, the white dwarf with a light-emitting disk around it and another small star. The two stars orbited around one another within a few hours. There were more of such ‘short period binary stars’ but if one of the stars was a white dwarf and if a light-emitting disk was seen, then the binary star was referred to as a ‘cataclysmic variable’, from the Greek word ‘kataklysmos’ that means ‘flood’ or ‘deluge’. In many cases, the disk was by far the brightest part of the system.

One of the very first questions was how such a cataclysmic variable could come about.

The first possibility was to compare the two with a single-egg twin. This suggestion could be rejected: one of the stars was a white dwarf, a highly compressed object with a prior history as a red giant. It was difficult to imagine how a white dwarf could be formed from a star in the making. Polish astronomer Paczynski (1976) sketched another scenario in 1975. Although he could not prove all of his speculations, his idea did take root. He proposed that both stars started together as a binary in which the two members were well separated. When the heavier of the two became a red giant, it grew so much in size that the second star entered the red giant and spiraled to the core: together they formed a ‘common envelope’ star. The red giant lost its mantle as a result of evaporation and the core became a white dwarf. A close binary star had now formed that was even smaller because they would emit gravity waves together and would thus lose energy. Paczynski even made it plausible that the shortest period they could reach was 81 minutes, which corresponded with the observations. Paczynski was not the first person to toy with the idea that the stars could creep closer together but his arguments were more convincing. 'Common envelope' objects will be discussed further in Chaps. 13 and 14.

There were many types of cataclysmic variables: classical novae, dwarf novae, recurrent novae, nova-like stars (Warner 1976; Robinson 1976). The explosions differed in strength and frequency from one type to another. Whether and how the types were related was unknown. Cataclysmic variables were eccentric beasts and they were rare. Why then was so much attention paid to this topic? The answer is that these binary stars were a modest prototype for other, more exciting objects discovered with X-ray telescopes in the 1970s and 1980s: close binary stars in which a neutron star or a black hole took the place of the white dwarf. It was this discovery that suddenly made studying cataclysmic variables more than a fun game.

Two separate, special, small groups of cataclysmic variables will be mentioned: the 'polars' (or AM Herculis stars) (Tapia 1977; Hearn and Richardson 1977; Stockman et al. 1977) and the 'intermediate polars' (or DQ Herculis stars). The letters DQ and AM indicate symbolically that the objects were discovered as variable stars, both (by chance) in the constellation of Hercules. DQ Herculis is what remained from a very bright nova in 1934 and has already been discussed in Chap. 2. In both cases the white dwarf contains a strong magnet. In AM Herculis stars the magnetic field is so strong that the matter ejected by the companion arrives at the white dwarf via the magnetic pole and there is no disk around the white dwarf. In DQ Herculis stars, the magnetic field is less strong and an accretion disk is formed, but in the inner region the gas is again channeled to the white dwarf via the magnetic poles. Both types of 'polars' are strong sources of X-radiation.

Neutron Stars and Stellar Black Holes[1]

Isolated Neutron Stars, Pulsars

The first neutron star was discovered as a pulsating source of radiation. It was an object without a companion star, an isolated object. Optically visible pulsars were known to exist: the first had been detected in the Crab nebula (see Chap. 4). In 1977 a pulsar was also seen optically in the remnants of a supernova in the Vela constellation (Wallace et al. 1977). But no more optically visible pulsars were found; if they existed, they were apparently very difficult to find. Radio telescopes continued to be the easiest way of tracing pulsars, and Jodrell Bank in England, Arecibo and Green Bank in the US and Parkes in Australia were the main radio telescopes. In 1977, there were 149 known radio pulsars and in 1985 this number had risen to 316 (Lyne et al. 1985). Still, it continued to be difficult to analyse these numbers statistically. Here, too, there was the perennial problem for astronomy: how many of the

[1] In this period people begin to realize that black holes occur in the center of quasars and other 'active' Milky Way galaxies. Such black holes have a mass millions of times that of the Sun and are referred to as 'massive black holes' or even 'supermassive black holes'. Black holes in binary stars have a mass just a few times that of the Sun; I will refer to these as 'stellar black holes'.

objects we are seeking have we missed? In many cases there was no good answer to this question. In an analysis from 1986 an estimate was made that there were 70,000 pulsars to be discovered in our Milky Way galaxy.

Besides radio and optical pulsars there were also X-ray pulsars. These were in all cases neutron stars in a close binary star system. They will be discussed shortly.

Measuring the extremely slow rate at which stars moved along the sky (slow by human standards) has always been an important aim of astronomical research. This started with Halley who discovered in the seventeenth century that the stars were no longer in the same place in the heavens as where they had been located around 1500 years earlier, according to Roman astronomer Ptolemy. Halley drew the conclusion that the stars indeed move across the heaven, which means they exhibit 'proper motion'. To measure this proper motion one measured the new position of the star in the sky and compared that with a position of long ago. The longer the interval of time between 'now' and 'long ago', the easier it was to measure proper motion accurately. It was only at the start of the twentieth century that photographic plates were made that were good enough to measure the proper motion of large numbers of stars. Radio astronomy offered some advantages here: much more precise positions could be determined using radio interferometers. This advantage proved useful for the study of pulsars. The radio measurements of the position of pulsars were indeed so precise that accurate proper motions were obtained over a time interval of only after a few years. It was an achievement that optical astronomy could not match (Lyne et al. 1982).

The most surprising aspect of these measurements was that the pulsars raced at much greater speeds through space than normal stars. For now, people accepted that the high speeds were explained by their birth in a supernova: they had probably been ejected with great velocity.

Millisecond Pulsars: A Revival

The Crab pulsar rotated 33 times around its axis in one second and was for a long time the fastest known pulsar. There was an easy explanation for this. The radiation that pulsars emitted meant a loss of energy, which was taken from the rotation energy: as a result of the radiation emitted, pulsars rotated more slowly on their axis. This could be measured: each known pulsar slowly but surely needed more time for each revolution. This occurred so slowly that it could take a pulsar a 100 million years to come to a stand-still. The pulsar in the Crab was just 1,000 years old and was probably the youngest pulsar. It could therefore be expected to rotate fastest. In addition, its rotation slowed down more rapidly than the other pulsars, a thing one expected.

It was an enormous surprise when in 1982 a pulsar was found that rotated on its axis 20 times faster than its relative in the Crab but its rotation speed, its spin, decreased much more slowly. The discovery of this 'superfast' rotator came unexpectedly after a dedicated study over several years of an unremarkable radio source that, however, had a number of contradictory characteristics. The source scintillated as a consequence of variations in the ionosphere of the Earth and this meant that it must be very compact[2]. But other measurements indicated that the source was much greater than the scintillation indicated. A group of five respected radio astronomers became intrigued by this enigmatic source. Observations followed with the Arecibo radio telescope in Puerto Rico, the Westerbork radio telescope and the large German radio telescope in the Eifel. It slowly became clear that there were two radio sources close together that probably had nothing to do with one another. The small, scintillating source was eventually found and the spectrum of the radiation appeared similar to that of a pulsar. Attempts to detect the pulses failed until researchers went all out in their efforts, eventually with some success: the small source was a pulsar that rotated on its axis no fewer than 642 times a second, 20 times faster than the Crab pulsar (Backer et al. 1982). The publication of this result in Nature was followed in a short time by no fewer than 16 other Nature articles that confirmed and discussed what was found.

The first idea was that the pulsar had been formed recently, even less than the 1,000 years since the Crab pulsar came to life. But there was nothing in the area around the pulsar that indicated a recent supernova: another explanation was needed. And apart from the high spin-rate there was the other peculiarity: this fast spinning pulsar slowed down much more slowly than a young pulsar would do. An explanation eventually came and at first it looked peculiar, but it was supported by calculations: These rapid pulsars were very old and dead neutron stars that had been revived; they had been 'spun-up'! This is the story: The pulsar was one star in a close binary. It became a supernova and that left behind a pulsating neutron star. The supernova had not removed the companion. The pulsar kept pulsing but after some time it slowed down its spin and ultimately it became extinct as a pulsar. But after another period of time the companion star aged, began to eject matter and formed a disk around the extinct pulsar. The rotation of the disk was transferred to the neutron star and its rotation was sped up again, the radiation process restarted and the pulsar came to life again: a 'spin-up pulsar' (Damashek et al. 1982; Alpar et al. 1982).

Figure 7.32 illustrated this process. Every dot is a pulsar. The location of each dot is determined by the rotation time (the logarithm) of the period P of the pulsar and (the logarithm) of the slowing down of the period, the 'period derivative'. A large concentration of dots is at (x, y) = (0, −14). The 'Crab' and 'Vela' dots, two young pulsars, are to the top left: they have a short period and their rotation is slowing down very rapidly. In a 100,000 years the Crab and the Vela pulsars will probably have joined the great concentration. A few new pulsars will have been added and several old pulsars will have shifted to the bottom right, on the 'death line', and, as the name says: a pulsar that crosses this line is 'dead' and has stopped emitting pulses. A few pulsars from that graveyard will 'spin up', rotate faster and again emit pulses thanks to the contribution of fast rotating gas from their companion. They will leave the graveyard and appear at the bottom left; 1937+21 is an example of this. The 'spin-up' line is a theoretical estimate of the shortest rotation period that is possible as a result of 'spin-up'.

[2] 'Scintillation' caused by the ionosphere of the Earth had been a proof around 1950 that the radio source Cygnus A was a small object in the sky.

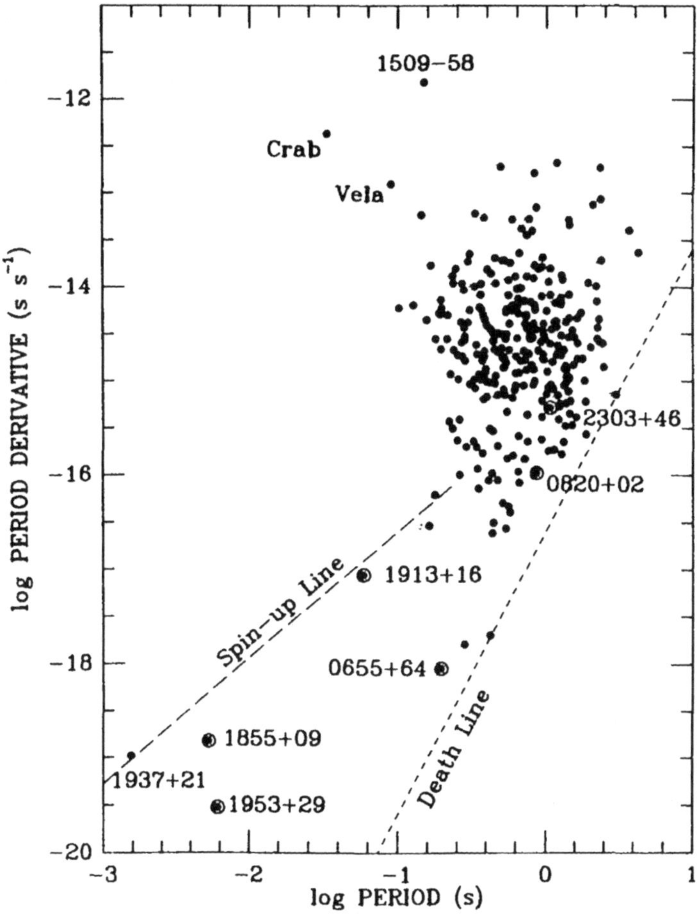

Fig. 7.32 Characteristics of radio pulses of neutron stars. The logarithm of the period is indicated horizontally and the logarithm of the speed at which the spin rate decreases is given vertically. The Crab and the Vela pulsars, together with pulsar PSR 1509-56, are the only pulsars that are found in the remnants of a supernova; all three of them are young. At the bottom left are six so-called millisecond pulsars. Further explanation is given in the text (Taylor and Stinebring 1986)

When this diagram was shown for the first time, it met with a lot of enthusiasm. The researchers received a new confirmation that they already understood a lot about the pulsar and its characteristics.

The Binary Pulsar and the Two Tests of the General Theory of Relativity

In 1975 American radio astronomers Hulse and Taylor (1975) discovered a pulsar, PSR1913+16, with a variable pulse period. The changes were regular and could easily be explained by the Doppler effect. This meant that the pulsar had a changing

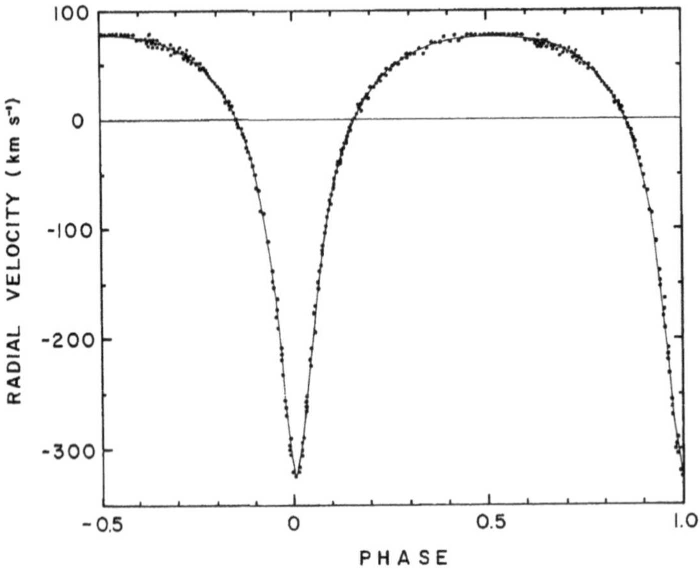

Fig. 7.33 The rotation period of the pulsar PSR1913+16 itself varies periodically. By ascribing the change to the doppler effect, it is possible to measure the changing speed of the pulsar in its orbit around another object. This diagram shows the points of the observations and the curve drawn is what is expected for the given orbit (Hulse and Taylor 1975)

velocity—like one component star in a binary. Thus the pulsar and its companion moved each in an orbit around the common point of gravity. By carefully tracking the periodic changes, the characteristics of the orbit of the pulsar could be determined very precisely. The orbit was a very flattened ellipse (eccentricity = 0.615). Figure 7.33 shows the variation during an orbit of the speed of the pulsar coming towards us and moving away from us. The velocity of the pulsar is indicated vertically. A curve has been drawn through the measurements; it shows the velocity variations according to Newton's laws. The correspondence between the measurements and the theory is remarkably good.

Six years of observations resulted in several important conclusions. To explain the reasoning, first think of the orbit of the planet Mercury around the Sun. This orbit is an ellipse and there is a point in the orbit where Mercury comes closes to the Sun, the 'pericenter' of the orbit. It had already been discovered in the nineteenth century that this pericenter showed a degree of 'creep'; slowly but surely it changed position. This pericenter shift was much greater than could be explained using Newtonian dynamics. The right answer came from Einstein's general theory of relativity (GTR). Such a change of the periastron, but much larger, was shown by the pulsar PSR 1913+15. Figure 7.34 showed this shift in the pericenter; the straight line at the top is what one would expect if Newtonian dynamics were applicable and the curve shows what the Einstein theory predicted.

There was also a second test for the GTR. The orbits of both stars and therefore also the mass of each star could be determined very precisely from the measure-

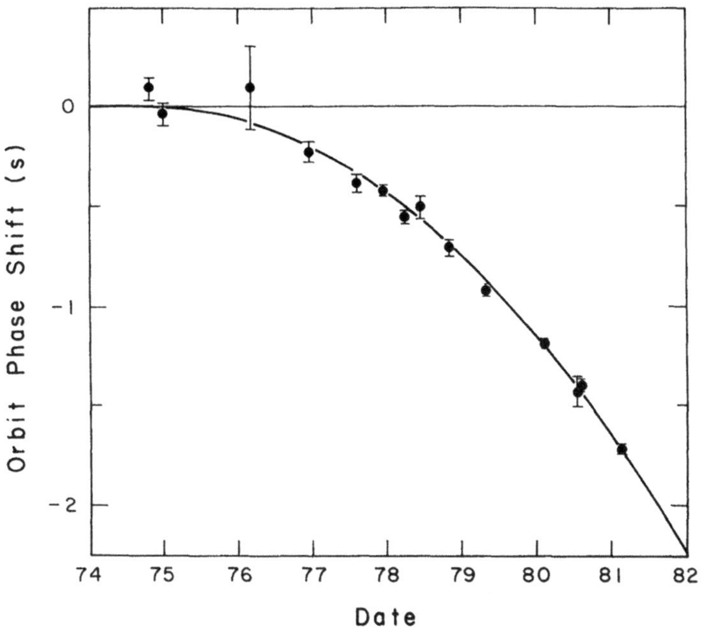

Fig. 7.34 When two bodies orbit around one another in an elliptical orbit, the point where the two bodies are closest to one another, the periastron, always remains the same, according to Newton's theory of gravity. According to Einstein's General Theory of Relativity, on the other hand, the binary emits gravitational radiation and the periastron moves around. This diagram shows that motion of the periastron. The curve is predicted by the theory of General Relativity and the straight line is what Newton's theory predicts (Taylor and Weisberg 1982)

ments, but also from the measurement of several relativistic effects: the neutron star was 1.42 ± 0.06 times as heavy as the Sun, and the companion star was 1.41 ± 0.06 times as heavy as the Sun. The researchers also saw that the two gradually orbited more slowly around one another; the period changed $-(2.30 \pm 0.22) \times 10^{-12}$ second by second. There was only one possible explanation for this decrease, and this was provided by the General Theory of Relativity (GTR): the two stars emitted gravitational radiation, a phenomenon that had been predicted but never observed. The GRT predicted a decrease of $dP/dt = (-2.403 \pm 0.005) \times 10^{-12}$, exactly as Hulse and Taylor had found. What a triumph! After this result, those in the know were no longer surprised that Jo Taylor was awarded the Nobel Prize for Physics in 1993 because 'this type of pulsar has brought new possibilities for the study of gravity'.

X-Ray Stars: A Summary from 1985

The basis for the research on X-ray stars had already been laid in the previous period. Nonetheless many new discoveries were made but most of them corresponded with the model of a close binary star with a donor and an acceptor. In most cases the acceptor has no or only a very weak magnetic field and then a disk forms around the acceptor where the gas from the donor is parked for a while and rotates around the receiving star. Slowly the gas moves inward and eventually settles on the receiver. In a small number of cases, the acceptor has a strong magnetic field. The magnetic field channels the overflowing gas to the receiver's magnetic north and south pool. The situation is similar to what we have encountered before with the cataclysmic variables.

In Chap. 4 the X-ray binary stars have been ordered according to the mass of the donor: HMXB (High-Mass X-ray Binary) have a heavy star as donor. LMXBs (Low-Mass X-ray Binary) consists of a donor star with low mass. Figure 7.35 again shows clearly that an LMXB is a 100 times smaller than an HMXB (1 versus 130 light seconds).

X-ray pulsars

X-ray pulsars are examples of close binaries in which the star receiving overflowing matter has a strong magnetic field. The overflowing gas is sensitive to a magnetic field because the gas is made up of charged particles (in professional jargon: the gas is a plasma): protons (the nuclei of hydrogen atoms) and electrons. The magnetic poles of the neutron star do not coincide precisely with the poles of the rotational

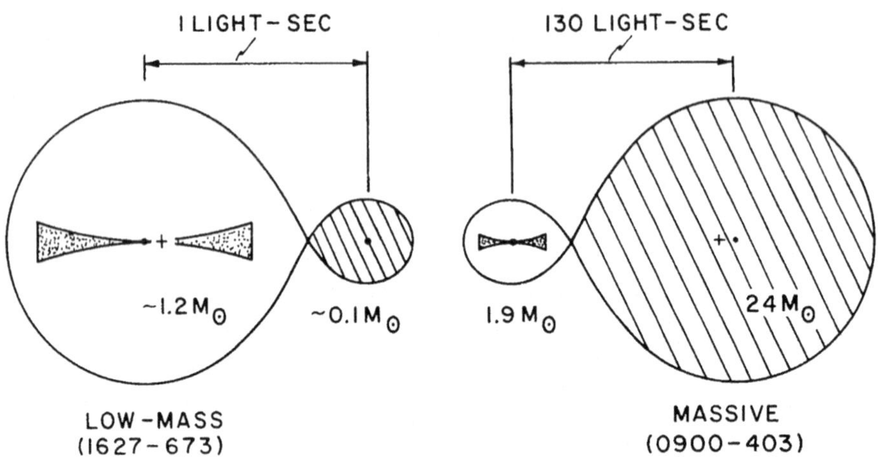

Fig. 7.35 Two sketches of an X-ray binary star with their Roche loop (see Chap. 3). The left-hand binary star (PSR 1627-6731) consists of two stars with a mass of ~1.2 M_{zon} and ~0.1 M_{zon} and the right-hand binary star (Vela X-1; an X-ray pulsar) is made up of stars with a mass of ~1.9 M_{zon} and ~24 M_{zon}; the symbol '~' means 'approximately'. There is a neutron star present in both binary stars, indicated by a dot in the left-hand loop. The point of gravity of each binary star is indicated with a cross; the cross is in the left-hand loop in the left-hand star, and in the right-hand binary it is in the right-hand loop. The left hand binary is 1 lightsecond across and the right hand binary 130 lightseconds. To compare these dimensions with those of the solar system: the diameter of the Sun is about 4.6 lightseconds and the innermost planet (Mercury) is at a distance from the Sun of 133 lightseconds (Bradt and McClintock 1983)

axis of the star. This is the same on Earth where magnetic North lies somewhere in North Canada and does not coincide with the North Pole, the point where the rotation axis of the Earth comes to the surface. The receiving star in an X-ray pulsar is always a neutron star and the donor is always a heavy star. The gas flowing inwards at the poles causes a strong X-radiation although in 1985 the mechanism was not known that produced the radiation. As in optical and in radio pulsars we see a flash that is repeated after a fixed and very stable period.

The period measured is in fact not constant but varies slightly and very systematically, just as in the case of the binary star discussed above (Fig. 7.33). This variation can also be ascribed to the movement of the neutron star around the donor star. It is even possible to deduce precisely the size of the orbit of the neutron star and its mass. In Fig. 7.36 a number of well-studied cases is shown and the mass of the

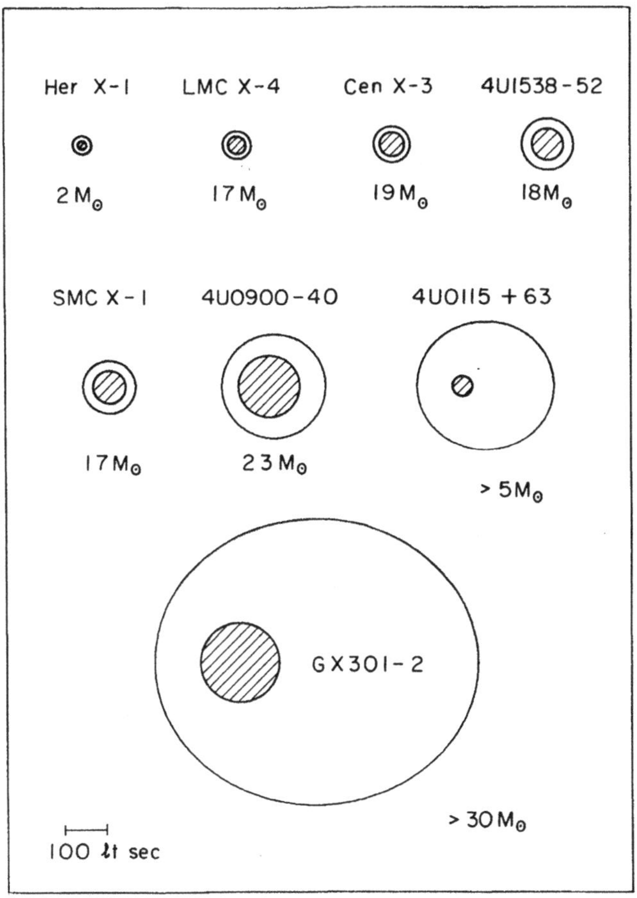

Fig. 7.36 Measurements of the orbit of the X-ray pulsar around the donor star. The circles and ellipses are drawn to the same scale. The circle with lines shows the size of the donor star and its mass is expressed in solar masses (Joss and Rappaport 1984)

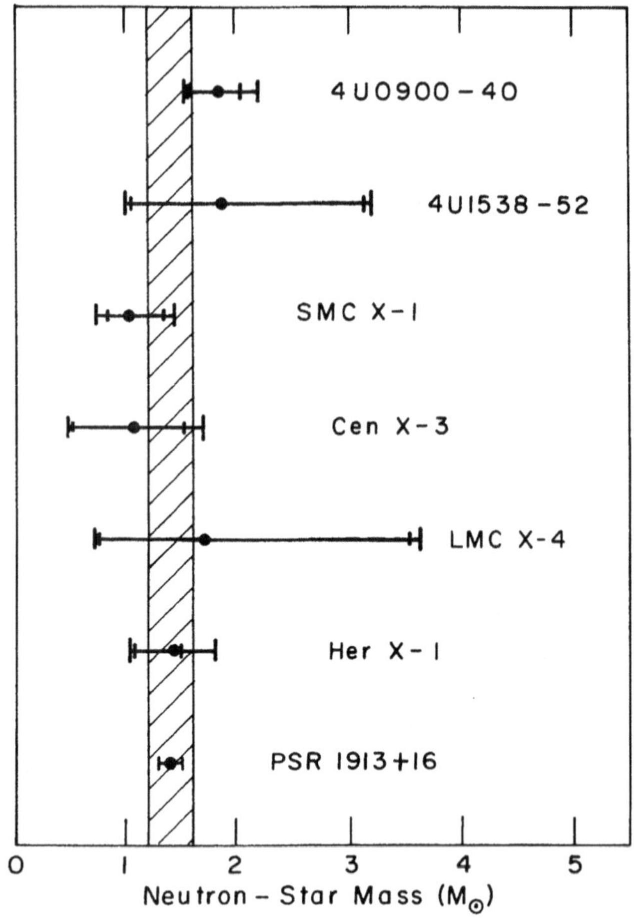

Fig. 7.37 The mass of seven X-ray pulsars deduced from the parameters of the orbit of the relevant neutron star. The length of the lines indicate the uncertainties in the measurement results. The vertical diagonally lined band is what would be expected on theoretical grounds (Joss and Rappaport 1984)

donor star is also given. From this it appeared that the lightest donor star was twice as heavy as the Sun: all pulsars are HMXBs. Figure 7.37 shows the mass of several neutron stars.

X-ray novae

In 1975 the British Ariel V X-ray satellite witnessed the appearance of what they called an 'X-ray transient' and that became for some time the brightest X-ray source in the sky. See Fig. 7.38.

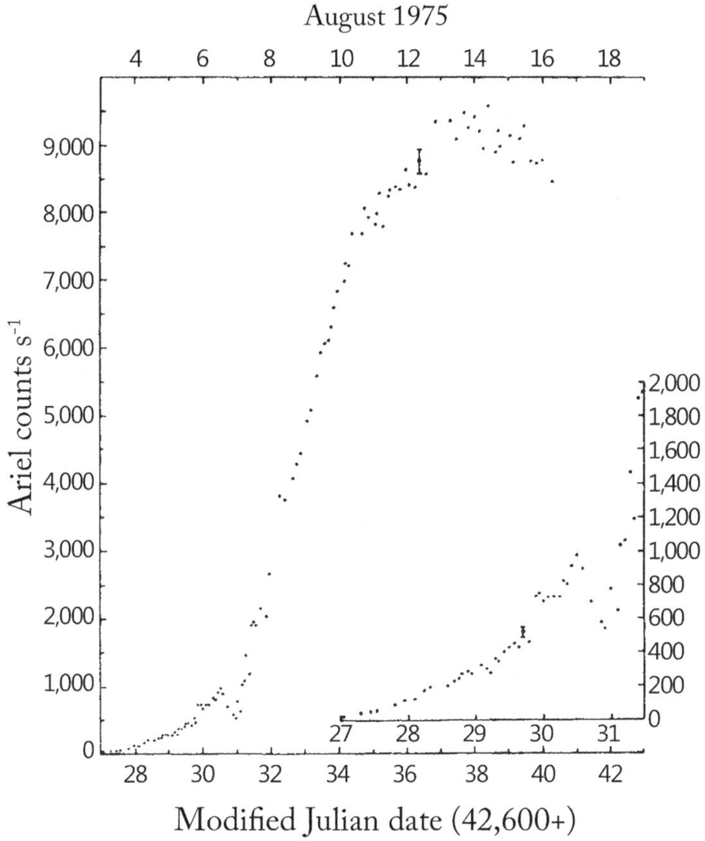

Fig. 7.38 The sudden appearance of a source of X-radiation that would become temporarily the strongest X-ray source in the sky. It was quickly identified with an optical nova. It received the X-ray name A0620-00, Nova Monocerotis 1975 and but also a variable star name, V616 Mon. Time is on the horizontal axis and the counts of the X-ray detector is on the vertical axis (Elvis et al. 1975)

The identification with a star was easy: at the position indicated a nova was observed, called Nova Monocerotis 1975, and located in the constellation Monoceros or Unicorn. The star had a magnitude 12, a hot spectrum and no spectral lines. Before the burst the star had had a magnitude 18.3 to which it returned 15 months later. It was a recurrent nova: an earlier outburst had happened in 1917. The object was known as a variable with name V616 Mon. The discovery had some strong similarity with that of Sco X-1. V616 Mon object was a spectroscopic variable with a period of 7.7 hours and thus a close binary with only one visible star and a compact object. Further optical observations showed that the red star orbited the compact object at a speed of 460 km/s from which it was derived that the compact object had a mass of 3.3 times that of the Sun. That was too heavy, as far as one knew, for a white dwarf or neutron star and the conclusion was that this must be a black hole (McClintock and Remillard 1986).This was not the first time that a black hole of such a mass had been encountered (Cygnus X-1 has been discussed earlier), but it was nonetheless a very important discovery. More of such X-ray novae were found later.

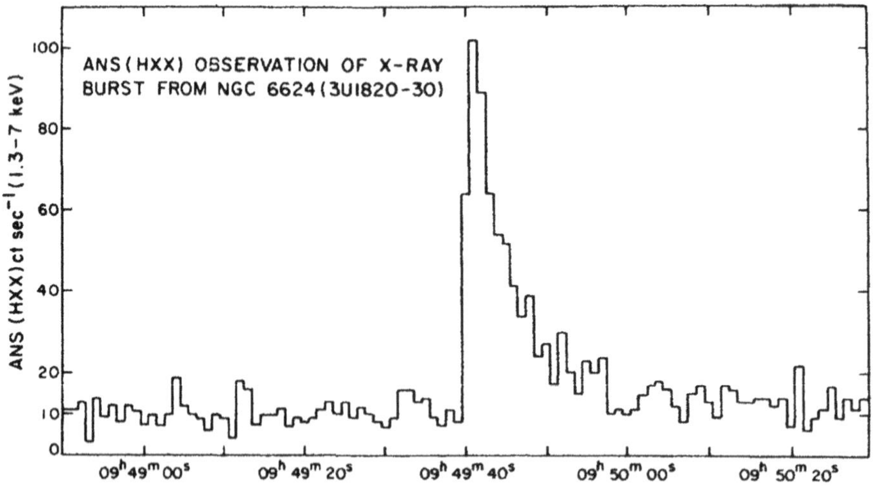

Fig. 7.39 The X-radiation received from the globular cluster NGC 6624 by the ANS satellite on 28 September 1975. The time is displayed horizontally and the number of X-ray photons received is displayed vertically (Grindlay 1976)

X-ray bursters

The X-ray nova V616 Mon was visible for months. This was not the case for the 'bursters', which were first discovered using the ANS satellite in 1975. The satellite had two instruments: one for far ultraviolet observations and an X-ray detector. Both instruments detected a short (5 seconds) burst in a globular cluster, at the same time, NGC 6624 (see Fig. 7.39) (Grindlay and Heise 1976; Grindlay 1976). Stars in a globular cluster are packed tightly together, which is why it could not be determined which star emitted the X-rays. Later satellites discovered similar explosions in a large number of other globular clusters. The association of these 'bursters' with a close double star and a light donor star was obvious: globular clusters are so old that all massive stars had decayed long ago.

The definite identification of the X-ray burster could not be made successfully. However, there was a pretty convincing argument that in such a globular cluster stars often come very near each other, and when a compact object then met another star there was a good chance that the two objects would form a double star (Fabian et al. 1975). Interestingly enough a short while later the same types of 'bursts' were noted in the area around the center of our Milky Way. This area, the thickening ('bulge') of our Milky Way, consists of many old, and thus light stars. The large distance indicated that there were strong X-ray sources.

A curious finding was that of a 'rapid burster', an object that has many bursts following one after the other—see Fig. 7.40. The source is not always active but will sometimes be dormant for months.

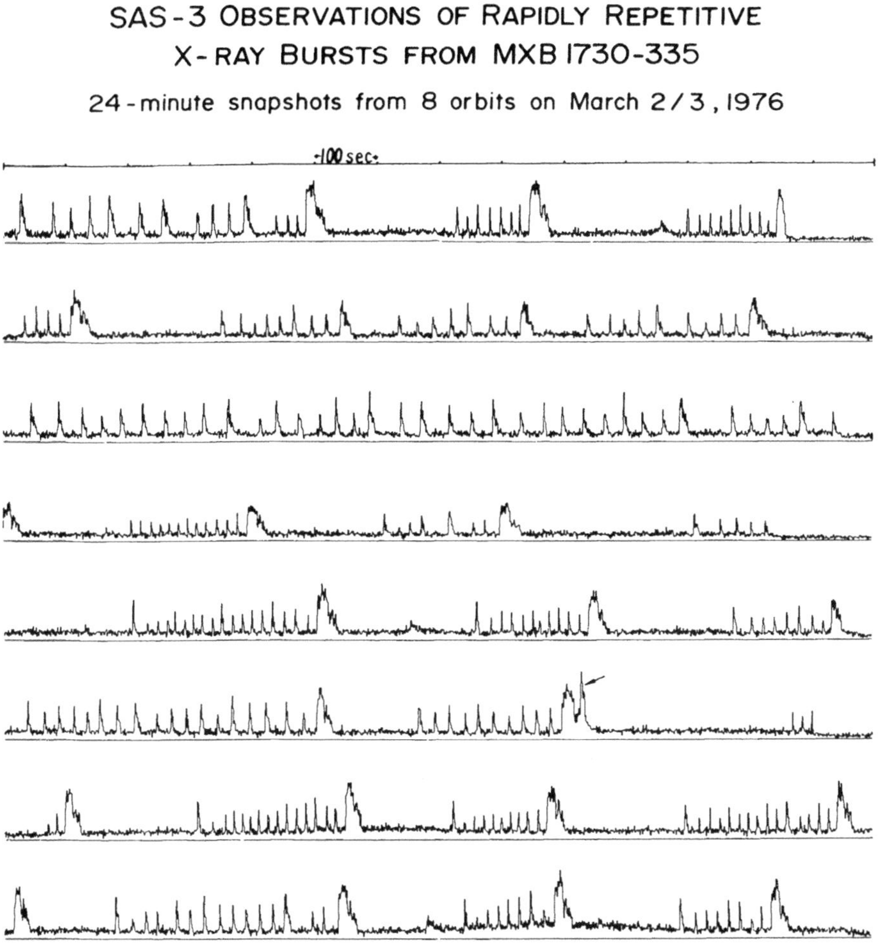

Fig. 7.40 The frequent outbursts of MXB1730-335 during an active period (Lewin and Joss 1981)

Accretion Disks

Chapter 4 reported on the first models of the accretion disc. These first models were obtained by pure reasoning and arithmetic, as the dimensions of the binary stars were too small to photograph or portray it in any way. Slowly but surely ways were found in which to test the models by means of observations. Good spectra were recorded at different times during an orbit of the two stars. Eclipses also occurred in some cases due to one part of the system being obscured by another part—the donor was obscured by the disk, for example. The accretion models became more detailed. I show two models, not to explain all of the details but to give an impression of the degree of responsible 'sophistication' that has already been achieved.

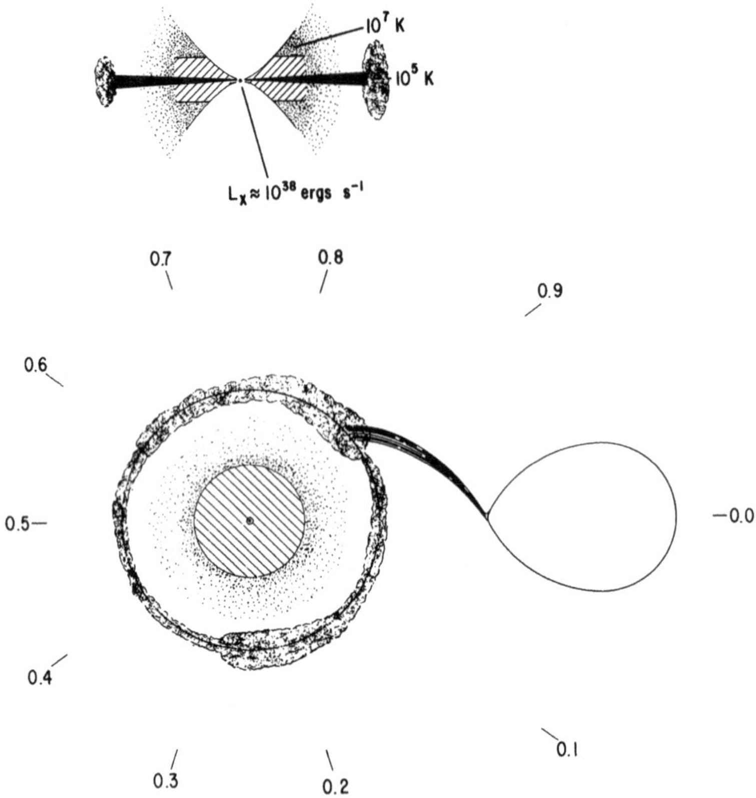

Fig. 7.41 A speculative model of an accretion disk around the X-ray binary star 4U 1822-37. Above is a side view, below is a view from above. In the bottom drawing, the donor star can be recognized, indicated with an object shaped as a tear. From the tip of the loop, the first point of Lagrange, matter streams to the receiving star, a neutron star. This matter falls on a ring of matter that lies around the neutron star at some distance. The neutron star is the point in the middle of a disc. The disk is surrounded by very hot gas that is heated by X-rays from the central object (White and Holt 1982)

One example is the Fig. 7.41.

Another example was a model for the disk around Cyg X-1, the object that caused a sensation in 1972 because it was suggested that it might contain a black hole. Although initially there was some reasonable doubt about this conclusion, the arguments for a black hole became increasingly more convincing. In 1975 the American theoretical physicists, Kip Thorne and Richard Price, presented a model for the disk with a black hole at its center (see Fig. 7.42).

New: The SS433 Microquasar

And then, unexpectedly, there was real news. It relates to the 433rd object in a catalogue from 1971 of hot stars with bright lines of hydrogen, published by Americans Stephenson and Sanduleak, and the star in question was indicated as SS433. The

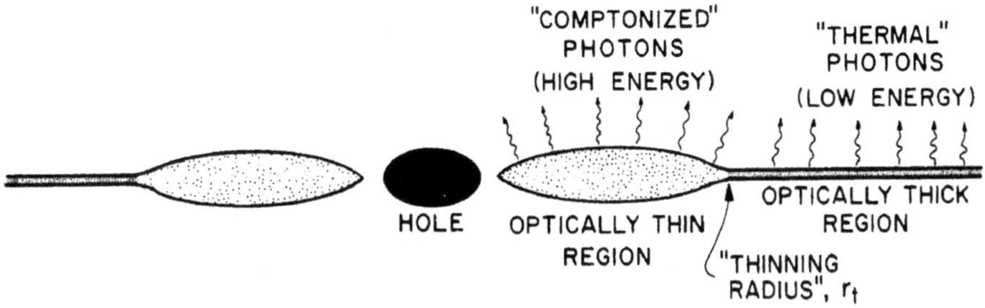

Fig. 7.42 A sketch of the disk around Cygnus X-1, a possible black hole. The terms 'optically thin' and 'optically thick' come from jargon and mean 'translucent' and 'opaque'. 'Comptonized photons' are photons that have collided with a high-energy electron; during such a collision energy is transferred from the electron to the photon; this is the so-called 'reverse Compton effect' (Thorne and Price 1975)

bright lines meant that ionized hydrogen gas (and plasma) were present just outside the star. In the summer of 1978, three radio astronomers (Seaquist et al. 1978) discovered that the star coincided with a source of radiation that varied in strength in an irregular manner. The (radio) radiation was stronger at longer wavelengths, which indicated synchrotron radiation (see Chap. 1). It was very uncommon to find such radiation in a star: previously such radiation always came from a galaxy. Within a few months it was also discovered that there was an X-ray source at the same position (Marshall, Mushotzky et al. 1978). And then, in April of 1979, Bruce Margon (Los Angeles) announced that lines in the optical spectrum 'shoot back and forth' in the star spectrum and in a short time. If this was a result of the Doppler effect, then the star was ejecting gas clouds at speeds between +50,000 and −30,000 km/s. And that was new. How was this possible?

Two answers were forthcoming, one from Cambridge (England) and one from Jerusalem, and both were the same: SS433 ejects plasma clouds, just like quasars, two jet streams expelled in opposite directions, at very high speed: 40,000 km/s, 13% of the speed of light. See Fig. 7.43.

Shortly after this, Margon came up with a more elaborate proposal: the source of the jet streams was a black hole or neutron star in a close binary star. Presumably the other star was heavier and younger. The two objects rotated around their center of gravity in 164 days. The axis of the jet stream was inclined to the orbital plane of the two objects, and thus one could observe a precession of the jet stream with a period of 164 days. This interpretation subsequently received strong support when two radio astronomers (Hjellming and Johnston 1981) used VLBI measurements to see two small, very bright spots move across the sky at the exact speed that Margon's model predicted. Furthermore, SS433 appeared to be at the center of a strong and well-known supernova remnant, W50 (see Fig. 7.44). Again this was a source with a supernova connection (Margon 1984) (Fig. 7.45).

The source was within our Milky Way and was the first (optical) star that was simultaneously an infrared, a radio and a X-ray source and with (almost) relativistic fast jet stream: SS433 was referred to as a 'microquasar'. More microquasars were

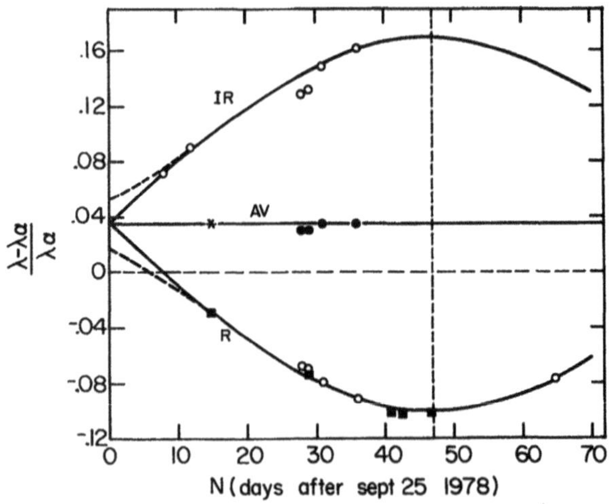

Fig. 7.43 The Doppler velocities that are derived from the doppler shift of bright lines in the SS433 spectrum are depicted vertically; the scale ((λ − λa)/λa) equals the speed as a fraction of the velocity of light. The horizontal axis shows the days on which the spectra were recorded. The horizontal line is the average of the two series of lines and the solid curve is what one can expect of a simple model in which the speeds in time change as a sine wave with a period of around 164 days (Milgrom 1979)

found in subsequent years, but they remained rare. Even so, the patent of extragalactic radio sources was broken: star-like objects in the Milky Way could also produce relativistic jet streams.

New: SGR, the Soft Gamma Repeater

On 5 March 1979, at 15.51 hrs Universal Time, a flash of gamma radiation was detected by two Soviet satellites that at that time were in the vicinity of the planet Venus. The flash was so strong that the satellites' detectors were saturated. Eleven seconds later the same happened to a NASA satellite. And there were more. It turned out that no fewer than ten satellites registered this burst: one satellite even stopped, completely automatically, all of its activities and shut down to 'safe mode'. The gamma burst started quickly but it took several minutes before the gamma photons disappeared again: during this decrease the strength of the burst oscillated with a period of 8 seconds. Fourteen and a half hours later this occurrence was repeated, although less strongly. This repetition gave the name to the object that had produced the phenomenon: 'Soft Gamma Ray Repeater', SGR (Mazets et al. 1979).

When the gamma rays arrived, each satellite was at a different location in the solar system and registered the signal at a slightly different time. The direction to the gamma source could be accurately reconstructed based on these differences:

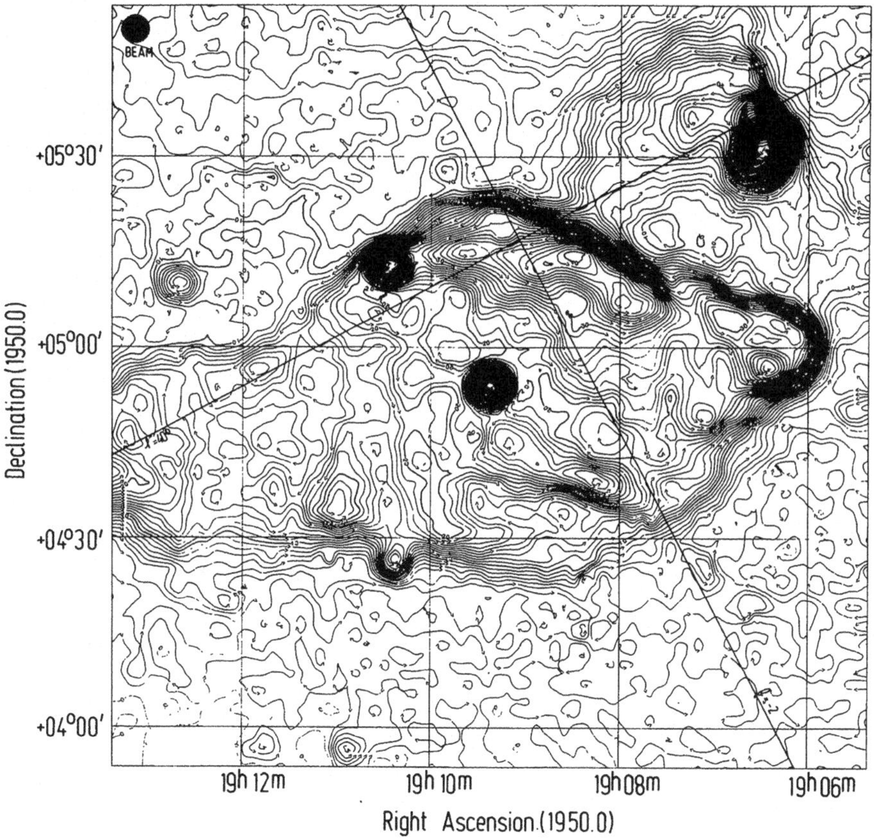

Fig. 7.44 Radio radiation at a wavelength of 11 cm measured using the 100 m Effelsberg telescope (see Chap. 6). The large structure of a flaccid flat bag containing a large ball marks the boundary of the supernova remnant. At the center of this figure is a small but strong source: the radio source that coincides with SS433. The strong source at the top right-hand side is a separate source and not part of W50 (Geldzahler et al. 1980)

the remnant of a supernova in the Large Magellanic Cloud—a satellite galaxy of the Milky Way at a distance of 180,000 light years away. The position made up the second part of the name: SGR 0520-66. Thus, SGRs were left behind after a supernova explosion. They might be related to, for example, the pulsar in the Crab nebula.

No explanation for this ‘event’ was forthcoming: there was too little information. Of course it was impressive that a flash at such a large distance could cause alarm conditions in equipment within the solar system. And then there was also the surprise at the speed in which the pulse reached it full strength: within one second. And that much energy! This was not a one-off event, as a short while later two other SGRs were discovered, but there was even less information on these. To illustrate

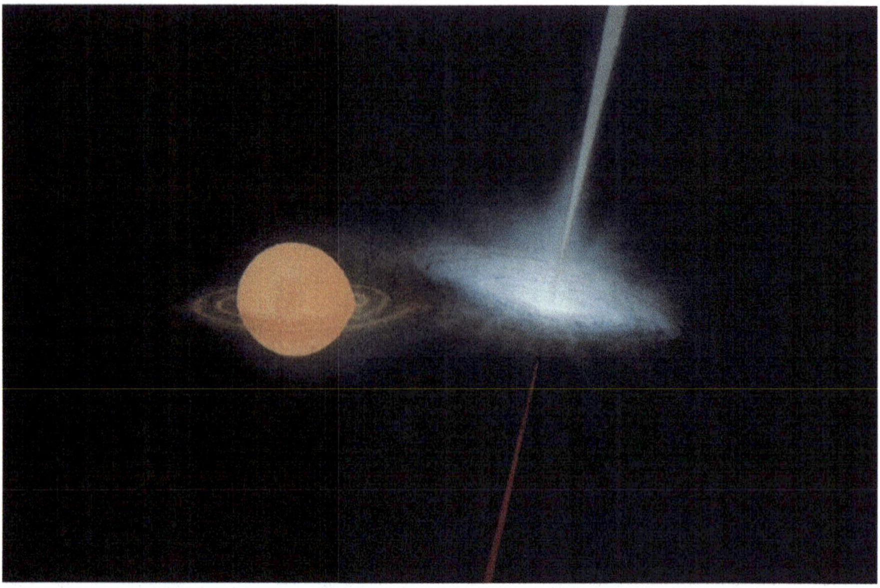

Fig. 7.45 Drawing of a model of SS433. To the right is a black hole with the jets and left is the heavy companion. Source: Astronomy Picture of the Day, 6 March 1996

the magnitude of the event: a similar gamma-ray outburst in 1996 was so strong that it affected some layers in the Earth' ionosphere so that radioamateurs could no longer make long distance connections.

References

Alpar, M.A., et al.: Nature. **300**, 728–730 (1982)
Angel, J.R.P.: Annu. Rev. Astron. Astrophys. **16**, 487 (1978)
Angel, J.R.P., Borra, E.F., Landstreet, J.D.: Astrophys. J. Suppl. **45**, 457–474 (1981)
Aumann, H.H., Gillett, F.C., et al.: Astrophys. J. **278**, L23–L27 (1984)
Backer, D.C., et al.: Nature. **300**, 615 (1982)
Baliunas, S.L., Vaughan, A.H.: Annu. Rev. Astron. Astrophys. **23**, 379–412 (1985)
Bradt, H.V.D., McClintock, J.E.: Annu. Rev. Astron. Astrophys. **21**, 13–66 (1983)
Branch, D., et al.: Astrophys. J. **270**, 123–139 (1983)
Castor, J.I., Abbott, D.C., Klein, R.I.: Astrophys. J. **195**, 157–174 (1975)
Cohen, M., Kuhi, L.V.: Astrophys. J. Suppl. **41**, 743–843 (1979)
Conti, P.: Int. Astron. Union Symp. **105**, 233 (1984)
Damashek, M., et al.: Astrophys. J. **253**, L57–L60 (1982)
Elvis, M., Page, C.G., et al.: Nature. **257**, 656–657 (1975)
Fabian, A.C., Pringle, J.E., Rees, M.J.: Mon. Not. R. Astron. Soc. **172**, 15–18 (1975)
Field, G.B.: HiA. **3**, 37–49 (1974)
Garmany, C.D., et al.: Astrophys. J. **250**, 660–676 (1981)
Geldzahler, B.J., Pauls, T., Salter, C.J.: Astron. Astrophys. **84**, 237–244 (1980)

Goudis, C., Maeburn, J.: Astron. Astrophys. **68**, 189–192 (1978)
Green, R.F., Schmidt, M., Liebert, J.: Astrophys. J. Suppl. **61**, 305–352 (1986)
Grindlay, J.E.: Comments Astrophys. Space Phys. **6**, 165–175 (1976)
Grindlay, J., Heise, J.: Int. Astron. Union Colloquium 2879 (1976)
Harnden Jr., F.R., Branduardi, G., et al.: Astrophys. J. Lett. **234**, L51–L54 (1979)
Hearn, D.R., Richardson, J.A.: Astrophys. J. **213**, 115 (1977)
Herbig, G.H., Jones, B.F.: Astrophys. J. **86**, 1232–1244 (1981)
Hillebrandt, W.: Prog. Particle Nuclear Phys. **17**, 215 (1986)
Hjellming, R.M., Johnston, K.J.: Astrophys. J. **246**, 141 (1981)
Hulse, R.A., Taylor, J.H.: Astrophys. J. **195**, L51–L53 (1975)
Humphreys, R.M., Davidson, K.: Astrophys. J. **232**, 409–420 (1979)
Joss, P.C., Rappaport, S.A.: Annu. Rev. Astron. Astrophys. **22**, 537–592 (1984)
Koenigl, A.: Astrophys. J. **261**, 115–134 (1982)
Koester, D., Schulz, H., Weidemann, V.: Astron. Astrophys. **76**, 262–275 (1979)
Kutner, M.L., et al.: Astrophys. J. **215**, 521 (1977)
Kwok, S., Purton, C.R., Fitzgerald, P.M.: Astrophys. J. **219**, L125 (1978)
Lewin, W.H.G., Joss, P.C.: Space Sci. Rev. **28**, 3–87 (1981)
Liebert, J., Dahn, C.C., et al.: Astrophys. J. **233**, 226 (1979)
Lyne, A.G., Anderson, B., Salter, M.J.: Mon. Not. R. Astron. Soc. **201**, 503 (1982)
Lyne, A.G., Manchester, R.N., Taylor, J.H.: Mon. Not. R. Astron. Soc. **213**, 613 (1985)
Margon, B.: Annu. Rev. Astron. Astrophys. **22**, 507–536 (1984)
Margon, B.: IAU Circ. No. 3345 (1979)
Marshall, F.E., Mushotzky, R.F., et al.: IAUC 3314 (1978)
Mazets, E.P., Golentskii, S.V., et al.: Nature. **282**, 587 (1979)
McClintock, J.E., Remillard, R.A.: Astrophys. J. **308**, 110 (1986)
McGraw, J., et al.: SAOSR. **385**, 125 (1979)
Milgrom, M.: Annu. Rev. Astron. Astrophys. **76**, L3–L6 (1979)
Myers, P.C., et al.: Astrophys. J. **220**, 864–882 (1978)
Nomoto, K., Thielemann, F.-K., Yokoi, K.: Astrophys. J. **286**, 644–658 (1984)
Noyes, R.W., Hartmann, L.W., et al.: Astrophys. J. **279**, 763 (1984)
Paczynski, B.: Int. Astron. Union Symp. **73**, 75 (1976)
Pallavicini, R., Golub, L.: Astrophys. J. **248**, 279–290 (1981)
Pravdo, S.H., Rodriguez, L.F., et al.: Astrophys. J. **293**, L35 (1985)
Robinson, E.L.: Annu. Rev. Astron. Astrophys. **14**, 119 (1976)
Seaquist, E.R., Gregory, P.C., Crane, P.C.: IAUC 3256 (1978)
Smith, B.A., Terrile, R.J.: Science. **226**, 1421–1424 (1984)
Snell, R.L., Loren, R.B., Plambeck, R.L.: Astrophys. J. **239**, L17–L22 (1980)
Stockman, H.S., Schmidt, G.D., et al.: Astrophys. J. **217**, 815 (1977)
Tapia, S.: Astrophys. J. **212**, L125 (1977)
Taylor, J.H., Stinebring, D.R.: Annu. Rev. Astron. Astrophys. **24**, 285–327 (1986)
Taylor, J.H., Weisberg, J.M.: Astrophys. J. **253**, 908–920 (1982)
Thaddeus, P.: Annu. Rev. Astron. Astrophys. **10**, 305 (1972)
Thorne, K.S., Price, R.H.: Astrophys. J. **195**, L101–L105 (1975)
Utrobin, V.P.: Astrophys. Space Sci. **98**, 115–147 (1984)
Vaiana, G.S., Cassinelli, J.P., et al.: Astrophys. J. **244**, 163–182 (1981)
Wallace, P.T., Peterson, B.A., et al.: Nature. **266**, 692 (1977)
Warner, B.: Int. Astron. Union Symp. **73**, 85 (1976)
Weaver, R., McCray, R., Castor, J.: Astrophys. J. **218**, 377–395 (1977)
Weaver, T.A., Zimmerman, G.B., Woosley, S.E.: Astrophys. J. **225**, 1021–1029 (1978)
Weidemann, V., Koester, D.: Astron. Astrophys. **132**, 195–202 (1984)
Wesemael, F.: Astrophys. J. Suppl. **45**, 177–257 (1981)
White, N.E., Holt, S.S.: Astrophys. J. **257**, 318–337 (1982)
Woosley, S.E., Weaver, T.A.: Annu. Rev. Astron. Astrophys. **24**, 205–253 (1986)
Zuckerman, B., Kuiper, T.B.H., Rodriguez Kuiper, E.N.: Astrophys. J. **209**, L137–L142 (1976)

Chapter 8
1975–1984: Galaxies and the Universe

8.1 New Physics

Gravitational Lenses

Einstein's General Theory of Relativity was first tested in 1919 when two English teams demonstrated that the rays of light from stars close to the Sun were no longer straight lines but were bent as a result of gravitational attraction by the Sun. Now consider a cluster of galaxies. Such clusters contain a great deal of matter, most of which is dark and invisible, as Zwicky had already proposed in the 1930s. Suppose that there is a galaxy far behind the cluster, and that it is, by chance, on the continuation of the line of sight from us to the cluster. Light rays from the distant galaxy are deflected by the cluster and purely by chance some rays of light now reach us, rays that we would have missed if the cluster had not been there. As a result, we receive more light from the galaxy; the light has been magnified. The image of the galaxy, though, will generally be distorted because the lens has an irregular shape. We may even see two different images of the same object. All this is the explanation and the justification for calling the cluster a 'gravitational lens'.

In 1979, Americans Walsh, Carswell and Weymann (Walsh et al. 1979) announced that they had discovered two quasars that were beside one another on a photo and, apart from a single detail, they had the same spectrum with a red shift $z = 1.41$ (see Fig. 8.1). The conclusion was that this was the effect of a gravitational lens: two different images of the same object.

The two images came from the same object, but there were different factors that could cause each of them to be distorted. In this case, the two quasar images A and B were too small to detect any distortion. However, thanks to long-baseline interferometry, radio astronomy was able to produce much sharper images. There were significant differences between the images, and the radio astronomers involved wrote in 'Science' that the two images could not come from one and the same source. However, less than a year later—and again in 'Science'—the same authors

H. J. Habing, *The Birth of Modern Astronomy*, Historical & Cultural Astronomy,
https://doi.org/10.1007/978-3-319-99082-8_8

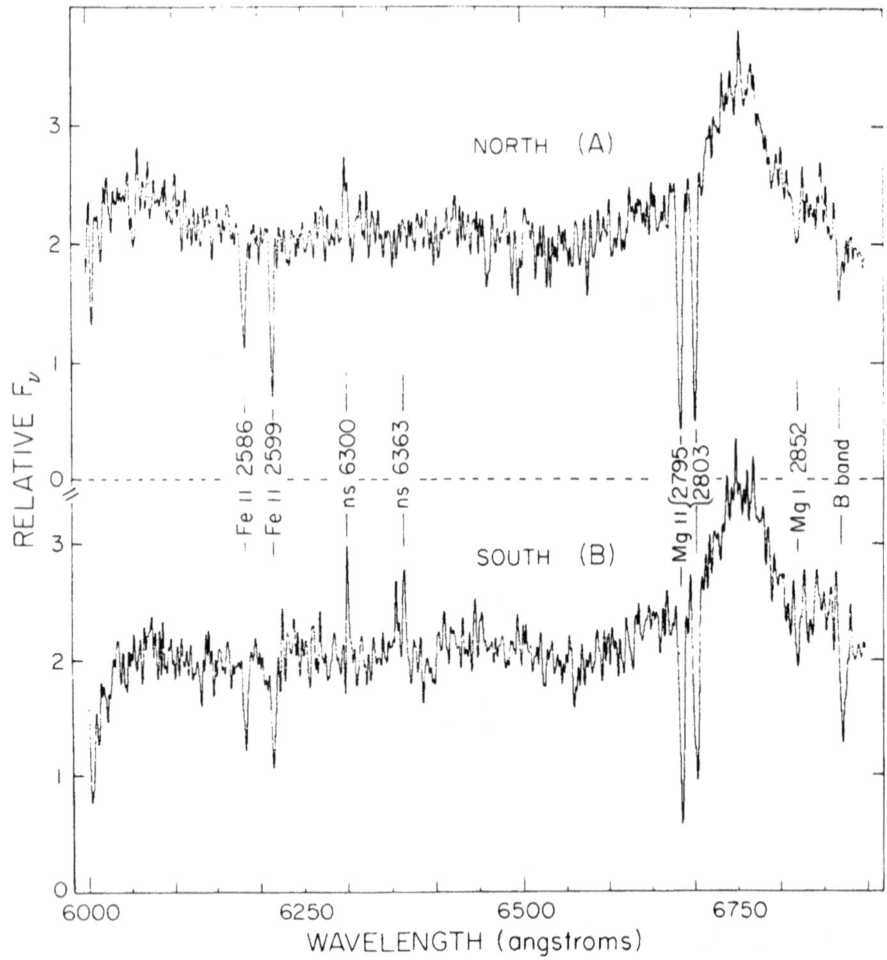

Fig. 8.1 The spectrum of each of the two images of quasar 05957+561 (Weymann et al. 1979a)

retracted their claims and stated that what one was seeing was the result of a gravitational lens. The discussion was cut short a few years later when the image of both objects was observed during a VLBI (Very Long Baseline Interferometry) experiment and much smaller details were observed. Figure 8.2 shows that in this experiment the two images did indeed look the same although there were minor differences.

The effect of the gravitational lens is explained in Fig. 8.3. This figure is a modern revision of the figure that appeared in a short article in 1924 in the then eminent journal '*Astronomische Nachrichten*', in which the author, Soviet astronomer Chwolson, concluded with the words: 'I have no idea whether this potential phenomenon will ever occur in reality.' In 1964, Norwegian astronomer Sjur Refsdal revised this idea and added that the discovery would offer completely new opportunities for cosmological observations. The Hubble constant could now be measured in a completely new way. Refsdal's work was more or less forgotten until the discovery of the double quasar.

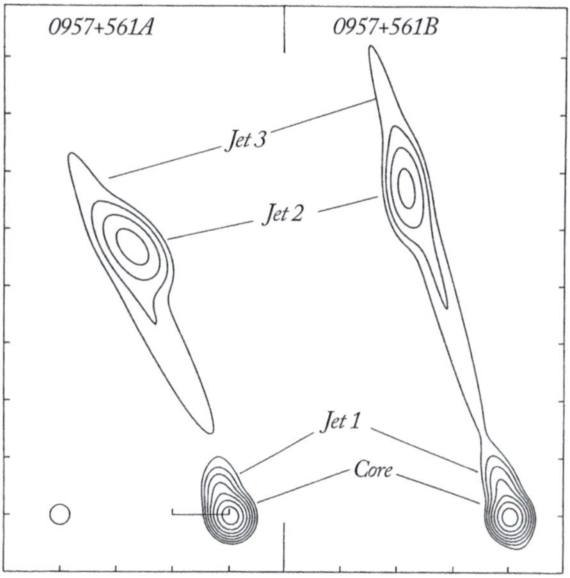

Fig. 8.2 Two images of the radio radiation from objects A and B of quasar Q0957+561 at a wavelength of 13 cm and obtained with VLBI measurements. The circle below left indicates the sharpness of the images; it has a radius of 3.5 milliarcseconds; the distance between two tickmarks in the frame of the figure is 10 milliarcseconds (Gorenstein et al. 1988)

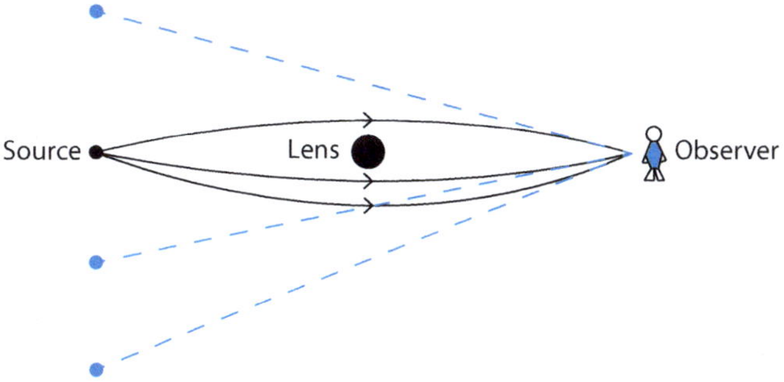

Fig. 8.3 The source emits rays of light in all directions. Three rays are drawn because they reach the observer after having been deflected by a lens, which is a massive galaxy, or a cluster of galaxies. The observer sees three images of the source because the rays of light are coming from different directions. Based on image from Citizendium

A nice example of a gravitational lens is the so-called 'Einstein cross': four images of a distant quasar produced by the gravitational attraction of a galaxy. The object was first recognized as a small object with a spectrum with spectral lines from two systems each with a different redshift. It took several years, and an excellently transparent and quiet night at 4200 m altitude, to show that there were four images within one arcsecond surrounding the quasar! (Huchra et al. 1985; Yee 1988). See Fig. 8.4.

There are several ways in which gravitational lenses can be used to help answer cosmological questions (Blandford and Narayan 1992). First, as explained above, a gravitational lens will brighten the image of the background galaxy. This could make galaxies detectable that would be too distant and too faint without magnification. Second, as Refsdal proposed as early as in 1964: when there are two images,

Fig. 8.4 Gravitational lens: the Einstein Cross: four images from the same quasar Q2237+0305 surrounding a galaxy. Source: NASA, ESA, and STScI

the light reaches us by two separate paths that differ slightly in length. Suppose that a supernova were to explode in the source, then the observer would see it appear in one image earlier than in the other. The distance to the source could be deduced from the time that elapsed between the two detections, as well as the Hubble constant, both in a way that was completely independent of other methods. Third, there is the possibility of calculating the total mass of a cluster by measuring the magnification or the distortion of a background galaxy by the gravitational field of the cluster. All three possibilities have been realised in subsequent years.

Invisible or 'Dark' Matter

In Chap. 5 a decennia-old, but rarely discussed, problem was finally addressed and resolved: the decision that 'invisible matter' existed—matter that could not be seen, and that emitted no visible light or electromagnetic radiation at any other

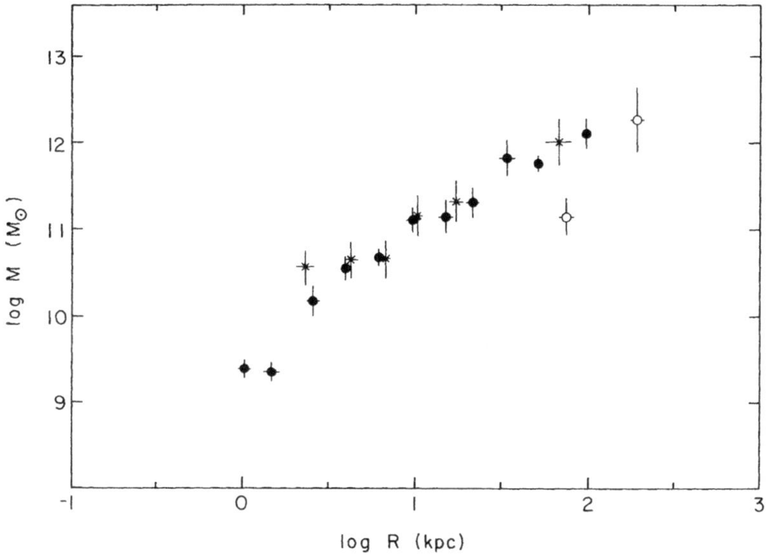

Fig. 8.5 The mass of the Milky Way within the radius R. Mass in the vertical direction, radius in the horizontal direction (Faber and Gallagher 1979)

wavelength. There is even much more of this invisible matter than the matter we can see. The existence of much invisible matter was of great importance because it had never been detected by the physicists, neither in theory nor in experiments. It was therefore essential for astronomers to collect as much evidence as possible to confirm the discovery and to rule out other, more profane explanations.

In a review paper in 1979 Faber and Gallagher (Faber and Gallagher 1979) showed measurements of the mass of the Milky Way through the velocities of globular clusters and satellite galaxies that were far away from the galactic center. Figure 8.5 shows that the mass of the Milky Way is still growing at a distance of 100–200 kpc (320,000–640,000 light years). At these distances there is hardly any luminous object to be found.

To show the presence of invisible matter, the density of the matter is compared to the amount of light that is produced. The Sun is representative of such a comparison. If we add up the mass of all the stars in the solar neighbourhood (a quantity called M) and if we also add up all the light emitted by these stars (call it L), then the ratio M/L is between 1.1 and 2.4 times the ratio for the Sun. From this we can conclude that invisible matter is present in a given object if M/L is much larger than 1.

The value of M/L has been determined for objects in these three categories. Globular clusters have M/L values around 1.6 and they are thus largely free of invisible matter (Illingworth 1976). For the Milky Way, the value calculated is M/L = 70 ± 20, which confirms the conclusion already drawn: the Milky Way is surrounded by a halo that contains much more invisible matter than the visible matter of the Milky Way (interstellar matter can be left to aside: its mass is a very small part of the mass of the stars) (Faber and Gallagher 1979).

Measurements of the 21 cm line radiation with the Westerbork Radio Telescope showed that a majority of spiral galaxies maintained a high speed of rotation far outside the visible galaxy (Bosma 1981). A very careful analysis has been made of the 21 cm line radiation from the spiral galaxy NGC 3198. Figure 8.6 shows a photograph of this galaxy and in white contours the distribution of atomic hydrogen gas. It is obvious that the gas layer extends to distances far from the visible galaxy.

Figure 8.7 shows the rotation velocity of the hydrogen gas as it varies in line with the distance from the center. A model of the mass distribution has been made, assuming that the mass is the sum of that of a flat, rotating disk and of a wide halo. As it turns out, there are various models for the disk and for the halo that reproduce the same sum—there is no unique answer. However, independent of what is the best model, the authors concluded that at a distance of 30 kpc from the center the value of M/L was 18.

Figure 8.8 is taken from a review in 1979. It shows values of M/L for various kinds of galaxies. In all cases, M/L is significantly larger than 1, indicating that invisible matter is present in all types of galaxies, and it always has more mass than

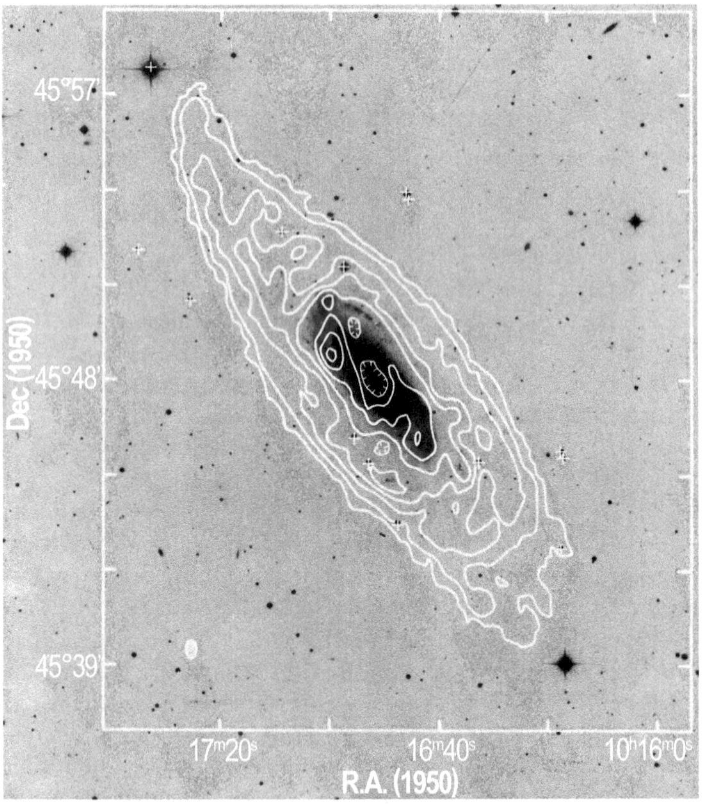

Fig. 8.6 A photograph of NGC 3198 and in white contours the distribution of the neutral hydrogen (van Albada et al. 1985)

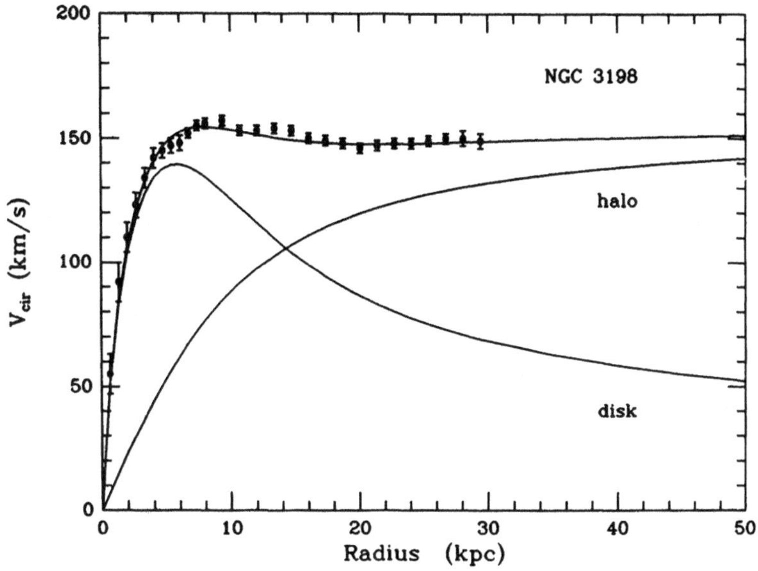

Fig. 8.7 Rotation curve of the galaxy NGC 3198. The velocity of rotation in the disk of the spiral galaxy has been measured using the 21 cm line. The points represent the measurements: the distance from the center is along the x-axis and the velocity along the y-axis. The curves are taken from model calculations. The upper curve through the measurements is the sum of the two components of the model: a disk and a halo. The halo consists of dark matter (van Albada et al. 1985)

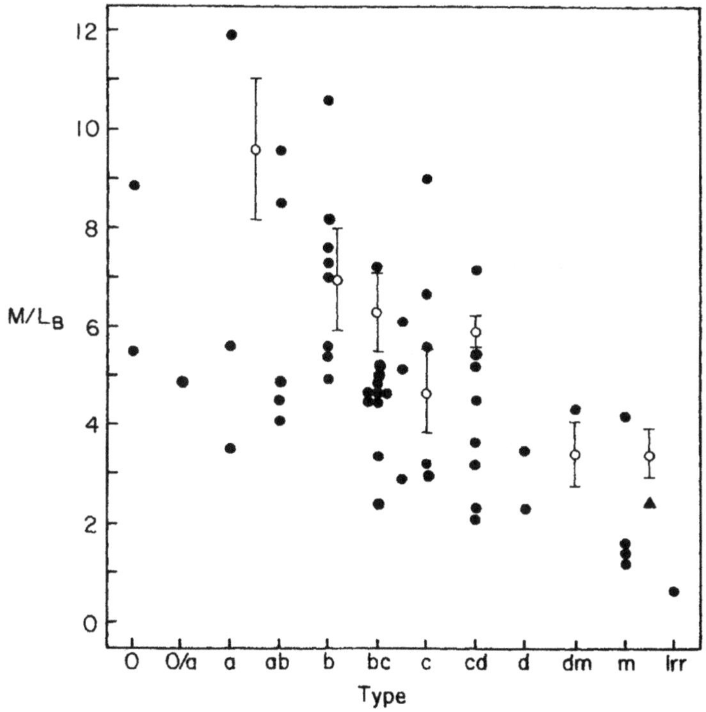

Fig. 8.8 The ratio between the total light output of a galaxy and its total mass, both compared to the Sun, is given vertically. The Hubble type of the galaxy is shown in the horizontal direction. 'O' are ellipticals; O/a are S0 galaxies; a, ab, bc, c, cd and d are spiral galaxies; dm, m and irr are various types of irregular galaxies (Faber and Gallagher 1979)

the visible matter. The critical reader will notice that the spiral galaxies in this diagram have M/L values lower than in the case of NGC 3198 discussed above. This reflects the uncertainties in the determination of M/L that still existed in 1979.

Elliptical galaxies contain little or no hydrogen and the mass has to be determined from optical spectra taken at various points in the galaxy. The derivation of the total mass from these measurements depends on various somewhat uncertain assumptions (see below). Nevertheless, the values obtained should be more or less right. In a study of 15 elliptical galaxies Schechter found M/L values between 5 and 20 (Schechter 1980). It thus appears that spiral and elliptical galaxies contain much invisible matter. In later years dwarf galaxies were also embedded in a halo of invisible matter (Aaronson 1983; Faber and Lin 1983).

Once the existence of invisible matter had been accepted, the speculation started about its nature. Could it be comprised of itinerant and undetectable black holes or almost undetectable brown dwarf stars? Or was it something more profound, a type of matter that physicists had never before measured, such as 'black neutrinos'? The theoretical physics of elementary particles did leave room for such speculations. Physics researchers took up the gauntlet, but without success, which was unfortunate, but for astronomy it was enough that sufficient characteristics of the dark matter were known to be able to explain their observations. In 1984, Blumenthal, Faber, Primack and Rees (Blumenthal et al. 1984) published an article in which they chose from three types of elementary particles: 'hot dark matter', 'warm dark matter' and 'cold dark matter', depending on whether the invisible matter comprised particles with a light mass (30 eV)[1], an average mass (a few keV) or a heavy mass (1 Gev). They concluded that the mass of these particles determined the mass of the first objects that they created: the greater the mass of the particles, the smaller the mass that contracted. In particular, the existence of light dwarf galaxies needed a heavy mass of invisible, elementary particles. 'Cold dark matter' (CDM) was the winning substance—for the astronomers at least, because the nature of this matter is still uncertain. A number of possibilities have been suggested, some of them requiring drastic changes in fundamental theories of physics.

The enormous impact of the article by Blumenthal et al. was not yet appreciated by the time a symposium on invisible matter that was held at Princeton in June 1985. ('Dark matter in the Universe', IAU Symposium 117). The first speakers raised all kinds of crucial questions, including what this invisible matter might consist of. Two possible answers were proposed for further study. One was that a lot of matter was stored in 'brown dwarfs', stars that were so light that no nuclear fusion took place at their center and they consequently radiated virtually no energy. The other answer was the storage of this matter in light black holes that floated around in the Milky Way without being noticed. These attempts to explain the invisible matter by normal matter in undetectable stellar objects did not take away the challenge for physicists: could types of elementary particles exist that would explain this invisible matter? There was no clear answer: this invisible matter was a fundamental problem. At the end of the symposium, Jim Gunn and Scott Tremaine, no

[1] The mass of one particle is expressed as a measurement of energy: $E = mc^2$.

lightweight astronomers, summarized the situation. Both started philosophically and concluded that 'invisible matter' represented a crisis in astronomical research. We were probably facing a 'paradigm shift' that, according to science historian Kuhn, could occasionally occur in a branch of the physical sciences that led to a completely new vision of the subject; the motto of the story could be 'Never waste a good crisis'. By the end of this chapter we will see that this crisis did indeed have a very useful effect: it became an essential part of a new theory on the formation of galaxies and of clusters of galaxies.

8.2 Galaxies Without an Active Nucleus

In the period from 1975 to 1984, a great many large new telescopes came into use, there were strong improvements in detection equipment and with the use of computers the quality of the data analysis reached new heights: in short, it was now possible to draw more precise conclusions based on the same measurements. Compared with photographic recordings (a chemical process), measurements of the light distribution of galaxies by CCDs were vastly improved as were measurements of the spectrum. As a result, the redshift of spectral lines of many more galaxies were measured, from which the distance of a galaxy could be determined using Hubble's law. It was also possible to deduce from the width of a spectral line how rapidly the stars swarmed around one another. As an analogy with the speeds of molecules in gases, the terms 'hot' or 'cold' were sometimes used when the line was broad or on the other hand narrow.

There were large spectroscopic differences between elliptical galaxies and spiral galaxies. The first contained no gas and consisted only of stars. The spectra were made up of very many different stellar spectra and had only absorption lines: dark lines against the weak background of the galaxy. The spectra of spiral galaxies, on the other hand, almost always contained clear emission lines from light-emitting nebulae, such as the Orion nebula. The presence of a lot of gas in spiral galaxies made it possible to measure the 21 cm line, but only in relatively nearby galaxies. Distant disk galaxies were too weak to be observed with radio telescopes even when they could be imaged easily using optical telescopes.

Elliptical Galaxies

Hubble had placed the elliptical galaxies in a sequence according to their degree of flattening (see the Prologue). The image of E0 galaxies was spherical, and images of E1, E2 to E6 galaxies were increasingly flatter ellipses. The flattening, it was assumed, was the result of a spherical galaxy revolving around its axis: the faster the galaxy revolved, the flatter the ellipse became. Theoretically, so it was thought, the elliptical galaxies were interesting but not fundamentally difficult. Of course, this should be checked observationally by measuring spectra of a few spots in these galaxies and by looking for the Doppler effect. However the elliptical galaxies were mostly too faint to obtain good spectra. Optical observers had to accept that the only observation within their reach was to measure the distribution of light.

In the 1970s, however, spectroscopy techniques had advanced so far that it was now possible to produce good spectra of elliptical galaxies. With two new, 4 m American telescopes, one in Arizona and one in Chile, and thanks to a special kind of television camera, a precursor of CCD cameras, a young Canadian-American astronomer, Gareth Illingworth (1977), was able to produce spectra of sufficient quality. He could deduce how rapidly a galaxy revolved around its axis (V) and at the same time the width of the line (σ) could also be measured. This parameter σ was an indication of how far the speed of individual stars differed from the average. The result, V was always smaller than σ, took observers and theoreticians as a surprise. The theory that elliptical galaxies were flattened by rotation was wrong. Illingworth had concentrated on relatively large elliptical galaxies. When measurements were made of intrinsically faint ellipticals, these rotated as expected for a flattened ellipsoidal object (Davies et al. 1983). This looked like an indication that faint, small ellipticals galaxies had been built in a fundamentally different way from how large elliptical galaxies had been put together (Fig. 8.9).

Within a year, British theoretician James Binney (1978) proved that elliptical galaxies were not flattened spheres, but rather that they were shaped like cigars. In mathematical terms, they were not oblate spheroids (objects that looked something like a pumpkin), but prolate spheroids (cigar-shaped). The first type came about when an ellipse revolved around its short axis, the second when it rotated around its long axis.

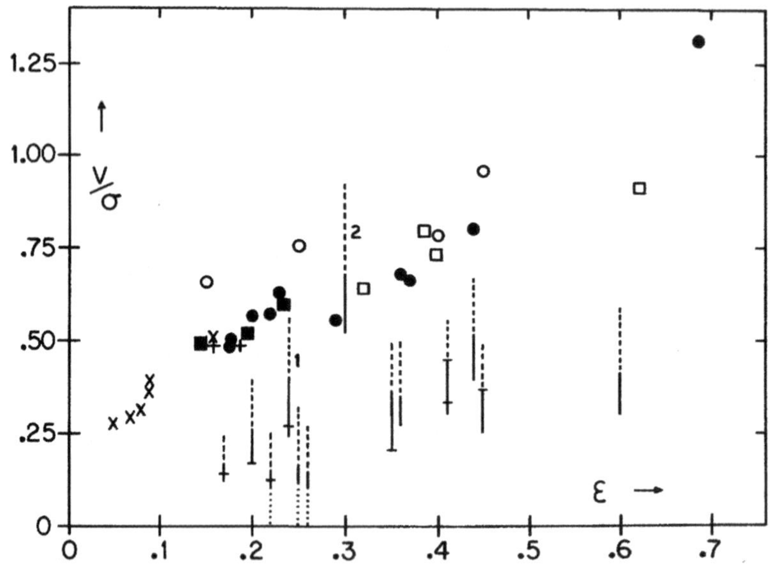

Fig. 8.9 Spectroscopically measured rotation velocities, V, of elliptical galaxies of different flattening. The width of the spectral lines, σ, has also been measured. The galaxies have an elliptical image and the flattening is given by the eccentricity of the ellipse ε. The eccentricity ε of the ellipse is given horizontally. The ratio V/σ is indicated in the vertical direction (Illingworth 1977)

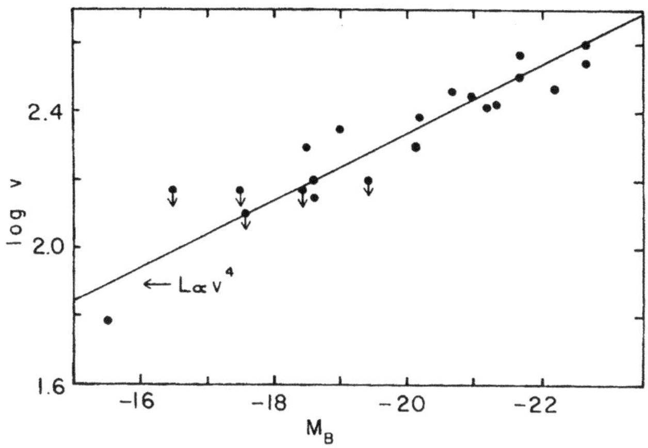

Fig. 8.10 Relationship between the breadth of the absorption lines and the absolute magnitude for elliptical galaxies (Faber and Jackson 1976)

Binney concluded that it had become a lot more difficult to construct models for elliptical galaxies. One no longer knew, for instance, along which orbits stars could move through such a galaxy. Mathematical equations were easy to write but they could not be resolved with pen and paper. Martin Schwarzschild, whose name we came across earlier, took up the challenge and used a computer to calculate possible orbits of stars, and eventually he managed to combine all these orbits and to construct a cigar-shaped galaxy (Schwarzschild 1979). A few years later, a Leiden PhD student, after a brief internship under Schwarzschild, found a family of elegant mathematical models with which Schwarzschild-like models could be approximated (de Zeeuw 1985).

Now, a very different and major discovery about elliptical galaxies: there was a strong correlation between the amount of light that an elliptical galaxy emitted and the width of the spectral lines—see Fig. 8.10 (Faber and Jackson 1976). Apart from any theoretical explanation there was a very practical application for this correlation. From now on, the total amount of light emitted by an elliptical galaxy could be calculated from the width of the spectral lines, and then the distance could be determined by comparing the amount of light produced by the galaxy with the amount of light received. This 'Faber/Jackson relationship' played the same role for elliptical galaxies as the Tully/Fisher relationship played for disk galaxies (see below). A few years later this relationship would be improved even further. See Chap. 15.

Spiral Galaxies

The most important characteristic of a spiral galaxy is the presence of a rotating disk of gas and young stars that have been formed in this gas. In the period from 1975 to 1984 the formation of stars became a topic of increasing importance and this remained so in subsequent years. The concept of the 'star-formation rate', the mass of newly formed stars per unit of time and abbreviated to 'SFR', was to become a household term. A subject of special attention is the 'starburst', the production of a large number of stars within a short period, a sudden almost momentaneous peak in the SFR parameter.

This has a deeper meaning: stars are born from interstellar matter and at the end of their lives they will return most of that matter to interstellar space. During each cycle of star birth and star death, some matter disappears in eternal graveyards such as white dwarfs, neutron stars and black holes. The continuous loss of interstellar matter will cause the Universe to grow old. The formation of stars increasingly became seen as a sign of the 'fertility' of galaxies and as a signal of the ageing of the Universe. From about the year 2000 we will see large efforts to measure the SFR in distant galaxies and in an earlier phase of the Universe.

An early and successful paper on starbursts in other galaxies appeared in 1978 (Larson and Tinsley 1978). They selected a set of 'normal' spiral galaxies, galaxies from the Hubble Atlas, that is used to define Hubble's classification. They also selected a set of 'peculiar galaxies' from Arp's Atlas (see Chap. 5). Larson and Tinsley collected the magnitudes of the galaxies through three filters, a magnitude U at ultraviolet wavelengths, B in blue light and V in yellow or visual light. Next they determined two 'colors' by taking U-B and B-V. Two diagrams of the (B-V) color versus the (U-B)-color are shown in Fig. 8.11. It is obvious that these peculiar

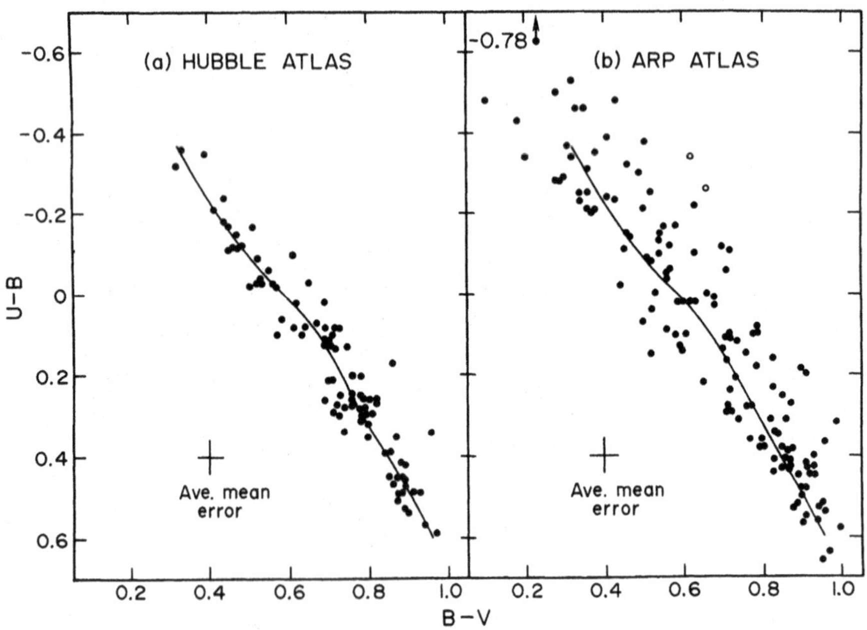

Fig. 8.11 The (B-V) and (V-U) colors of two samples of galaxies: from the Hubble Atlas (the 'normal' galaxies) and from Arp's atlas of peculiar galaxies. In the left diagram the points show a strong correlation between the two colors whereas the correlation is much poorer in the sample of peculiar galaxies (Larson and Tinsley 1978)

galaxies have a different, wider spread in the diagram than the normal galaxies. The authors ascribe the wider spread to 'starbursts' and they draw the conclusion that the violent events going on in peculiar galaxies stimulate the formation of new stars: a result that at that time was somewhat speculative but that later will be confirmed over and again.

The Larson-Tinsley approach of characterising a galaxy by two colors only is, of course, rather crude. A more subtle way of determining the SFR of a disk galaxy is by measuring the strength of the Hα-line that is emitted by ionized nebulae. The reasoning behind this is that the strength of the Hα-line measures most of the light from stars of spectral types O and B, thus of massive, luminous, young stars that live only a brief moment (astrophysically speaking). The strength of the Hα-line appears to be a reliable proxy for the value of the SFR (Kennicutt 1983; Kennicutt and Kent 1983). Making use of other indicators Kennicutt estimated the average value of the SFR over the last several billion years. In Fig. 8.12 the ratio is given between the present-day value of SFR and <R>, its value integrated over a long, previous period. This is a first indication that the evolution, the history of spiral galaxies is measurable.

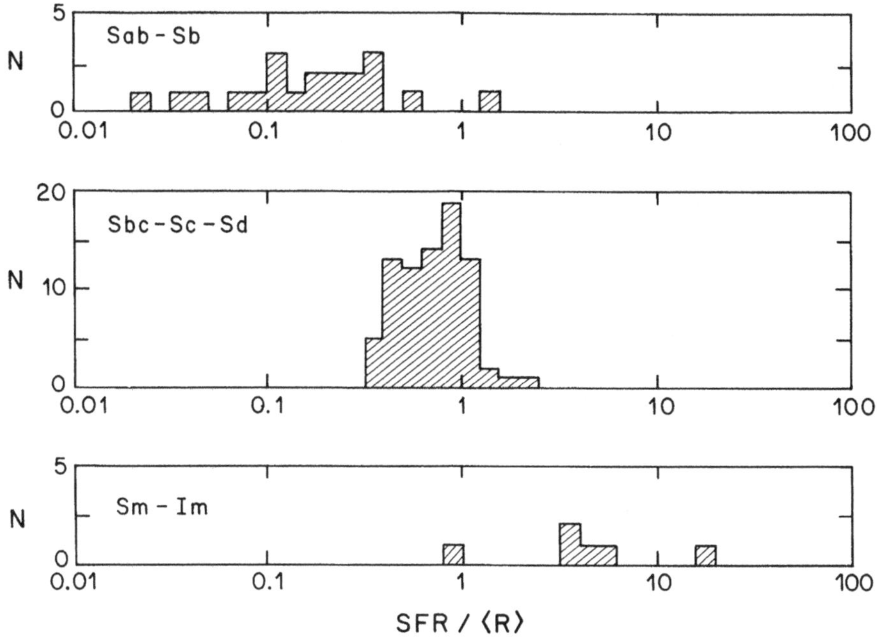

Fig. 8.12 Three bar diagrams of the ratio between the present-day star formation rate (SFR) and the same rate measured over a long previous period, R. The three different groups contain galaxies with tightly wound spiral arms, intermediate open spiral arms and very open spiral arms (Kennicutt 1983)

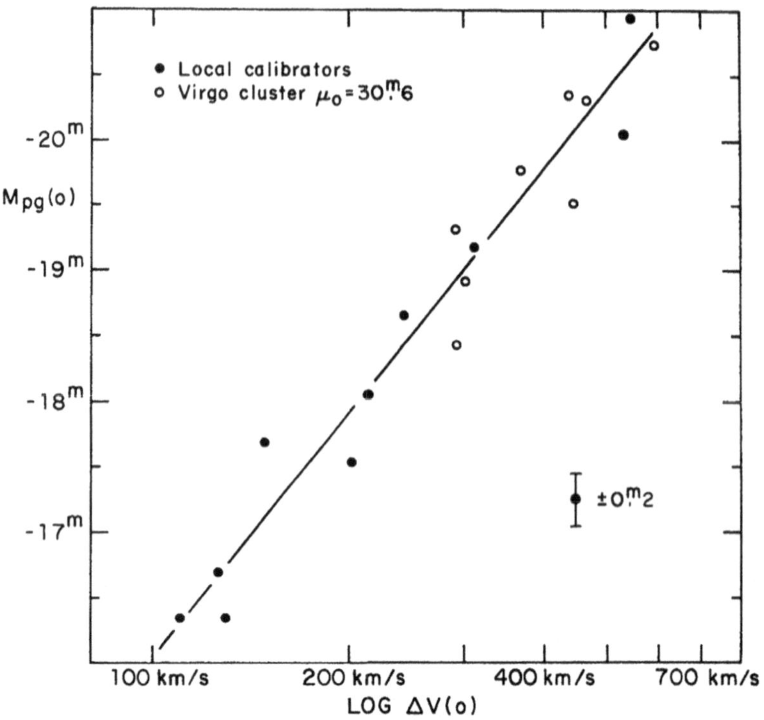

Fig. 8.13 The width of the 21 cm line is shown horizontally and the absolute magnitude vertically, that is the (logarithm of the) amount of light emitted by the galaxy (Tully and Fisher 1977)

The discovery of the Faber-Jackson method for measuring the distances of elliptical galaxies is described above. An equally important discovery was that a somewhat similar relation existed for spiral galaxies. Measuring the total width of the (radio) spectral lines at 21 cm, it became apparent that this width correlates strongly with the amount of light that a galaxy emitted, the so-called Tully-Fisher relation (Tully and Fisher 1977; see Fig. 8.13). By measuring the width with a radiotelescope you derived how much light a galaxy radiated and (in the same way as for the Faber-Jackson relation) from the amount of light that you actually received it was possible to derive the distance of the galaxy. A few years later, the distances were found to be even more precise if one measured the total light of the galaxy at near-infrared wavelengths: the stars were then less obscured by interstellar matter.

8.3 Radio Galaxies: Galaxies with an Active Nucleus

The first 'radio galaxies' were very strong radio sources (1950). Later on, optical detections from active sources in Seyfert galaxies (1963), from the quasars (1963) and from the blazars (1972) were added. The two jet streams emanating from the core of the galaxies made impressive photographs. The amount of energy deposed by the jets in the surrounding was measured to thousands of supernovae. Radio telescopes were not alone in studying galaxies with an active nucleus. From the start of 1970s, there were also X-ray and infrared observations of these galaxies.

Systems with an active core, AGN systems (AGN = Active Galactic Nucleus), were observed in many different ways and these observations yielded all kinds of relationships between different quantities that were measured. There were numerous conferences and many reviews. All measurements were put together and discussed in different diagrams and more and more knowledge was gathered but not compressed. More and more different types were distinguished; in a review article from 1993, which we will discuss later, there are more than ten different acronyms introduced. One has to ask whether people were still able to see the wood for the trees. An example is the two Fanaroff-Riley types of radio galaxies (FRI and FRII) from 1974: observationally it was still a convincing distinction 40 years later, but its meaning is still unclear. One had the impression that there was too much detail veiling the essence.

I do not want to bore the reader with too much information and will limit myself to a few key issues.

Speeds Greater than the Speed of Light?

Chapter 5 reported that radio astronomers very accurately determined the position of the bright spots in a radio system, using the VLBI technique. When they repeated the measurements several years later, a number of radio systems showed that the location of these bright spots on the sky had changed: they had moved (Zensus and Pearson 1988). Such movements had been known for stars for centuries by the term 'proper motion'. The 'proper motions' of these VLBI spots created a problem: if the movements were combined with the distance, the bright radio spots seemed to move at a speed greater than that of light. This was, of course, a fundamental problem: Einstein's two theories of relativity, the special and the general, are based on the assumption that no object can move faster than the speed of light, c. Fortunately, this discovery of 'superluminous motion' had already been anticipated by British theorist Martin Rees (1966) but the radio astronomers had not read Rees's article and they made an issue out of the 'superluminal motion'. There is a report in the journal Nature of a study week on superluminal motion in which the title of one of the presentations was: 'Super luminous motions; astronomers still puzzled'. Despite the puzzle, the crisis was already over. In the English version of Wikipedia there is an illuminating explanation in the lemma 'superluminal motion'.

Massive Black Holes

In Chap. 5 the suggestion was presented that there was a very massive black hole in many galaxies, but that this hole was silent, dormant and only active in a few galaxies. True to their name, black holes do not emit radiation and are 'silent'. Only when matter from the outside is fed into the hole does this matter produce (much) radiation just before it enters the black abyss. For a long time there was no direct evidence for the suggested dormant 'massive black holes', but the idea continued as an important hypothesis. Black holes had been found, for example in the X-ray binary Cyg X-1, discussed in Chap. 4. But in these cases they had the mass of a star, and not of millions of stars. Galaxies were diligently searched for direct evidence of massive black holes, and every time the measurements were found to be suggestive but not compelling—the smoking gun was missing. A much-discussed example at the time was the source in the center of the large elliptical galaxy M87, the origin of the strong radio source Virgo A (see earlier chapters). The spectrum of the inner parts of the core was measured using a new photon detector built in London and linked to the 4 m-telescope in Arizona. I quote the last sentence from the abstract of the relevant article (Sargent et al. 1978): 'We conclude that the observations of M87 are entirely consistent with the presence of a black hole of about 5 billion solar masses.' The observations were consistent, but not compelling.

Even though no definitive proof could be found, the concept of a massive black hole as the heart of an AGN remained: there was no alternative for the source of a flow of so much energy. Figure 8.14 can be found in a review article from 1984 on supermassive black holes as the true source of an AGN. It is an illustration of the large number of paths by which, according to the author, matter eventually must end up in a black hole of large mass. The origin of such a hole, he says, is the natural outcome of a large number of different processes. In summary, it would be highly surprising if massive black holes did not exist.

In 1982 a Polish astronomer (Soltan 1982) published a bold but very interesting estimate of the mass of such a—still hypothetical—massive black hole in quasars. An estimate could be made of how much matter disappeared into the black hole per million years on the basis of the amount of light a quasar produced. By counting the number of quasars, it was possible to estimate how long the average quasar existed and thus how much matter it ingested: about 100 million times the mass of the Sun. This, too, was a good indication for the existence of massive black holes.

In Chap. 5, Fig. 5.20 showed the energy that the different sorts of galaxies emitted: spirals, elliptical systems, quasars, etc. The figure suggested

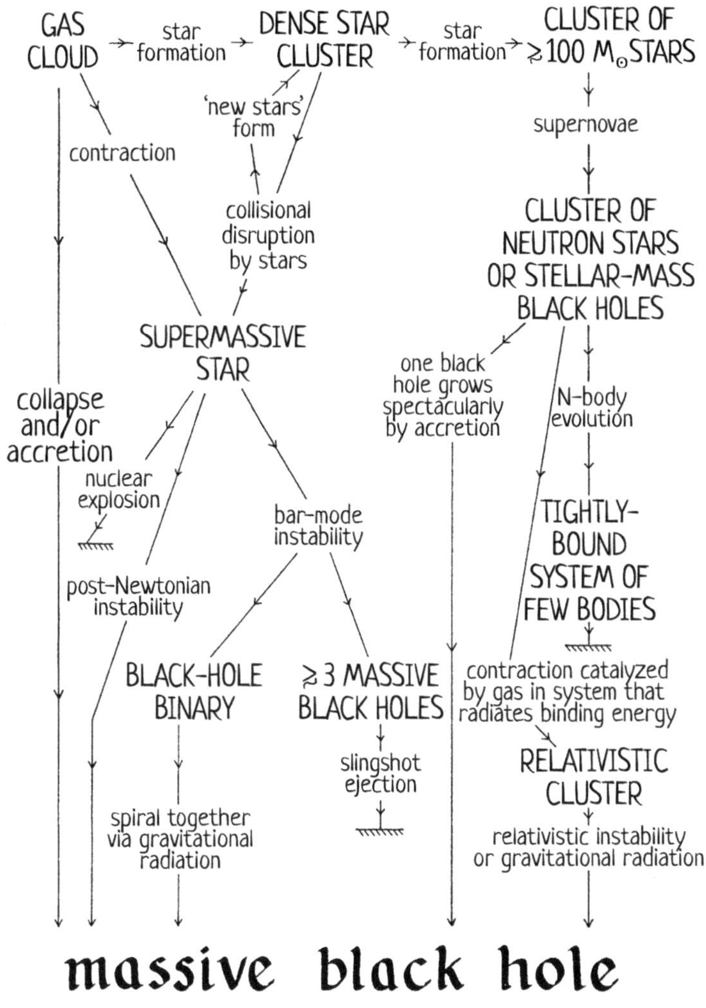

Fig. 8.14 Starting at the top left with a gas cloud, possible developments have been given of which most paths lead to a massive black hole. Please note, not all roads lead to Rome (Rees 1984)

(perhaps unintentionally) that there was a certain relationship between all these different types. 'Unification' was an idea that many researchers harbored but could not prove. Another attempt at unification can be seen in Fig. 8.15.

In 1980 a young American astronomer, Tim Heckman, published several articles in which he posed the question of whether there was a significant difference between normal and special galaxies. He recorded spectra of the cores of numerous ordinary systems and found clear indications that many of them had a core with some activity. In other words, there could also be a black hole there. Research by others confirmed the conclusion: the boundary between 'active' and ordinary galaxies is fluid.

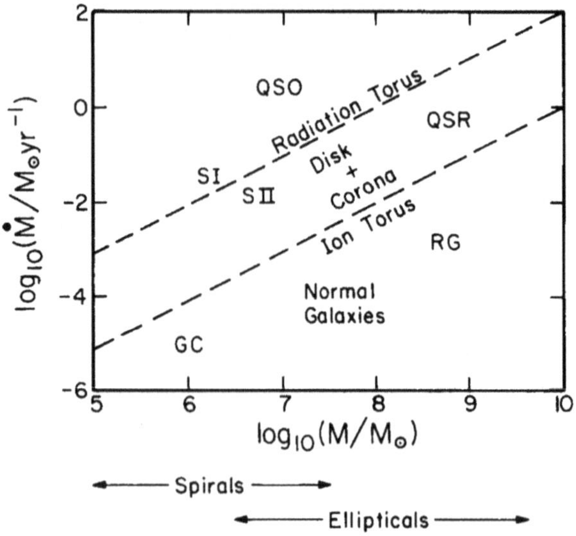

Fig. 8.15 'Attempt' at unification of all sorts of active galaxies. It is assumed that the properties of all types of AGNs can be explained from only two variables: the mass of the black hole (horizontally plotted) and the amount of matter added to the hole in a year (vertically plotted). A QSO is a quasar that is silent at radio wavelengths, a QSR is a quasar with radio radiation; SI and SII are the two types of Seyfert galaxies; RG is a radio galaxy (a galaxy with strong radio radiation) and GC is the compact radio source in the centre of our Milky Way. A 'torus' is a kind of donut around the center, which is large and thick (radiation torus), but can also be thin (ion torus). The corona is a tenuous sphere of very hot gas around the center that protrudes above the donut (Blandford 1984)

8.4 Mergers, Cannibalism and Massive Star Formation

In 1972, brothers Alar and Juri Toomre published an article on close encounters between galaxies and their consequences (see Chap. 5). In hindsight, their article proved to be a fruitful beginning of research on the merging of similar galaxies and about 'cannibalism' whereby large galaxies swallow little ones. Alar Toomre (1977) claimed that collisions occurred more often than one might think and illustrated this using several nearby galaxies that all exhibited 'feelers' (see Fig. 8.16). In 1994, a small galaxy was even found that today is being devoured by our own Milky Way, but that's for another chapter.

This led to the idea that there must be a connection between merging galaxies on the one hand, and starbursts on the other. When two galaxies collide, the stars do not themselves collide; they are too small, compared to the empty space in which they move but gas clouds are a different matter: they take up much more volume and do collide with one another. Such a collision between nebulae would cause the gas to compress, which would lead to the formation of many new stars. These new stars would heat up the cloud and this heat would then need to be released as infrared radiation.

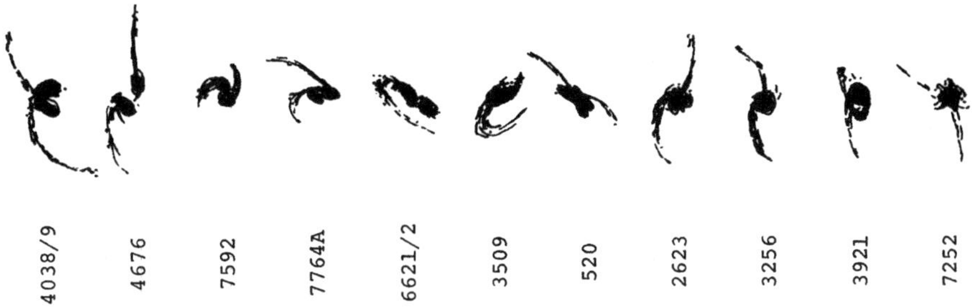

Fig. 8.16 Eleven sketches of entanglements of galaxies that will most likely lead to a merger. The numbers refer to a well-known catalogue (NGC) (Toomre 1977)

Infrared astronomy made a major leap forward in 1983 with the American-Dutch-British infrared IRAS satellite. For the first time, the infrared radiation of the whole sky, both the northern and the southern hemispheres, was mapped by wavelengths where both galaxies with an active core were reported as well as the starburst galaxies.

One of the strongest IRAS sources at 60 and 100 μm was a galaxy that occurs as number 220 in Arp's 'Atlas of Peculiar Galaxies' from 1966. After the publication of the IRAS measurements, a number of different research groups simultaneously started a more in-depth examination at those infrared wavelengths where the radiation can be measured from the ground. It was confirmed that Arp 220 was actually a case of colliding nebulae and that it was a starburst galaxy of exceptional strength: in the infrared alone the same amount of radiation is emitted as 10^{12} (= a thousand billion) times that of the Sun. Arp 220 turned out to be another, very rare source of radiation: it emitted maser radiation at 18 cm. There was no easy explanation for the strength of this maser line but it indicated at least very dense matter and an intense infrared radiation field. Arp 220 was a very bright galaxy in the infrared and the prototype of a new class: LIRG, 'Luminous Infra Red Galaxy'.

8.5 Deeper in the Universe

The distance to nearby galaxies could be measured directly using variable stars—as was done by Hubble. The detection of the Faber-Jackson and the Tully-Fisher correlation (see above) made it possible to go deeper and still measure the distance directly. Beyond the limits of these new methods the generally accepted method was to measure z and to derive the distance d. As long as z is small (say <0.3) the Hubble relation can be used ($zc = Hd$) but for larger values of z a more complex relation is needed. This will be discussed in Chap. 16. For the impatient among us: the mathematical relationship is well explained by Wikipedia under the keyword 'Lambda-CDM model'. To put the value of z in perspective, let us calculate the value of z when 'recombination' took place, the point when the CBR temperature had dropped to 3000 K and protons and electrons combined to become hydrogen atoms. Nowadays, the background radiation has a temperature of 2.75 K. Using the relation $T = 2.75\ (1 + z)$, you will find that the background radiation was formed at a distance of z = 1000. For the Big Bang it was (and is) true that: z = ∞ (infinity). How deep could one actually see into the Universe? Naturally, this depended on the furthest objects that one could study, such as quasars; as we have seen, quasars had been detected with $z > 3$. No normal galaxies with $z > 0.5$ were known.

More Red shift Measurements and the Discovery of the Cosmic Web

CCD cameras, better spectrographs and the help of computers made it possible to obtain more precise measurements of the spectra of many more galaxies than previously, even with telescopes of (by now) modest sizes, 1.5 m telescopes, for example. This realisation spurred a number of groups in the US to action, groups at the Mount Palomar observatory, at Harvard and in Tucson, Arizona. Spectra were first measured in dozens of galaxies but later in more than a thousand. This chapter is bound by the year 1985; many more redshifts were measured in the following period. Here only the beginning of a revolutionary development is described.

When studying the literature of this period, it is noticeable how obstructive two major uncertainties were in explaining the results of the measurements: uncertainty about the value of the 'Hubble constant' H and the ' density', ρ_c. There were two camps: one that proved H to be 50 km/sec/Mpc and the other proved H to be 100 km/sec/Mpc, that is: a galaxy that was at a distance of 1 Mpc (3.26 million light years) was moving away at a speed of 50 km/s or 100 km/s. Reversing the argument: if one calculated the distance by applying the Hubble rule $d = V/\mathrm{H}$, where V was the observed speed away from us, this distance was uncertain by a factor of 2. The meaning of the critical density, ρ_c, has been explained in the prologue: if this was the density of the Universe, then its expansion would eternally slow down and reach value zero at time infinity. Observers consistently measured a lower value for the density, but they were never sure about their results. As a reminder: the critical density, ρ_c, is usually used as a standard. In cosmological questions the parameter Ω is generally used, defined as $\Omega = \rho/\rho_c$. If $\Omega < 1$ then the density ρ is less than the critical density ρ_c; if $\Omega = 1$ the Universe is critical. The value of Ω will be a recurring theme in the discussions.

Coming back to the redshift surveys, they caused quite a surprise. An early but very clear example can be seen in the fan in Fig. 8.17. We are at the top of the fan and each point is a galaxy in the direction recorded and at a distance deduced using Hubble's law and the redshift measured. The figure is two-dimensional but in reality it is a small piece of three-dimensional space that has been compressed into two dimensions. What was surprising was the irregular distribution: galaxies in small groups and in large groups (clusters or superclusters) with a void between them.

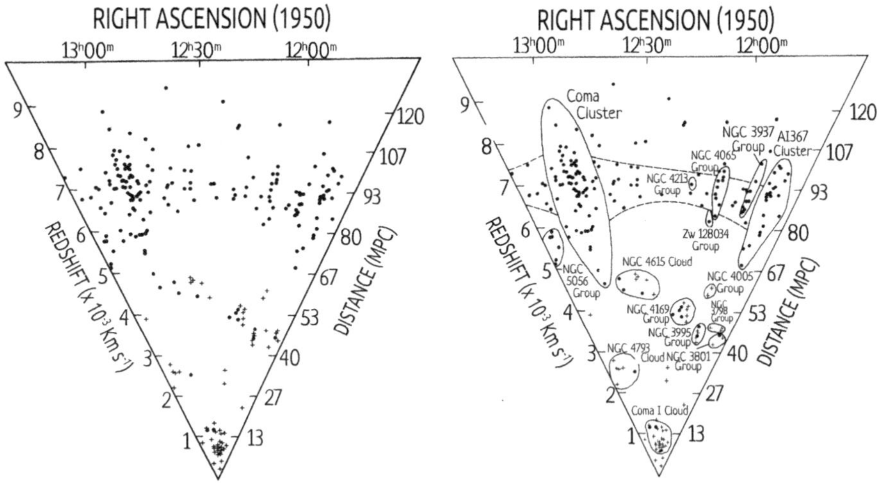

Fig. 8.17 Two maps in the form of fans from a small section of the sky around the Coma Cluster, a large cluster of galaxies. This fan is a section of space compressed into two dimensions. At the top of the fan is in the Milky Way and along the line of sight within the fan the points represent those galaxies of which a spectrum has a been recorded. Along the radius of the fan is the distance d from the Milky Way to the galaxy, which is deduced from the speed V using Hubble's law: $d = V/H$. The right ascension (=α) of the galaxy is indicated at the top. In the left-hand fan, only the position of the galaxies is shown and in the right-hand triangle they are divided into groups. The authors used the value $H = 75$ km/s/Mpc (Gregory and Thompson 1978)

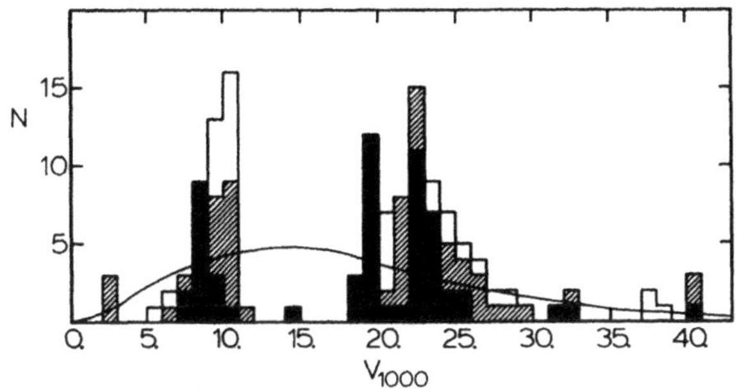

Fig. 8.18 Bar diagram of the number of galaxies at a given velocity in the direction of the constellation of Boötes; horizontally the velocity is given in units of 1000 km/s. The curve shows what was expected. There is clearly a void between 10,000 km/s and 19,000 km/s: the Boötes void (Kirshner et al. 1981)

The authors were highly surprised by the result: firstly, there was so much empty space between the concentrations of galaxies and secondly at the top of the fans there was a long trail of galaxies with two large concentrations, the Coma cluster and the A1367 cluster. Trails and voids occurred not only around the Coma cluster but were also confirmed elsewhere. There was general surprise at the discovery of a very large void seen in another part of the sky (see Fig. 8.18)

Figures 8.17 and 8.18 showed a completely different distribution in space than had previously been assumed. New words were needed to describe this distribution: ‘a honeycomb without honey’, ‘foam’ and ‘sponge-like’ were some of the terms used. It seems that this turnaround in the thinking took place during a major symposium held in 1977 in Talin (Estland) on the topic of ‘The large-scale structure of the Universe’. Experts from both sides of the Iron Curtain met and on both sides it was clear to many participants that old descriptions were no longer adequate. The title of the contribution by Estlanders Jôveer and Einasto was ‘Has the Universe the cell structure?’ and the contribution of Americans Tully and Fisher for the first time contained the word ‘void’. The word itself filled a (linguistic) void; it had been looked for and it was found at this symposium. At the end of this chapter I will discuss the outline of a theory that tried to explain this structure of walls and voids and that was to be successful decades later.

Now that so many redshifts had been measured, questions started to be asked about to what extent galaxies reflected Hubble’s law. The general assumption was that this law applied to large distances and on average to large volumes, but deviations could occur in small volumes, as in our direct environment. A remarkable example was already known in the 1920s: the Andromeda nebula is moving not away from us, but towards the center of our Milky Way at a speed of 120 km/s. The usual and generally accepted explanation is that the Andromeda nebula and our own Milky Way are a binary galaxy. Together they form the core of the ‘Local Group’, where they rule over a whole string of smaller galaxies. The largest of these, M33 or the Triangulum galaxy, is a few times lighter than our Milky Way; the others are many times lighter.

There are a number of other groups in the vicinity of the Local Group, each associated with a large galaxy: M81 at 31 million light years, M101 at 24, M51 at 31 million light years. By far the biggest concentration of matter is the Virgo cluster at 65 million light years. It contained so many galaxies and so much dark matter that it could easily attract our Local Group with its gravity and this possibility could be explored. Using the Faber-Jackson relation or that of Tully-Fisher (see above), one could measure directly the distance d for a large number of galaxies, then calculate the expected velocity $V_{H} = H \times d$, followed by the difference Δv between V_{H} and the measured velocity V. Having done this for many galaxies, one could study the statistical properties of the values of Δv. If the Local Group moved in the direction of the Virgo cluster, one would expect to see a systematic effect in the values of Δv. Such studies were made, examples being (Yahil et al. 1979; Aaronson et al. 1980; Tonry and Davis 1981a, b) but the results did not correlate well. The research was continued and will be discussed in Chap. 15.

Another important strand of continuing research related to the statistical description of the distribution of galaxies. As the redshift of many galaxies had been measured, this information could be used to determine the distances involved, which in turn meant that the three-dimensional distribution of the galaxies was known. This made it possible to derive the autocorrelation of this distribution, an important statistical tool that allowed a comparison to be made of simulated and observed distributions in space. This was described in a publication by Davis and Peebles (1983) and will be discussed at greater length in the following chapter on galaxies and the Universe, Chap. 15.

It had long been realized that if our Milky Way moves in relation to 3K background radiation, the Doppler effect would cause this radiation to be slightly stronger in one direction and fainter in the opposite direction. Several groups tried to measure this, with some success. The measurements could not be made from the ground because the Earth's atmosphere has a large, disrupting influence on the signal. Using U2 spy planes that can fly at very high altitudes and balloons at even higher altitudes a systematic deviation was found with an amplitude of about 1/1000 (Smoot et al. 1977; Cheng et al. 1979; Boughn et al. 1981; Lubin et al. 1983). These measurements were largely surpassed in quality when NASA's COBE satellite became operational in 1989 and further discussion is therefore postponed to Chap. 15.

Quasars

Quasars had large redshifts. Two years after the first quasar was discovered, another was found with $z = 2$. Chapter 5 contains a description of a violent controversy between renowned astronomers on the question of whether that redshift was the result of the large distance or whether quasars were fragments of galaxies that were expelled during an explosion. The battle over this was never really resolved but, over the years, the skirmishes, as we might call them, decreased in number and intensity. Moreover, there was a new generation of researchers for whom it was obvious that the quasars were overly bright galactic cores. There were two ways in which this became evident.

First of all, the improvements in the electronic detection of light led to new research on the question of whether quasars were truly the very bright nuclei of galaxies. Could this galaxy be detected? The question had been answered positively (see Chap. 5) but it was apparently felt that better data could now be obtained and were still necessary. Several groups reported on the matter between 1981 and 1983 and agreed in their conclusion: quasars were surrounded by faint nebulosity ('fuzz') with a spectrum that contained absorption lines with the same redshift as the quasar. Figure 8.19 has images of stars (none had fuzz) and quasars (all had fuzz).

The second argument came from the absorption lines in quasar spectra, attributed to clouds of hydrogen between the quasar and us. The higher the redshift of the quasar, the more of these lines were visible, which corresponded with the interpretation that the speed at which a quasar recedes from us is proportional to the distance of the quasar: the line of sight to a more distant quasar will pass through more clouds and will thus give more absorption lines: higher redshift, larger distance (Sargent et al. 1980).

Quasars promised to offer the means to study the distant and thus the early Universe. Searches were conducted, hoping to find quasars with an even larger redshift than had previously been found. One of the methods used was to photograph the same field twice, once through an ultraviolet filter and once through a blue filter. The observers then looked for objects that shone more brightly in the ultraviolet than in the blue. In some instances a prism was placed in front of the telescope, so that every star image was a spectrum. Quasars distinguished themselves by strong clear lines. In 1980, these various surveys resulted in a catalogue of 1550 quasars and related objects such as the blazars; see Chap. 5. A bar chart (histogram) Fig. 8.20 shows how many quasars were found at different redshifts.

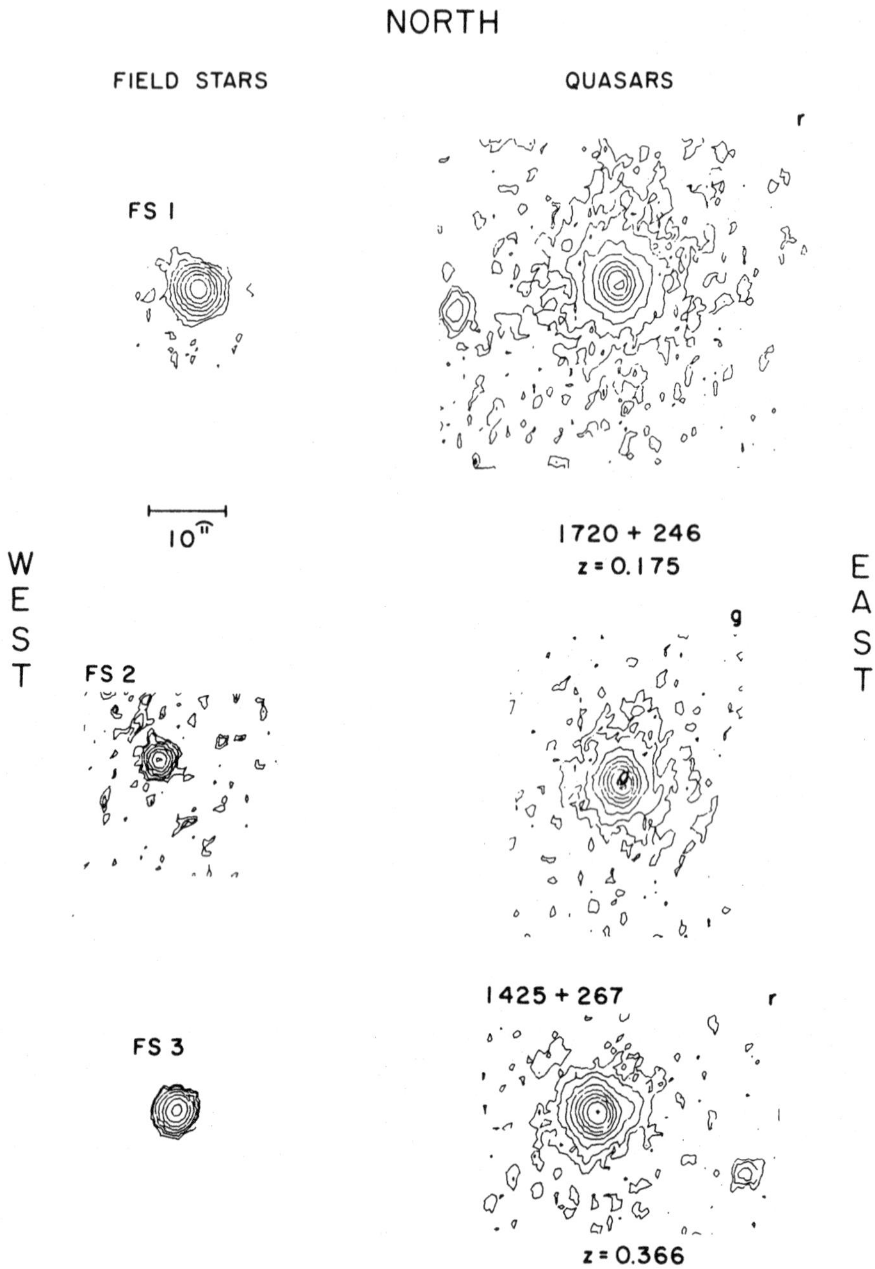

Fig. 8.19 On the right from top to bottom are the light contours of three quasars. On the left are the contours of three equally bright stars. The author concluded that every quasar image contains a star-shaped object plus light from a surrounding nebula. This nebula is the light from the galaxy that has the quasar as its very bright nucleus (Malkan 1984)

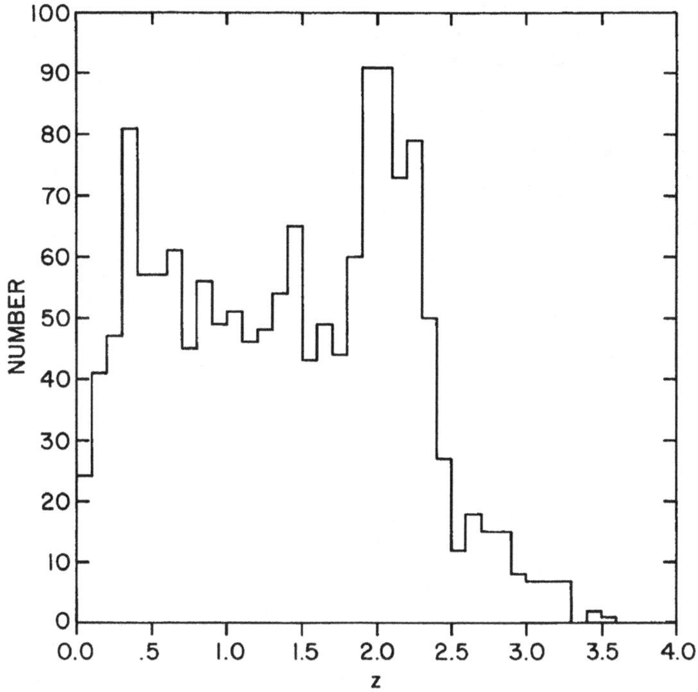

Fig. 8.20 Histogram (bar chart) of the numbers of quasars within intervals in z, the redshift parameter (Hewitt 1980)

The quasar with the largest redshift had z = 3.5, which equates to a look-back time of about 11.5 billion years (and a distance of 11.5 billion lightyears). The distance was too great to detect an ordinary galaxy. A point to bear in mind is that when we see quasars at a distance of 11 billion light years, we are seeing them as they were 11 billion years ago. We are witnessing directly what happened in the past. How happy historians would be if they could look back over a distance of 2050 light years and see the murder of Julius Caesar!

In this bar chart there is a peak in the distribution between z = 1.8 and z = 2.2. This has no statistical significance because quasars were looked for in so many different ways that an unknown number could have been missed. Neither could the highest redshift be taken as definitive. In 1985 a quasar with a redshift z = 3.78 was detected by Parkes, a large Australian radio telescope.

The Quasar Population Past and Present

Chapter 5 mentioned Sandage and Luyten's optical research, which focused on weak blue objects: could these be white dwarfs or quasars? That research was disappointing: the blue objects were mostly white dwarfs. Sandage threw in the towel but the exploration was continued by Schmidt and Green. They launched a search program based on blue and ultraviolet photos of large portions of the sky, as discussed in Chap. 7. It was a long-term program of which the results were published in 1983. In Chap. 7 the peculiar star PG1159 was discussed that was discovered during this survey. Now the quasars will be considered. The most important conclusion was that there used to be many more quasars in the past than there are today. Schmidt had already determined this in 1968, but this new number of quasars made the conclusion even more convincing: the quasars were a little bit like dinosaurs: they had been much more important in the past but now they are largely extinct.

For the critical questioner: in these discussions one must take into account that the Universe is expanding. The volume of a cubic light-year NOW was a factor z^3 smaller at a redshift z. In calculating densities, this change in volume has to be taken into account. The technical term is that one always assumes a 'co-moving volume'.

Between Galaxies: the Lyman Forest

Chapter 5 introduced the term 'Lyman-alpha Forest': many absorption lines present in spectra of very distant quasars and these contributed to intergalactic hydrogen clouds passed through by the line of sight to the quasar. The absorption lines were narrow and often shallow and could only be studied with a large telescope, a modern spectrograph and a modern detector. Fortunately, more and more such telescopes were being commissioned and appropriately equipped. An example (mentioned earlier) was an ultra-sensitive, revolutionary new electronic camera built in London. The camera and its makers visited large observatories in the US and Australia every year to make measurements of unparalleled spectroscopic accuracy in collaboration with local astronomers.

Let us take a look at an example of such a spectrum. Figure 8.21 shows the spectrum of a quasar recorded in segments using the Anglo-Australian 4.2 m telescope, on eight nights between December 1976 and February 1978. A simple look at the figure is shows a bewildering collection of absorption lines. How to explain all that one sees?

A study of some 30 similar spectra led to the conclusion that in spectra like these one can recognize several different 'systems' of lines, apart from the emission lines that belonged to the quasar: lines from galaxies in the same cluster as the quasar, lines from gas ejected by the quasar and, thirdly, lines from galaxies and clouds somewhere along the line from the quasar to us (Weymann et al. 1979b).

Clusters of Galaxies

Galaxy clusters formed large structures in the Universe. Take the Coma cluster with a diameter of 36 million light years and compare it with the Local Group, our 'home', that measures about 2 million light years. Moreover, the Coma cluster

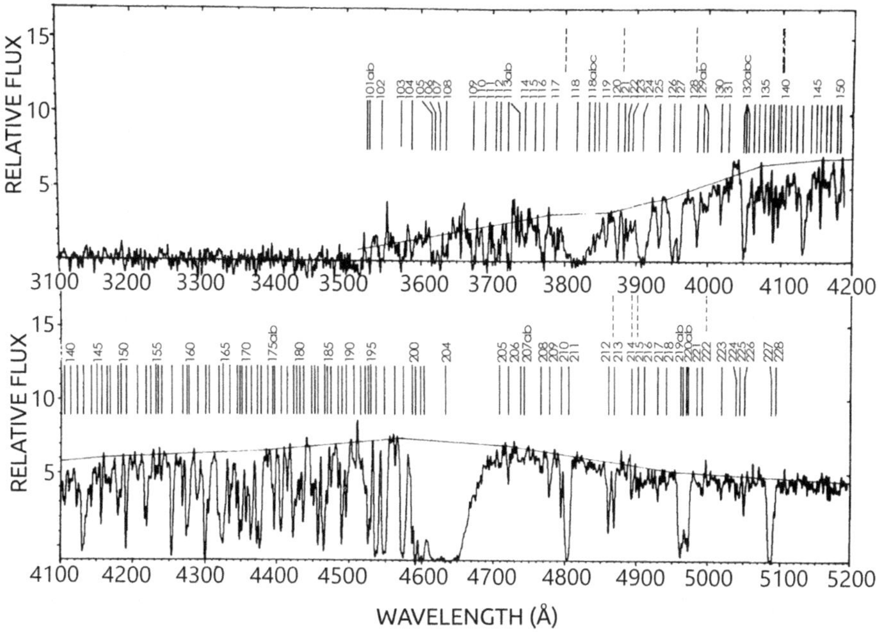

Fig. 8.21 The spectrum of quasar PKS 0528-250. The wavelength is indicated in Ångstrom units; 1 Å = 10 nm. A large number of absorption lines (intensity lower than that of the surrounding wavelengths) can be seen but no significant emission lines (higher intensity). At wavelength 4630 and to a lesser extent at 3829 broad absorption lines can be seen, the so-called damped Lyman-α hydrogen line, all at redshifted wavelengths. See the text (Morton 1980)

consisted of 3000 times as much matter as our own Milky Way (Rood et al. 1972); see also the Prologue. Clusters were looked for mainly on survey plates taken with Schmidt telescopes, initially in the Northern hemisphere (Abell 1958) and later in the Southern hemisphere (Abell et al. 1989). This yielded a total of around 4000 clusters out to a distance of three billion light years. It was very difficult to determine the shape of a galaxy at a distance of more than 1 billion light years: was this an elliptical or a spiral galaxy? Color measurements could offer a solution (red: elliptical galaxies, blue: spiral galaxies).

The member galaxies of a cluster could be reliably identified as long as they were close to the center, but not in the outer areas: there it was difficult to distinguish interlopers from real members. Because galaxies of the same cluster have similar redshifts, a cluster's membership could be established by measuring the redshift of potential members. This required large telescopes and improved detectors, such as became available after (roughly) 1980.

The progress in cluster research also came from simulations of clusters made by computers. An early simulation is shown in Fig. 8.22. In a simulation a 300 point masses was taken and distributed in space (left figure). Then the strength of gravitational attraction between the points was calculated and how this changed the motions of the galaxies. It was meant to explain the velocities and positions of galaxies in the Coma cluster. After a while (middle figure) all galaxies have contracted to the center

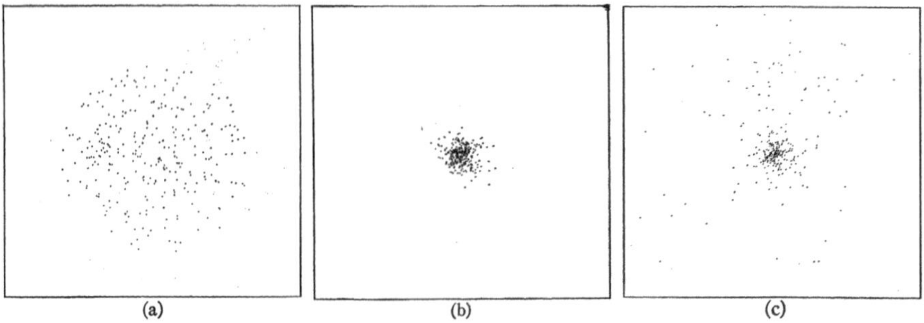

Fig. 8.22 Three hundred point masses are distributed in a large volume (image a) and follow the expansion of the Universe (image a). The force of gravity is then calculated that each point mass exercises on the others and then the subsequent motion by Newton's laws. The collection of point masses collapses to a small volume (see image b), which is followed by a redistribution of the kinetic energy, a process called relaxation. After this, what remains is a central core with most galaxies surrounded by a halo of galaxies that remain bound to the whole cluster (Peebles 1970)

of the cluster. This results in leads to an equalisation of the energies of the galaxies, a process that is referred to as relaxation. In the right-hand figure, an equilibrium distribution has been reached. The expansion of the Universe was taken into account in the calculations.

The three situations pictured in Fig. 8.22 correspond with the observational classification of clusters in two main categories: on the one hand the irregular clusters that are so young that they have not yet relaxed and on the other hand the regular clusters that are older and have relaxed: see Fig. 8.23.

There is an important difference between unrelaxed and relaxed clusters: in unrelaxed clusters the percentage of spiral galaxies is much higher. The opposite is true for elliptical galaxies: a low percentage in unrelaxed clusters, and high in the relaxed ones. Apparently spiral galaxies are turned into elliptical galaxies following the collapse, a conclusion for which there is no satisfactory explanation. Several mechanisms had been devised, none of which worked very well (see e.g. Oemler 1974)).

Is the existence of irregular, unfinished clusters 13 billion years after the Big Bang not surprising? No. In 1974 Americans Press and Schechter published a very influential article on the construction of the Universe being made up of small units—the so-called hierarchical structure formation; this is further elaborated on later. In the spirit of this article galaxy clusters are formed last—and the time for that is now, see the statement: 'The present is very much the epoch of cluster formation' (Gunn and Gott 1972). In short: the build of clusters is ongoing in this Universe.

In Chap. 5, clusters of galaxies were shown to be strong sources of X-radiation. This radiation was generated by a thin gas that filled the inner part of a cluster and had a temperature of between one and one hundred million degrees. A new discovery confirmed this: the X-ray spectra contained spectral lines that originated from atoms that had lost all but one or two of their electrons, iron lines, magnesium, silicon and sulphur (see Fig. 8.24). The presence of such seriously damaged atoms indicated that they were inside a gas with very high temperatures, where the atoms moved so fast that collisions between atoms caused a lot of damage. The discovery of these lines also implied another conclusion: the gas that was produced shortly

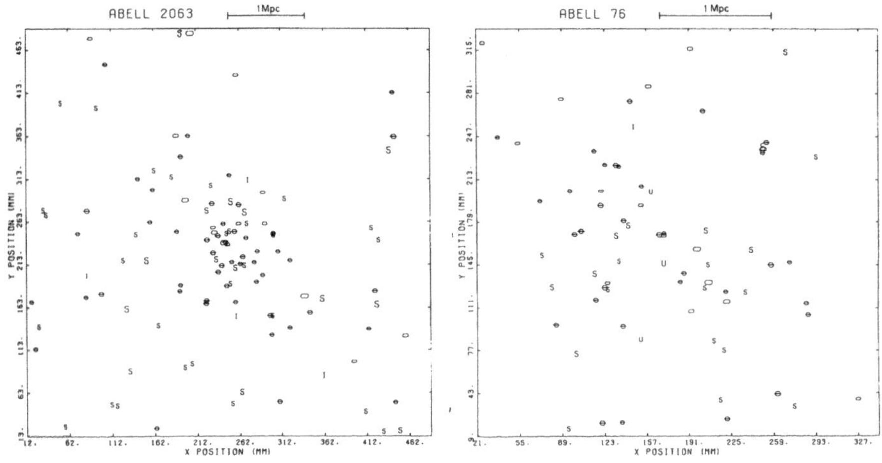

Fig. 8.23 The image on the left is of a relaxed cluster. The central condensation and the halo of galaxies around them can easily be recognized. The image on the right is of a (not yet relaxed) cluster prior to its collapse (Dressler 1980)

Fig. 8.24 X-ray spectrum with lines of different, strongly ionised atoms in the M87 galaxy. The horizontal axis shows the energy of the photons, not their wavelengths. The unit of energy used is the kilo-electronvolt, keV (Mushotzky and Smith 1980)

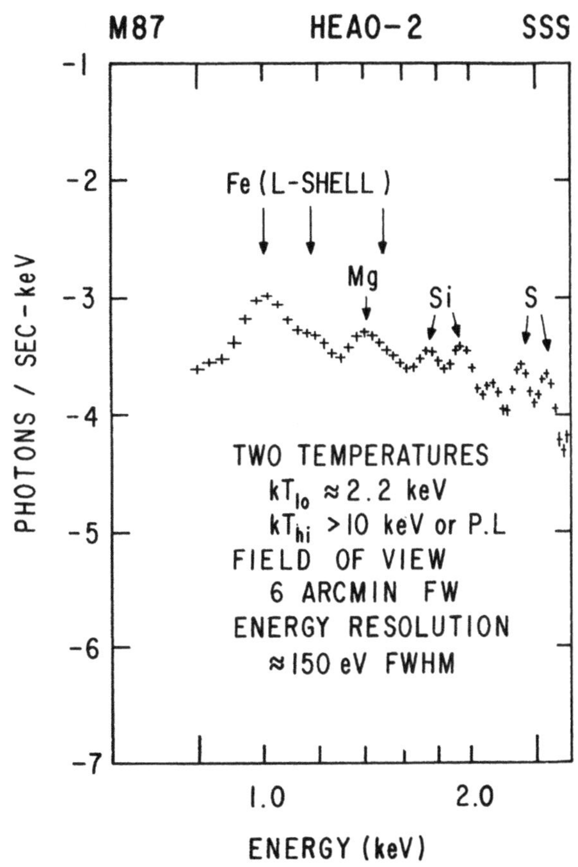

after the Big Bang consisted only of hydrogen and helium gas. Because the gas inside the clusters contained other atoms, this gas was not 'primordial' but had already been enriched by an older generation of stars and supernovae.

The presence of these other atoms had an important consequence: as a result of their radiation the hot gas would cool off and then compress. Deep in the cluster the gas was denser and the cooling was more rapid. There one should find a flow of cool matter, a so-called 'cooling flow'. Could this be detected? It became an important, almost vexing question, to which there was as yet no answer.

Immediately after the discovery of quasars, astronomers started looking at what types of galaxies coincided with radio sources (see Chap. 5). Among them were mainly D-type galaxies, a type that did not appear in Hubble's classification. The galaxies seemed related to elliptical systems but were surrounded by a diffuse light. These were massive systems that occurred mainly in the center of a cluster of galaxies. It was assumed that they were formed there due to accretion from mergers, but there was so far no good theory to support this. Now X-radiation images of clusters of galaxies added some more information. It had sometimes a strong peak at its center. In that case one usually found there a very massive galaxy—often a D-system, sometimes an elliptical system. If there was no D or E system in the center, the X-ray was more widespread and less concentrated (see Fig. 8.25). This was another indication of the difference between unrelaxed and relaxed clusters.

All in all, it was remarkable that the clusters of galaxies showed the ongoing construction of the Universe, whereas the quasar population was already past its peak.

The Butcher-Oemler Effect

When looking at distant clusters, one is looking back in time. This yielded some surprises: in 1976/1978 two young American astronomers used a 'vidicon', an early example of an electronic camera, and focused it on two clusters with galaxies with a redshift of about 0.4 (a look-back time of 5 to 6 billion years). The galaxies in these clusters were weak, between the 19th and 21st magnitude. However, there were good images of the cluster, made through a yellow and a red filter. To their surprise, the researchers found many blue galaxies in these clusters, in contrast to what was seen in nearby clusters (see Fig. 8.26a, b). Butcher and Oemler analysed more clusters and concluded in 1984 that the blue galaxies were blue because they were actively forming new stars: a collection of young stars is always blue in color because of the presence of very hot and luminous stars with strong emissions in the ultraviolet. Thus the final conclusion was that the number of blue galaxies in clusters has greatly decreased in the time since $z = 0.4$, that is in the last 6 billion years. Perhaps these blue galaxies had now become red. It was an attractive, but speculative conclusion that not only the quasars but also the more normal galaxies had evolved significantly in the recent history of the Universe.

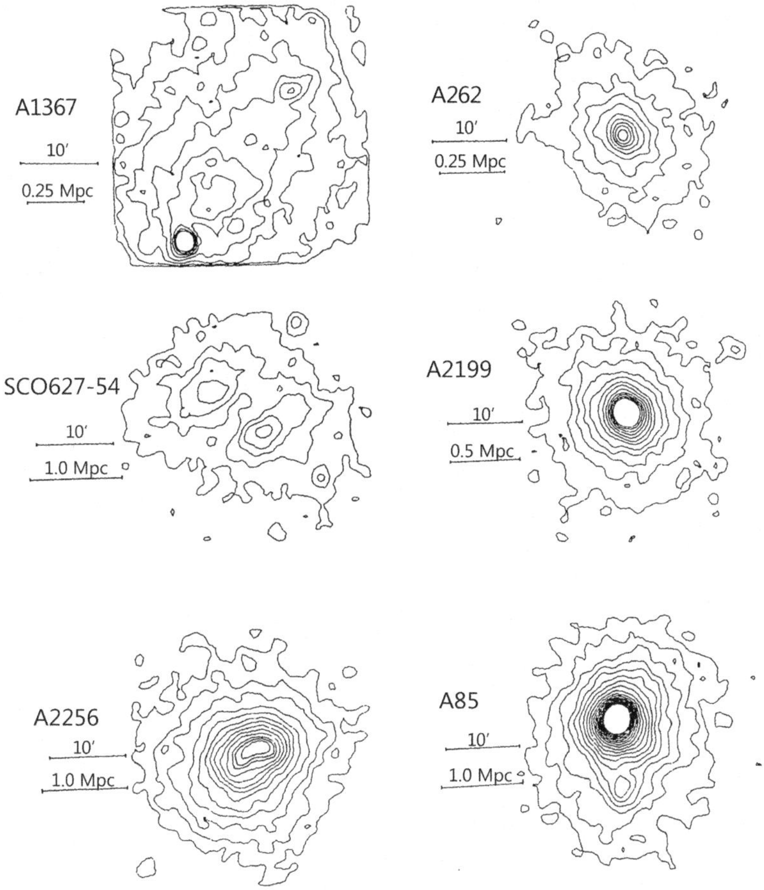

Fig. 8.25 Six different clusters, the three on the left without and the three on the right with a heavy central galaxy (Forman and Jones 1982)

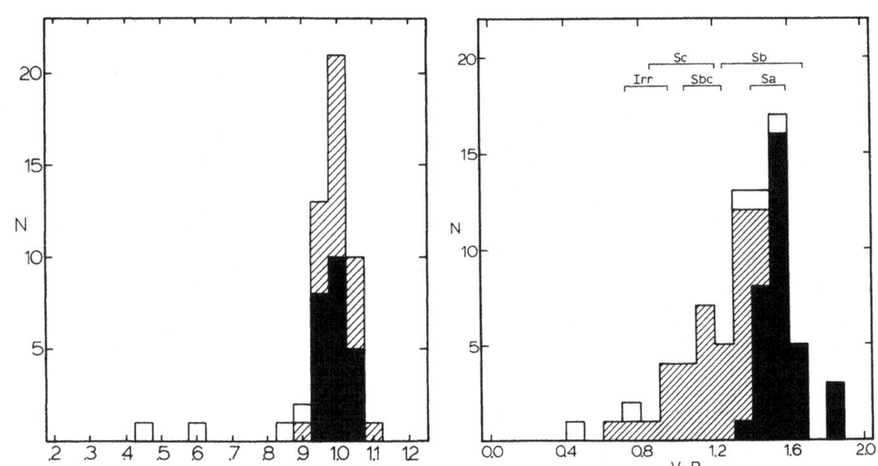

Fig. 8.26 Left: A bar chart of the number of galaxies in the nearby Coma cluster (redshift z—0.023) ranked according to 'color' 'B-V': more specifically: the difference in magnitude seen through a blue filter and through a yellow filter. Galaxies with B-V = 1.0 are red. Right: Bar chart of the number of galaxies in the distant cluster CI 0024 + 1654 (z = 0.39) ranked just as in the left figure. This distant cluster looks structurally similar to the Coma cluster, but contains many more blue galaxies (Butcher and Oemler 1978)

8.6 Structure in the Universe

Background Radiation

The discovery of cosmic background radiation (CBR) was discussed in Chap. 5: the map of this radiation was an image of the Universe 300,000 years after the Big Bang. The radiation was weak and difficult to measure; it was at its strongest at wavelengths around 0.9 mm and at this wavelength all radiation from outer space is strongly affected by the Earth's atmosphere. But the topic was so important that observers continued their attempts, for example from balloons and aircraft (e.g. the spy-plane U2) that could fly at very high altitudes. One important discovery concerned the so-called 'dipole': a light decrease in intensity (a few millikelvin) in one direction and an equally small increase in the opposite direction. It was explained by the assumption that the Milky Way, together with the rest of the Local Group, did not remain in its position in the Universe, but moved at a speed of several hundred kilometers away in some direction, for example that of the Virgo cluster. Another important result came from a ground-based radio telescope in 1984. Measurements of the cosmic background at a wavelength of 1.5 cm showed that the radiation was very uniform and homogeneous; any variations were less than 1 in 20,000 (Uson and Wilkinson 1984). This was still more than the theoreticians expected (see Chap. 5) and more accurate measurements were needed.

Inflation!

The background radiation was isotropic all over the sky: equally strong in all directions. Observations of weak radio sources showed the same effect. The large-scale isotropy bothered the theoreticians. Directly after the birth of the Universe, differences should exist between different parts as a result of quantum effects. These differences could not have been smeared out: there was not enough time for homogenizing existing differences. So, why was the Universe so homogeneous and isotropic on the largest scales? A possible solution for this theoretical problem came in 1981 from American theoretical physicist Alan Guth in 1981. He proposed that very early, between 10^{-36} and 10^{-32} seconds after the Big Bang, the Universe expanded enormously (faster than the speed of light!) and was smoothed out thanks to a mechanism he proposed. You can read about this mechanism in Wikipedia; I will not attempt an explanation. Guth wrote a book about it for non-specialists that is eminently readable: 'The inflationary universe'. It was a brave attempt and Guth was duly praised, but theoretical physicists are still not actually convinced that he is right. However, these reservations have not stopped astronomers from using Guth's inflation theory as a cornerstone for the 'cosmological standard model', the model that explained all—well, at least the most important—characteristics of the Universe. The cosmological standard model will be discussed in Chap. 16.

First Attempts at an Explanation of the 'Structure'

This beautifully smooth Universe expanded and the contents—photons and gas consisting of protons, alpha particles and electrons—expanded uniformly. Theoretically it was expected that very small quantum fluctuations during the expansion continued to exist and slowly started to grow: the start of 'structure'. The fluctuations should be present in the background radiation, but too small to be observable—see Chap. 5. The highly eminent Soviet cosmologist B. Zel'dovich argued that the matter would first condense into flat, thin 'pancakes' that would then later break up into fragments. His arguments were not necessarily accepted, but they were taken seriously, The word 'pancake' appealed to the imagination: so much so that it appeared in dozens of articles. Ultimately they were incorporated in the dominant hierarchical theory (Fig. 8.27).

As the observations offered no help here, maybe computer simulations might assist. A simulation reported in 1974 by Americans Press and Schechter (Press and Schechter 1974) became the starting point of an ultimately successful theory of the formation of stars, galaxies and clusters of galaxies. Press and Schechter suggested that after the recombination of protons and α-particles with electrons, the matter condensed in small objects that drifted on the expansion of the Universe. Thanks to gravity, the objects attracted one another. In a computer simulation, Press and Schechter were able to follow individual objects when they clumped together, and the authors saw how, gradually, increasingly larger objects were formed. The authors showed that the relationship between the number of lighter and heavier galaxies (technically, the mass distribution) quickly reached a set value and strongly resembled what was seen in the masses of the galaxies. The article by Press and Schechter, quoted in other papers over 3,000 times, proved to be a highly stimulating start for a theory on the 'hierarchical build-up' of structure in space.

In 1978, Brits White and Rees (White and Rees 1978) introduced a new element into these calculations and in so doing established the basis for a theory on the origin of visible and invisible structures in space. As we all know, counting citations of an article has become a modern plague for science, but if the number of citations exceeds 2000 it is worth mentioning; the articles by Press and Schechter and by White and Rees more than meet this criterion. Over a period of 20 years the theory evolved to the 'ΛCDM' theory that in the twenty-first century became the paradigm for the origin of galaxies. CDM stands for 'cold dark matter': the dark, invisible matter. The Greek capital letter Λ was added in 1998; it is the symbol for the cosmic pressure that Einstein had introduced to his General Theory of Relativity in order to prevent the collapse of the Universe and that he later removed—see the Prologue for why it was taken out and see Chap. 16 for why it was reinstated.

White and Rees suggested that the contracting matter was comprised 80–90% of invisible matter (also called 'dark matter') while the remaining 20–10% was made up of protons (the nucleus of hydrogen atoms), alpha-particles (the nucleus of the helium atom) and electrons. Protons and alpha-particles are called baryons; the electron was a lepton. The mass of the electron is almost 2000 times smaller than that of the proton and it played no role in gravitational effects; thanks to the electrical force, the electrons always loyally follow the protons and alpha-particles.

Fig. 8.27 Two eminent Soviet astrophysicists in a characteristic pose. Left is Iosif Samuilovich Shklovskii (1916–1985) who among other things discovered that cosmic radio radiation must be synchrotron radiation. To the right is Yakov Borisovich Zel'dovich (1914–1987), a theoretical physicist who had a profound influence on the development of cosmology. Source: Wikimedia Commons

The proposal started with 'dark' objects, known as 'halos' with a bit of baryon density at the center. The readers will in their youth have seen frog spawn: a transparent ball with a small black dot in the centre: this was the image that was evoked. Just as Press and Schechter, the halos meet one another and sometimes would fuse and form larger halos: in short, they would cluster hierarchically. Now the importance of the fundamental difference between dark matter and the baryons became evident. The dark matter could not radiate any energy and its energy was thus preserved. The baryons, however, could emit light and similar radiation, and as a result they lost energy. The consequence was that they ended up in the centre of the halo, just like the dark spots in frog-spawn. To use technical jargon: they finally settled in the bottom of the potential well. There, deep in the halo, this matter could form stars and galaxies, elliptical and disk-like. Because the halos moved freely, they collided and sometimes merged with other halos. The galaxies of both halos would also collide and merge. In this way, large galaxies were built up from smaller ones. To follow this process of collision and merging N-body computer calculations were necessary—calculations that soon became called 'simulations'. White and Rees made estimates of how long it would take before the baryons had formed a proto-galaxy and whether the bigger structures that we know corresponded with the observations. The result was favourable. 'N-body' calculations made with N equal to a few tens of thousands of 'bodies' provided what the authors had predicted. The project had a future.

This future was confirmed 5 years later in publications (Frenk et al. 1983; Klypin and Shandarin 1983) in which the results were presented of many simulations with tens of thousands of particles. This irregular structure with galaxies in filamented structures and with empty voids could arise through 'hierarchical clustering'. In Fig. 8.28 from 1985 it is possible to see how well the theory managed to predict the division of galaxies.

The simulations have produced wedge diagrams that resembled the observations very closely. A controversial assumption, however, had been made: in the simulations,

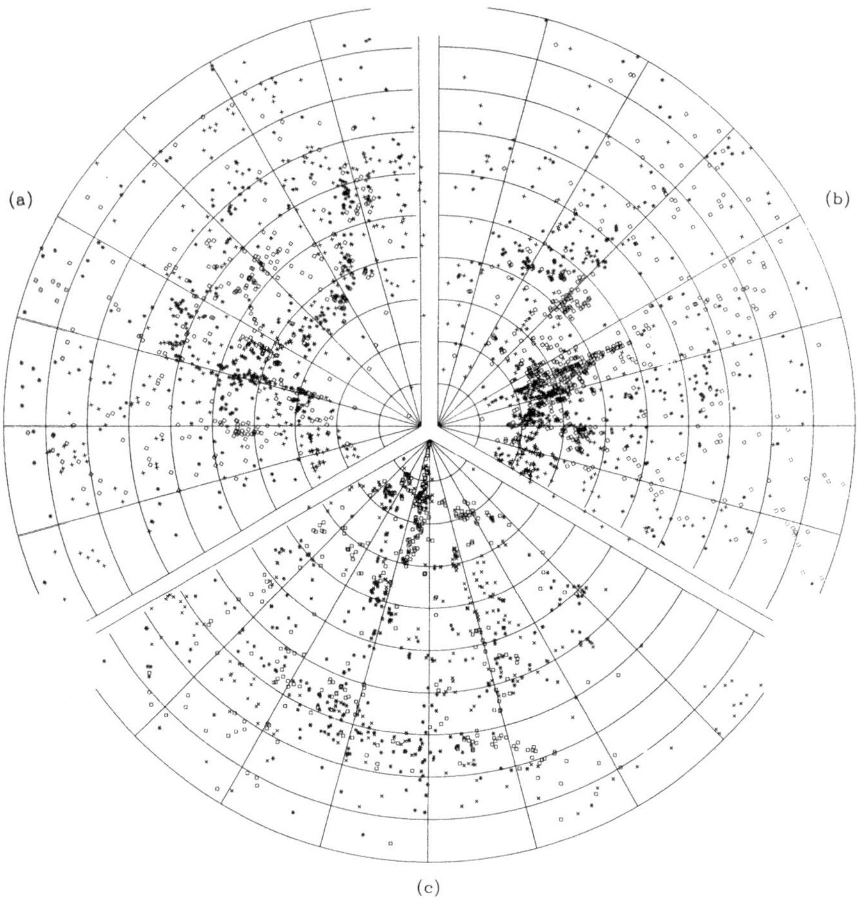

Fig. 8.28 The distribution of galaxies in the near Universe proposed in three wedges just as in Fig. 8.17 earlier in this chapter. (a) and (b) are the result of various N-body simulations with almost 32,000 objects, and (c) represents the observations (Davis et al. 1985)

the matter in the Universe had a density equal to the critical density, or, in other words: $\Omega = 1$. Observers kept finding that Ω was much smaller; the difference between these two assumptions about the density of matter in the Universe constituted a fundamental problem for which there was no solution in 1985.

References

Aaronson, M.: Astrophys. J. **266**, L11 (1983)
Aaronson, M., Mould, J., Huchra, J.: Astrophys. J. **237**, 655 (1980)
Abell, G.O.: Astrophys. J. Suppl. **3**, 211 (1958)
Abell, G.O., Corwin Jr., H.G., Olowin, R.P.: Astrophys. J. Suppl. **70**, 1 (1989)
Binney, J.: Mon. Not. R. Astron. Soc. **183**, 501–514 (1978)
Blandford, R.D.: Ann. N. Y. Acad. Sci. **422**, 303–318 (1984)

Blandford, R.D., Narayan, R.: Annu. Rev. Astron. Astrophys. 30, 311 (1992)
Blumenthal, G.R., et al.: Nature. **311**, 517 (1984)
Bosma, A.: Astron. J. **86**, 1825–1846 (1981)
Boughn, S.P., Cheng, E.S., Wilkinson, D.T.: Astrophys. J. **243**, L113–L117 (1981)
Butcher, H., Oemler Jr., A.: Astrophys. J. **219**, 18–30 (1978)
Cheng, E.S., et al.: Astrophys. J. **232**, L139–L143 (1979)
Davies, R.L., et al.: Astrophys. J. **266**, 41–57 (1983)
Davis, M., Peebles, P.J.E.: Astrophys. J. **267**, 465 (1983)
Davis, M., et al.: Astrophys. J. **292**, 371–394 (1985)
de Zeeuw, T.: Mon. Not. R. Astron. Soc. **216**, 273 (1985)
Dressler, A.: Astrophys. J. **236**, 351–365 (1980)
Faber, S.M., Gallagher, J.S.: Annu. Rev. Astron. Astrophys. **17**, 135–187 (1979)
Faber, S.M., Jackson, R.E.: Astrophys. J. **204**, 668–683 (1976)
Faber, S.M., Lin, D.N.C.: Astrophys. J. **166**, L17 (1983)
Forman, W., Jones, C.: Annu. Rev. Astron. Astrophys. **20**, 547 (1982)
Frenk, C.S., White, S.D.M., Davis, M.: Astrophys. J. **271**, 417–430 (1983)
Gorenstein, M.V., et al.: Astrophys. J. **334**, 42–58 (1988)
Gregory, S.A., Thompson, L.A.: Astrophys. J. **222**, 784–799 (1978)
Gunn, J.E., Gott, J.R.: Astrophys. J. **176**, 1–19 (1972)
Hewitt, B.G.: Astrophys. J. Suppl. **43**, 57–158 (1980)
Huchra, J., et al.: Astron. J. **90**, 69 (1985)
Illingworth, G.: Astrophys. J. **204**, 73 (1976)
Illingworth, G.: Astrophys. J. **218**, L43–L47 (1977)
Kennicutt Jr., R.C.: Astrophys. J. **272**, 54–67 (1983)
Kennicutt Jr., R.C., Kent, S.M.: Astron. J. **88**, 1094 (1983)
Kirshner, R.P., et al.: Astrophys. J. **248**, L57–L60 (1981)
Klypin, A.A., Shandarin, S.F.: Mon. Not. R. Astron. Soc. **204**, 891–907 (1983)
Larson, R.B., Tinsley, B.M.: Astrophys. J. **219**, 46–59 (1978)
Lubin, P.M., Epstein, G.L., Smoot, G.F.: Phys. Rev. Lett. **50**, 616 (1983)
Malkan, M.A.: Astrophys. J. **287**, 555–565 (1984)
Morton, D.C.: Mon. Not. R. Astron. Soc. **193**, 399–413 (1980)
Mushotzky, R.F., Smith, B.W.: HiA. **5**, 735–740 (1980)
Oemler Jr., A.: Astrophys. J. **194**, 1–19 (1974)
Peebles, P.J.E.: Astron. J. **75**, 13 (1970)
Press, W.H., Schechter, P.: Astrophys. J. **187**, 425–438 (1974)
Rees, M.J.: Nature. **211**, 468 (1966)
Rees, M.J.: Annu. Rev. Astron. Astrophys. **22**, 471–506 (1984)
Rood, H.J., et al.: Astrophys. J. **175**, 627–647 (1972)
Sargent, W.L.W., Young, P.J., et al.: Astrophys. J. **221**, 731 (1978)
Sargent, W.L.W., Young, P.J., et al.: Astrophys. J. Suppl. **42**, 41 (1980)
Schechter, P.L.: Astron. J. **85**, 801 (1980)
Schwarzschild, M.: Astrophys. J. **232**, 236 (1979)
Smoot, G.F., Gorenstein, M.V., Muller, R.A.: Phys. Rev. Lett. **39**, 898 (1977)
Soltan, A.: Mon. Not. R. Astron. Soc. **200**, 115–122 (1982)
Tonry, J.L., Davis, M.: Astrophys. J. **246**, 666 (1981a)
Tonry, J.L., Davis, M.: Astrophys. J. **246**, 680 (1981b)
Toomre, A.: Conference on Evolution of Galaxies, vol. 401, (1977)
Tully, R.B., Fisher, J.R.: Astron. Astrophys. **54**, 661–673 (1977)
Uson, J.M., Wilkinson, D.T.: Astrophys. J. **277**, L1 (1984)
van Albada, T.S., et al.: Astrophys. J. **295**, 305–313 (1985)
Walsh, D., Carswell, R.F., Weymann, R.J.: Nature. **279**, 381 (1979)
Weymann, R.J., et al.: Astrophys. J. **233**, L43–L46 (1979a)
Weymann, R.J., Chaffee Jr., F.H., et al.: Astrophys. J. **233**, L43 (1979b)
White, S.D.M., Rees, M.J.: Mon. Not. R. Astron. Soc. **183**, 341 (1978)
Yahil, A., Tammann, G.A., Sandage, A.: Astrophys. J. **232**, 352 (1979)
Yee, H.K.C.: Astron. J. **95**, 1331 (1988)
Zensus, J.A., Pearson, T.J.: Int. Astron. Union Symp. **129**, 7 (1988)

Part II
1985–2015: Long-Planned Exploration

The successes described in the first part led, the whole world over, to plans to explore further all these new discoveries, often with the hopeful thought that new surprises would come this way—as they sometimes did. These large and expensive plans led to a reorganisation in how the scientists worked. Large telescopes were planned, in space and on the ground, and groups of researchers were formed to watch the construction of these costly instruments and to make certain that they were used to their full potential. These groups came from all of the participating countries. In the European Union these developments led to large-scale international collaboration. In its scientists Europe is more united than in most other areas.

To describe what went on I collected the material in several themes that are being followed over the whole period after 1985. I admit that writing comprehensive contemporary history is almost impossible, and I accept that the book is less complete about the years 2005–2015 than I had hoped for.

Chapter 9
1985–2015: Plans, People, Organizations and Instruments

9.1 People and their Plans

What the Astronomers wanted

Since 1945 many great discoveries had been made, culminating in the period 1960–1975, discoveries that each led to a new research field: active galactic nuclei, interstellar molecules, neutron stars, black holes and cosmic background radiation. Radio, infrared, X-ray astronomy had been established as new subdisciplines. The question was, how to proceed from here on? From having read several of the documents about the future of astronomy that appeared in the 1980s, it appears that generally the first goal was to explore the newly detected objects: Can we find more to see how different or how similar these are? How and where did they first form and what do they tell us about the Universe? For example, it had been suggested that huge black holes existed at the centres of active galaxies: could one prove their existence? Should one know how many galaxies contained such a supermassive black hole?

In 1985 it was clear that, apart from finding more and weaker sources, one should study these objects in more detail and at all suitable wavelengths; for example, X-ray sources had to be studied also in the optical and radio windows of the electromagnetic spectrum. However, it takes a long time between the formulation of the requirements of an instrument and the moment when the instrument effectively comes into operation. In practice this delay was often around 10 years or more. The large X-ray telescopes Chandra (NASA) and XMM-Newton (ESA), both launched in 1999, were first planned in the 1970s. Perhaps the most astounding fact about this long-term planning is that the instruments turned out to do what was needed and this more than a decade after the first plan had been formulated. Planning over a period of 20 years had been largely successful: no mean feat and a huge compliment for the organizations that did the planning, such as NASA, ESA, ESO, etc. Let us make a brief summary of the wishes of the astronomers around 1985. The order in which these wishes are formulated here is not of importance.

H. J. Habing, *The Birth of Modern Astronomy*, Historical & Cultural Astronomy,
https://doi.org/10.1007/978-3-319-99082-8_9

The first wish was to find more and weaker sources. Larger samples should lead to better statistical conclusions. In extragalactic research weaker sources are in general at larger distance and thus farther back in time: we see them as they were long ago. An example was the Butcher-Oemler effect: fainter (more distant) galaxies were bluer than nearby galaxies. This was an effect of evolution: we saw the more distant galaxies at a time when the Universe was younger. To find more sources, one had to use larger telescopes and more sensitive receivers.

The second wish was to detect more details about the new objects. Sharper images and more precise spectra were necessary to find out which of the competing theoretical explanations should be disregarded. Another example: one cannot look into a black hole: no information from within the black hole will reach the outside world. However the theory predicts some interesting out-of-the-box effects immediately outside the black hole. Could those be detected?

Third, the importance of studying the same object at as many wavelengths as possible was realized. The Crab nebula is the example par excellence. It has been shown in earlier chapters that the Crab was detectable at all wavelengths of the electromagnetic spectrum and that observations in each wavelength interval contained new information on the object. The Crab is undoubtedly a somewhat extreme example, but nevertheless 'multi-wavelength studies' were essential in most astronomical objects (Fig. 9.1).

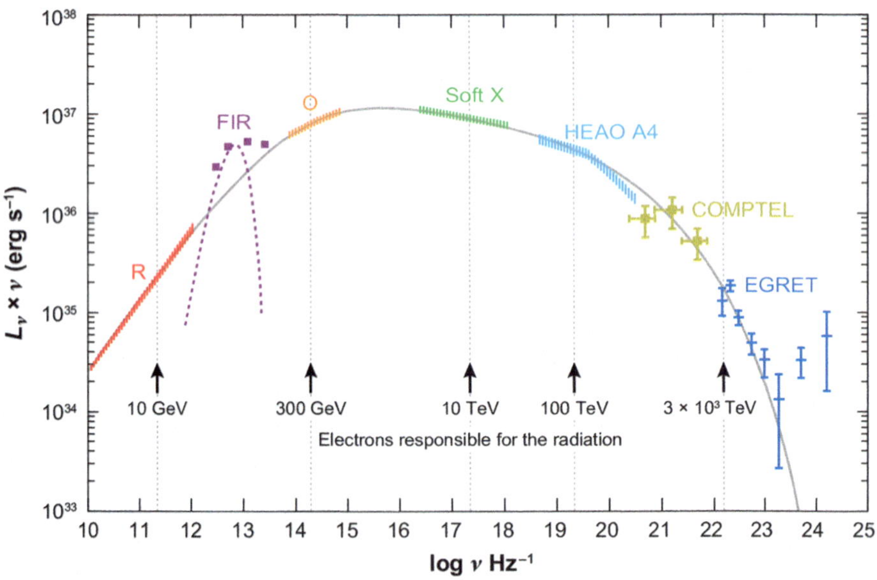

Fig. 9.1 Electromagnetic radiation from the Crab nebula with frequencies between 10^{10} and 10^{24} Hz. 10^{10} Hz corresponds to a wavelength of 3 cm. The radio observations are indicated with the letter R; the symbol FIR means 'Far InfraRed', O = optical; HEAO4 was a NASA X-ray satellite; COMPTEL and EGRET were gamma-ray telescopes on NASA's Compton Gamma Ray Observatory. The curve drawn is a theoretical spectrum based on a combination of synchrotron and inverse Compton radiation. The latter means that in the event of a low-energy photon colliding with a high-energy electron, the photon becomes high-energy. At the highest energies, EGRET observations seem to deviate from the curve drawn. This deviation will be confirmed later on in this chapter (Hester 2008)

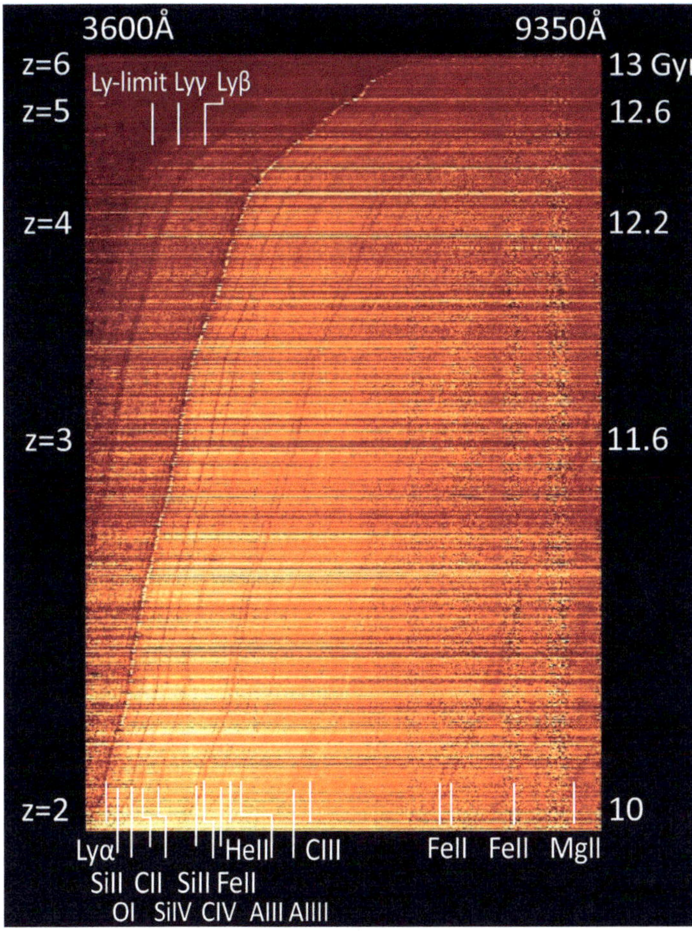

Fig. 9.2 A large number of spectra of similar galaxies have been ordered from bottom to top by the redshift (indicated in the left-hand margin). In the right-hand margin is the 'look-back' time: how long ago was the radiation emitted that we now receive? The main goal of showing this picture is to demonstrate that for a comparison of the spectrum of two similar galaxies with different redshift one needs observations at very different wavelengths (Le Fèvre et al. 2015)

Redshifted galaxies permit us to look in the past. We see galaxies as they were billions of years ago and the significance of measurements of such galaxies is obvious. Redshifts, however, also pose a practical problem: for redshifts larger than $z = 2$ the spectrum shifts to wavelengths in another detection range. Consider for example the strong, important line of Hα that has a wavelength of 656.4 nm in nearby galaxies. The line is easily measured. In a galaxy with $z = 7$ it is seen at a wavelength $(z + 1)$ or 8 times larger, thus at 5.3 μm. This is in the infrared where the Earth' atmosphere is nontransparant and the measurement difficult. The problem is demonstrated in Fig. 9.2 that shows a stack of spectra of galaxies with redshift between 2 and 6.

The demand for multiwavelength studies directly led to the wish to explore the sky at sub-millimeter wavelengths: very little had been done so far. IRAS had made a survey in four broad wavelength bands between 10 and 100 μm; this called for more detailed follow-up observations at these wavelengths. And the wavelength range from 100 to 1000 μm had hardly been explored at all. Observations from planes had shown that in this wavelength range several strong sources existed, but the observations had had a very limited sensitivity. In the same sub-millimeter wavelength range, the Cosmic Background Radiation reached its maximum intensity. Closer research of the CBR was very necessary and had to be done in this wavelength range. Another neglected range of wavelengths were those of the photons of the highest energies: the gamma rays and the even more energetic TeV photons (Fig. 9.1).

Organizations and Finances

In 1985 there were plenty of arguments to support the large investigations needed for further astronomical observatories, on Earth and in space. The impressive thing is that the money was provided, but by whom, and why?

Major sources of funding were the space exploration organizations, in the US, in the URSS, in Japan and in Europe. In the US and in the USSR, these organizations had largely been founded because of their military-strategic importance, but as time went on there were more and more important civil, non-military applications. In the 1960s and not even in the 1970s there is no mention of the usefulness of the GPS system. Nobody dreamt of a little phone that fits in your trouser pocket and that can tell you where to go in any city on the world. Space exploration had become inextricably connected with our daily life.

One of the well-established tasks of the large space organizations is to carry out astronomical research from space. This requires only a small fraction of their total budget, usually no more than a few percent, and making this available has remained an obligation over the years. It reflects perhaps the fact that worldwide astronomical discoveries are interesting to so many people, even though they have little effect on daily life. Is it the sense of relativity, the sense of being part of a larger unity? I have encountered in my life so much warm appreciation for astronomy that I think it a general human attitude, existing not only in this small West European country I live in.

The other major source of funding for the new investments has been the science research councils. They owe their existence and their budgets to the worldwide, positive attitude towards science. Scientific research had to be stimulated, partially because it was good for industry, but partially because of the assumption that it is culturally healthy to advance knowledge. And in the specific case of Europe, one of the positive consequences of the Second World War was the scientific collaboration it engendered. CERN in Geneva has been a great success for physics, and CERN's counterpart for astronomy, the European Southern Observatory, is also a success story.

The third category of suppliers of money has been the philanthropists, but almost exclusively in the US. The two 10 m 'Keck telescopes' on Hawaii are good examples. In Europe, and I expect everywhere outside the US, philanthropic generosity in the US has at least given astronomers strong arguments to ask for the same money from their governments.

Sociological Observations

Increasingly, groups of astronomers conducted joint research. Slowly but surely, groups of specialists arose, spread over several countries. They wrote proposals together. The ownership of data also changed: in the 1960s and 1970s collaborating groups asked for a space instrument; the group that received it was responsible for the construction of the instrument and once the instrument was launched, received all the data. But when the telescopes became bigger and ever more complex, something more than an instrument group was needed and the data became public property. A single group of local specialists was no longer capable of building all the instruments for a satellite (or large telescope) and international consortia were formed that grew in size over the years.

In 1960, one single observer working at night at a telescope, together with the professional night assistant, was responsible for the whole observation, from making charts for the object he wanted to study, to pointing the telescope; for installing and later developing the photographic plate, for analysis of the plate and for writing the paper. Gradually, and particularly after the 1970s, much of this work was taken over by computers, but at the same time the observer was replaced by a number of experts each of whom had knowledge about a particular aspect of the observation.

The growth in the number of researchers involved in a program speeded up after 1990. This can be seen in the numbers of authors of an article that at times grew to enormous proportions. Major projects such as the SDSS, 2dFGRs and GAIA had a team of hundreds of astronomers. The publication of the second detection of a gravitational wave had over 3,700 authors (Abbott et al. 2017). Hopefully this remains an anomaly.

For an inquisitive person it is interesting to look at the names of the authors at the top of a scientific paper. In European and American scientific journals more and more names appear of a clearly Slavonic, Indian or Chinese origin and often they are connected to a science institute in the US or Canada, in Australia or in Europe. In the next paper on the subject, many of these names are connected to another institute. These names belong to the growing population of 'itinerant postdocs'. It is obvious that it has become much harder for a young PhD-graduate to obtain a more or less permanent position. Nowadays a lot of talent for research disappears prematurely.

Internet, Computers, Software

In around 1985 'internet' entered the world and scientists were among the first to use it to its full advantage (and disadvantage). It stimulated cooperation between institutes far away from each other. It is becoming gradually easier and easier to exchange notes, preliminary graphs, and drafts of articles. SKYPE and FACETIME were increasingly used for informal contact without the need to travel.

Another product of the internet was electronic publication. Subsequent issues of a science journal appeared more quickly because of the increasing number of researchers and the higher speed at which data could be collected. The cost of printing, collating and mailing resulted in subscription costs that were much too high for a single individual. The proper answer was to publish electronically, and that is what happened. Reading an issue electronically is now the standard. Paper versions of scientific journals are still bought, but, I guess, only by libraries, probably for documentary reasons.

All major astronomical surveys (including the POSS) have been made electronically available free of charge. A large database at Harvard allows easy access to most scientific journals and a comparably large data base in Strasbourg provides elaborate data on practically all astronomical objects. NASA and ESA have similar databases for practically every space observatory they have launched.

The internet makes it easy to obtain data, but do you know how to use these data?. In many instances, only an instrument specialist is able to reduce the data to a usable set and often it is only this individual that can judge the many subtleties needed to fully understand the data. Sometimes these instrument specialists have made themselves superfluous by developing software packages that allow the user to analyze the data and arrive at scientific conclusions.

Ultimately, one wants to compare observational results with theoretical predictions. Here computers have increased the quality of statistical analysis. It becomes easier to compare data to complex theoretical predictions. The remaining problem is: do I control the computer, or does it control me? This is a problem that is unlikely to go away. A critical attitude remains an absolute necessity.

Making a survey has always been of importance. A large number of surveys have been made in this period of 1985–2015 as will be illustrated in this and in following chapters.

Legacy Programs

Atlases and catalogues are legacies for future generations. A new type of heritage was born in 1995 when the 'Hubble Deep Field' made its entrance. As compensation for his administrative duties, the director of an observatory is usually given some observing time that he can use freely. In 1996 the director of the Science Institute of the Hubble Space Telescope, Robert Williams, decided to use all his observing time making as many observations as possible of a small field far away from the Milky Way. This would allow him to achieve a census of galaxies deeper in the Universe, and thus deeper into the past, than anybody had made so far. He chose a square of 5.6 arc minutes (1% of the size of the full moon) and made observations using various color filters; in the end he had spent 46 hours per filter. He made the data of this Hubble Deep Field publically available as soon as possible and without a further analysis of the data (Fig. 9.3).

Fig. 9.3 The Hubble Deep Field. Practically all objects are galaxies, and some are at a distance of billions of light years. Only one in 250 objects is a star inside the Milky Way. Source: NASA, ESA

Williams published the results together with two co-authors (Ferguson et al. 2000). The field contained fewer than 20 stars and about 5,000 galaxies. Redshifts had been measured for only 150 of those and about 30 of them had a redshift $z > 2.0$, implying a distance of several billions of light years. These data, I repeat, were made publically available.

Williams' initiative had many followers. In the programs of later space telescopes similar deep fields were defined and observed in depth, e.g. in the cases of the large X-ray telescope Chandra, the infrared space telescope SPITZER and the sub-millimeter space telescope HERSCHEL. Further down I will discuss the so-called 'Legacy Programs' for which the inspiration came partially from the success of the Hubble Deep Field.

Another, related activity came quite naturally when a telescope and its receiver had been optimized over the years and when the flow of new and exciting observing proposals became smaller. Then, at several radio telescopes, one realized that a more fruitful way of using the instrument was to let it make a large survey, for example of the central parts of the Milky Way. The end result of a survey is a catalogue, formerly in the form of a book and now in the form of an electronic database.

9.2 Optical Telescopes

Observational optical astronomy made great progress in this period. Many telescopes became operational with mirrors between 8 and 10 m and at the same time larger arrays of light detectors with increased sensitivity became available. This was a success story, that was, however, beaten by a 2.4 m telescope, because it was placed in the ideal location for any telescope: in space.

Hubble Space Telescope

The advantages of an optical telescope in space must have been known for centuries: it meant that there was no atmosphere that worked as a dirty window. Sharper images and the detection of fainter sources were two of the obvious advantages. It remained a dream, until in 1957 Sputnik opened up space. Nevertheless, it took about 30 years before plans with a reasonable chance of implementation were made—and of course by NASA, the leading space agency. It was only after the original plan of a 3 m telescope was reduced to 2.4 m and a contribution from ESA had been obtained did the US Congress allocate the money. The Large Space Telescope was later renamed the 'Hubble Space Telescope'.

The launch in May 1990 was a success but the telescope was not. The images were blurred and diffuse. The cause was found pretty soon: the mirror had been ground to perfection but to the wrong specifications. It was a painful conclusion and the public reacted accordingly. Numerous cartoons appeared, and many comedians on TV made terrible jokes. The NASA management must have had a hard time. But NASA came back.

The fact that the telescope had been ground to perfection made it possible to design a small instrument (later named COSTAR) that, placed in the light path of the HST, corrected the error and would bring all radiation back into the desired focus. When a special SHUTTLE mission had delivered COSTAR to the telescope in 1993, some astronauts went outside and installed the savior. With success: HST produced the sharp images that had been promised and from now on NASA could make propaganda that their astronauts were capable of making complicated reparations in space (Fig. 9.4).

The SHUTTLEs have visited the HST a few times to replace instruments with new and more appropriate light receivers. One of these was to replace an optical instrument by an instrument operating in the infrared: the replacement reflected the evolution of astronomical observations brought about by the HST itself. The last 'service mission' was in 2009 because the SHUTTLE program was terminated. Nevertheless the HST has continued to produce fundamental astronomical data of the highest quality ever since. At the same time the community is waiting for HST's successor, the James Webb Space Telescope, known in Europe as the 'James Who

Fig. 9.4 Hubble Space Telescope. Source: NASA

Space Telescope', because there they do not remember that Webb was the administrator (director) of NASA in its glorious days of the first moon landings.

In almost each of the following chapters, the enormous impact is mentioned that the HST has had on astronomical research in almost all subfields.

More and Larger Telescopes in Dry Conditions

Chapter 1 mentioned the opening of the 5 m 'Hale telescope' in 1948. It was the largest telescope in the world, a position it held for some 45 years. Nominally a 6 m telescope began operations in the Soviet Union but it never functioned as had been hoped. In the following decades some fifteen 3–4 m telescopes were built at much better sites than Mount Palomar and with the introduction of CCD light collectors these telescopes surpassed the performance of the 5 m back in 1948.

The need for larger telescopes had, however, increased: more and more research reached its limits and required a deeper view than could be provided for. In Chap. 1 it was explained why the 5 m telescope was the largest telescope ever to be built: one required a rigid mirror that would not bend or otherwise deform when it was pointed in all directions in the sky; the rigidity required a mass of glass that for a telescope larger than 5 m would be too heavy to be handled by any mechanical construction.

The solution to this problem came in the 1980s when it was realized that computers could be introduced to keep a much lighter mirror in its proper shape by controlling a set of activators, magnetic rods as support for a thin mirror. The concept was successfully used in the Nordic telescope (1988) and especially in the New Technology Telescope (NTT) that ESO opened in 1989 and that had been a pilot project for the design of much larger telescopes. The actuator concept was called 'active optics'. The NTT was a success right from the beginning and an excellent demonstration that the way to much larger telescopes lay open.

At the same time, comparable plans were made across the Atlantic. The California Institute of Technology (CalTech) and the University of California started the construction of a 10 m telescope along the same principles. Helped by a donation by the W.M. Keck Foundation (a first gift of 70 million dollars, later increased to 140 million dollars) two 10 m telescopes were built on Mauna Kea (Hawaii) and put into operation in 1993 and 1996. The telescopes functioned as expected, right from the start.

The opening of the first Keck telescope marked the beginning of a decade in which 12 telescopes began operations each with a diameter in excess of 8.0 m, four on the equator (Hawaii) two in the northern hemisphere (Texas and one of the Canary Islands) and six in the southern hemisphere (Chile and South Africa).

Sharpening the Images: Adaptive Optics

The possibility of deforming the mirror of a telescope at will using a computer stimulated optical engineers to use this capacity to correct for irregularities in the atmosphere. A guide star was chosen close to the object to be observed. The variations in the image of the star were converted by a computer into the necessary changes of the mirror by the actuators. The development of the technique was far from easy but as Fig. 9.5 shows, ultimately it was successful.

Extremely Large Telescopes

The success of the 8 and 10 m telescopes inspired astronomers to make plans for still larger instruments, and the scientific progress in various fields made clear that much was to be gained by such larger telescopes. ESO made the most progress. It is presently building an optical telescope with a mirror that will have a diameter of 39 m, a collecting area of 978 m^2 and will be located in the Atacama desert in North Chile. It is planned to be operational in 2024. See Fig. 9.6.

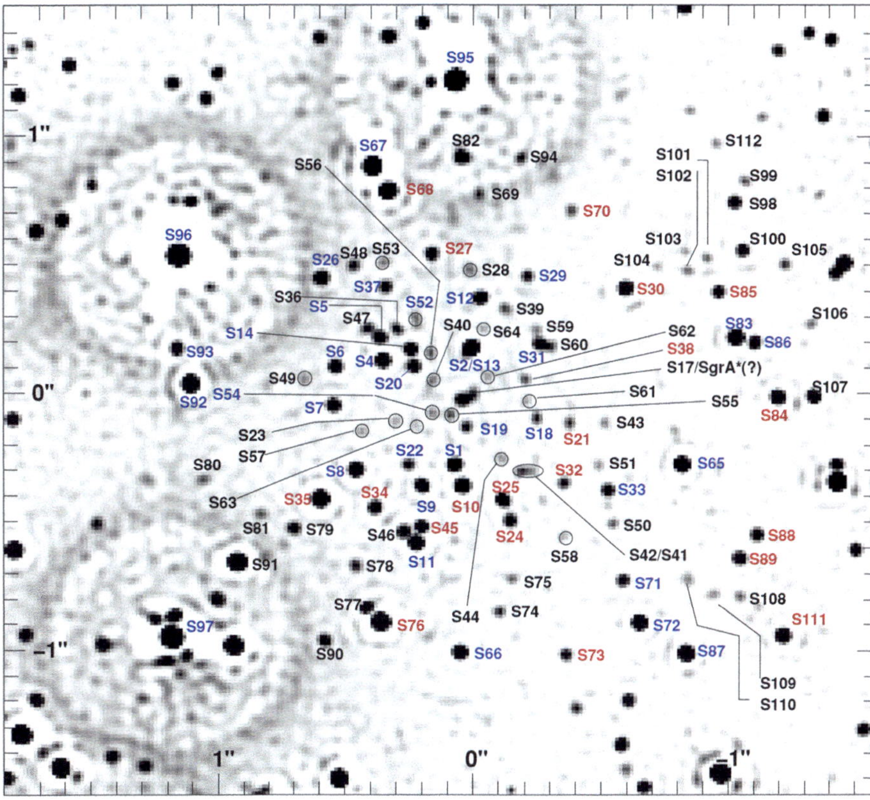

Fig. 9.5 This area, close to the centre of the Milky Way, near the radio source SgrA*, has been observed through a filter at near-infrared wavelengths and has been sharpened using active optics. To judge the result, consider the fact that this image is about 2 arcsec by 2 arcsec (see the scale at the bottom). Without sharpening, the image of each star would have been 1 arcsecond. In Chap. 16 it will be explained that images like this taken during a number of years have shown that the stars follow orbits around a heavy object that coincides with SgrA*: a proof that SgrA* coincides with a massive black hole (Gillesen et al. 2009)

In the US, a 30-meter telescope is being planned as a collaborative project together with Canada, China and Japan. It will have a collecting area of 655 m^2. The intention is to build it on Mauna Kea (Hawaii) and the building did start. However, serious objections were raised of a religious nature by native Hawaiians, bringing the building work to a temporary halt. It is uncertain what the outcome of the process will be. Another project is the Giant Magellan Telescope, to be completed in 2025. Seven 8.4 m telescopes are planned with in total a collecting area of 368 m^2. Participants are the US, Australia, Brazil, South Korea and Chile. The location will be in Chile.

Fig. 9.6 The ELT, the Extremely Large Telescope, with a 39 m diameter mirror now under construction in Chile. Note the size of the cars, and in particular the size of the large truck. Source: ESO/L. Calçada

Fig. 9.7 The 2.5 m robot telescope used to make the SDSS survey. Left: the location of the telescope; the movable white housing covers the telescope in the daytime. Right: an image of the telescope (Gunn et al. 2006)

Robot Telescopes and Surveys

Special astronomical observations require the presence of a human observer, but many observations follow a routine. In these cases a computer can take over much of the work of an observer. In fact, in routine work a machine will make fewer mistakes than a human. The ultimate consequence is a robot telescope that is operated from far away.

Robot telescopes have been used to make routine observations and also for monitoring the sky in search of 'transient objects' (see Chap. 14). The most impressive survey is certainly the Sloan Digital Sky Survey, which will be discussed extensively in Chap. 16. Figure 9.7 shows the location and the 2.5 m telescope that was used.

9.3 Satellites for Astrometry: Hipparcos and GAIA

The oldest existing catalog of stars was compiled by the Greek astronomer Hipparchus and dates from about 150 BC. It has survived because it was included in a large compendium (now called the 'Almagest') by the Roman astronomer Ptolemaeus (about 150 AD). It contains about 100 stars and gives the magnitude of each star, thus the brightness as we see it, and the position, two numbers on the celestial sphere. In around 1718, English astronomer Halley noticed that several stars in Ptolemaeus' catalogue now had somewhat different positions. He attributed this to the fact that stars had a 'proper motion': each moved, very slowly, over the sky. Measuring proper motion became an important occupation for astronomers, an activity that was called astrometry. It required the measurement of highly accurate positions for all stars. There was another driver for such accuracy: during one year the stars had to move along a circle in the sky: a reflex of the motion of the Earth around the Sun. The size of the circle was called the parallax of the star and it revealed its distance.

In practice, proper motions were measured for nearby stars but the greater accuracy needed to measure the parallax was achieved between 1837 and 1840 when three astronomers, independently, measured it for the first time and so determined for the first time the distance of a few stars. In the remainder of the nineteenth century more parallaxes were measured and many more proper motions but progress was slow and hard. The application of photography (after 1880) helped a bit, but astrometry remained a technically very demanding—and ultimately unsatisfying—task. I remember a question from a highly-experienced astrometrist who, at a colloquium where the positional accuracy of radio sources was discussed, asked the cynical question: 'Nice result you got; have you already measured it for the second time?' In the second half of the twentieth century astrometrical progress ground to a halt, in spite of enormous efforts by die-hard astrometrists. Proper motions and parallaxes remained of fundamental value, but the technique of astrometry had reached its natural limit.

In 1958 space research opened many new avenues for astronomical research, but not for astrometry. How could one pinpoint the direction of a star from the unstable platform that space research offered? And then came a brilliant solution, invented in 1967 by French astrometrist, Lacroute, who expanded an idea of his teacher, Danjon. Go back to early astronomy, when astronomers used an instrument called a Jacob's staff, an instrument that measured the angle between the directions of two stars. Now measure the angles between thousands of stars, and then you may take one star as a reference and knowing the angle between all stars you can put them all on a celestial sphere. How accurate your celestial globe is depends on the accuracy that you achieved in measuring the angles. Lacroute found a simple technique to measure this angle very accurately from space and he was convinced that a space telescope could deliver the data for a catalogue that was many times more accurate than any catalogue that existed.

Lacroute wrote a proposal that made a long journey via the slow desks of several organizations. It was a proposal that differed so much from all others that the engineers and physicists at the space agencies rejected the idea out of hand.

There was no astrophysical problem to be solved and it was therefore not within the domain of space research. But finally the proposal received backing from a united front of astronomers that managed to convince the ESA bureaucracy of the enormous value of this purely astrometrical project. In 1980 ESA decided to follow the advice and started a project called HIPPARCOS. It was an acronym, but there was a clear reference to Hipparchus. The satellite was launched in August 1989 and ended its activities in August 1993, having 'accomplished' its mission. A few years later two catalogues were delivered. One contained 120,000 stars with a positional accuracy of 0.001 arcsec and the second had a catalog with one million stars with a positional accuracy of 0.03 arcs. These catalogs revitalized astrometry and found applications in other fields such as radioastronomy.

While the HIPPARCOS measurements were being processed, ESA already decided that there should be a follow-up mission, with higher accuracies and many more stars, a project that was given the name GAIA. This second mission also reached its goals. The data are slowly being distributed from 2018 onwards. The data should make it possible to construct a 3D map of the Milky Way (e.g. Gaia Collaboration, Babusiaux et al. 2018).

9.4 Infrared Telescopes

The atmosphere of the Earth is transparent for radiation at a wavelength below 1 μm (0.001 mm) and again above 1 mm. Between 1 and 20 μm there are a few 'windows', small wavelength intervals where the atmosphere is transparent, at least at high observatories and during dry nights: water molecules in the Earth's atmosphere are a major cause of absorption. At the other end, near 1 mm, the atmosphere, when it is in a friendly mood, allows observations, especially at (second rate) windows around the wavelengths 0.45 and 0.85 mm, but only in places that are both very high and very dry. It is customary to divide the range of wavelengths from 1.0 μm (0.001 mm) to 1 mm in two parts: from 1 to 100 μm they are referred to as 'infrared wavelengths' and from 100 μm (0.1 mm) to 1 mm are called 'sub-millimeter'. But the boundary at 100 μm (0.1 mm) is not used consistently, and in practice there is some overlap in spectral range; in spite of the overlap HERSCHEL will be called a 'sub-millimeter' telescope and SPITZER an infrared telescope.

Infrared and sub-millimeter radiation is emitted by dust particles, that re-emit the energy they get from the absorption of light and ultraviolet radiation produced by stellar sources or by black holes—more precisely: from the neighborhood of black holes. Hot dust emits infrared radiation; cool dust emits sub-millimeter radiation. The total amount of infrared and submillimeter radiation provides an estimate of the amount of starlight produced inside a dark cloud.

Radiation by dust is always continuous in wavelength. Spectral lines are another important form of radiation. Molecular lines are the most important because they reveal what happens deep inside dark clouds. Some atomic lines are also present, the foremost a line at 158 μm produced by ionized carbon, [CII]. Between 1 and 20 μm there are several atomic lines to be observed in bright nebulae such as the Orion Nebula.

The presence of much dust and the presence of molecular gases inside clouds of gas and dust makes measurement of infrared/sub-millimeter radiation important. Radiation between 10 and 800 μm was the last of all electromagnetic radiation to be explored in astronomy; all other wavelength ranges were 'opened' earlier. The cause of this delay was largely technical; it took until around 1990 before suitable detectors were available.

A quite different argument in favor of infrared radiation was the detection of objects with a large redshift. In Chap. 5 the quasar 3C9 has been mentioned that has a redshift $z = 2$. This implies that its far-ultraviolet lines (for example, the hydrogen line Ly α) is observed at a wavelength three times larger, and thus at optical wavelengths, and lines that are normally in the optical window now appear in the infrared. To study such objects as 3C9, it is mandatory to make observations in the infrared. The argument was so strong that NASA approved an expensive SHUTTLE visit to HST to replace one of the optical instruments with one for the infrared.

Ground-Based Infrared Telescopes and Surveys

The first infrared observations were made from the ground using existing (optical) telescopes. The need to eliminate infrared radiation from the telescope itself made it necessary during observations to alternate quickly between looking at the proper object and an area of nearby sky. Because optical telescopes were not equipped to do this at the required speed, a mirror was placed close to the prime focus and this was tilted and back and forth a few times per second. The problem was overcome with new telescopes equipped specially for infrared measurements. In the early seventies, a 60″ telescope became available at Mount Lemon in Arizona. US observers made many of the early discoveries using this telescope. At the same time a number of discoveries (e.g. of the galactic centre) were made with the telescopes on Mount Wilson and Mount Palomar.

A new era started when new infrared telescopes with large mirrors were commissioned. The 2.3 m Wyoming Infrared Telescope came into operation in 1975. NASA's 3 m InfraRed Telescope Facility, 'IRTF', dates from 1979 as does UKIRT, a then British telescope with a diameter of 3.8 m and now owned by NASA. IRTF and UKIRT are both located at 4200 m, at the summit of Mauna Kea in Hawaii. A major product delivered by UKIRT were the five UKIDSS surveys carried out between 2009 and 2012. The latest near-infrared telescope is ESO's 4.2 m VISTA (Visible and Infrared Survey Telescope for Astronomy) that saw first light in December 2009. It is the UK's contribution in kind for the country's accession to ESO. VISTA is intended to make surveys; first results have been published and the data have been released.

The largest of all infrared telescopes is the Japanese 'Subaru' on Mauna Kea (Hawaii) with a dish measuring 8.2 m in diameter. The first observations were made in January 1999. It appears to be working very well. See Fig. 9.8 to get an impression of the size of this 8 m telescope.

Fig. 9.8 The Subaru telescope viewed from inside the dome. The equipment leaning against the wall of the dome gives an impression of the size of this enormous instrument. Source: NAOJ

Big telescopes are impressive and they produce unique results. Nevertheless, small telescopes can also be of great value if there is a good plan and the time is ripe. An example is the successful all-sky survey made with two 1.3 m telescopes in wavelength bands in the near-infrared and centred on 1.2, 1.7 and 2.2 μm; one telescope was in Arizona and the other in Chile. Together they provided the data for the 'Two-Mass' survey. It filled a gap between optical and near-infrared observations. The paper explaining the set-up and the execution of the survey has been cited more than 6,400 times (Skrutskie et al. 2006). 'Two-mass' was adopted by NASA and was financially well funded. A European counterpart, called DENIS, lacked such funding and was consequently less successful, although the first discovery of brown dwarfs was some consolation, as was the first complete near-IR mapping of the Magellanic Clouds (see Chap. 10).

Telescopes in Space

The IRAS survey (see Chap. 6) had demonstrated convincingly the need for more infrared space telescopes with a sharper view and a spectroscope for the many lines that had to be detectable. NASA, Japan and ESA responded to this need. In 1995 ESA successfully launched the Infrared Space Observatory (ISO), a medium-sized mission, with a mirror of 60 cm diameter. The instruments were cooled by liquid helium, and the amount of helium set the lifetime of the telescope: 28 months. ISO had four instruments on board, covering the wavelength range from 2.4 to 240 μm. Judging by the number of times that ISO has been quoted in scientific papers, it was a successful mission. Figure 9.9 shows a remarkable result obtained with the short-wavelength spectrograph.

In December 2003 NASA launched SIRTF, the last of its four 'Great Observatories', the satellite for infrared studies. After launch it was renamed 'SPITZER' in honor of the great astrophysicist who had done much for space research on astronomy. SPITZER had a telescope mirror with a diameter of 85 cm and three main instruments, a camera, a spectrograph and multiband photometer, that together covered the wavelength range from 3 to 180 μm. In May 2009 the telescope ran out of coolant

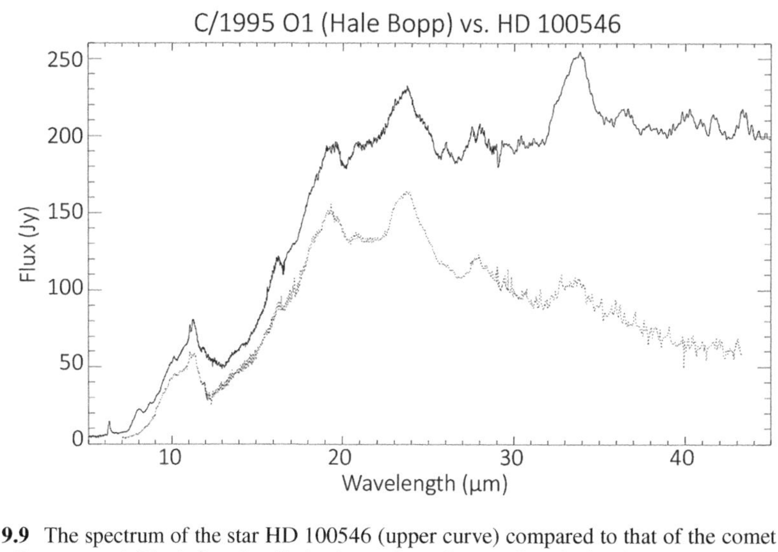

Fig. 9.9 The spectrum of the star HD 100546 (upper curve) compared to that of the comet Hale-Bopp (lower curve). The infrared radiation is not from the star directly, but from a surrounding disk that is expected to form planets. The similarity between the material of the comet and that of the planet-forming disk is striking; both contain identical species (Malfait et al. 1998)

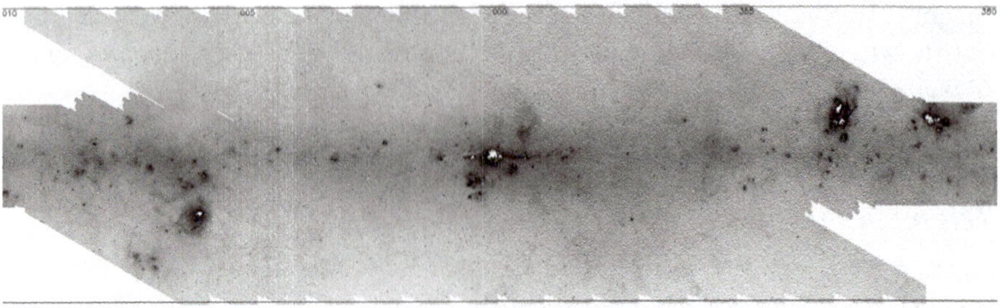

Fig. 9.10 The inner 20 degrees of the Milky Way at a wavelength of 24 μm. At the rightmost point the galactic longitude l = −10°; at the other end it is l = +10° (Carey et al. 2009)

Fig. 9.11 A dark cloud with the appearance of a winding hose that contains so much darkening matter that it absorbs radiation even at the long wavelengths 24 and 70 μm (Carey et al. 2009)

liquid, which brought most of the observations to a halt, although a limited range of observations at short wavelengths is still possible. SPITZER was undoubtedly a great success. One example of successful observations is the central piece of a full picture of the Milky Way (Fig. 9.10) whereas Fig. 9.11 shows an elongated and exceptionally dark filament.

A special feature of the SPITZER program was the introduction of 'legacy programs'—programs that were expected to make measurements that would continue to be important after the satellite had died. NASA formulated three criteria by which such a program would be judged: it should be large and coherent; it should be of general importance; the data should be released as soon as they are fit to be used by an experienced but non-specialist astronomer. Behind the institution of the legacy programs were several considerations, one of which was that observers often request short observations in order to have a high probability of obtaining observing time; this often leads to fragmentation of time available for observations. Observing time on a satellite is very precious as it is a one-time-only possibility. The concept of the legacy programs has become a standard part of the use of major instruments, in space but also on Earth.

The third major space satellite is AKARI (Japanese for: light). It was operational from February 2006 until November 2011. It had a 67 cm diameter mirror

Fig. 9.12 A Boeing 747SP equipped with SOFIA, a 2.5 m telescope for infrared observations. Source: NASA/USRA

and was sensitive to radiation from 2 to 200 μm. It made a number of surveys (Ishihara et al. 2010).

A few smaller projects should be added. One is the MSX satellite (=Midcourse Space eXperiment) carried out in 1996 and 1997 by an infrared camera as a 'piggy back' on a military satellite; successful surveys were carried out in wavelength bands between 4.3 and 21.3 μm. The next successful satellite is WISE, which performed an all-sky survey with images in 3.4, 4.6, 12 and 22 μm and with an angular resolution of 6.5″ at 3.4, 4.6 and 12 μm and 20″ at 22 μm. The satellite was remarkably successful. A promising recent addition is the new airplane observatory, SOFIA (Stratospheric Observatory For Infrared Astronomy), a 80/20 US/German collaboration that uses a Boeing-747SP and carries a 2.5 m telescope. It is the successor to the Kuiper Airborne Observatory discussed in Chap. 6 (Fig. 9.12).

9.5 Millimeter and Sub-Millimeter Telescopes

Radiation at Wavelengths between 1 cm and 1 mm

Telescopes on the Earth's surface are good enough, although you do want to locate them in a high, dry place with no clouds. Telescopes are sensitive to wind and sun and are therefore generally placed inside a dome with a ceiling that is transparent for radiowaves. The most important single-dish telescopes are a 45 m telescope in Nobeyama, Japan, a 30 m French/German telescope in Spain, a 15 m Swedish/ESO telescope in Chile (now no longer active), a 20 m telescope in Sweden, and a 12 m telescope in Arizona. This last is an old instrument that was build in 1969 as the first telescope to detect millimeter radiation.

Interferometry is do-able and is also highly desirable because a lot of research is focused on star formation and that takes place in very small objects. On the Plateau de Bure in the French Alps, at an altitude of 2,550 m, an array of nine telescopes is owned by IRAM, a Franco-German-Spanish radio observatory. It is now called NOEMA (NOrthern Extended Millimeter Array). In the US, for the period of a decade there were two successful millimeter arrays, which at some point in time were merged and ultimately between 2000 and 2010 lost their financial support due to the US participation in ALMA. In Japan VERA is a very ambitious project: an array of millimeter telescopes at large distances (a baseline of over 2000 km) to determine with high accuracy the parallaxes of natural masers at large distances in the Milky Way.

Radiation at Wavelengths between 1 and 0.1 mm

As already mentioned, the atmosphere does not allow this radiation to pass through very easily. Satellite telescopes are therefore ideal, even if they are expensive. Fortunately, the atmosphere is transparent at some wavelengths (called 'windows') where you can, so to speak, look outside. Water vapor in the atmosphere in particular causes much absorption. Places that are high and dry (the Andes, the dormant volcanoes in Hawaii) are therefore the places most frequently chosen. A little less good, but still excellent are the dormant volcanoes on the Canary Islands. And then there is Antarctica, where the atmosphere is thin and very clean.

The largest single telescope has a parabolic mirror with a diameter of 50 m and is located in Mexico; only in use in May 2013. (Website: lmtgtm.org). Then, since 1984, there has been the 30 m IRAM telescope in Pico Veleta, making observations at 0.9, 1, 2, 3 mm: very widely used and very successful. There is also a 15 m telescope that has been on Mauna Kea since 1987: the James Clark Maxwell telescope (JCMT). It was a British-Dutch-Canadian project, but in 2015 it was transferred to an organization heading a partnership between China, Taiwan, Japan, Canada, the UK, South Korea and the University of Hawaii.

Perhaps the most important result obtained with the JCMT is the survey with the SCUBA-2 sub-millimeter array detector; 0.45–0.85 mm. There appears to be a significant category of galaxies with a strong sub-millimeter radiation. These are galaxies with a very bright core, the radiation of which is, however, completely absorbed in a surrounding sphere with dust particles that absorb the radiation and emit it again in the infrared. Another single-dish sub-millimeter mirror with a diameter of 10 m is the Heinrich Hertz telescope in Arizona. A remarkable instrument is the South Pole Telescope, SPT, located at latitude 90° at the Amundsen-Scott South Pole Station at an altitude of 2800 m, owned by a large collaboration of US universities and one university in Germany. Its goal is to measure the background radiation and to find the faint signature of the Sunyaev-Zeldovich effect that shows the presence of a cluster of galaxies.

Since 1992 there has been a further array of telescopes on Mauna Kea called the 'Sub-millimeter Array' (SMA) that is used and managed by the Smithsonian Astrophysical Observatory in Boston and the Academia Sinica Institute of Astronomy and Astrophysics in Taiwan. The research is strongly focused on star formation. For example, flat rotating disks have been found around very young stars. The expectation is that planets form in these disks and the current goal is to detect structure in the disks (cavities, empty rings) from which it can be concluded that planets are actually formed there.

The two most important developments in sub-millimeter astronomy were the HERSCHEL space telescope and the construction of ALMA, the Atacama Large Millimeter Array. Let us start with HERSCHEL: an ESA satellite telescope with a diameter of 3.5 m. Telescope and detectors were placed in a cooling vessel that kept them at a temperature close to absolute zero, thanks to the evaporation of helium. The mission lasted from 2009 to 2013, 1 year longer than planned. The satellite mainly carried out research on spectral lines of molecules. One of the molecules, O_2, was expected to be the most common after H_2 and CO, but it was very difficult to detect, although a Scandinavian satellite, Odin, made the discovery in 2007 (Larsson et al. 2007).

Finally, there is the most remarkable sub-millimeter telescope, ALMA, the Atacama Large Millimeter Array. The instrument stands on the 5,000 m high Chajnantor Plain in northern Chile and it is a synthesis telescope consisting of 66 twelve-meter and seven-meter telescopes. It has been working successfully since the second half of 2011. Its cost was approximately 1.4 billion euros and it was financed by national research funds in the US, in Europe through ESO, Canada, Japan, South Korea and Chile, the latter as host. Like any project of this size, its materialization is a long and complex history in which the most astonishing thing is that all the planning finally resulted in a concrete object. This was proof of the significance that astronomers attributed to sub-millimeter astronomy and of their ability to convince funding organizations to provide the necessary finances. The telescope is used for research into star and planet formation, but also for extragalactic research.

9.6 Radio Astronomy

Single-Dish Telescopes

In earlier chapters it has been argued that the primary obstacle in the progress of observational radio astronomy was a lack of sharpness of view, technically named 'limited angular resolution'. This problem had been overcome by the discovery of aperture synthesis, of the possibility of obtaining a sharp image of a radio source and its environment by making many observations with an interferometer and converting these in an image through Fourier transformations (see Chap. 6). But aperture synthesis was only successful if the radio source was strong enough and, of course, there always more faint sources than strong ones. There remained a strong case for large, single telescopes, for example in the search for faint pulsars or faint quasars and for the study of extended clouds of atomic or molecular gas.

This as an introduction to the unique story of the building of the 100 m diameter 'Robert C. Byrd Telescope' at the site of the national Radio Observatory in Green Bank (West Virginia).

Telescopes are not built for eternity although astronomers like to think they are. In the early 1960s the National Radio Observatory was established with headquarters in Charlottesville, Virginia, and a site for telescopes in a valley in the rural, quiet surroundings of West Virginia where disturbances by various radio emitters would be minimal. The first plans foresaw the construction of a 140 ft telescope as the most important instrument. The construction lasted longer than was hoped, and to keep the science going a rather cheap 300 ft telescope was constructed. It was never a very good telescope, but was nonetheless very useful and there was always sufficient demand for the telescope to keep it going, including after the 140 ft had been successfully completed.

In the morning of 15 November 1988 the telescope was in use. At lunchtime everyone left and the telescope continued under computer control. But when the staff and visitors returned after lunch, they saw a pile of metal. The telescope had totally and silently collapsed into a heap. For the engineers this was a miracle: how could such a large structure collapse almost silently and in such a controlled manner? It was an enigma for quite some time, but in the end the engineers found a satisfactory answer for their mechanical problem (Fig. 9.13).

Fig. 9.13 300 ft before (left, image courtesy of Carleton College) and after (right, image courtesy of NRAO/AUI) on 15 November 1988

Then there was the question of rebuilding. Was the building of another large single dish the right thing to do? The cost would be tremendous and would it be money wisely spent? But while the director, Paul Vandenbout, was still undecided he received a phone call from Robert C. Byrd, senator for the state of West Virginia. Byrd explained that the 300 ft had been a very visible symbol of progress in the otherwise poor state of West Virginia. That large telescope had to be rebuilt, no question about it. He, Senator Byrd, was chairman of the Senatorial appropriations committee and he could certainly help to obtain the necessary funding. And this is what happened. A new 100 m telescope was built, much better than the old. The telescope was named the 'Robert C Byrd Green Bank Telescope'. It complemented many other telescopes and is still being put to good use (Fig. 9.14).

Fig. 9.14 The Robert C. Byrd Green Bank Telescope. Source: Wikimedia Commons

For many years the 300 m telescope in Arecibo (Puerto Rico) was the largest single-dish telescope. That has now changed. In September 2016 a 500 m telescope was opened in the province of Ghizou in China. It fills a valley and has therefore some of the same shortcomings as Arecibo (only observations close to the zenith are possible) but apparently less so: measurements can be made to 40° from the zenith. The mirror is parabolic and will be good enough to make observations at wavelengths between 6 cm and 1 m.

Aperture Synthesis Telescopes

The opening of the large Australian synthesis telescope (ATCA or 'Australia Telescope Compact Array') in September 1988 was an important event: finally there was now a synthesis telescope in the southern hemisphere. In 1998 the array was expanded to make it sensitive to observations at millimeter wavelengths. Another synthesis telescope was successfully built in India: thirty 45 m telescopes with a largest baseline of 25 km: the 'Giant Metrewave Radio Telescope' of 'GMRT'. It was operational in 1995 and made measurements at wavelengths between 21 cm and 60 m.

You do not need new telescopes to set up an aperture synthesis telescope; the clever use of existing telescopes will yield excellent observations. This has been the principle of Merlin, a system consisting of a set of existing radio telescopes in England and of the EVN, the European VLBI network using telescopes spread throughout the European Union and even outside. Although VLBI had been pioneered in the sixties and seventies, major progress in sensitivity, accuracy and flexibility became possible at the turn of the century by benefiting from advances in the ICT industry; the introduction of real-time, e-VLBI in the European VLBI Network (EVN) enabled high-resolution detections of transient phenomena: short bursts of radiation that have recently been detected at radio (FRB), optical, X-ray and gamma-ray (GRB) wavelengths.

1994: SKA, a Grand New Plan

The acronym SKA appears for the first time in 1994: the 'square kilometer array': a new concept for a radio telescope with effectively a square kilometer collecting area. Such a telescope is desired for making new breakthroughs, notably in cosmology and pulsar research. Technological advances make it possible to construct larger telescopes by combining many more antenna elements electronically, instead of making a few large parabolic dishes. In a complex process led by the SKA office in Manchester, more than ten countries finally decided to first design a more modest SKA phase-1 telescope, consisting of a low frequency array in Australia and mid-frequency dishes in South Africa.

Resulting from this global ambition, many initiatives emerged to build precursors to SKA. The Australians have the MWA and ASKAP at their Western Australian site, while South Africa commissions KAT7 and later MeerKAT in the Karoo desert. A large pilot project at low frequency is LOFAR, operational initially in the Netherlands and expanded with stations across Europe. It demonstrates that—with some computational difficulty—it is possible to calibrate observations made with simple antenna stations at frequencies where the atmosphere and interference environment are harsh. It also pioneers the processing of gigantic, peta-byte[1] data sets into scientific results.

9.7 Astronomy of the Extreme Ultraviolet and X-Radiation

Extreme Ultraviolet Telescopes

Radiation at wavelengths shorter than about 330 nm is absorbed by the Earth's atmosphere. The radiation is called extreme ultraviolet. Stars with a surface as hot as that of the Sun or hotter emit a lot of radiation at these wavelengths. In Chap. 6 the IUE satellite, launched in 1978, was mentioned that was dedicated to observations of radiation down to a wavelength of 110 nm. It was operational until September 1996.

In 1992 the EUV-explorer (of EUVE) put into orbit by NASA, an imaging telescope that detected some 1000 sources in the 7–76 nm wavelength range. Radiation at these wavelengths is strongly absorbed by interstellar hydrogen, but apparently the hydrogen distribution is sufficiently irregular to allow the detection of nearby objects. The detections were not very surprising as only well-known kinds of objects were detected. The exception is the discovery of a new class of objects, the so-called 'supersoft X-ray sources'. They are probably white dwarfs in a narrow binary system. Their existence does not pose any major new astrophysical problems.

In 2003 the GALEX satellite became operational. The plan was to make maps of the radiation in the wavelength range of 135–280 nm of over one million galaxies. The strongest stellar sources of this radiation are young high-mass stars, and GALEX provided an overview of these stars in various types of galaxies. This was important for the study of star formation in distant galaxies where the redshift has changed the wavelengths of ultraviolet radiation into visible, optical radiation.

X-Ray Telescopes

X-ray and gamma-ray studies are best characterized by the energy of the photons. The usual unit is eV (electron volt; 1 eV = 1.6×10^{-19} Joule). By definition, X-ray photons have an energy between 100 eV and 100 keV.

This chapter deals with instruments in the period from 1985 to 2015. Let us start with the first half of that period: 1985–2000. The two dominant X-ray telescopes have been the German ROSAT satellite (1992–1999) and the Japanese ASCA satellite (1993–2001). Both were highly successful and for a large part complementary. Simplifying the comparison between the two: ROSAT detected many sources, and ASCA took spectra of the brightest. ROSAT is an acronym for 'Röntgen Satellite'; in German, X-radiation is called Röntgen radiation, referring to the discoverer of the radiation, Wilhelm Röntgen. The Japanese Astronomical Satellite for Cosmology and Astrophysics has the acronym ASCA.

[1] 'peta' equals 10^{15} or one million times a billion.

In a review of ROSAT, after the launch, the Principal Investigator, Truemper (1993), stressed the enhanced sensitivity shown by the much larger numbers of detected sources: from 840 to 60,000. Also, a large number of extended sources were found, such as supernova remnants. The ROSAT database turned out to be a treasure trove for many later studies. The catalog with some 19,000 sources (Voges et al., 1999) gained the reputation in X-ray astronomy that the 3C-catalogue had for radio astronomy. More than 200 clusters of galaxies were detected, making this an important database for studies of the evolution of the Universe (see Chap. 16).

The Japanese 'Astronomical Satellite for Cosmology and Astrophysics', ASCA, was equipped with several instruments of which the spectrograph appeared to be the most successful part. This is well illustrated in Fig. 9.15. Another important ASCA result was the spectroscopic proof of the existence of supermassive black holes (see Chap. 16).

A surprising, and yet not unexpected result was the discovery of synchrotron radiation from the rim of a supernovae (see Figs. 9.15 and 9.17). This led to the suggestion that in this rim electrons were accelerated to relativistic velocities and that this rim is one of the places where the mysterious 'cosmic rays' were produced. The suggestion was strongly supported by the discovery of TeV-photos (photons with an energy 1000 times that of gamma rays) from the same area (Tanimori et al., 1998).

Two other, smaller, but still very successful X-ray satellites were first NASA's RXTE (X-ray Time Explorer; 1995–2012). The letter R was later added to the acronym XTE to honor Bruno Rossi who greatly stimulated the birth of X-ray astronomy (see Chap. 3). The satellite was designed to study the variability of X-ray sources and to search for periodicities in the variations. It was sensitive over a very broad range of photon energies (2–250 keV) and functioned as expected. It made some important contributions.

The second satellite, called originally SAX (Satellite per Astronomia a raggi X) was an Italian project with a significant contribution from the Netherlands. Later, the name 'Beppo' was added to commemorate the colorful and important Italian physicist Guiseppe ('Beppo') Occhialini. BeppoSAX operated from 1996–2003. Like RXTE, the instrument detected photons over a wide range of energy (0.1–300 keV). With its sensitivity to gamma rays as well as to X-rays, it made its most important contribution: the first identification of a Gamma Ray Burst (see Chap. 14) with a very distant galaxy.

The new millennium began with the 'flagships' of NASA and ESA: CHANDRA and XMM-Newton launched late in 1999. Both were the result of long-term planning that began in the 1970s; NASA decided to build four 'Great Observatories', and ESA had four 'cornerstones'. CHANDRA was the third 'great observatory' and XMM-Newton the third cornerstone. Number four, to be built within a few years, was for both agencies an infrared telescope: SPITZER for NASA and HERSCHEL for ESA.

Both telescopes were called 'observatories' and that is what they were. CHANDRA made maps of X-ray brightness with details as small as in ordinary optical astronomy, 1 arcsec; XMM-Newton produced spectra and that made the two telescopes somewhat complementary, although that had not been planned. Both worked as intended and this is still true 15 years after the launch.

One of the very first detections made by CHANDRA was a neutron star at the centre of the supernova remnant Cassiopeia A. The neutron star did not pulse and

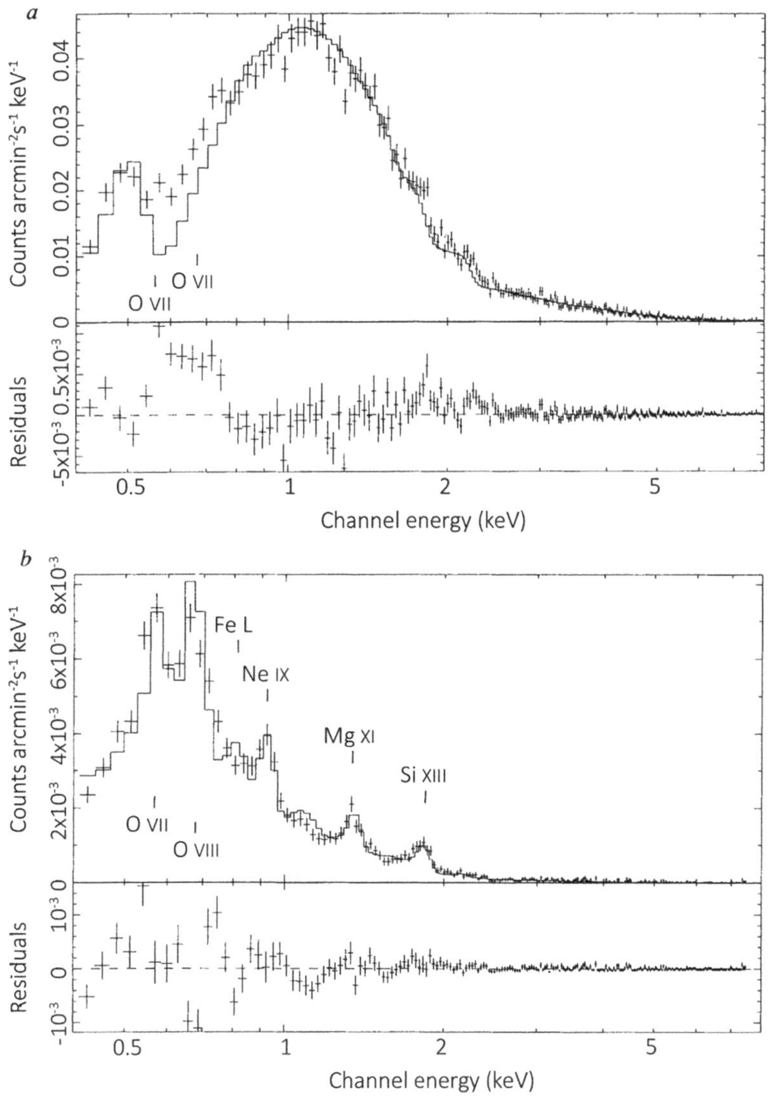

Fig. 9.15 Spectra of the remnants of supernova SN 1006 obtained with ASCA. The upper figure shows the spectrum of the northern rim (see Fig. 9.17, farther down) that is explained by synchrotron radiation. The lower figure refers to the centre of the remnants. Several lines are present that have been produced by highly ionized atoms (Koyama et al. 1995)

this made it exceptional. More stars, but still only a few 'naked neutron stars', have since been detected (see Chap. 13). An important item of news was the detection of 'pulsar winds': very hot, very rarefied gas emitted by pulsars. The first detection had been the blue light in the Crab nebula, detected in around 1940 and explained in the late 1950s. Figure 9.16 shows the blue light of the pulsar winds in four supernova remnants obtained with CHANDRA and Fig. 9.17 has been produced from data obtained by XMM-Newton.

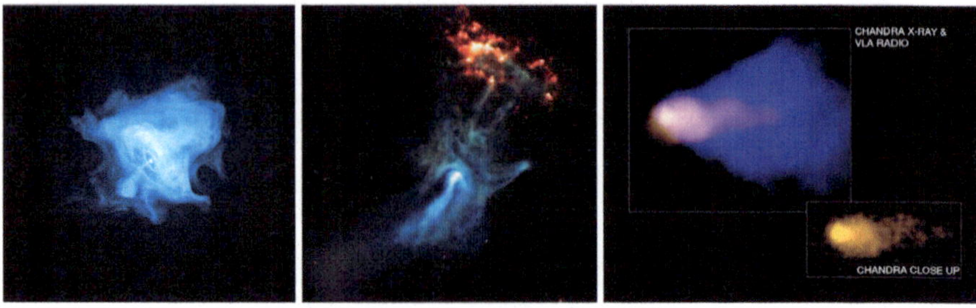

Fig. 9.16 Left is the Crab Nebula (size 5′); in the middle is the nebula associated with the pulsar PSR B1509-58 (20′); to the right a pulsar moving at high speed through the interstellar matter leaves behind a bow wave; the image is 1.2′ wide; this object is known as the 'Mouse' (Kaspi 2010)

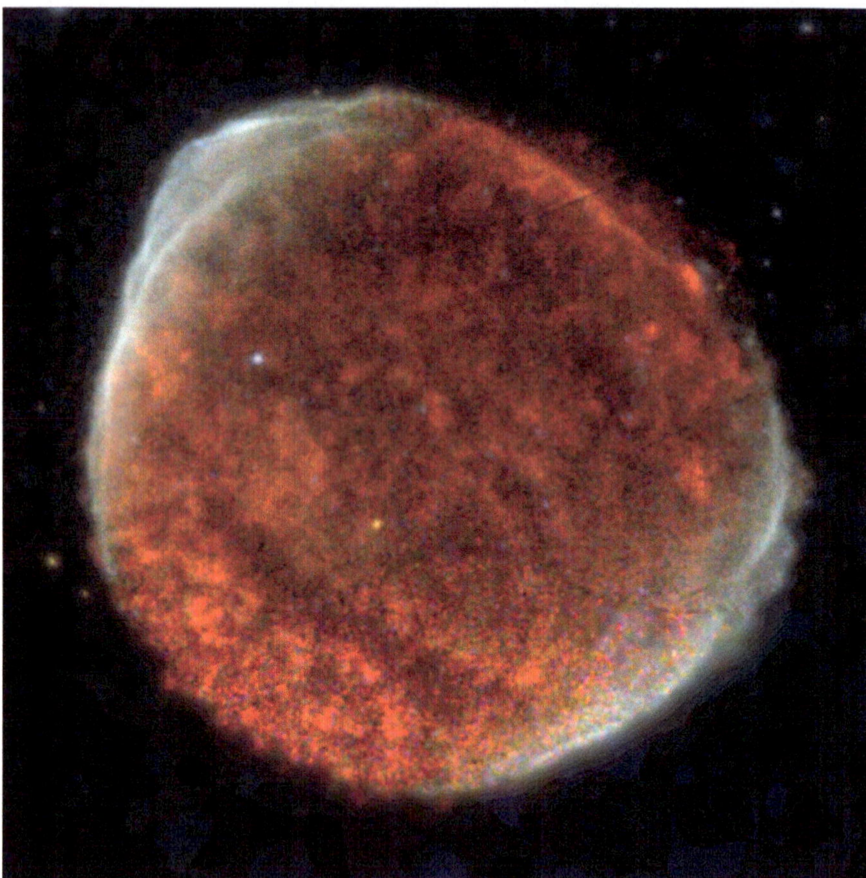

Fig. 9.17 SN 1006. This image is a composite or recordings each taken with a differently colored filter. The white edges occur where the radiation is equally strong in all filters; this is an indication of synchrotron radiation (Rothenflug et al. 2004)

9.8 Detecting Photons with the Highest Energy

Detecting Gamma Radiation

Photons of gamma radiation have (by definition) an energy that is greater than 100 keV and, as an illustration and for those who for whom this means something: 100 keV is approximately 7000 times the energy of a photon needed to ionize a hydrogen atom—and such an ionization is a fairly energy-rich process in astronomical terms. But gamma photons are far from being the most energy-rich. As we will see, there are galaxies that emit photons with millions of times more energy. The processes in which gamma photons originate are still unknown and the search for them is a challenge, but is proceeding with some difficulty. The biggest challenge in recent years has been the GRBs, those mysterious Gamma Ray Bursts, which, as described above, have taken place at great distances from us, releasing a great deal of energy that has been inexplicably produced in a matter of mere seconds.

In December 1989, the Soviet Union successfully launched the GRANAT telescope, which was to remain active for 9 years; substantial instrumental contributions were made from France, among others. The device detected photons with energies between 2 and 100,000 keV. Initially the telescope was focused on specific objects, but later it served as an all-sky detector for GRBs, Gamma-Ray Bursts.

In November 1991, NASA launched a large space platform, CGRO, which remained operational until 2004. CGRO, the Compton Gamma-Ray Observatory, was one of four ‘Great Observatories’ provided for in NASA’s long-term plan. The name ‘Compton’ referred to Arthur Compton, an American physicist who was awarded the Nobel Prize in 1927 for research on the scattering of photons after colliding with an electron. This process plays a very important role in AGNs, active nuclei of galaxies.

CGRO had four major instruments on board, one of which was built by a German-Dutch partnership. All four instruments worked as expected and produced significant results. (1) The detection of two spectral lines from radioactive atomic nuclei (^{44}Ti and ^{26}Al) in the Milky Way, which must have been dropped there recently in supernova explosions. (2) The detection of photons with energies of up to 100 MeV from about 30 active galaxies and in particular from ‘blazars’, a special underclass of quasars. (3) A catalog of 271 sources of gamma photons with more than 100 MeV of energy. One hundred of them could be identified with known objects, including surprisingly many blazars, but the identity of some 170 different sources could not be established.

In 1973 a spectral line with a photon energy of about 500 keV was discovered by a gamma ray telescope transported to a high altitude by a balloon (Johnson and Haymes, 1973) and a few years later, again by a balloon flight, when the photon energy was determined more accurately as 511 keV. This confirmed what had been guessed: these photons are produced when an electron (e^-) collides with its antiparticle, the positron e^+. The detection was significant because it showed the presence of positrons, and thus of ‘anti-matter’. This was a theoretical concept that

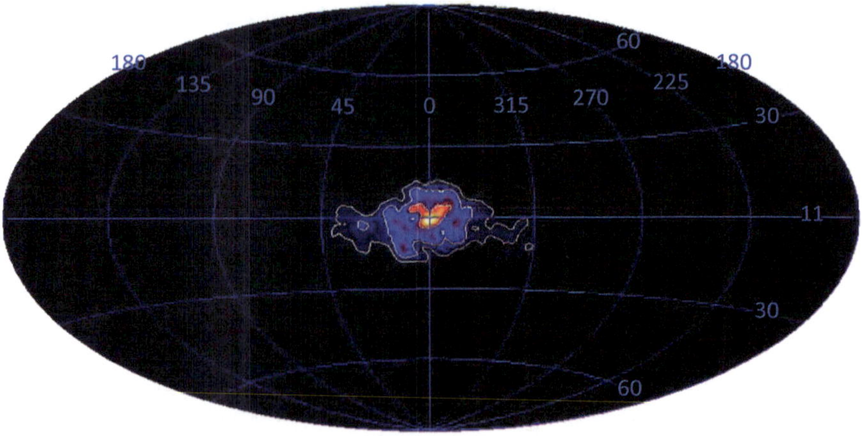

Fig. 9.18 A map of the full sky that shows the strength of the 511 keV gamma-ray line. The centre of the map coincides with the centre of the Milky Way, but there is some radiation in the disk of the Milky Way. Results obtained with the INTEGRAL satellite (Knödlseder et al. 2005)

was later verified in the laboratory, but that had never been seen in astronomical objects. In fact, the absence of much anti-matter is a fundamental problem for theoretical cosmology.

The same line was detected by OSSE, one of the four instruments on CGRO, but only in the direction of the galactic centre. ESA's gamma ray telescope, INTEGRAL, launched in 2002, made a survey of the line in the Milky Way (see Fig. 9.18). The line is strongest in the neighborhood of the galactic centre but there is an extension along the Milky Way equator. There is as yet no explanation for the origin of the positrons.

A very successful telescope, called the FERMI Gamma-Ray Space Telescope, was launched by NASA in 2008 and has been in operation since then. So far it has produced a catalogue with about 1800 gamma-ray sources, 70% of which have been identified as objects known to produce gamma rays. 30% are unexplained. Fermi also discovered over 100 gamma pulsars whereas only a handful had been known. What was most surprising and spectacular was the discovery of two enormous gamma-ray bubbles, one on each side of the Milky Way and touching one another at the galactic centre. This most likely points to a very energetic explosion in the centre some millions of years ago. It is a mystery (Fig. 9.19).

VHE: Photons of Very High Energy

Astronomers considered gamma photons to have a very high energy. Practically all the processes that have been discussed in this book do not produce gamma photons and yet nature does it. And gamma photons were not the end of the story. Another phenomenon pointed at the existence of photons with energies at least a

Fig. 9.19 Observation with the Fermi-satellite showed that there are two large balloons of gamma radiation symmetrically placed on both sides of the centre of the Milky Way. The balloons are most likely the product of an explosion. Source: NASA

thousand times more, and thus energies of a few times 10^{12} eV or 1 TeV (Tera electronvolt).[2] This evidence was the observation of short light flashes at night that could be attributed to 'Cherenkov radiation': radiation caused by a particle that travels through the Earth's atmosphere at a velocity greater than the local speed of light (this does not contradict the theory of relativity, because that demands that no object moves faster than the speed of light *in vacuum.* Inside some dense medium, light may have a smaller speed, as it has, for example, in water: the lower speed of light in water explains the well-known fact that a stick looks as if it has broken on the transition of air to water). Cherenkov was a Soviet scientist who studied the radiation produced by super-relativistic particles in such media; he received the Nobel Prize for his studies in 1958. The effect is sometimes referred to as an 'air shower'—an appropriate name.

What produced the high-velocity particles that were the cause of the Cherenkov radiation? Ultimately two different sources were recognized: one was the collision of atoms in the Earth's atmosphere with particles from the cosmic rays, that is: atomic nuclei and electrons with relativistic velocities that travel in interstellar space and whose origins were mysterious. The other source of Cherenkov radiation was atoms in the atmosphere hit by TeV-photons from outside the Earth. Both cosmic ray particles and TeV photons caused a cascade of particles in the atmosphere, but the shape of the two cascades produced were different (see Fig. 9.20).

[2] The prefix 'T' implies a multiplication by a factor of 10^{12}, much as 'k' implies multiplication by a factor of 1000 (or 10^3) and G a multiplication by 10^9.

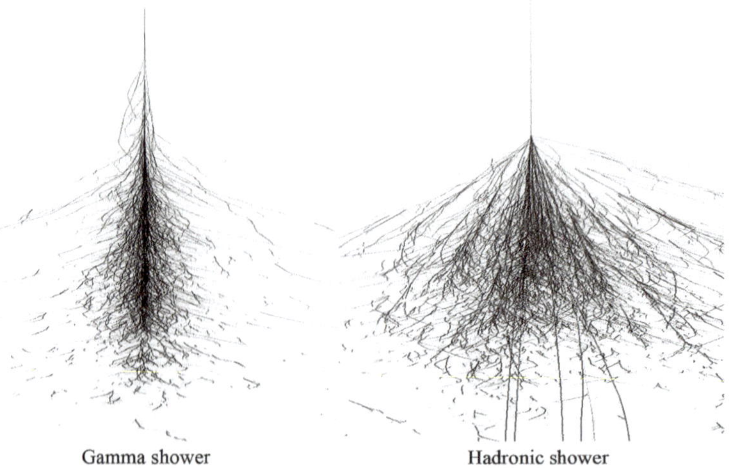

Fig. 9.20 The cascade (or air shower) in the atmosphere caused by a high-energy photon (left) or by a particle of the cosmic rays (Völk and Bernlöhr 2009)

Fig. 9.21 Veritas TeV telescope at the Whipple Observatory in Arizona. Four 12 m dishes collect the flashes of Cherenkov radiation caused by TeV photons from outside the Earth. Source: VERITAS

Various groups undertook the systematic measurement of Cherenkov radiation. One particular breakthrough was the detection by the Whipple telescope of TeV photons coming from the active galaxy Markarian 231 (Punch et al. 1992). Another success was the detection of TeV photons from the rim of the supernova of SN1006. In 2015 a number of TeV observatories were active, including the German telescope, H.E.S.S. in Namibia a French telescope CAT in the Pyrenees and the Whipple observatory in Arizona (Fig. 9.21).

References

Abbott, B.P., Abbott, R., et al.: Astrophys. J. **848**, L12 (2017)
Carey, S.J., Noriega-Crespo, A., et al.: Publ. Astron. Soc. Pac. **121**, 76 (2009)
Ferguson, H.C., Dickinson, M., Williams, R.: Annu. Rev. Astron. Astrophys. **38**, 667 (2000)
Gillesen, S., Eisenhauer, F., et al.: Astrophys. J. **692**, 1075 (2009)
Gaia Collaboration; Babusiaux, C.; van Leeuwen, F.; et al.; A&A v616 A10 (2018)
Gunn, J.E., Siegmund, W.A., et al.: Astron. J. **131**, 2332 (2006)
Hester, J.J.: Annu. Rev. Astron. Astrophys. **46**, 127 (2008)
Ishihara, D., Onaka, T., et al.: Astron. Astrophys. **514**, A1 (2010)
Johnson III, W.N., Haymes, R.C.: Astrophys. J. **184**, 103 (1973)
Kaspi, V.M.: Proc. Natl. Acad. Sci. U. S. A. **107**, 7147 (2010)
Knödlseder, j., Jean, P., et al.: Astron. Astrophys. **441**, 513 (2005)
Koyama, K., Petre, R., et al.: Nature. **378**, 255 (1995)
Larsson, B., Liseau, R., et al.: Astron. Astrophys. **466**, 999 (2007)
Le Fèvre, O., Tasca, L.A.M., et al.: Astron. Astrophys. **576**, A79 (2015)
Malfait, K., Waelkens, C., et al.: Astron. Astrophys. **332**, L25 (1998)
Punch, M., Akerlof, C.W., et al.: Nature. **358**, 477 (1992)
Rothenflug, R., Ballet, J., et al.: Astron. Astrophys. **425**, 121 (2004)
Skrutskie, M.F., Cutri, R.M., et al.: Astron. J. 131, 1163 (2006)
Tanimori, T., Hayami, Y., et al.: Astrophys. J. **497**, L25 (1998)
Truemper, J.: Science. **260**, 1769 (1993)
Voges, W., Aschenbach, B., et al.: Astron. Astrophys. **349**, 389 (1999)
Völk, H.J., Bernlöhr, K.: Exp. Astron. **25**, 173 (2009)

Chapter 10
1995–2015: Main-Sequence Stars and Failed Stars

As mentioned previously, the pressure and temperature at the core of a star are so great that nuclear fusion, hydrogen combustion, ensues spontaneously, generating the energy that the star emits throughout its life. The pressure and temperature at the core of lighter stars are lower; they have less mass and weaker gravity. If compression is too low, nuclear fusion will not take place: we then have what is known as a failed star. Estimates indicate that this happens when a mass of gas contracts less than one-tenth of the Sun. The contraction itself provides a little bit of energy that is radiated away mainly at infrared wavelengths. These 'failed stars' have a very low luminosity; it took until the 1990s before the first examples were found and they were always within a few light years from the Sun.

This is an appropriate place to show two very recent (2018) Hertzsprung-Russell diagrams of large numbers of stars for which the distance has been measured very accurately by the GAIA satellite. They show the progress made in the basic field of astrometry (Figs. 10.1 and 10.2).

10.1 Brown Dwarfs or Failed Stars

Predicting, Naming and Discovering Brown Dwarfs

Stars that are lighter than the Sun emit less light and are redder in color. What are the lightest and reddest stars? Until the 1980s the star called VB10 was known as the star with the lowest luminosity; it was 2,000 times fainter than the Sun; the temperature on the surface was estimated to be 2,200–2,500 K and its mass was just one-tenth that of the Sun (Greenstein et al. 1970). The 'VB' in this stellar name stands for 'Van Biesbroeck', a legendary Belgian astronomer working in the US who was an active observer of binary stars up to the age of 97; he observed many of these stars himself, by eye and without taking any photographs. He was

H. J. Habing, *The Birth of Modern Astronomy*, Historical & Cultural Astronomy,
https://doi.org/10.1007/978-3-319-99082-8_10

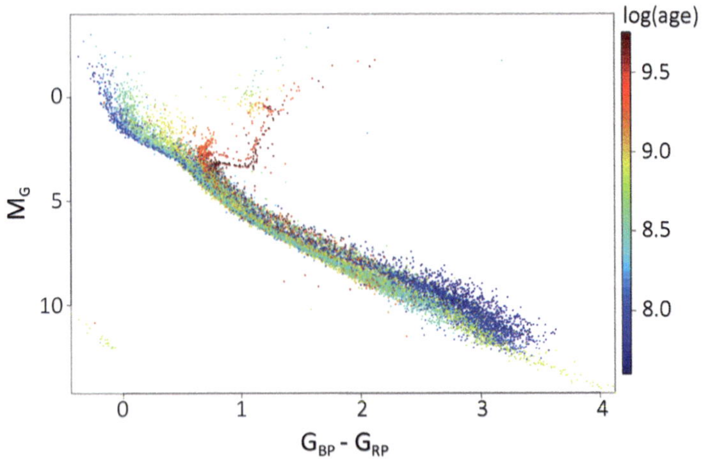

Fig. 10.1 Hertzsprung-Russell diagram of a number of star clusters. The horizontal scale gives the color of the star. The vertical scale shows the (absolute) magnitude. The precise location of the stars depends somewhat on the age of the cluster; see the bar on the right-hand side of the figure (Babusiaux et al. 2018)

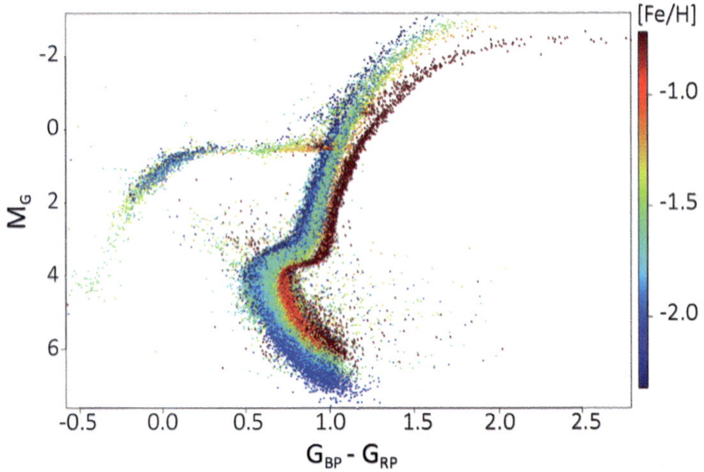

Fig. 10.2 The Hertzsprung-Russell diagram of stars in globular clusters. As these clusters are old they contain many red giant stars and no blue main-sequence stars. The precise location of the stars depends on the 'metallicity'—see the bar on the right side of the diagram: [Fe/H= –1.0 implies that the star contains 10 times less iron atoms (compared with the n umber of hydrogen atoms) than the Sun; [Fe/H]= –2 implies a hundred times lower presence (Babusiaux et al. 2018)

probably the only professional astronomer still doing so. After he retired, at the age of 65, he took on a new job as an observing astronomer, a role in which he continued to be active for a further 30 years. His final publication appeared two years after his death (Hardie 1974).

What was the smallest mass that a star could have and still emit energy obtained from nuclear fusion? Theoreticians had something to say about this. As has already been mentioned in Chap. 4, a young, light star before it arrived at the main sequence was a contracting ball of gas; 'TTau-stars' were such objects. A 'protostar' is a better term than 'star' for these objects. The contraction continued until the density and temperature at the core were so high that spontaneous combustion occurred and hydrogen atoms started to fuse to form helium atoms. From that point onwards, the object was no longer referred to as a 'protostar' but as a 'star'.

Calculations had already shown in the early 1960s (Hayashi and Nakano 1963; Kumar 1962) that a contracting protostar some 20 times less massive than the Sun would not emit energy obtained from nuclear fusion: the core would never be hot enough nor the density large enough for hydrogen combustion. The protostar would never become a true star. Its fate was clear: it would continue to contract until the process was halted by 'degeneration pressure', as occurred in white dwarfs. Degeneration pressure does not occur in classical physics but is the consequence of quantum laws, in particular Pauli's exclusion principle. Once the contraction stopped, it left behind an object that emitted less and less infrared radiation and that consequently cooled down strongly. For some time these hypothetical objects were called 'black dwarfs' but in her 1975 dissertation, Jill Tarter[1] in Berkeley proposed another name that came up during a discussion with her supervisor, Joe Silk: 'brown dwarfs'. Brown does not occur in the solar spectrum and Silk and Tarter agreed that it therefore was appropriate to call these then hypothetical stars 'brown'. It was a name that was immediately embraced by everyone.

These objects now had to be found. That took a while and, looking back, it was only to be expected: brown dwarfs emitted mainly infrared radiation and only very little, at that. Large telescopes and new infrared detectors were needed. The first detections of an infrared star at a short distance from another faint star were made in 1987 and 1988 (Forrest et al. 1988; Becklin and Zuckerman 1988). For an image of one of the two, see Fig. 10.3. In both cases the objects could be brown dwarfs, but the authors were very cautious and dared not exclude the possibility that they were ordinary red dwarfs. Certainty only came several years later (Rebolo et al. 1995) when Spanish astronomers found two very faint stars in the Pleiades that emitted mainly infrared radiation. A spectrum recorded with one of the two 10 m Keck telescopes on Hawaii showed that the element lithium was present in the atmosphere of the star. This was significant: the presence of lithium proved that no nuclear fusion had taken place in the object because lithium would disappear as collateral damage almost immediately after hydrogen fusion started. Without nuclear fusion, the object was by definition a brown dwarf.

[1] Jil Tarter would later become famous as the powerful motor behind the SETI project, the 'Search for Extra Terrestrial Intelligence', that has searched for over 30 years for intelligent signals of extraterrestrial life, a search that has so far not produced any results.

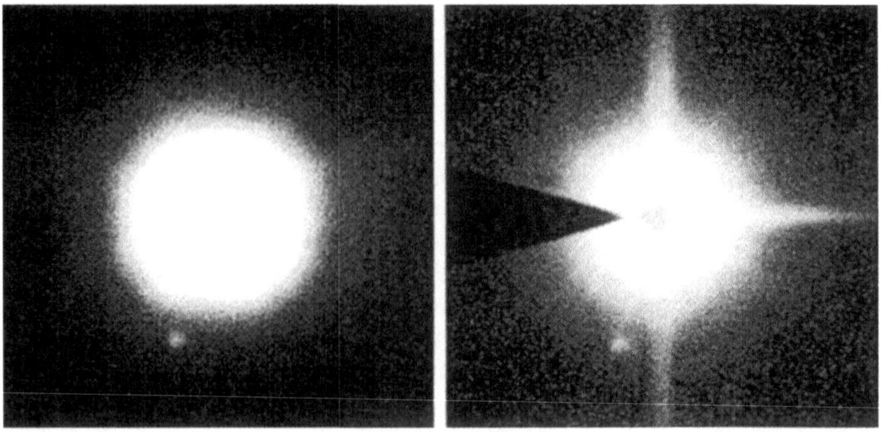

Fig. 10.3 Two images of the star Gliese 229 with the small companion Gliese 229B. The relatively huge size of the image of Gliese 229 is not true to life, but is an effect of the over-exposure needed to detect its faint companion. The left-hand photo was taken through a very red filter (0.9 μm) using the 60-in. telescope on Mount Palomar, and the right-hand photo with the Hubble Space Telescope through an infrared filter (2.2 μm). The black triangle in the right-hand figure points to a deeper image that is larger than the star. The authors believe this is a disk around the star; more of these disks will be found near to other stars (Nakajima et al. 1995)

In 1995, when the spectrum of the Gliese 229B star was recorded, it appeared to contain an absorption band[2] of methane (CH_4), something that was previously seen only in the atmospheres of the major planets, but never in a stellar spectrum. It indicated that the 'little star' had a surface temperature of around 1,000 K; it was definitely not a star but a brown dwarf (Oppenheimer et al. 1995).

Brown dwarfs had less mass than a star. The *highest* mass had been estimated to be one-fifth of the mass of the Sun (see above). What was the *lowest* mass a brown dwarf could have? This question arose when a free-moving object was detected that was eight times more massive than Jupiter, and the mass of the planet Jupiter is 1,000 times smaller than that of the Sun. Was this free-floating object a planet or a brown dwarf? Was it a brown dwarf trying to form a planetary system, or was a system of moons being created around a planet, like those found near to Jupiter and Saturn? Was there actually a fundamental difference between massive planets and brown dwarfs? There are as yet no answers to these questions (Luhman et al. 2005; Kellogg et al. 2016).

More Brown Dwarfs

The very fact that brown dwarfs existed caused something of a sensation. More examples were soon found, with most of them appearing after two new surveys had been conducted at an infrared wavelength of 2.2 μm; see Chap. 9. A similar survey

[2] A absorption band is formed in a spectrum if a large number of absorption lines are close together. Only molecules cause such series of closely packed lines.

had already been carried out in the 1960s but it was now possible to repeat it with a 100-fold increase in sensitivity. The first of these two surveys, DENIS, was a poorly financed, rough-and-ready European organization that was kept afloat on the enthusiastic support of many—mainly young—astronomers. In spite of all the disasters it was the first to produce good candidate brown dwarfs (Delfosse et al. 1997). But, fair's fair, the second all-sky infrared survey, the 2MASS project, produced much more valuable results. NASA generously funded the project that brought to light not only brown dwarfs but numerous other interesting objects as well. After 2005, a number of infrared surveys were carried out from space (see Chap. 9). On 12 November 2012 the 'DwarfArchives.org' website announced almost thirteen hundred brown dwarfs.

Theoretical Considerations about Brown Dwarfs

In 2000, French astronomers Gilles Chabrier and Isabelle Baraffe published a very successful summary article in which they discussed the models of the lightest stars and the brown dwarfs and how the stars evolved over time (Fig. 10.4).

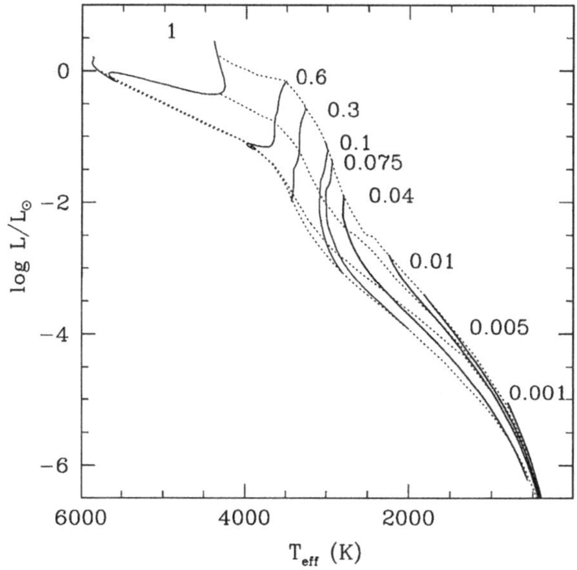

Fig. 10.4 This diagram shows the evolution paths of light stars and brown dwarfs. The surface temperature is indicated along the horizontal axis and the amount of light and infrared radiation are shown along the vertical axis. The mass of each object is indicated at the start of each path. The upper black line explains how, having contracted, the Sun (mass 1.0) arrives at the main sequence, where it will remain for billions of years. Stars in the making, with a mass of 0.6, 0.3 and 0.1, share the same destination. But stars in the making that have a mass below 0.1 have a different fate. They slide, as it were, downwards along a path that is practically the same for all of them. The dotted lines are isochrones, curves that mark the passing of time. From right to left, the dotted line furthest to the right indicates that 1 million years have passed; the next ones indicate a period of 10 million, 100 million and 5 billion years. (Chabrier and Baraffe 2000)

This diagram shows why it was observationally practically impossible to distinguish these young and light brown dwarfs from old, more massive ones: in the diagram their development paths run close to one another.

The difference in life-time between brown dwarfs and almost equally massive stars is dramatic. Brown dwarfs take millions of years to cool down, but the lightest protostars that start hydrogen fusion have a life-time of more than 10,000 million years.

Spectra and the Types of Brown Dwarfs

The spectrum of a star provided a lot of information, and the same applied for brown dwarfs. From around 1900, stellar spectra could be photographed in large numbers. On the basis of the dark lines in the spectrum, researchers were able to categorize almost all stars into 'spectral classes', which were indicated by a capital letter. After the discovery in the 1920s that the color indicated the surface temperature of the star, it was concluded that the sequence of spectral classes (OBAFGKM) indicated stars from hot to cold. The temperature of the coldest stars was 3000–3300 K (2730–3030 °C); the star was dark red and the class, the 'spectral type', was named M; M0 were the hottest M stars and M9 the coldest.

The surface of brown dwarfs was much colder than that of the coolest known stars. One of the first brown dwarfs discovered, Kelu-1, had a surface temperature of 1,900 K, far below that of the coldest star (Ruiz et al. 1997). The same applied to other brown dwarfs. At an international conference in 1999 it was therefore decided to introduce a new spectral class, which was assigned the letter L. But that was not enough. Later, another class of brown dwarfs—the T-class—was discovered; this class had strong bands of methane (CH_4) that indicated a surface temperature of 1,000 K or less. And, ultimately, a further class, Y, was found that comprised even colder stars with spectra that included ammonia (NH_3).

At the end of 2015, the coldest brown dwarf known (WISE 0855-0714) had a surface temperature between 220 and 270 K, or, in degrees Celsius, between: −43° and −3°. The star was very close by, just 2.2 pc (7.2 light years) away. The surface was so cold that water (if there were any present there) would be frozen to the surface (Luhman and Esplin 2016) (Fig. 10.5).

The Atmosphere

The presence of H_2O, CH_4 (methane) and NH_3 (ammonia) molecules had a major influence on the spectra. Japanese astronomers put forward an even more unusual hypothesis, namely that clouds could be formed in such cold atmospheres (Tsuji et al. 1996). The existence of these clouds has not yet been proven convincingly.

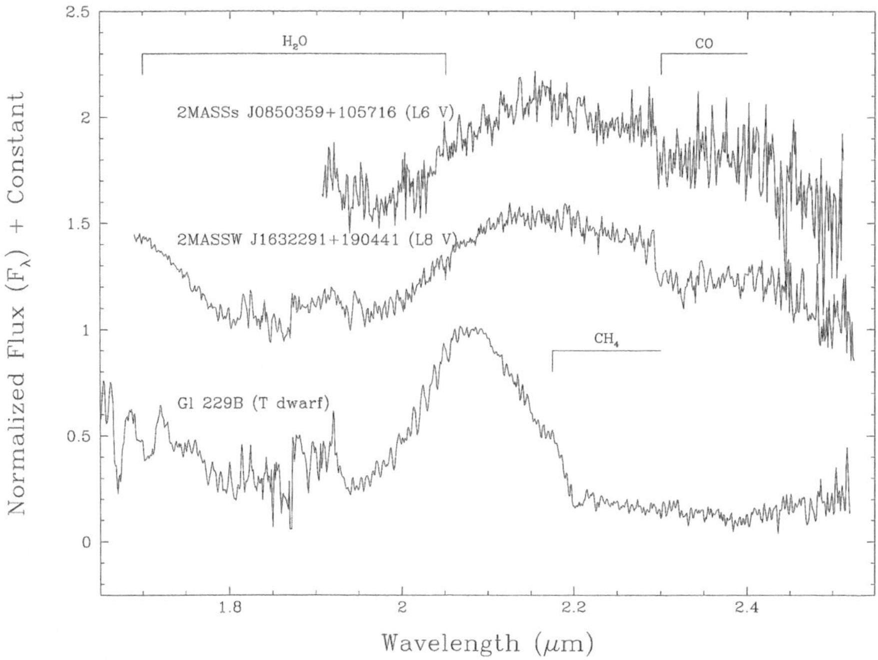

Fig. 10.5 Two spectra of L dwarfs and the spectrum of the Gliese 229B T dwarf. The wavelength is plotted along the horizontal axis and the strength of the radiation along the vertical axis. The lowest star radiates very faintly, between 2.3 and 2.5 μm and has a very strong absorption band of CH_4 (methane). (Kirkpatrick et al. 1999)

Circumstellar Disks and Planets

Another surprise was the finding of disks of gas and dust around some young brown dwarfs, even though these discs were a 1000 times 'thinner' than the disk around a typical TTau star. Planets were also found around a number of brown dwarfs. More about this subject in Chaps. 11 and 12.

Activity

M stars had been known for some time to be 'active' (see Chap. 4). They had chromospheres, and emitted both radio radiation and X-radiation, sometimes in the form of 'bursts'. L and T stars were also active; radio radiation was detected in six out of seven L and T dwarfs studied using the American 'Karl Jansky VLA' telescope (Kao et al. 2016). The measurements surprised the observers because the characteristics of this radio radiation were different from the norm. It was similar to that of the Northern lights and Jupiter. M dwarfs also experienced bursts of X-radiation, but the L dwarfs did not (Berger et al. 2010).

How Brown Dwarfs were formed

Ultimately there was the question of how brown dwarfs were formed. There were two different views on this issue, and there was no way of telling which was correct. According to the first, brown dwarfs were formed in the same way as normal stars, and according to the second, brown dwarfs were formed in the same way as planets, but had escaped the mother star.

10.2 Light Stars

'There's little reason to be satisfied with our physical knowledge of the stars.' J. Cristensen-Dalsgaard, 2002

The most important fact about a star is its mass. Massive stars explode as supernovae, leaving behind a neutron star or a black hole. Light stars eventually peacefully fall apart and leave behind a white dwarf. Earlier chapters have already described how, as they age, light stars twice reach the stage of red giants. Each time they have a hard core surrounded by a very tenuous, very large mantle of hydrogen. The second time, when they are referred to as AGB stars, they gradually shed this mantle, after which the core remains as a white dwarf. The borderline between massive and light stars is poorly defined, but it is generally assumed that the cut-off point is somewhere between eight and ten times the mass of the Sun.

I will first look at advances in our knowledge about the Sun.

The Exterior of the Sun

The solar atmosphere comprises a number of layers: the photosphere, where most sunlight comes from; above this a hotter layer that is red in color—the chromosphere—and outside this the thin and even hotter corona that surrounds the Sun like a crown, but that is so faint that it is only visible during a solar eclipse (see Fig. 10.6).

Space telescopes have collected a great deal of visual material about the Sun and in particular spectral areas (UV, X-radiation), of which the radiation does not penetrate the Earth's atmosphere. These satellites include TRACE (Transition Region and Coronal Explorer, NASA), the Japanese satellite YOKOH, and SOHO (Solar and Heliospheric Observatory; ESA and NASA jointly). Besides these, a number of Soviet and later Russian satellites have also been active.

Figure 10.7 shows an image of the Sun taken with the SOHO telescope in a spectral line of hydrogen. At the bottom left, a large solar flare can be seen that starts on the surface of the Sun in two thin rays that curl into one another further up. The rays show how invisible magnetic fields transport the hot gas far above the surface of the Sun before returning it to the surface.

Fig. 10.6 A recording of the Sun during an eclipse in 1999. The black object is the Moon. The chromosphere can be seen, in red, as a thin, hotter layer of gas just above the photosphere. Outside this the corona surrounds the Sun. Both the chromosphere and the corona exhibit large irregularities. Chromosphere and corona are only visible from Earth during an eclipse of the Sun by the Moon. Coincidentally the image of the Moon is just as big as that of the Sun. Source: Luc Viatour (lucnix.be)

Figure 10.8 shows a similar image but this time of radiation emitted by iron atoms, from each of which 11 electrons have been removed. Such 'ions' indicate that the gas is at a temperature of a few million degrees.

The Coronal Mass Ejections or CMEs are a newly discovered phenomenon. In such an ejection a million times a million kilograms of solar gas are expelled into space at a speed of between a hundred and several thousand kilometers per second. When a CME passes the Earth, some of it is caught by the magnetic field of the Earth and can then be perceived as Northern and Southern lights. Figure 10.9 shows the emission of a CME and Fig. 10.10 shows, as it were, the opposite: a 'coronal hole': a 'hole' in the corona.

Solar flares and certainly CMEs are the result of magnetic activity; to be able to explain this precisely calls for some knowledge of 'magnetohydrodynamics', the theory of the behavior of plasmas, that is gas made up of electrically charged

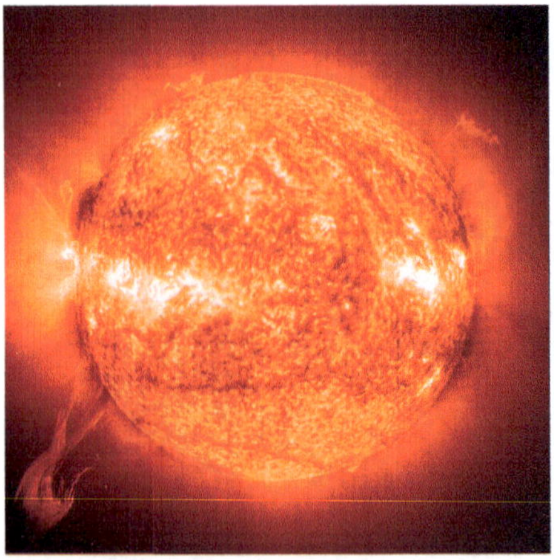

Fig. 10.7 An image of the Sun photographed through a filter that only allows the radiation in an important spectral line of hydrogen to pass. Recording made in 2000 with the SOHO satellite. Source: Astronomy Picture of the Day, 1 June 2008, SOHO Consortium, EIT, ESA, NASA

Fig. 10.8 The 'iron Sun'. An observation made with the SOHO satellite on 22 September 2001. The image was made using only the light from 11 times ionised iron; such iron ions only occur in the hot gas of the corona. Source: Astronomy Picture of the Day, 25 September 2004, SOHO—EIT Consortium, ESA, NASA

particles (protons and electrons) under the influence of magnetic fields. Experts still differ in their opinions about the exact explanation for these phenomena. The increasing complexity of our knowledge of what happens above the photosphere is splendidly illustrated in Fig. 10.11.

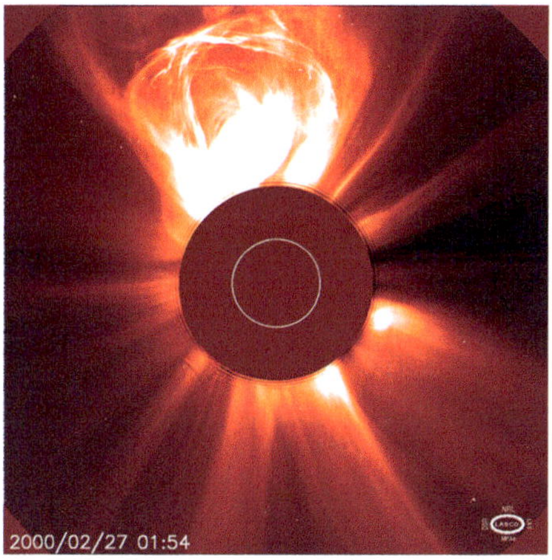

Fig. 10.9 A 'Coronal Mass Ejection' (CME): a large bubble of gas is ejected by the Sun. Source: Astronomy Picture of the Day, 9 March 2000, SOHO Consortium, ESA, NASA

Fig. 10.10 Below left in the solar image is a large hole in the corona through which the surface of the Sun can be seen. Such a hole contains 'open' magnetic lines along which hot plasma can escape, giving rise to such spectacles as the Northern or Southern lights. Source: Astronomy Picture of the Day, 18 March 2003, SOHO—EIT Consortium, ESA, NASA

Below the Surface of the Sun

Figure 10.12 shows the general construction of the Sun. It contains nothing new that was not already known in 1985. Still, a number of important advances have been made, thanks both to computer simulations of the convective zone and to seismic observations.

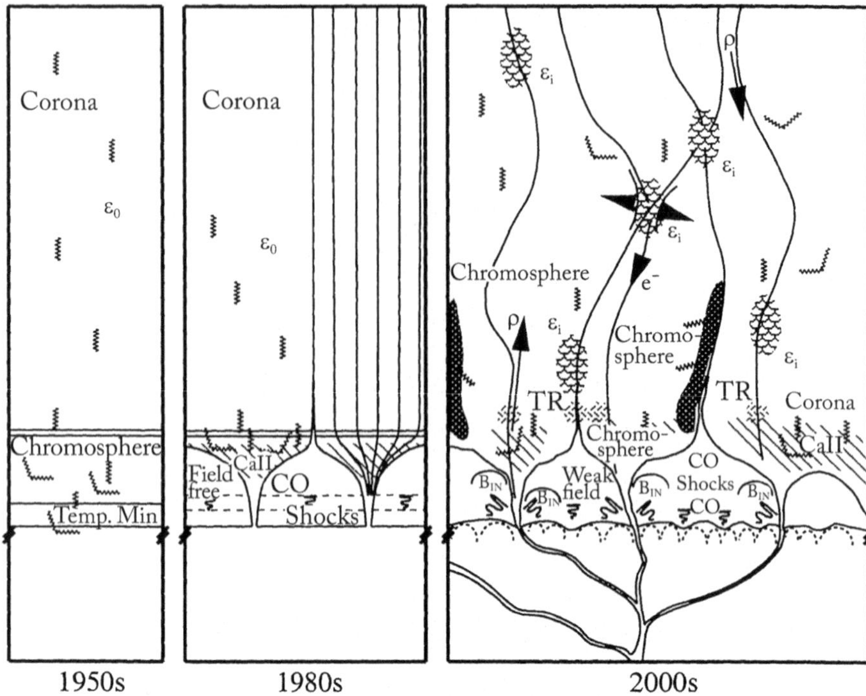

Fig. 10.11 Photosphere, chromosphere and corona of the Sun as described in the years 1950, 1980 and after 2000 (Schrijver 2001)

The transport of heat through convection is difficult to calculate accurately. The process involves flows of heat that change constantly although statistically they remain the same. Figure 10.13 shows a small part of the surface of the Sun with an umbra, the slightly less dark penumbra and the granules of the surface. Three-dimensional and time-dependent calculations are needed to understand the characteristics of these granules. Computers capable of making these calculations became available in the 1990s. Figure 10.14 shows the results of such a calculation in which it is not the umbra that is estimated but the granules on the surface. A start had been made.

Helioseismology

The other important step forward was achieved thanks to seismology: measuring the Sun's vibrations. Information thus obtained provides insight into the internal structure of the Sun, just as seismology generates knowledge about the structure of

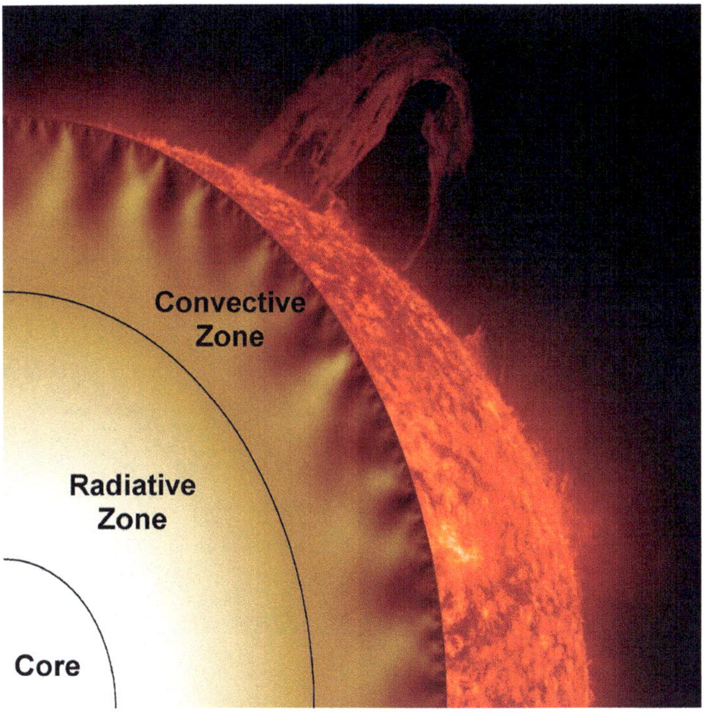

Fig. 10.12 This figure shows the overall construction of the Sun. There is a core in which the fusion of hydrogen atoms produces the energy that is emitted by the Sun. Around the core is a radiative zone in which the energy from the core is transmitted as radiation, that is, in the form of photons. Around this is a convective zone in which the energy is transported to the exterior via convection: hot bubbles of gas rise and cooled bubbles descend. Source: NASA

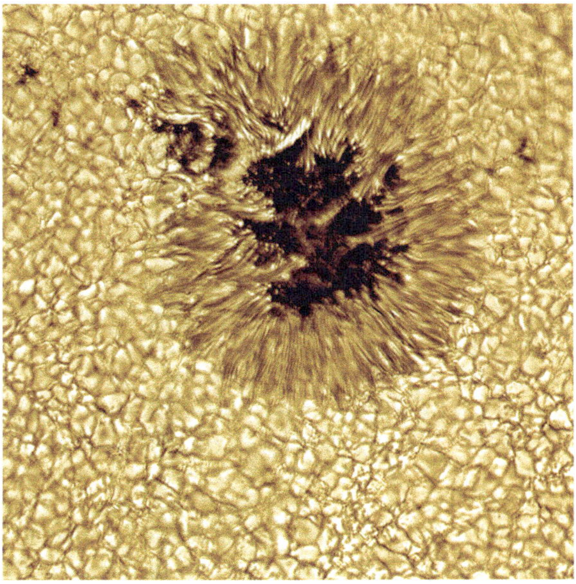

Fig. 10.13 A section of the surface of the Sun showing granules and an umbra surrounded by the penumbra. Source: Astronomy Picture Of the Day, 6 November 2005, Vacuum Tower Telescope, NSO, NOAO

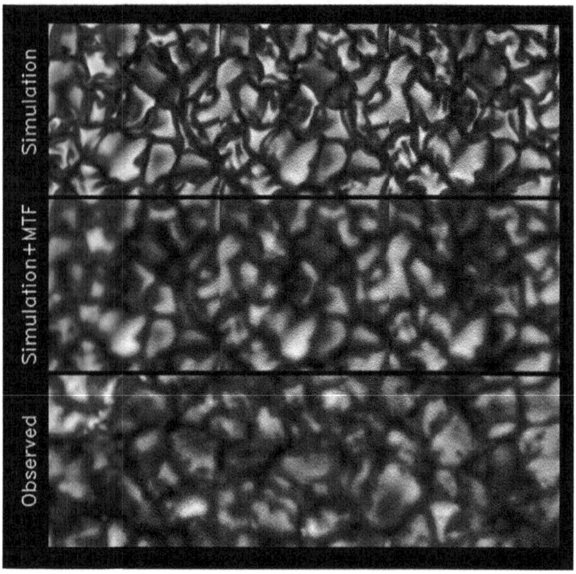

Fig. 10.14 Granules on the solar surface produced by convection: as calculated and as observed. The image consists of three rectangles placed one above the other. In the uppermost one, the direct result can be seen from a present-day simulation. Below this, the result of the calculation is re-shaped to add the effect of the Earth's atmosphere. The third rectangle is the field as it was observed (Stein and Nordlund 1998)

the Earth. An earthquake in Peru causes measurable vibrations all over the world and by measuring the time at which the vibrations arrive at other locations it is possible to plot how the vibrations spread over the Earth. Seismological research has produced a very detailed image of the structure of the Earth, and seismic measurements of vibrations of the Sun are now also being used to determine its internal structure. Figure 10.15 shows how vibrations advance through the Sun.

Vibrations are a fundamental phenomenon in physics, and by studying these vibrations a lot can be learned about the object that is vibrating. It is rather like music played on a piano. Listen to a piece of music on a piano and you know how often and with what force the different keys are struck and which keys are not played at all, or only seldom. In the same way, it is possible to deduce from extended observation of the Sun what vibrations occur and how strong they are, and what vibrations do not appear at all. The result is a diagram with the frequency of a vibration shown horizontally, and the strength of the vibration vertically (see Fig. 10.16). Various models can then be made of the Sun and the vibrations that each model would produce can be calculated. Comparing the vibrations that are observed with those predicted makes it possible to select the best model.

An important step forwards in helioseismology was made with the formation of GONG (Global Oscillations Network Group) in 1995. The network comprised six identical telescopes that together made a continual series of observations of the Sun from different locations on Earth. Later the SOHO satellite was added to the network. This satellite, orbiting at a great distance from the Earth, was directed at the Sun and used some of its instruments to make seismic measurements. The eventual results—and all the researchers involved from many different countries are in agreement on this - provided a very reliable model, of which the global

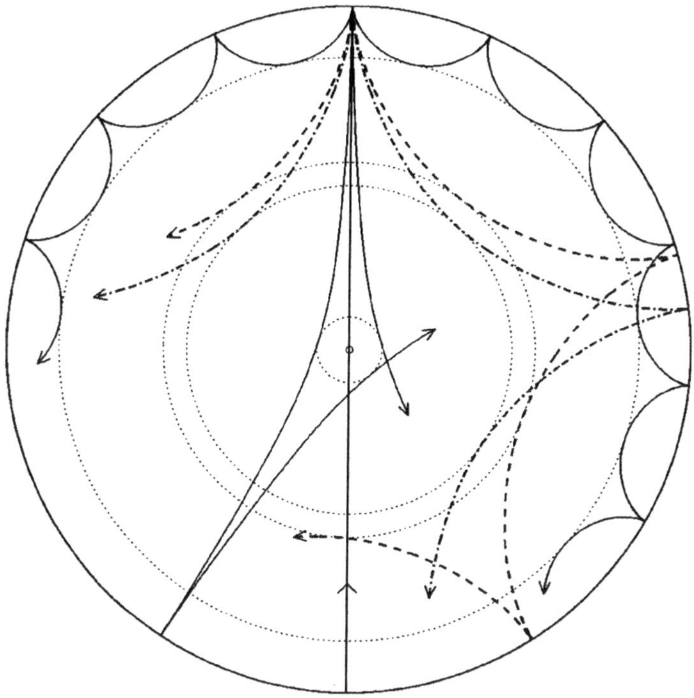

Fig. 10.15 Transmitting soundwaves in the interior of the Sun. Soundwaves precisely directed at the center penetrate throughout the Sun, all other waves are deflected somewhere in the interior of the Sun from and are redirected to the surface, where they are reflected once again (Christensen-Dalsgaard 2002)

image is shown in Fig. 10.12. The findings generate new, detailed knowledge of the systematic flows of matter within the Sun and at different distances from the rotation axis.

There is a review dating from 2013 (Gough 2013) in which the renowned author reports that helioseismology has answered all the questions that had been posed in around 1995. Like the truly critical scientist that he is, he comments that numerous new questions have arisen that have not yet been answered. The conclusion that I draw is that we know a lot about the Sun and this knowledge is reliable, but not complete.

Neutrinos from the Sun

Another difficult problem was also resolved, but not by the astronomers. For years the flow of neutrinos from the center of the Sun had been measured in subterranean detectors, large tanks filled with purified water. The measurements showed a flow of

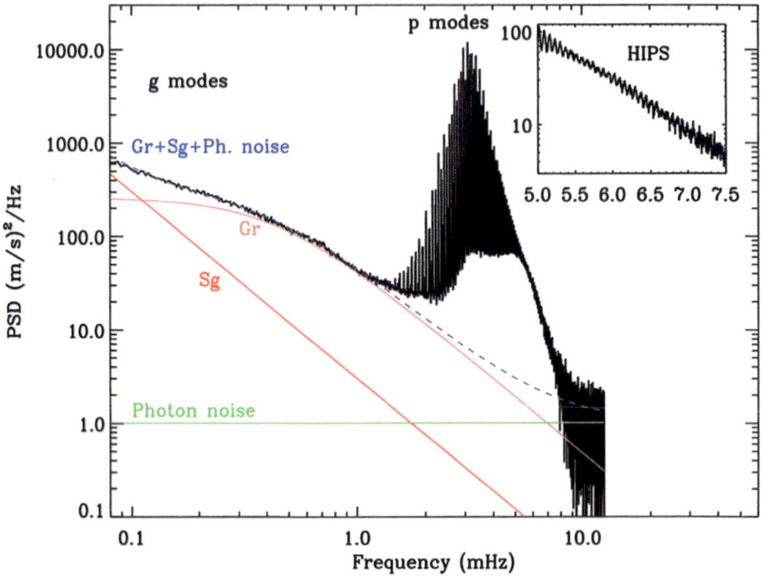

Fig. 10.16 Diagram of the strength of solar vibrations at different frequencies. The black curve shows measurements made by the GOLF instrument on the SOHO satellite. Different contributions to the curve have been plotted: in purple-red ('Gr') the contribution of the granulation, the small cells on the Sun's surface that come and go in a short period of time; the red straight line indicates 'supergranulation'. At frequencies higher than 1 mHz, the character of the curve changes: the pulsations of the Sun are now visible ('p-modes'). As an explanation: a frequency of 1 mHz corresponds with vibrations that repeat every 1/0.001 second = 1000 seconds = 16.6 minutes (García 2009)

neutrinos that was much lower than expected. The neutrinos are a by-product of the nuclear fusion that takes place in the Sun and the predicted value was very reliable; astrophysicists could find no fault in the theory. But eventually theoretical physicists discovered that neutrinos could spontaneously change type. On Earth not all types could be detected and once this was corrected for, theory and observation agreed very well.

Other Light Stars than the Sun: Asteroseismology

Previous chapters have discussed the general evolution of light stars, their path in the Hertzsprung-Russell diagram: their rise up the first giant branch, their descent to the horizontal branch and finally their rise to the Asymptotic Giant Branch (or AGB) and their subsequent decomposition. Ultimately, they end as white dwarfs following a brief phase as a planetary nebula.

In the period after 1985 little progress was made with resolving the remaining important questions although a lot of effort has been expended. There was, for example, as yet no good explanation for why an AGB star disintegrated: why does the star expel the extensive and thin mantle of gas that lies around its hard hot core? There was also no definitive study of which atoms are created in the two combustion zones and how many of them eventually find their way into interstellar space.

But to offset the lack of progress on these difficult subjects, there was a surprising increase in the knowledge gained from seismic research on light stars. In an earlier chapter it was mentioned that seismic research requires long periods to be spent staring at stars in order to make a recording of how starlight varies. Because it was necessary for stars should be watched 24 hours per day, an association was established between observatories all around the world called the 'Whole Earth Telescope' (WET). Figure 10.17 shows the locations from 1990 of the telescopes that made up the WET.

The first successes were achieved with observations of white dwarfs and later of other known variable stars. But the big success came not from the WET but once again from a small observatory in space: a small French-Brazilian-German-ESA telescope, CoRoT (Convection-Rotation-Telescope; diameter of the telescope 27 cm) launched in October 2006 and the much bigger NASA Kepler satellite, with

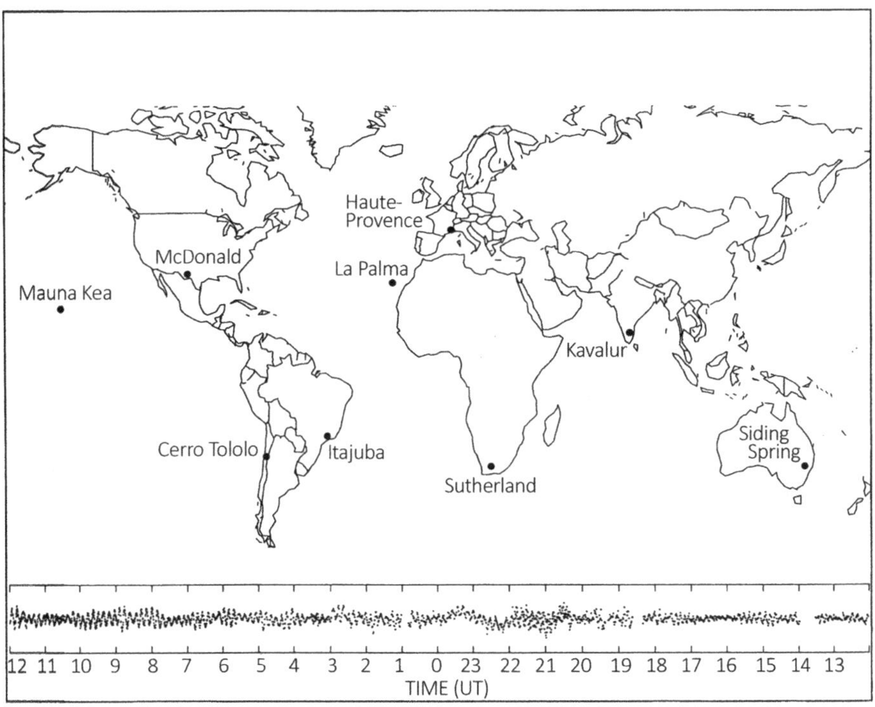

Fig. 10.17 Locations of telescopes that originally took part in the Whole Earth Telescope (Nather et al. 1990)

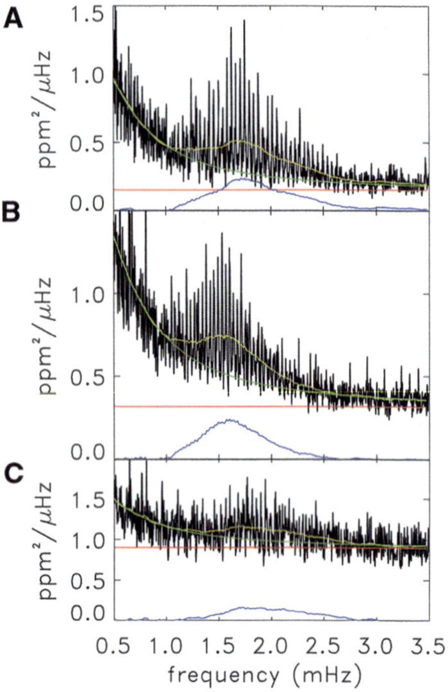

Fig. 10.18 The diagram of the strength of elementary vibrations of different stars. The diagram strongly resembles that of figure 10.16 (Michel et al. 2008)

its 95 cm diameter mirror, launched in March 2009. Both telescopes observed a selected number of stars as frequently as possible, and with some success: seismic signals were measured of 'normal' stars, stars like the Sun, stars on the main sequence. The first, excellent results came from CoRoT (see Fig. 10.18) but were quickly followed by measurements with the Kepler satellite.

The big difference between observations of the Sun and 'normal' stars lay in the large image: the image of the Sun is a circular disk of a half a degree and it was possible to measure vibrations at different places on the disc. Because of the great distance, the disk of a star was much too small for such an analysis. The spectrum of the stellar vibrations was therefore necessarily less rich than that of the Sun and the analyses were not as deep. Nonetheless, seismic measurements produced important data. Consider the case of star HD5265 that had been observed almost daily for 4 months by CoRoT. The analysis of the data showed that the star was between 1.14 and 1.32 times as massive as the Sun, had a diameter of 1.30–1.34 times that of the Sun and was between 2.1 and 2.4 billion years in age, making it slightly younger than the Sun (Lebreton and Goupil 2014); this kind of precision in measuring stellar properties had not been reached before. These improvements in precision may become significant in explorations of our Milky Way.

Besides, the successes achieved through seismological research were at least in part supported by the strong improvement in more conventional research techniques, such as better spectroscopy. The much-improved distance measurements by the Hipparcos and Gaia satellites were also important.

Finally, I should mention that seismological observations of red giants also provide a lot of information about the core of these stars; in terms of density, the core differs dramatically from the rest of the star.

10.3 Massive Stars

Where are they?

Massive stars are rare, which makes systematic studies difficult. Moreover, in our Milky Way their distances are often unknown. On the other hand, massive stars are such powerful light sources that they can be studied in other galaxies. The advantage of such a study is that the distance is well known. Studying stars in another galaxy even has an additional bonus: they often have a slightly different chemical composition, which could have a considerable effect. In the Small Magellanic Cloud, for example, there are four times fewer heavy atoms per hydrogen atom than in our Milky Way, which has a big influence on the disposal of matter by massive stars and consequently on the whole further course of the stellar development.

Stars are born in clusters and the most massive stars, which live for only a very short time, are therefore also found there. A good example is the R136 cluster in the Large Magellanic Cloud, the largest satellite of our Milky Way. The stars of this cluster illuminate the bright 30 Doradus nebula. Fig. 10.19 shows R136 in a part of the nebula and Fig. 10.20 shows the center of R136 and its division into a few individual stars. 'Adaptive optics', as described in Chap. 9, is used to separate the stars from one another.

The complex is around two million years old, which, in astronomy, is quite young; the Sun and the Earth, for example, are 2,000 times older. The stars a1 and a2 in Fig. 10.20 have both already evolved to Wolf-Rayet stars (type: WN5).

The Development of Massive Stars is still not fully understood

Successful models for heavy main sequence stars (six or more times the mass of the Sun) had already been calculated in the 1960s (see Chap. 4) but the evolution away from the main sequence was problematic. The techniques for model calculations had since been much improved. Figure 10.21 shows newer results in which the rotation by the stars had been taken into account and the comparison with observational results is now quite good. This success is also due to improved insight into the mass ejected by these massive stars (Vink et al. 2001; Puls et al. 2008). Strong magnetic

Fig. 10.19 Part of the nebula known as '30 Doradus' is visible with a large number of bright white stars arranged around a bright patch. This patch is object R136, a concentration; only recently been it has become possible to distinguish the individual stars from one another. Source: NASA/ESA

fields have not been included and still form the Achilles' heel of the model calculations (Langer 2012).

On the observational side, four types of stars have been known for a long time that apparently lie on the development path of massive stars: blue and red super giants, LBVs (Luminous Blue Variables) and the Wolf-Rayet stars. Even though they have been observed extensively using the best techniques and by the best observers, it is still not possible to state what role each of these objects plays in the evolution of massive stars as was shown yet again in the proceedings of a well-attended conference in 2015 (Hamann et al. 2015).

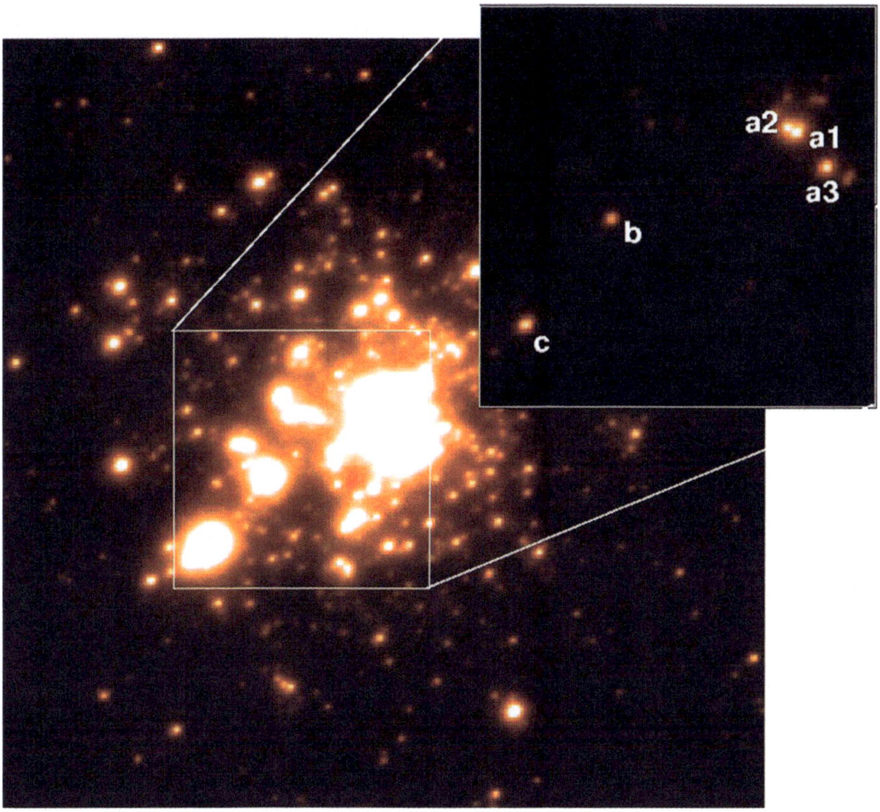

Fig. 10.20 The central part of the object R136 and the most important individual stars in it. The star a1 is around 300 times as massive as the Sun, a2 200×, a3 100×, b 90×. They emit respectively nine million, four million, two million and two million times more light than the Sun (Crowther et al. 2010)

Massive stars occur frequently (one estimate says, in two-thirds of all cases) in a close binary star with an equally massive star as companion. In such a construction there will at some time be a strong flow of matter from one star to the other (Sana et al. 2012). This means that the evolution of such a star can follow a very different path than if the star were on its own. According to estimates, around one-third of all massive binary stars will end with both objects merging. No direct evidence of such a merging of two massive stars has yet been found, although such evidence does exist for light stars (see Chap. 14).

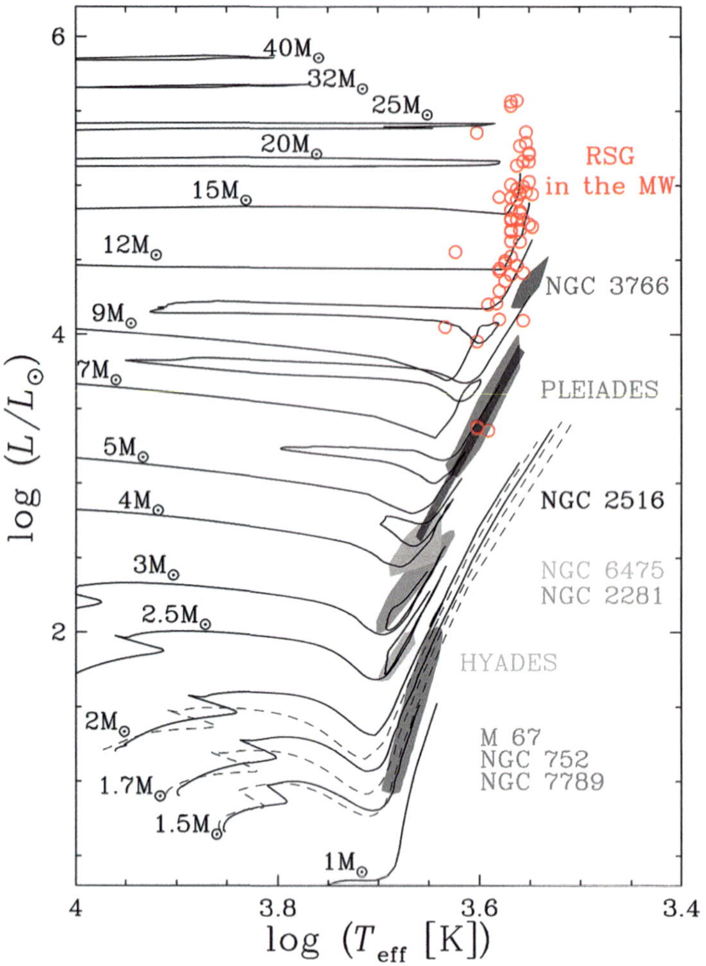

Fig. 10.21 The Hertzsprung-Russell diagram with evolutionary tracks for stars of mass between 1 and 40 solar masses. Observational data are given in grey scale and in red circles. The names and identification numbers of the groups to which the evolved stars belong are shown. By 'RSG in the MW' are meant the (rare) red supergiant stars in the Milky Way (Ekström et al. 2012)

References

Babusiaux, C., van Leeuwen, F., et al.: Astron. Astrophys. **616**, A10 (2018)
Becklin, E.E., Zuckerman, B.: Nature. **336**, 656 (1988)
Berger, E., Basri, G., et al.: Astrophys. J. **709**, 332 (2010)
Chabrier, G., Baraffe, I.: Annu. Rev. Astron. Astrophys. **38**, 337 (2000)
Christensen-Dalsgaard, J.: Rev. Mod. Phys. **74**, 1073 (2002)
Crowther, P.A., Schnurr, O., et al.: Mon. Not. R. Astron. Soc. **408**, 731 (2010)

Delfosse, X., Tinney, G.C., et al.: Astron. Astrophys. **327**, L25 (1997)
Ekström, S., Georgy, C., et al.: Astron. Astrophys. **537**, A146 (2012)
Forrest, W.J., Skrutskie, M.F., Shure, M.: Astrophys. J. **330**, L119 (1988)
García, R.A.: AIPC. **1170**, 560 (2009)
Gough, D.: Sol. Phys. **287**, 9 (2013)
Greenstein, J.L., Neugebauer, G., Becklin, E.E.: Astrophys. J. **161**, 519 (1970)
Hamann, W.-R., Sander, A., Todt H.: Wolf-Rayet Stars: proceedings of an international workshop, Universitätsverlag Potsdam (2015)
Hardie, R.H.: J. R. Astron. Soc. Can. **68**, 202 (1974)
Hayashi, C., Nakano, T.: Progr. Theor. Phys. **30**, 460 (1963)
Kao, M.M., Hallinan, G., et al.: Astrophys. J. **818**, 24 (2016)
Kellogg, K., Metchev, S., et al.: Astrophys. J. **821**, L15 (2016)
Kirkpatrick, D., Reid, I.N., et al.: Astrophys. J. **519**, 802 (1999)
Kumar, S.S.: Astron. J. **67**, 579 (1962)
Langer, N.: Annu. Rev. Astron. Astrophys. **50**, 107 (2012)
Lebreton, Y., Goupil, M.J.: Astron. Astrophys. **569**, A21 (2014)
Luhman, K.L., Esplin, T.L.: Astron. J. **152**, 78 (2016)
Luhman, K.L., Adame, L., et al.: Astrophys. J. **635**, L93 (2005)
Michel, E., Baglin, A., et al.: Science. **322**, 558 (2008)
Nakajima, T., Oppenheimer, B.R., et al.: Nature. **378**, 463 (1995)
Nather, R.E., Winget, D.E., et al.: Astrophys. J. **361**, 309 (1990)
Oppenheimer, B.R., Kulkarni, S.R., et al.: Science. **270**, 1478 (1995)
Puls, J., Vink, J.S., Najarro, F.: Astron. Astrophys. Rev. **16**, 209 (2008)
Rebolo, R., Zapatero Osorio, M.R., Martin, E.L.: Nature. **377**, 129 (1995)
Ruiz, M.T., Leggett, S.K., Allard, F.: Astrophys. J. **491**, L107 (1997)
Sana, H., de Mink, S.E., et al.: Science. **337**, 444 (2012)
Schrijver, C.J.: Sol. Phys. **198**, 325 (2001)
Stein, R.F., Nordlund, Å.: Astrophys. J. **499**, 914 (1998)
Tsuji, T., Ohnaka, K., et al.: Astron. Astrophys. **308**, L29 (1996)
Vink, J.S., de Koter, A., Lamers, H.J.G.L.M.: Astron. Astrophys. **349**, 574 (2001)

Chapter 11
1995–2015: Something Completely New: Exoplanets

11.1 How to Find Planets of Stars other than the Sun?

The Initial Discoveries

The distance from the Sun to a star was measured for the first time in 1838. From this result, researchers calculated how much light the star radiated, and the outcome was very comparable to the amount that the Sun emitted. This confirmed once and for all what had been suspected for centuries: the Sun was a star like all the others that people saw in the night sky. When the distance of more stars was measured, the Sun was seen to be a 'medium' star; not big, not heavy, not hot, not cold. However the Sun did have something very uncommon: planets, moons, comets. It was obvious to assume that in this respect, too, the Sun was no different from other stars, but there was no direct evidence at all for this assumption.

An important indication that planets did occur was discovered in 1983. Measurements taken by the IRAS infrared satellite showed that a few ordinary nearby stars had a disk of dust particles around them, similar to the disk of the planet Saturn but many times larger and more massive. The disk was assumed to be a remnant of the birth of a star. The inference was that a short while before one or more planets had been formed from this disc. Once again, this was not hard evidence, but the discovery reinforced many astronomers in their belief in the existence of 'exoplanets'.

Why did the observers fail to detect planets? The answer was: the light of the planet would be reflected starlight and a minuscule amount in comparison with the light of the star itself. A planet is such a faint light directly next to the star, that it would be completely and unrecognisably buried in the imprint of the star on the photographic plate. To demonstrate the presence of a planet it must be done in some other, indirect way.

The first exoplanet was discovered in 1992 and totally unexpectedly. Polish astronomer Alexander Wolszczan and his American colleague Dale Frail recorded the arrival times of radio pulses from a newly discovered pulsar, PSR1257+12.

H. J. Habing, *The Birth of Modern Astronomy*, Historical & Cultural Astronomy,
https://doi.org/10.1007/978-3-319-99082-8_11

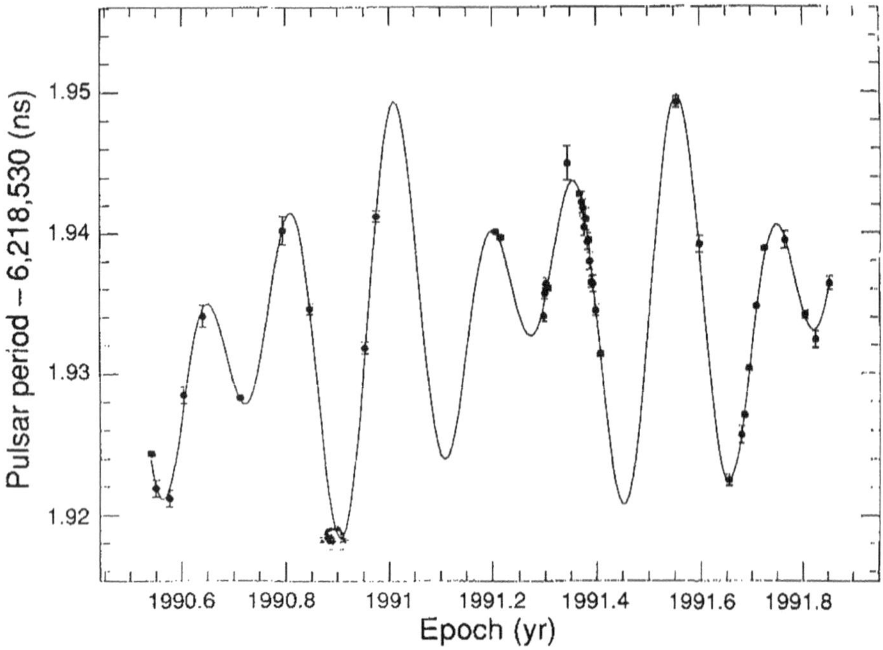

Fig. 11.1 The time is indicated horizontally and the delay in the period of the pulse is shown vertically. Through the measurement points a curve has been drawn that corresponds with the presence of two planets around the star (Wolszczan and Frail 1992)

The pulses arrived at intervals of 6.2 milliseconds, but there were very small differences between the consecutive arrival times. This difference increased and then decreased periodically, and when this was plotted in a graph the shape strongly resembled that of the radial velocities of a triple star (see Fig. 11.1). The discoverers concluded that the pulsar was encircled by two planets with masses of 2.8 and 3.4 times the mass of the Earth (Wolszczan and Frail 1992). The conclusion was not doubted but still it met some unbelief: a pulsar is the remnant of a massive star after a supernova explosion. If the star had had a planet before the explosion, this would almost certainly have disappeared during the explosion. The discovery was therefore highly remarkable, but it did make researchers think that if planets could be formed in such a bizarre, hostile environment, it could almost certainly occur during the formation of a normal star.

This discovery showed the direction in which astronomers should look for exoplanets near to normal stars: do not search for the planet itself, but for the effect the planet has on the star. Take the example of an adult swinging a child around by its arms (fun, but frowned on by paediatricians). As we all know, the adult has to bend backwards: while the child makes a large circle, the upper body of the person doing the swinging moves in a small circle around the shared centre of gravity (the 'reflex circle').

Dutch astronomer Pieter van de Kamp, who spent his scientific life in the US, looked for decades for such a circle around the 'star of Barnard'—a small star at a distance of a few light-years from the Sun. Van de Kamp photographed the star together with a number of other stars in almost the same direction; he repeated photographing the same area thousands of times and measured the position of the star in relation to the surrounding stars. He published the results of all this work in 1963, concluding that Barnard's star had a planet (van de Kamp 1963). Unfortunately, he saw something that was not there. His findings were not confirmed. Decades later it became clear that his method was right but the observations were not precise enough. From the 1990s, star positions had been measured much more accurately thanks to a revolutionary new technique used in Hipparcos, an ESA satellite. The Hubble Space Telescope was also capable of much better measurements. These improvements meant that in 2002, 40 years after van de Kamp's doomed publication, the motion of a star as a reflex on the presence of a planetary companion was measured by the HST in the way Van de Kamp intended to follow (Benedict and McArthur 2002).

However, there was a very different way of measuring this same 'reflex circle'. When a star went around its little reflex circle, it moved one half of the orbit towards us and the other half away from us. As a result of the Doppler effect, the spectral lines of the star had to shift at regular intervals: in one half of the period, the lines would be shifted towards the blue side of the spectrum and in the other half towards the red side. If the shifts could be measured and if they were regular, this would indicate that a planet had been found. It was a good theory, but estimates showed that the expected variations in velocity were below 100 m/s and were much too small to be measured, even in the 1980s. Nonetheless, there were a few instrument specialists who believed it might be possible to resolve this difficulty.

The most successful of these was Michel Mayor at the observatory in Geneva. He built a spectrograph that used computer support and that was indeed accurate enough. He made a list of suitable stars and, together with Didier Queloz, over a period of a year and a half spent many nights making measurements of the spectrum of each star. They used a 1.5 m telescope at the French 'Observatoire de Haute Provence'.[1] Finally Mayor and Queloz were sure of themselves and on November 23, 1995 they published their findings for the star 51 Pegasi in the journal *Nature*. The measurements had shown that the changes in the wavelength of the spectral lines increased and decreased precisely like a sine with a period of 4.42 days (see Fig. 11.2). Analysis also showed that the planet was roughly as massive as Jupiter.

The discovery was met with enthusiasm worldwide. Finally the assumption was confirmed that had been around so long: other stars had planets, just like our star, the Sun. We would now find out just how special or ordinary our solar system was and ideas about the origin of the solar system could be verified! Within a few years the subject of 'exoplanets' became an important field of research among all organisations that supported astronomical research, in the US as well as in Europe, Australia, South Africa, Japan, South America and other countries.

---

[1] It is ironical that such an outstanding discovery was made at an observatory that higher authorities had decided on more than one occasion to close down

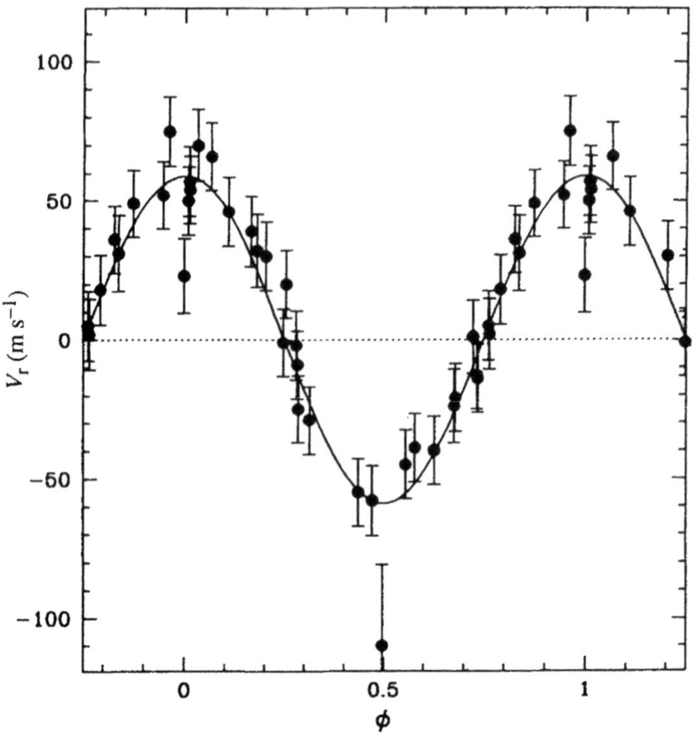

Fig. 11.2 The regular shift in the lines of the spectrum of star 51 Pegasi; the period is 4.42 days. The time of the measurements at a particular moment is indicated horizontally, and the shift is shown vertically. The curve drawn through the measurement points closely mirrors the shape of a sine; this means that the star tracks a circle around the centre of gravity of the star and the planet (Mayor and Queloz 1995)

Two astronomers were deeply disappointed: Geoffrey Marcy and Paul Butler from the University of San Francisco. Completely independently of Mayor and Queloz, they had constructed their own similar instrument and had already recorded some positive results when the article by Mayor and Queloz appeared in Nature. A few months later they published an article on planets close to two other stars—an equally convincing result. Their achievement was no less significant, but unfortunately they were not the first. Their result is shown in Marcy-Butler-96. The curve through the data points is in this case not a sine because the orbit of star and planet around the centre of gravity is elliptical rather than circular (Fig. 11.3).

Another type of curve arose later: that belonging to a star with two planets (see Fig. 11.4). Compare this with the above curve shown in Fig. 11.1.

In the box an explanation is given of how the mass of the planet and the distance between star and planet can be deduced from the measurements made by Mayor and Queloz.

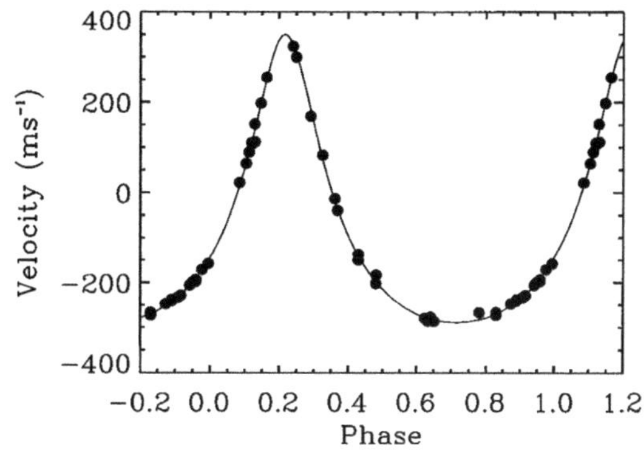

Fig. 11.3 The regular shift in the lines in the spectra of star 70 Virginis. The curve is not a sine, but can be fully explained by assuming an elliptical orbit instead of a circle (Marcy and Butler 1996)

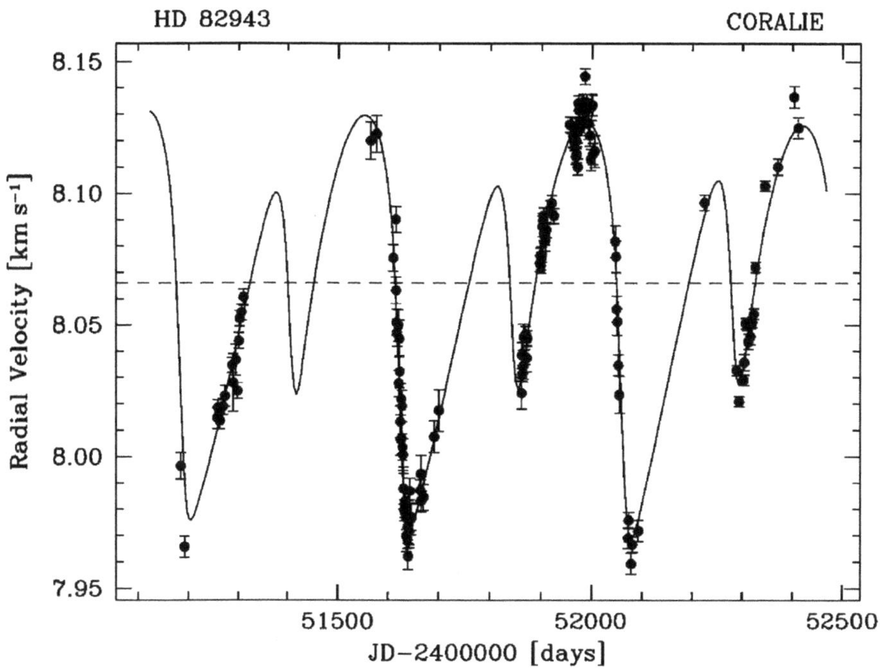

Fig. 11.4 The radial velocity curve of the star HD 82943. The curve can be explained by assuming that there are two planets orbiting around the star (Santos et al. 2002)

How to Deduce the Mass of the Planet

51 Pegasi is a common type of star, whose distance and mass (M_S) have already been measured in the past. Mayor and Queloz determined the period of the orbit of the planet. Using this information and applying Kepler's third law, it was possible to calculate the distance, a, between star and planet.

(continued)

(Strictly speaking, Kepler's equation requires knowledge of the mass of the planet, but this is so small that it can be ignored.) The period and the measured velocity of the star are then used to calculate the size of the circular orbit of the star, and from this the diameter, a_S, of the orbit. From a and a_S one calculates the distance, $a_{p,}$ from the planet to the centre of gravity of star plus planet. Because $M_S/M_p = a_p/a_S$ one calculates M_p, the mass of the planet. Mayor and Queloz had the good fortune that there was only one planet to the star, and its orbit was circular and not an ellipse.

The discovery of the first planets was followed by the inevitable question: can there be a planet somewhere else where there could be life? This question was, of course, premature—so far practically nothing was known about planets. In our solar system there was only life on planet Earth, although there could be bacterial life on Mars. The question of life elsewhere continually plagued and stimulated planet researchers, particularly when looking for new financial support. About extraterrestrial life there is nothing certain to report even in 2015.

More Detection Techniques

The technique of finding planets developed by Mayor and Queloz was of course taken over by other researchers. Mayor was not (yet) quite honored with a statue, but he was given all possible financial scope to improve his instruments, which he did successfully and continuously. He now had easy access to large telescopes, which meant he was able to reach fainter and thus more stars. The same happened to Marcy and Butler in the US. As there were by now a large number of other astronomers at all the big observatories, there was a rapid growth of the community of specialists in the search for 'exoplanets', planets around other stars. The number of exoplanets kept increasing, although slowly. The technique was very labour-intensive: you could only identify a star with a planet after spending many nights taking spectra of always the same star to find that in only a few cases you were successful.

Another new method of finding planets was devised, in fact a very simple method: if a planet passes before our eyes in front of a star, it will take a few percent of the light away from the star: a (minuscule) eclipse. Such eclipses are known from the solar system: Venus and Mercury pass, from our perspective, in front of the Sun a few times a century. Finding a temporary decrease in the light of a star was not yet the guarantee that you had found a planet; only measurements repeated many times gave you this certainty.

Measuring an eclipse had an additional advantage, a bonus: if the amount of starlight that was obscured by the planet was measured (say 4%), the surface of the planet was 4% of that of the star, and the diameter of the planet was one-fifth that of the star (diameter of planet 1/5 of the star; surface of planet 1/25 = 4% of the star). And because the diameter of the star was known, the eclipse could give an indication of how big the planet was and how large the volume. If you had derived the planetary

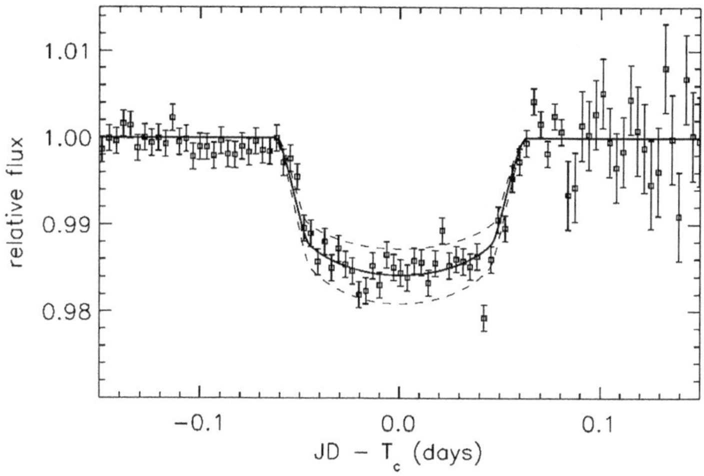

Fig. 11.5 The reduction in the light of star HD 208458 when a planet passes in front of the star (Charbonneau et al. 2000)

mass as in the box above, you could work out the average density of the planet by dividing the mass by the volume. Density was a factor that contained very important information in our solar system. Earth, for example, had an average density of 5,520 kg/m^3, and Mercury, Venus and Mars were similar. These planets must consist of matter heavier than water (1,000 kg/m^3) as is the case: iron and silicon are the most important materials inside the Earth. Jupiter, by contrast, has a density of 1330 kg/m^3 and consists primarily of frozen, light gases like carbon dioxide, methane and water.

In 2000 the first tiny eclipse was seen of the star HD 208458. Using the technique of the shifting spectral lines, one knew that the star had a planet. Figure 11.5 shows the eclipse. The planet had a diameter 1.27 ± 0.02 times that of Jupiter, while the mass of the planet was 0.73 times that of Jupiter. The density was just 380 kg/m^3, which was less than the density of water and just over half the density of Saturn. Conclusion: in terms of measurement and size, the planet near to HD208458 was comparable with the large planets in our solar system.

Figure 11.6 shows an eclipse of the same star, but measured a year later using the Hubble space telescope. Notice the higher quality of the second measurement.

A third possibility for detecting a planet was a direct image, a possibility that had previously been dismissed. After around 1990, it became possible to freeze the stellar image during measurement, that is, compensating for the oscillations in the Earth's atmosphere while making the recording (see Chap. 9). Once this could be done successfully, it was possible to image planets directly. This led to the splendid discovery of a planet in the disk around the very bright and near star Fomalhaut - see Fig. 11.13. This direct method was very labour intensive and needed large and modern telescopes, on which it was difficult to obtain observation time. It was therefore impractical to search for planets using this method, but it was at the same time fantastic to be able to see a planet directly whose existence had been demonstrated with one of the other indirect methods.

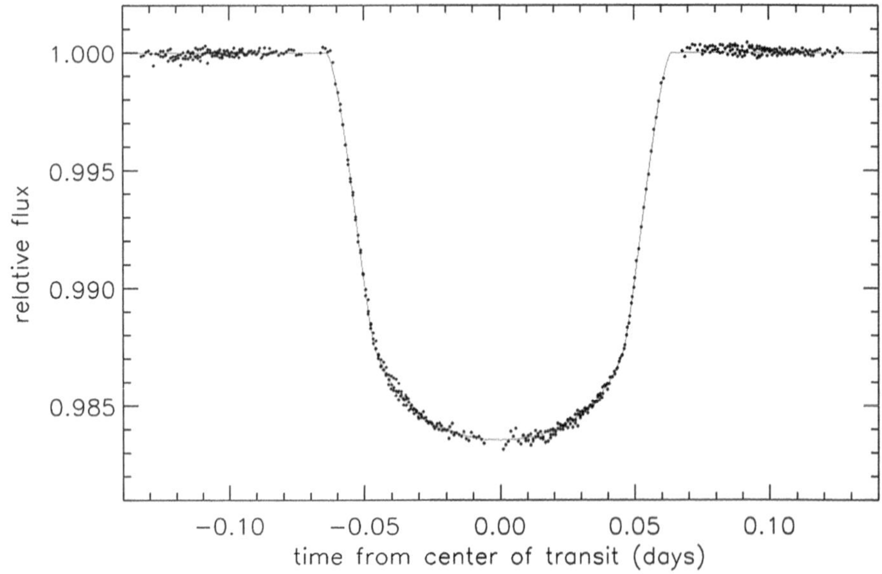

Fig. 11.6 A repeat, more than a year later, of the measurement of the eclipse of HD 208456 by a planet. This time the measurement was made with the Hubble Space Telescope. Compare the quality of the two measurements (Brown et al. 2009)

Another—fourth—method of finding planets was by using microlenses, as described in Chap. 14: when star A passed closely behind star B, B could intensify the light of A (see Alcock et al. 1993). If A also has a planet, this planet could cause an extra peak in the light curve (see Fig. 11.7).

What was even more interesting was the detection of a free-moving planet passing in front of a star and acting as a lens. This was a short-lived event, lasting a few days at most (see Fig. 11.8).

If a star had two planets, the light curve would be more complex, as shown by Fig. 11.9. The data for this magnification curve has been obtained by a large network of astronomers and observatories around the Earth, which made it possible to observe the object 24 hours a day.

Space Telescopes: CoRoT and Kepler

Finding exoplanets requires patience: many stars are observed many times before one of the stars shows signs of a planet. The observations are all made in the same way and are suitable for robot telescopes: the robot decides whether the sky is clear, opens the dome and directs the telescope toward the first on its list of targets. Such robot telescopes are in operation, but their productivity in space would be many times higher: observations 24 hours a day without any interference from the Earth's

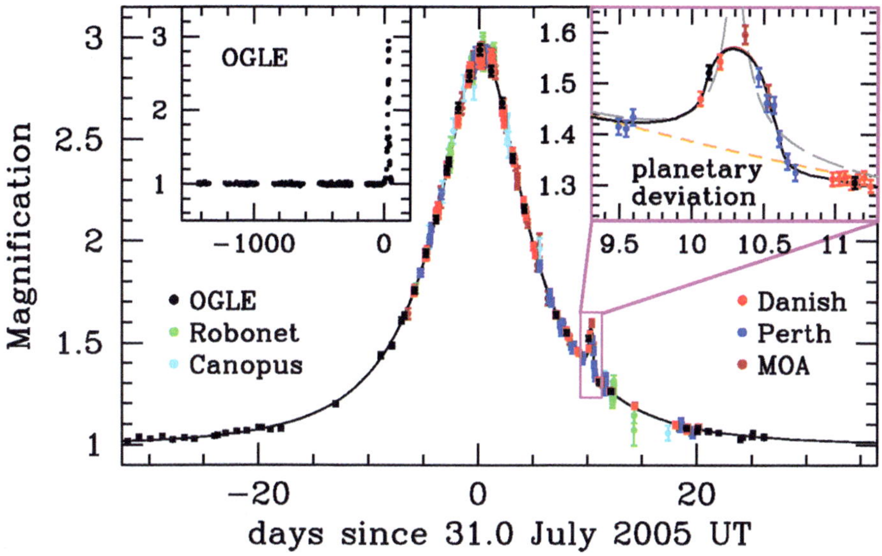

Fig. 11.7 The intensifying of the light of a star and the small deviation caused by a second source, the planet. The different colors indicate the contributions by different researchers at different observatories (Beaulieu et al. 2006)

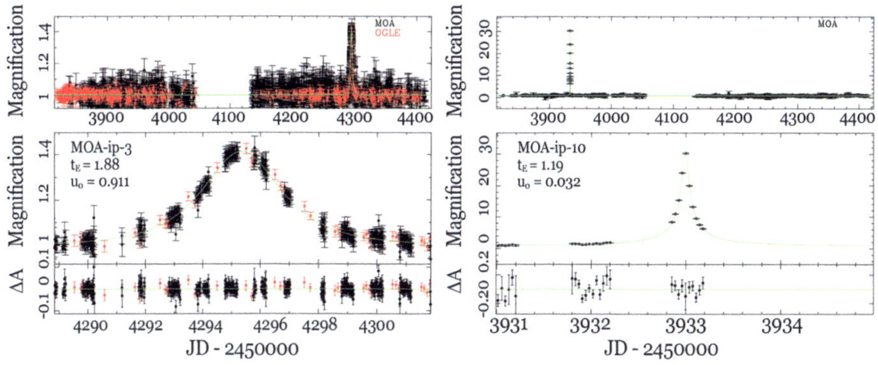

Fig. 11.8 A magnification event that lasted only a few days. The time in days is given horizontally. The intensification of the light is shown vertically. The short duration indicates that the gravitational lens is a planet (Sumi et al. 2011)

atmosphere. It goes without saying that satellites are very expensive and outside the reach of smaller countries. Two of these satellites were developed, the first a small French satellite (mirror diameter 27 cm) that was constructed with the support of ESA and launched with a Russian rocket. The satellite, known as CoRoT ('Convection Rotation et Transits planetaires'), was operational from February 2007 to June 2013 and discovered some 30 exoplanets. There were a further several hundred detections, for which it still has to be determined whether they corresponded with an exoplanet.

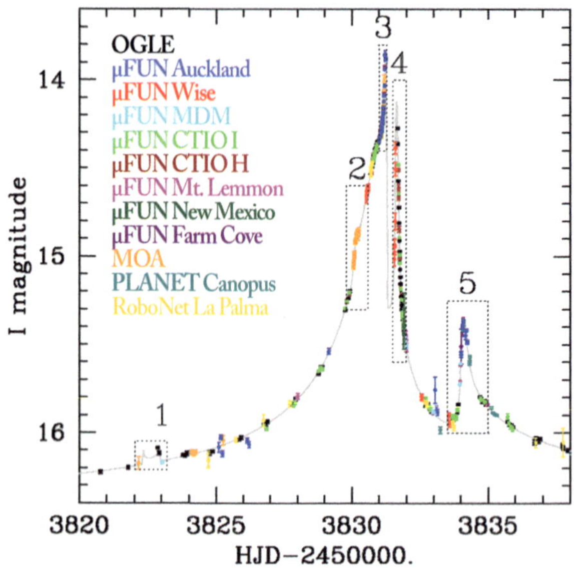

Fig. 11.9 The magnification of the light of a star while another star passes in front and acts as a gravitational lens. Vertically the increase in magnitude, horizontally the day of observation indicated by the so-called Julian number. The main rise in magnitude is because of the lensing effect of the star; the two planets cause small, sharp, extra magnifications. The two rectangles A and B contain additional information that is not relevant here (Gaudi et al. 2008)

The second was a NASA satellite and it was large: the Kepler satellite, with a mirror diameter of 1 m. It was in operation from May 2009 to August 2015. More than 140,000 stars were monitored continuously. There were numerous false positives: a convincing eclipse was reported but there was no repetition of the signal; these cases had to weeded out. The mission resulted in a total of 1,284 newly discovered exoplanets. NASA was rightly proud of this achievement.

Both satellites, CoRoT and Kepler, will be discussed again in Chap. 12, because as well as identifying planets the measurements also provided seismic information on the stars.

11.2 The Solar System as an Example

The Kant-Laplace Theory of the Origin of our Solar System

Let us take a look back. In 1450, Copernicus published his famous work, in which he claimed that the Sun was in the center of the Universe, with the planets orbiting around it. After Galilei's discoveries in 1609/1610 and those of Kepler around the same time, the scientific world was sure: Copernicus was right, even if the Pope believed that this should not be spoken aloud. How the solar system was formed remained an unanswered question for centuries. The first person to put forward a modern vision was German philosopher Immanuel Kant. In 1755 and before writing his major philosophical works he proposed an explanation for the origin of our Universe and a specific explanation for the origin of the Sun and its planets. This was well argued without any mathematical equations and yet it was a source of inspiration for more recent studies. A slightly rotating cloud contracts; at the centre a star forms, but much of the gas is contracted into a rotating disk and it is in this disk that the planets form. Kant's vision explained a number of important facts: all planets move in almost the same plane as the Sun and revolve in (almost) circular orbits, all in the same direction. Kant was no mathematician and he presented his arguments in language. A few decades later, French mathematician-physicist Laplace added mathematical formulae, but he did not change the main thread of Kant's argument. When it became known that first the star and then the planets were formed in the disks of gas, the Kant-Laplace model became the basis for the modern theory of the origin of planetary systems.

Kant did not explain everything; he did not, for example, explain how it came about that Jupiter and Saturn (he did not know of the existence of Uranus, Neptune and Pluto) were so much larger and more massive than the inner planets Mercury, Venus, Earth and Mars. The current view is that these giant gas planets started as rocky planets (like Earth) and that gas subsequently froze on them. The inner planets were too close to the Sun.

There were more questions that the Kant-Laplace theory failed to explain: the existence of asteroids, rocky debris particularly between Jupiter and Mars (the biggest chunks are known as 'dwarf planets'). There were also frozen dwarf planets far distant from the Sun (Jewitt and Luu 1993) with Pluto being the longest known of these. Thousands of times more distant there was a large reservoir of comets, the so-called 'Oort cloud' (Oort 1950). And there was still no answer to the obvious question: 'How did Earth get its Moon?'

Late Heavy Bombardment; Migration; Water on Earth; the Nice Model

According to the Kant-Laplace model, planets were formed at their present distance from the Sun. This proposition has undergone some serious change in recent decades as a result of three developments. The first of these was the very detailed microscopic study of Moonstones that were brought back by astronauts and also of meteorites—space debris that landed on Earth. The second development was the computer calculations that were used to study how the early solar system developed, when it comprised mainly debris—see below. The third development was the discovery that some host stars had a large planet, nicknamed a 'hot Jupiter', dangerously close to them.

Let us start with the analysis of that rocky debris. Moon material proved that there has been a small period in time when many of the rocks had been molten: 3.92–3.85 billion years ago. This was long after the start of the formation of the Sun and after the formation of the planets and the Moon. This led in 1974 to the hypothesis of a 'heavy bombardment' of the Moon by asteroids at a late stage during the formation of the solar system (Tera et al. 1974). Additional evidence came from the study of craters on the Moon and on Mars. Although there were objections to the hypothesis, the conclusion itself was not disputed. In 2005 an explanation was given why this planetary late summer storm was stirred up by the publication of the Nice model, as will be shown below (where—English-speakers please note—Nice refers to the French city on the Mediterranean coast) (Tsiganis et al. 2005; Gomes et al. 2005; Morbidelli et al. 2005).

The second development followed from calculations of the orbits of planets and of the debris. This led to the insight that the planets Jupiter and Saturn have shifted back and forth in the early history after the planets were formed. Jupiter first moved inwards and arrived at a distance of one and a half times the distance from the Earth to the Sun. Then Jupiter made a wide detour and moved outwards again. During this migration, the orbit time of the planets changed and there were times when Saturn orbited once around the Sun and Jupiter twice. This '2:1 resonance' had dynamic effects; one of these was that a great number of asteroids were propelled inwards, thus causing the 'Late Heavy Bombardment' (Raymond et al. 2014).

Two major problems have not yet been explained. The Moon is almost certainly part of the debris of the Earth: the rock of the Moon is very similar to that of the Earth. The Moon can be the result of a frontal collision of the Earth and a very large asteroid, but there is still little or no information on this. The second problem is the question of where the water on Earth comes from. During the formation of the Earth and also during the Late Heavy Bombardment, the Earth was very hot. At that time, it was made up mainly of melted rocks. Any water that was present must have evaporated during this melted stage; the Earth was boiled dry, so to speak. The water that is now so abundant and that was needed for life on Earth must have arrived later, perhaps via a comet, or an asteroid.

Now, 2016, it is too early to give a final conclusion on these new developments. But some fascinating ideas are doing the rounds and there is a lively debate.

Warning: the Solar System is Small—at least in Astronomical Terms

The Sun is around 150 million kilometers away; a distance known as the Astronomical Unit, AU. Expressed in these units, Jupiter is by far the biggest and most massive planet of the solar system, at a distance of 5 AU. Further outwards there is a belt of ice planets and comets, the so-called Kuiper belt, of which Pluto is a good example. Pluto has average distance of 39 AU. Much further outwards there is only the cloud of comets. Remember this number: 40 AU is roughly the limit of our solar system. Outside this there is very little more to discover, apart from the cloud of comets.

Forty AU is a very small length in the distance seen from the close and important star-forming region, the molecular cloud in Taurus. This is around 20 million times further away from us than the Sun. At this distance, 40 AU is around 0.4 arc seconds, which is very small. In optical astronomy, a visual acuity of 1 arc second was achieved, although from around 1990 special techniques had been developed that could produce much sharper images. But light was not the most important form of radiation that observers were able to receive: objects in molecular clouds could only be observed at wavelengths that were not affected by their clouds: in the far infrared and at millimetre and sub-millimetre wavelengths. At these wavelengths the smallest details that a telescope could see were around 4–10 arc seconds in size: at such a distance the analogue of the solar system was immeasurably small. With radio interferometers (e.g. ALMA) one can achieve an acuity below 1 arc second. But observers had to live with the idea that in most molecular clouds solar-system-like objects were too small for significant details to be perceived.

11.3 Planet Formation

According to Shu's original paradigm (see Chap. 12), disks were created during the formation of a star. Material from the disk crept slowly towards the star while rotating, and was swallowed by the star. But apparently not all matter disappeared in the star, because observations by the IRAS satellite had already shown in 1983 that after a star was formed, the disk could still be present although it was strongly emaciated—see Chap. 7. Such residual discs consisted mainly of grains of dust—the matter from which the Earth and the other inner planets (Mercury, Venus, Mars) were formed. A disk of such material offered a good start for planet formation.

From Dust Particles to Planetesimals

Planets such as the Earth were ultimately formed from the dust particles. These dust particles were much smaller than a micron (one millionth of a meter) and the Earth has a diameter of almost 13,000 km. This means that the particles grew to an object that is 10^{13} times bigger (a 1 with 13 zeros). How this came about is a puzzle that has still not been resolved. Ice formation could help: grains with a layer of ice stick to one another more easily. A few HAeBe stars that had almost reached the main sequence had spectra showing particles formed from silicates like olivine or another mineral. Moreover, these particles were much larger than in the cloud from which they had originated (Van Boekel et al. 2004).

Almost from the beginning of the exoplanet era there has been the suggestion that planets are found more frequently around stars with a higher abundance of heavier atoms like Fe and Mg. Further studies have confirmed the suggestion, but unfortunately every sample of planets is incomplete because of the limitations of all the detection methods. Still, the answer is important because interstellar grains are formed from such atoms and the grains are essential for the building of planets (Adibekyan et al. 2009).

Grains of dust grew in the disks from which planets were formed, but even in 2015 it was still not known how the blocks of rock (planetesimals) together constructed a planet. All suggested explanations had inadequacies and there was no solution in sight.

The Structure of Planet-Forming Disks

Planet-forming disks produced planets. This process can be described theoretically, but the theory was approximate and highly mathematical. I will therefore focus on observations of protoplanetary discs, a subject that has developed rapidly in recent years. The reason for this surge in interest lay in the new possibilities of making measurements of objects smaller than 1 arc second so that disks as large as our solar system can be seen in some detail.

It has been mentioned above that the star β Pictoris is in the centre of a disk and that there is also a planet. Another example (there are many more than I will show) related to protoplanetary discs with a gap in the middle where there is a planet under construction. In Fig. 11.10 notice the blue ellipse at the bottom-left. This gives the size of the smallest detail. All blue ellipses are each much larger than the solar system.

The ALMA telescope in Chile has been made operational in recent years (see Chap. 9). With its new receivers, the number of mirrors and its location at an altitude of 5,000 meters, this radio telescope surpasses all previous telescopes. It is possible to see details smaller than 0.1 arc second. Figure 11.11 shows a protoplanetary disk with empty rings that have probably been swept clean by whole planets. This is the state of the art in the summer of 2016.

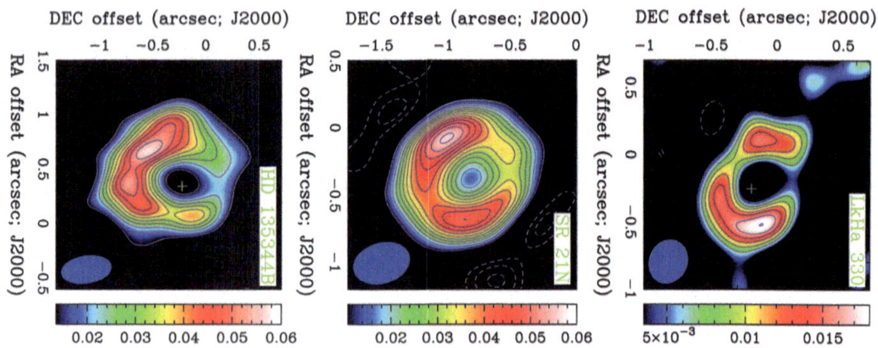

Fig. 11.10 The images of three protoplanetary discs at a wavelength of 0.88 μm. In all cases the disk has a gap in the centre; in this gap there is a star under construction, which is not depicted. The objects shown have a diameter of around 1–2 arc seconds, and the central gap is clearly recognised (Brown et al. 2009)

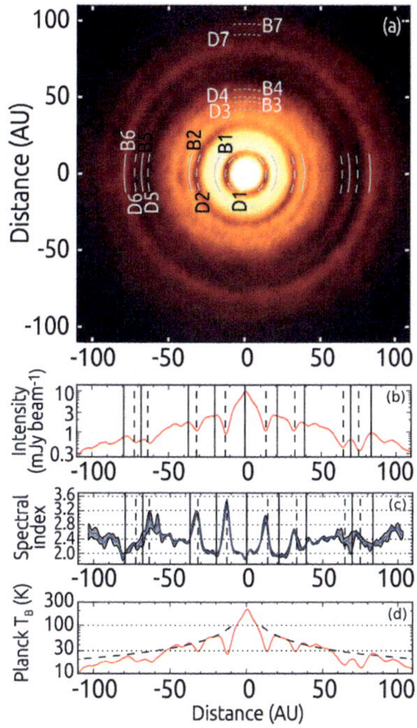

Fig. 11.11 An image of the protoplanetary disk around the star HL Tau obtained using the ALMA radio telescope. The whole image is around 2 arc seconds in size (Alma Partnership et al. 2016)

The Problem of the Hot Jupiters

In 1995 Mayor and Queloz reported the discovery of the first exoplanet. This immediately caused a problem: the planet was roughly as massive as Jupiter but was very close to the star, at a smaller distance than Mercury was from the Sun. It was like unsuspectingly opening the front door and seeing an enormous elephant in your garden with its front feet on the doorstep. How did that enormous beast get there? Assuming that the elephant is a gas planet like Jupiter, the planet must have originated further outside and then emigrated to this orbit. This was a shock: there had been speculation on the migration of planets in the early solar system, but what had been only a remote possibility, suddenly turned out to be real.

After 1995 more 'hot Jupiters' were found, even though they were in the minority. Their proximity to the mother star continued to be an unexplained problem for the time being.

Planet Atmospheres

Planets were difficult to find and it was much more difficult to observe their atmospheres. The best chance was during the observation of an eclipse when a planet disappeared behind the star during the last moments of its transit.

Then the atmosphere of the planet will be the last thing you can see. Various attempts have been undertaken (starting in 2002) and they were also successful but in 2015 all this had not generated very much information.

Summary: What have we learned about the Planetary Systems?

It is remarkable that in 1994 nobody knew how you would ever find out whether exoplanets existed and 20 years later there are almost 4,000 known exoplanets—see for example exoplanet.eu/catalog/, openexoplanetcatalogue.com or exoplanet-archive.ipac.caltech.edu. It is time to do some statistics.

First, look at the way the exoplanets were detected. The diagram in Fig. 11.12 shows the exoplanets distributed according to the distance from their host star (actually the ratio of distance and snowline; see the legend) in which the colors and sizes of the dots indicates the method by which the planets have been found; see the legend to the figure. Most planets have been found by the eclipse method–this is largely the result of the Kepler space mission. This diagram shows also something else: the dots cluster in some parts of the diagram. Heavy planets at short distance from the host stars are preferentially found through radial velocity measurements of heavy planets. The microlensing technique discovers planets at large distances, but over a large range in mass. The eclipse technique finds preferentially heavy and thus large planets: they give the largest eclipses. The conclusion is that each detection technique has strong limitations and that makes statistical analysis a dangerous undertaking.

The spectral type of the stars with planets are almost all of type A or later, implying that the stars have a surface temperature below 10,000 K. But stars hotter than these, B-type and O-type stars, have not been examined, probably because there all methods to detect planets fail. The hottest star with a planet may well be kappa Andromedae, spectral type B9V, surface temperature 11,400 K that carries with it a planet 13 times as massive as Jupiter (Carson et al. 2014). Figure 11.13 shows the planet around another hot star, Fomalhaut, with a surface temperature of 8600 K.

We thus do not know whether the high mass stars have any planets or whether protoplanets are destroyed during the formation of these stars. For low-mass stars the opposite is true: planets have been found around brown dwarfs (Chauvin and Lagrange 2004; Joergens and Müller 2007). A suggestion had already arisen early in exoplanet studies, namely that planets are found more often around stars with a larger abundance of heavy atoms. The answer is still somewhat uncertain but it appears that heavy planets, with a mass comparable to Jupiter, do preferentially occur when the abundance is high. As said before, it is an important piece of information because more heavy atoms should lead to the formation of more, or of greater grains and both factors would facilitate the formation of planets.

Though the statistical analysis is difficult and dangerous, different groups have expressed the somewhat intuitive conclusion that practically every star has planets.

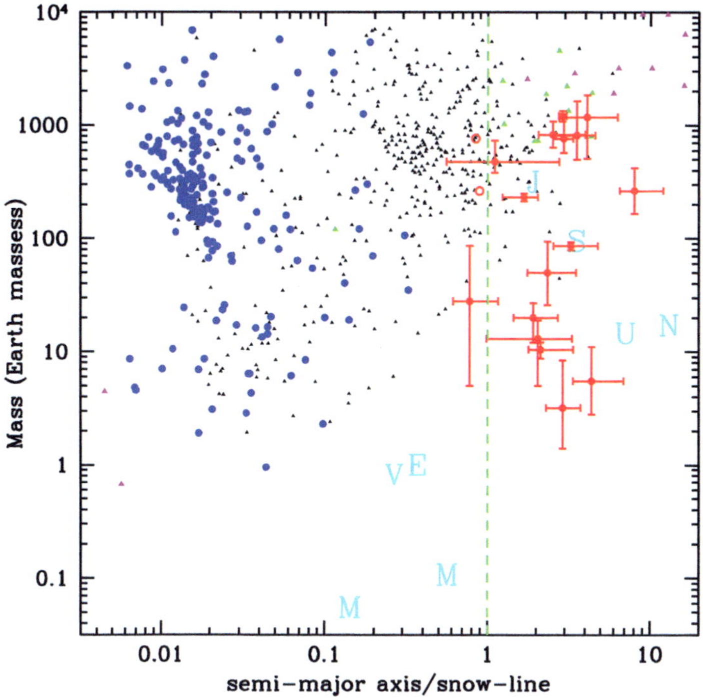

Fig. 11.12 A diagram of known exoplanets. Horizontally is the ration of the distance from star to planet to the distance from the star to its 'snow line' (the snow line is the distance from a star to the point where water would freeze). The blue dots indicate planets discovered using the Mayor-Queloz method; the black dots (small triangles) are planets discovered using the eclipse method. Planets discovered using direct imaging techniques are shown in light blue and green. The red symbols indicate the planets discovered via the microlensing effect (Mao 2012)

The conclusion got some support when in 2016 a planet was found near to Proxima Centauri, the star that is closest to the Sun at a distance of a few light-years (Anglada-Escudé et al. 2016).

Let us look at multiplicity of planets. The Sun has eight planets around it. Is the Sun an exception? No: host stars were found with five or more planets, and even then there was still no certainty that all planets close to this star had been discovered. It frequently occurred that first one planet was found, then a second and a third, and later a few more. The planetary system of the different stars would in the future appear to have more planets in close proximity. Figure 11.14 shows distances and measurements of a number of planetary systems with five or six planets.

A few binary stars were found with a planet in a wide orbit around the binary star. Such a wide orbit was to be expected. The 'three body problem' is very old and research has shown long ago that if three bodies move around one another there is a high chance that at a certain point in time one of the three will suddenly disappear in space. Planets around close binary stars were always at a safe distance.

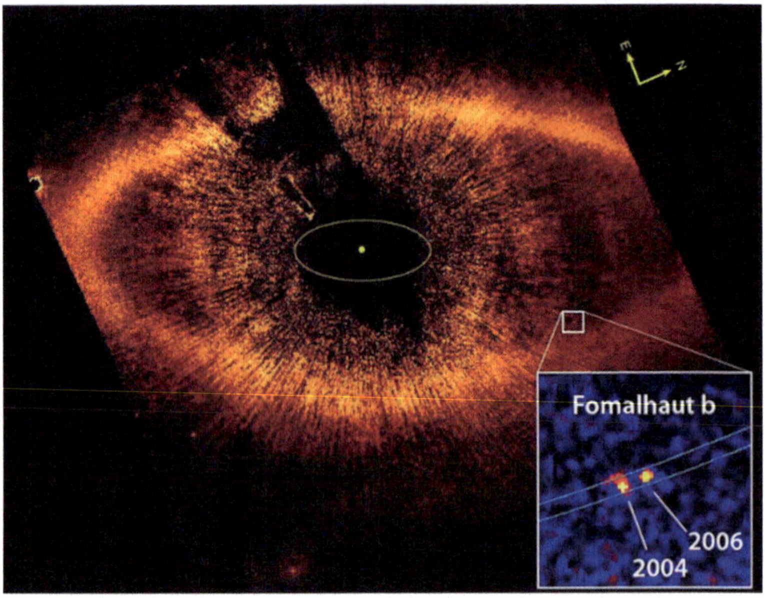

Fig. 11.13 Image of the immediate surroundings of the star Fomalhaut made at a wavelength of 600 nm with the Hubble Space Telescope. The direct stellar light has been blocked off by a coronagraph. The position of the planet Fomalhaut-b is shown as it was in 2004 and in 2006 (Kalas et al. 2008)

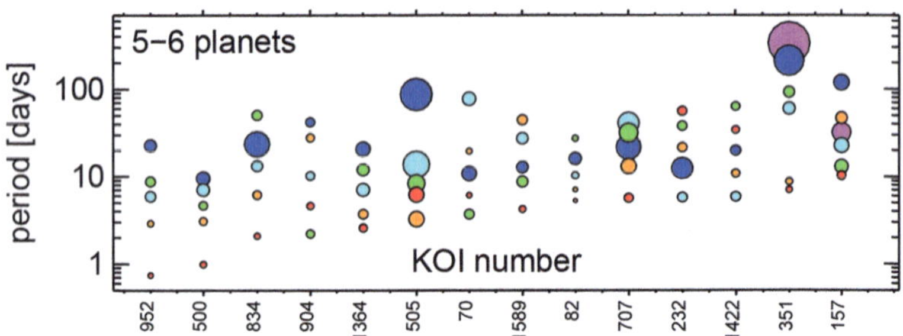

Fig. 11.14 A number of planetary systems with five and six planets around the same star. Each vertical line represents the planets around a specific star. The horizontal axis shows an identification number that refers to all the planets detected by the Kepler satellites. The size of each circle indicates the size of the planet. The significance of the color within the circles is not mentioned, but it probably indicates the method of detection (Winn and Fabrycki 2015)

Finally there is the centuries-old question: is there life elsewhere? The meaning of the 'snowline' has been explained before; another expression for a similar thing is the 'habitable zone', the zone in which water could be present in liquid form. Too close to the star there is no water (it would evaporate in a very short time; Mercury is a good example). Nor could water be too far from the star where it would be frozen (Jupiter is a good example). Looking for planets in the habitable zone was an important motivation for NASA to launch the Kepler satellite. It seems to me that it would have been better to postpone the search for habitable planets and first wait for the detection of planet atmospheres that could support life. After all, life on Earth was very primitive for a very long time until a special kind of bacteria had developed that produced enough free oxygen to kick start evolution.

References

Alcock, C., Akerlof, C.W., et al.: Nature. **365**, 621 (1993)
Anglada-Escudé, G., Amado, P.J., et al.: Nature. **536**, 437 (2016)
Alma Partnership, Brogan, C.L., et al.: Astrophys. J. **808**, L3 (2016)
Adibekyan, V.Z., Santos, N.C., et al.: Astron. Astrophys. **543**, A89 (2009)
Beaulieu, J.-P., Bennett, D.P., et al.: Nature. **439**, 437 (2006)
Brown, J.M., Blake, G.A., et al.: Astrophys. J. **704**, 496 (2009)
Benedict, G.F., McArthur, B.E.: Astrophys. J. **581**, L115 (2002)
Charbonneau, D., Brown, T.M., et al.: Astrophys. J. **568**, 377 (2000)
Gaudi, B.S., Bennett, D.P., et al.: Science. **319**, 927 (2008)
Carson, J., Thalmann, J., et al.: Astrophys. J. **763**, L32 (2014)
Chauvin, G., Lagrange, A.-M.: Astron. Astrophys. **425**, L29 (2004)
Gomes, R., Levinson, H.F., et al.: Nature. **435**, 466 (2005)
Jewitt, D., Luu, J.: Nature. **362**, 730 (1993)
Joergens, V., Müller, A.: Astrophys. J. **666**, L113 (2007)
Kalas, P., Graham, J.R., et al.: Science. **322**, 1345 (2008)
Mao, S.: RAA. **12**, 947 (2012)
Marcy, G.W., Butler, R.P.: Astrophys. J. **464**, L147 (1996)
Morbidelli, A., Levison, H.F., et al.: Nature. **435**, 462 (2005)
Mayor, M., Queloz, D.: Nature. **378**, 355 (1995)
Oort, J.H.: Bull. Astron. Inst. Neth. **11**, 91 (1950)
Raymond, S.N., Morbidelli, A., et al.: IAU Symp. **310**, 194 (2014)
Santos, N.C., Mayor, M., et al.: Astron. Astrophys. **392**, 215 (2002)
Sumi, T., Kamiya, K., et al.: Nature. **473**, 349 (2011)
Tsiganis, K., Gomes, R., et al.: Nature. **435**, 459 (2005)
Tera, F., Papanastassiou, D.A., Wasserburg, G.J.: E&SPL. **22**, 1 (1974)
Van Boekel, R., Min, M., et al.: Nature. **432**, 479 (2004)
van de Kamp, P.: Astron. J. **68**, 515 (1963)
Wolszczan, A., Frail, D.A.: Nature. **355**, 145 (1992)
Winn, J.N., Fabrycki, D.C.: Annu. Rev. Astron. Astrophys. **53**, 409 (2015)

Chapter 12
1985–2015: The Formation of Stars

12.1 Successive Stages in the Formation of Low-Mass Stars

Finally: A Comprehensive Theory

The preceding years had shown that star formation took place in molecular clouds and that infrared observations were essential in finding the newborns. In the 1980s ground-based, observational infrared astronomy had become a well-established discipline. Large-scale programmes in search of young and very young stars could be planned and carried out. Numerous locations were known where they could be sought. IRAS had surveyed almost the complete sky at four wavelengths out to 100 μm. Radio telescopes could routinely make observations of many molecular lines, and the lines of ammonia (NH_3) in particular proved to point towards very dense parts of molecular clouds. Even the photographic plates of the good old Palomar Observatory Sky Survey were a rich and fruitful source of very dark areas, and thus potential places where stars may have formed.

Between 1980 and 1990, several 'safaris' were made and hundreds of sources were identified that emitted almost completely at wavelengths longer than 1 μm. The total amount of radiation was usually equal to that of the Sun, or a few times more. The infrared spectrum made it possible to separate the sources into three classes, called I, II, III (see Fig. 12.1 (Lada 1987)). This scheme was interpreted by a basic mental scheme supported by hydrodynamical calculations: after a spherical cloud started to collapse, a very small protostar appeared very quickly in the centre; thereafter the protostar grew in mass by accreting matter from the rest of the cloud. In this scheme infrared stars of class I were still accreting at a high rate from a surrounding, thick disk of gas and dust. Stars of class II had passed this stage and were accreting at a much smaller rate. Finally class III objects had become main sequence stars, often weakened through a foreground with much dust.

H. J. Habing, *The Birth of Modern Astronomy*, Historical & Cultural Astronomy,
https://doi.org/10.1007/978-3-319-99082-8_12

Fig. 12.1 Infrared spectrum of protostars in three consecutive configurations. In I the protostar is still accreting surrounding gas at a fast rate. In II the accretion has gone down and the configuration is that of a star surrounded by a disk of dust and gas. In III the star is no longer embedded in much material; the object is a fully-grown star (Lada 1987)

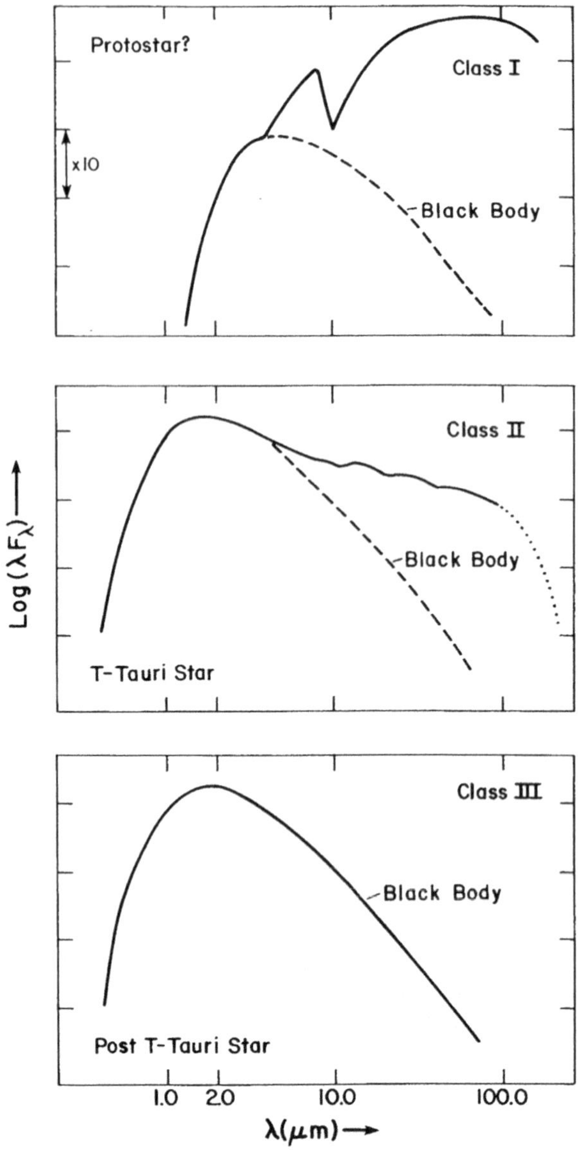

Lada's scheme appealed to many; it was the formulation of insight that already existed in the minds of many. In fact Lada's paper appeared contemporaneously with another article setting out theoretical considerations about the various stages a star went through from beginning to maturity. The article in question was published in 1987 by Shu, Adams and Lizano, all three from Berkeley, and became a cornerstone of the knowledge about star formation (Shu et al. 1987). It is one of the most cited papers in the 1980s, a 'classic'.

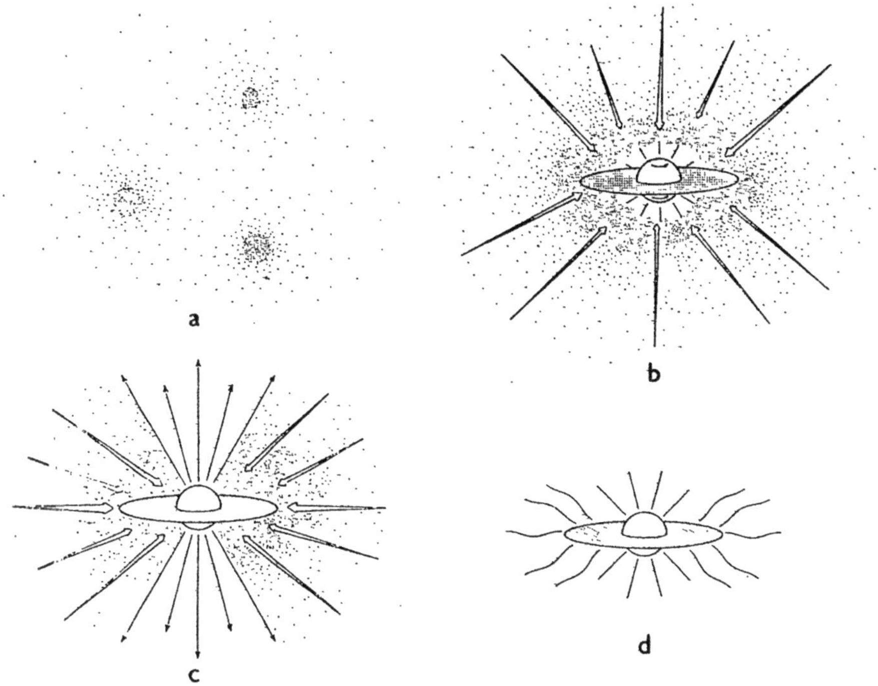

Fig. 12.2 Four consecutive stages in the formation of a star, starting top left with the first stage, with a cloud that has three cores (**a**). A new star will be formed in each of these cores. In (**b**) there is already a 'protostar' in one of the three cores, but gas is still flowing from all directions to the protostar and to the disk that has formed around the protostar. In (**c**) matter is still falling onto the disk, but the star (the former protostar) expels gas upwards and downwards. Finally, in (**d**) the young star makes its appearance but it is surrounded by a disk with a mixture of gas and dust; the tortuous lines indicate the final magnetic field. Planets can form later in the disk (Shu et al. 1987)

Shu and his researchers came to the conclusion that there are four different consecutive stages in the formation of stars. The paradigm is shown in graph form in Fig. 12.2.

The paradigm differentiated four consecutive stages in the birth and development of light stars such as the Sun. TTau stars coincided with the last two stages. The infrared observations had shown that TTau stars fitted Lada's class I as well class II. The class to which they belonged was determined by their radiation at wavelengths longer than 10 μm (see Fig. 12.1): some TTau stars emitted strongly there while others did not. Clearly, the type I emitters were less evolved than those in class II. In Class III the star had left its mother cloud. Together the TTau stars represented stages (c) and (d) in Fig. 12.2 (Bertout 1989). There were no objects known that fitted into stage (b).

The Shu-Adams-Lizano model was to become the paradigm for formation of the light stars. It is still largely valid today, 30 years later. There were modifications and the model was developed further, but it held its ground. This having been said it must be admitted that the main omission of the model—seen with the benefit of hindsight!—related to the first two stages: what caused the first cores to form and why had the objects in phase (b) escaped detection? The first omission (how did the star form?) will be discussed below. Now consider the second question: what about objects in phase (b)?

The answer came in 1993 from observations at wavelengths longer than 20 μm. It has been said several times that radiation with wavelengths between 20 and around 800 μm do not reach the surface of the Earth: they are stopped in the atmosphere, especially by water molecules. But in the late 1980s it was discovered by trial and error that there existed a number of 'windows' that were sometimes open to observers at very high altitude. The peak of Mauna Kea in Hawaii has this quality. A successful survey was made using the UKIRT (UK Infrared Telescope) at wavelengths of 0.35, 0.4, 0.8 and 1.1 mm of a small area of the dark cloud near the star ρ Ophiuchi (Ward-Thompson et al. 1989). A new, truly submillimeter source was found, SM1. A few years later the same area was observed again, now with a larger telescope at Mauna Kea, the 15 m JCMT (see Chap. 9). Now, with higher resolution and a larger telescope, four separate submillimeter sources were found at the location, one of which was SM1. But it was the source SM4(=VLA1623) that was most interesting: it coincided with a radio source already detected with the VLA in New Mexico. Actually, it was a source with a strong outflow of molecular gas. Additional data and careful considerations convinced the authors (Andre et al. 1993) that they had detected the first example of a source in stage (b): a protostar significantly younger than the TTau stars detected earlier. They called it 'a class 0 star' (Fig. 12.3). Their conclusion was not disputed. The sequence of events described graphically in Fig. 12.2 was confirmed.

The chart below (Fig. 12.4) is a summary of the different stages that Shu et al. proposed with some additional properties (Feigelson and Montmerle 1999).

Jets are common Phenomena

Jets are a common phenomenon in areas where new stars form or have formed, as Fig. 12.5 shows.

TTau Stars have an Active Surface and often emit lots of X-Radiation

When the IUE space telescope measured the spectrum of the star TTau, it proved to have the same lines (but much brighter) as the chromosphere of the Sun. These lines indicated that there were strong convective streams under the surface of the star. TTau stars must also have very strong 'solar' spots and 'solar' flames. This assumption received support when it appeared that the TTau stars generally were strong sources of X-radiation (Walter et al. 1988). This was attributed to a 'corona' comparable with a solar corona. Spots were also seen and from time to time a very strong

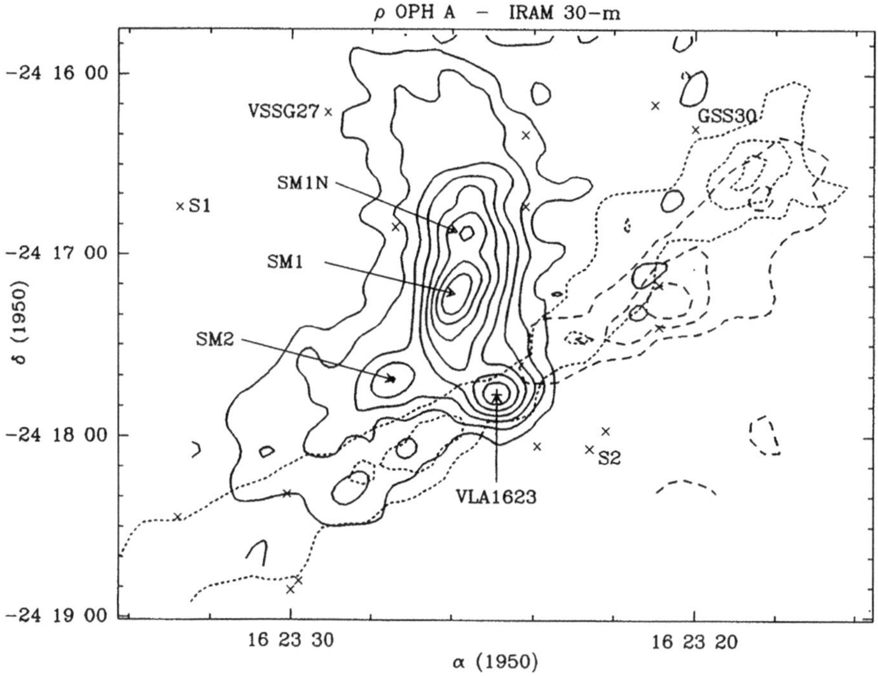

Fig. 12.3 Contours of the 1.3 mm radiation from a small area in the Ophiuchus molecular cloud. Several submillimeter sources are indicated. The source name VLA1629 is a submillimeter source that coincides with a faint radio source. The source was the first of a class called '0'. Matter is still accreting onto the source, but it has already developed a molecular outflow. All the radiation is emitted by warm and hot dust particles. The short-dashed contours indicate a molecular outflow from VLA1629 (Andre et al. 1993)

flame occurred. TTau stars were active, even more active than the Sun. Finally, several TTau stars were found with a jetstream (see the diagram in Fig. 12.4). To cite from a review paper (Bachiller 1996): 'A violent outflow of high-velocity gas is one of the first manifestations of the formation of a new star.'

Silhouetted Disks

While observers did their best to prove the presence of disks around young stars, in 1993 American astronomers O'Dell et al. (1993) examined several maps of the Orion nebula made with the refurbished Hubble telescope. They saw something that, with hindsight, could have been expected: black disks around young stars that were clearly silhouetted against the bright background of the Orion nebula (see Figs. 12.6 and 12.7). The discoverers named the objects 'proplyds', an abbreviation for 'protoplanetary disks' but this ugly word was fortunately replaced in time by a name that better described what was observed: 'silhouetted disk'.

PROPERTIES	*Infalling Protostar*	*Evolved Protostar*	*Classical T Tauri Star*	*Weak-lined T Tauri Star*	*Main Sequence Star*
SKETCH	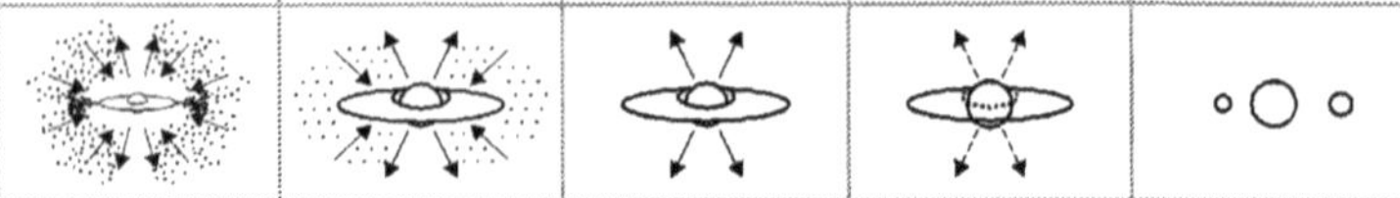				
AGE (YEARS)	10^4	10^5	$10^6 - 10^7$	$10^6 - 10^7$	$> 10^7$
mm/INFRARED CLASS	Class 0	Class I	Class II	Class III	(Class III)
DISK	Yes	Thick	Thick	Thin or Non-existent	Possible Planetary System
X-RAY	?	Yes	Strong	Strong	Weak
THERMAL RADIO	Yes	Yes	Yes	No	No
NON-THERMAL RADIO	No	Yes	No ?	Yes	Yes

Fig. 12.4 A table with properties of the various stages in the life of a proto-star (Feigelson and Montmerle 1999)

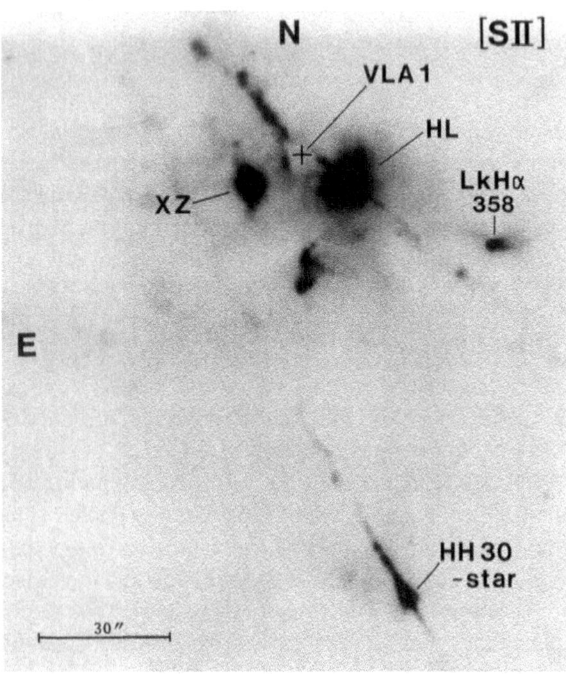

Fig. 12.5 A small part of the sky around the HL Tau star. A large number of jetstreams can be seen expelled by Young Stellar Objects (YSOs): HL Tau, XZ Tau, HH30, VLA1 and LkHα. The recording was made with a filter that only allows light to penetrate at the same wavelength as a spectral line S⁺ (or SII) at 674 nm. The filter strengthens the contrast between the jetstream and the background (Mundt and Ray 1990)

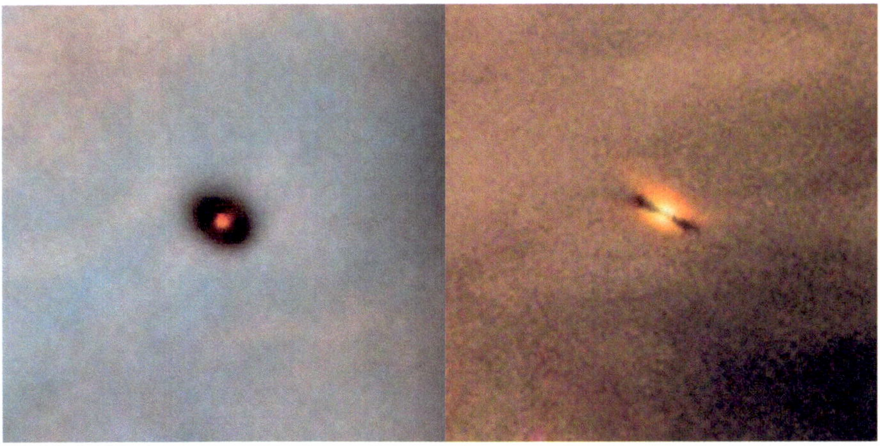

Fig. 12.6 Silhouettes of two star-forming disks seen against the background illuminated by the Orion nebula. Source: NASA

Fig. 12.7 A recording of the large nebula in Orion showing several silhouetted disks. Source: Astronomy Picture of the Day, 22 December 2009, NASA, ESA, M. Robberto, L. Ricci

Stages of Star Formation indicated by Molecular Species

Let us now look at a confirmation of Shu's paradigm more than 10 years later when much more was known about the molecules that occurred in a molecular cloud (Fig. 12.8).

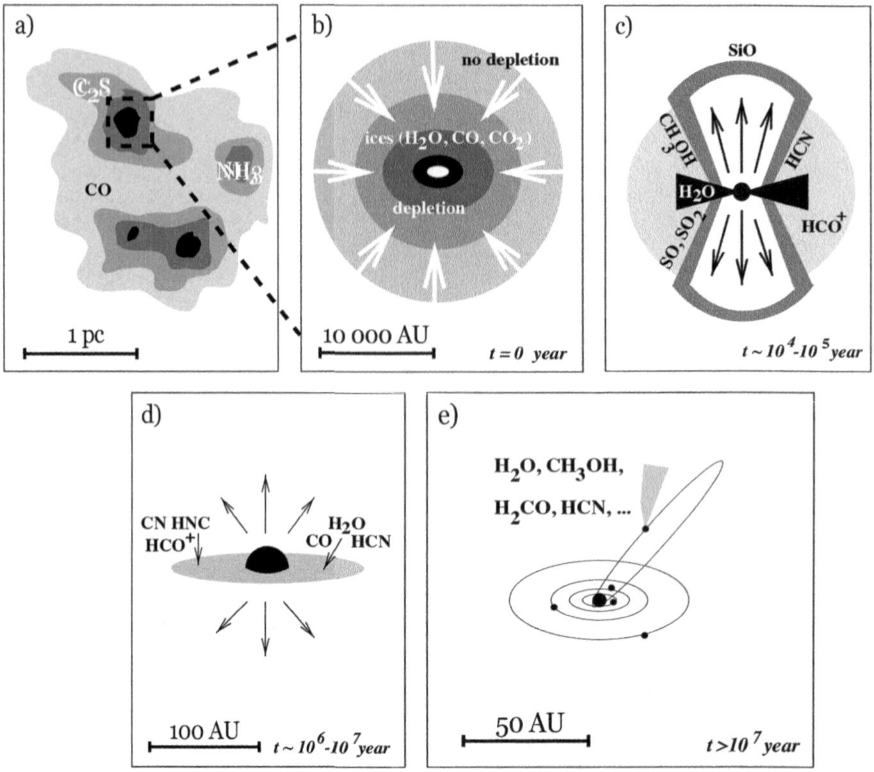

Fig. 12.8 A diagram displaying the four different stages of the formation of a light star. It is similar to Fig. 12.2 but this time as displayed by molecular lines. Figure (**a**) shows three cores in a cloud. The lightest grey indicates where the CO molecule (carbon monoxide) is found. The slightly darker grey indicates an area where C_2S is found, and the darkest grey is mainly characterized by the presence of ammonia molecules (NH_3). A line at the bottom of each image shows the size of the object. Figure (**b**) is a small area of (**a**); the scale in (**b**) is around 1/20th of the scale in figure (**a**). The central part (**b**) of the cloud is so cold that gases are frozen onto the omnipresent dust particles. This means that little CO, CO_2 and H_2O are found in the core of the star under construction. Figure (**c**) is the core of (**b**) but 10 to a 100 thousand years later. The star is surrounded by a disk that contains water vapour (H_2O) among other substances. The embryonic star emits gas in two directions. The core that is formed has edges in which such molecules as SO, SO_2, CH_3OH and HCN occur. The HCO^+ molecule characterizes the remains of the original material (light grey). The configuration in figure (**c**) lasts 100 times shorter than the following one in which almost the whole star can be seen. The star is still embedded in a rotating disk in which molecules such as CO, CN, HCN, HCO^+, H_2O and CNHCN occur. The disk and the star are a 100 times smaller than the object in figure (**b**). Finally, there is figure (**e**) in which we see the star in the middle of its planets with a comet also making its entrance. The scale is two times smaller than in figure (**d**). During the time of the appearance of Shu's figure (1987) and this one (1998), the first exoplanets had been discovered; Shu et al. did not know of their existence. From the small size of figure (**d**) compared with figures (**a**), (**b**) and (**c**), a stellar system is apparently a small object compared with the objects from which it was created (van Dishoeck and Blake 1998)

12.2 Molecules Galore

Since 1968 many molecules have been detected in dark clouds. They had a twofold use: they indicated locations where stars form and they could be used to measure temperature and pressure where they were present.

The Treasures of the Collector

In around 1985, radio telescopes were fitted with new receivers that could record very sharp spectra at very short wavelengths around or below 1 mm over a broad wavelength interval. Two groups, one at Onsala in Sweden and the other in California, pointed their telescope at a known, rich source of molecular lines, made a scan over a significant wavelength range and identified all lines they could detect (Johansson et al. 1984; Sutton et al. 1985). Figure 12.9 shows a small part of the Swedish results. Some rather complex organic molecules were present.

Small Grains of Dust, the Formation of Ice and Complex Molecules

Figure 12.10 shows the infrared spectrum of a very young star deep in a molecular cloud. It is an absorption spectrum, which means that gases had removed the radiation of the young star selectively at different wavelengths. These gases are located in

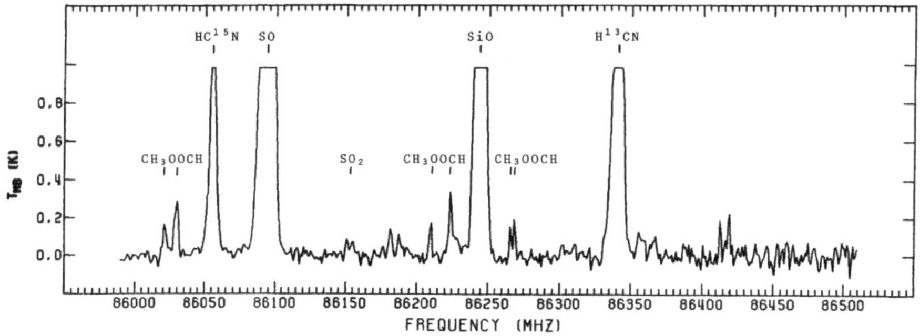

Fig. 12.9 A small (3%) segment from a spectrum that extends from 72 to 91 GHz (4.1 to 3.3 mm) measured with a radio telescope in Onsala, Sweden. In the figure, the strongest lines have been clipped. The full spectrum contains about 170 lines from 24 known species. A few lines could not be identified (Johansson et al. 1984)

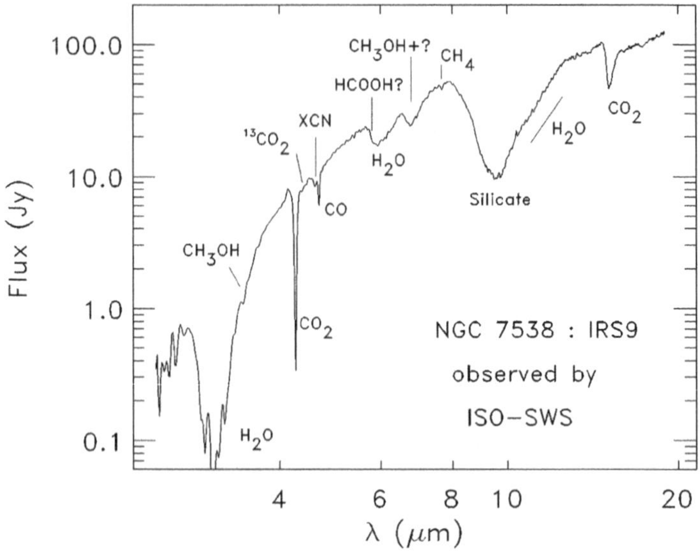

Fig. 12.10 Spectrum of the strong infrared source IRS 5 in the molecular cloud near to the bright nebula NGC 7538. Deep valleys (absorption lines) can be seen, where various molecules have absorbed the radiation. The two deepest valleys are caused by frozen water (at around 3 μm), and by a silicon-containing mineral at 9.7 μm. There are also absorption lines of all kinds of molecules, such as, for example, CH_4 (methane) and CH_3OH (methanol) (van Dishoeck and Blake 1998)

the cloud between the star and us. The most remarkable absorptions were at wavelengths of around 3 and 10 μm. The deep absorption at 10 μm was caused by silicate particles in the dust while that at 3 μm could be attributed to water, H_2O, but in a frozen state: ice.

Molecules exhibit different absorption lines when they are in a gas than when they are inside a solid body. This is illustrated in Fig. 12.11 that shows the absorption spectrum of the CO molecule in three different objects: in the uppermost spectrum the CO gas—like CO_2—is frozen to a particle of solid matter, and in the lower image the CO has become a gas but the CO_2 is still frozen solid.

In Chaps. 4 and 7, solid interstellar particles have been discussed deep inside a molecular cloud that provide the cold surface to which individual atoms may adhere and then form a molecule with another atom. The molecules can be changed later by infrared radiation. It is clear that the composition of the mantle of a grain contains the history of the grain: the radiation to which it had been exposed in the past. That history becomes apparent when the grain is heated and the molecules are set free, such as happens when a new star is born deep in the cloud. Figure 12.12 shows how the successive changes in form of the dust particle were imagined.

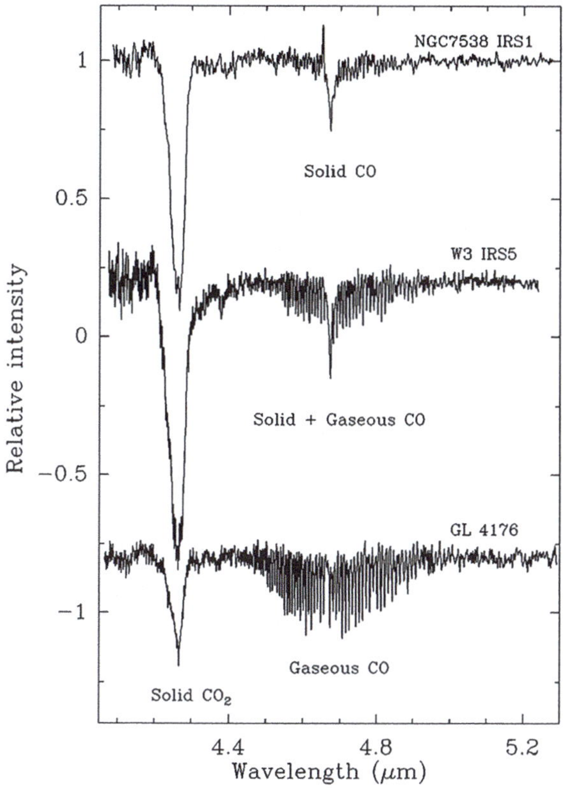

Fig. 12.11 Spectral lines of CO in solid form and in gaseous form. See text (van Dishoeck and Blake 1998)

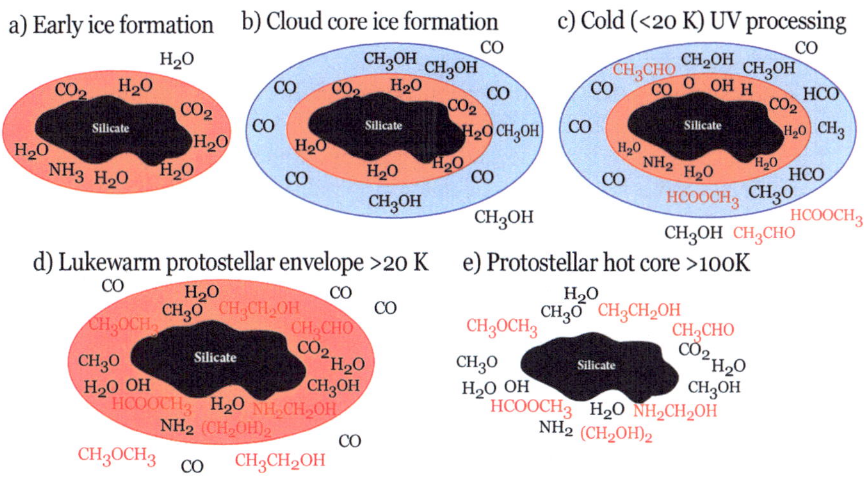

Fig. 12.12 A portrayal of the different forms that a dust particle in a molecular cloud can take. If the formula for a molecule is outside the red or grey area of the particle, the molecule is 'free' and has all the characteristics of a free molecule (van Dishoeck 2014)

12.3 The Formation of High-Mass Stars

The discovery of pre-main sequence stars by Walker in 1952 showed that even after the star had accreted all its matter, nuclear fusion had not yet started. This happened because the Kelvin-Helmholtz contraction time of the star (see Chap. 2) was long enough. For high-mass stars, however, the situation was reversed: nuclear fusion started before the star had been fully assembled. A high-mass star produced so much radiation that its radiation pressure would prevent any further influx. An estimate in 1974 predicted that stars more massive than 40 times the mass of the Sun could not be formed (Kahn 1974). A high-mass star was stopped too early in its growth, which meant that the existence of high-mass stars was a problem.

That high-mass stars disturbed the 'scene of the crime'—and even destroyed it—is clear if we look at the clouds of gas in which high-mass stars were formed. Figure 12.13 shows the chaos that hot stars can cause in their direct environment.

The high-mass star, destroying its cradle, led to the general expectation that high-mass stars could not be formed in the way the lightweight stars were formed. Eventually, after around 2010, observations showed that, contrary to expectations, the formation of high-mass stars proceeded in stages that resembled what happened with the birth of light stars. Accretion disks have been detected even for young stars with a mass as high as 30 M_{sun} (Beltrán and de Wit 2016). An example is presented in the next section. Accretion disks for stars of still higher mass have so far not been detected.

How a High-Mass Binary Star was discovered in Statu Nascendi

The road to an important discovery can be long and on the way new directions are taken without a solid plan and without any certainty about what the end result will be. A good example is the story of how 50 years of accidental discoveries led to the detection of a high-mass binary star in statu nascendi.

The story starts in the 1950s in radio astronomy. A number of large sky surveys had been made and a great variety of interesting radio sources had been detected. All these sources had the remarkable property that they were stronger at longer wavelengths. This went against all existing explanations in astronomy for the emission of electromagnetic radiation and was solved by the discovery of synchrotron radiation. This discovery was felt by all radio astronomers as a major triumph. One might therefore think (as many radio astronomers did) that the future of radio astronomy was at long (radio) wavelengths. However, that was not the only argument for preferring long radio waves. A more down-to-earth argument was that shorter radio waves had higher frequencies and high-frequency receivers were rather experimental. In the fifties the state of the art of receiver technology was at wavelengths of a decimeter. The discovery in 1951 of the 21 cm line (see Chap. 1) was a tribute to the technical skills of the physicists who built the receivers.

Fig. 12.13 A recording with the Hubble Space Telescope. In the upper part of the image a cluster of stars can be seen, Pismis 24, and in the lower part there is light-emitting and dark matter at the same distance as the cluster. The ultraviolet radiation of the hot stars in Pismis 24 has destroyed the cloud. In the lower center, there is a new-born blue star visible in a cavity that it has excavated for itself in the gas cloud. Source: Astronomy Picture of the Day, 27 March 2016, NASA, ESA and Jesús Maíz Apellániz

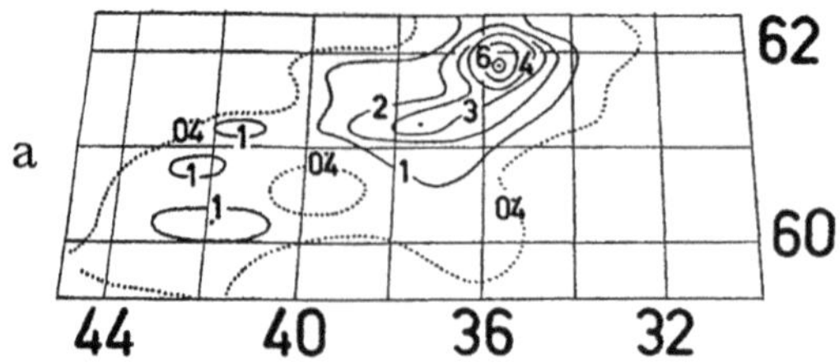

Fig. 12.14 Radio emission at 22 cm wavelength in the direction of three optically-bright nebulae. Right ascension is shown to the right horizontally (expressed in degrees) and declination vertically. Contours of the same brightness have been plotted. The brightest object is W3 (Westerhout 1958)

It was therefore pioneering work when Gart Westerhout in 1957, against advice, started his PhD research with an exploration of the Milky Way at a wavelength of 22 cm using the brand-new 25 m radio-telescope in Dwingeloo. When setting up his measurements, Westerhout gave priority to the part of the Milky Way that was closest to the galactic center, where the most important objects were expected. But at the northern latitude of the Netherlands the central part of the Milky Way is only above the horizon a few hours a day. So what could one do in the 'empty' hours? Westerhout chose a few nebulae in the Milky Way far away from the galactic center and by luck he chose the optically bright nebulae IC1795, 1805 and 1848. He discovered that there was a strong source near to IC 1795 while the other nebulae were fainter sources. The three sources were given the names W3, W4 and W5. They were accidental discoveries and not much attention could be attached to them: bright nebulae were predicted to emit radio radiation and detecting it would add no new knowledge—so one thought (Fig. 12.14).

The story continues with two different discoveries some 8 years later. First, totally unexpectedly a strong OH maser was found near W3 (Weaver et al. 1965). Because the spectral line was so strong, the maser became a useful test object for a VLBI experiment between Berkeley and Boston (a baseline of thousands of kilometers). The outcome was surprising: the source consisted of a large number of bright points distributed over an area of 2 arcseconds × 2 arcseconds. Even at the largest baselines no detail was seen in the bright spots (Moran et al. 1968).

One year later the new radio 140-foot telescope in the US was used to study several HII regions in the galaxy at 2 cm wavelength at much higher resolution, almost 20 times better than Westerhoud achieved. The radio source W3 was found to consist of three smaller sources (see Fig. 12.15) (Schraml and Mezger 1969). The smallest, most compact of the three coincided with the OH maser. Comparing the radio results with a photograph of the same area made it clear that the sources were actually inside or behind a dark (molecular) cloud and could not be seen optically.

The source in the middle has been discussed in Chap. 4, where a more detailed map is shown in Fig. 4.10 in Chap. 4. Here I continue with the object in the lower left corner, known from now on as W3(OH), because it coincided with the OH maser.

In the same year, 1969, the position of the OH source was measured more precisely using a radio interferometer in California (Raimond and Eliasson 1969). The maser did not precisely coincide with the 2 cm radio source, but was located

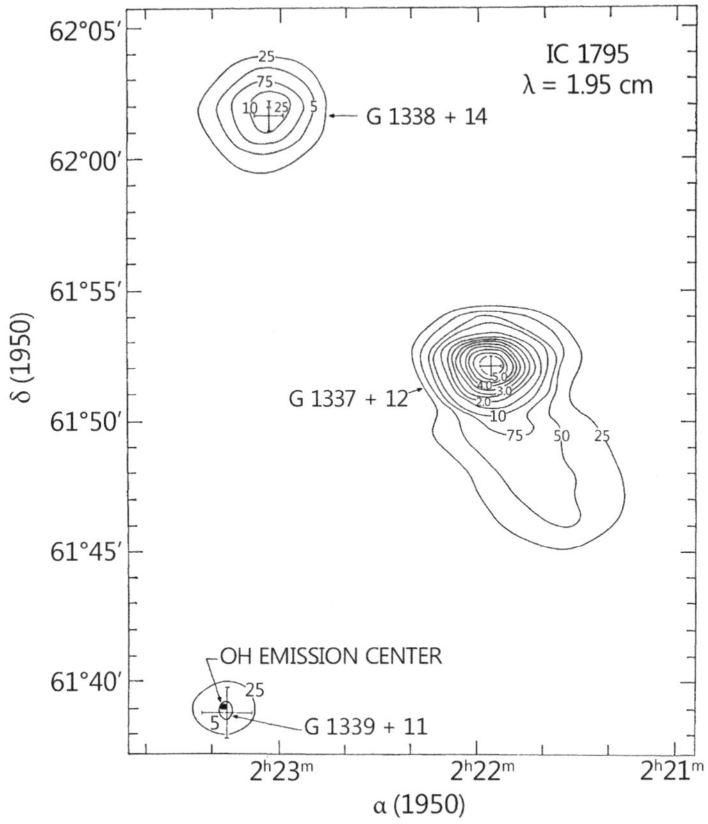

Fig. 12.15 Contour of the radiation from W3 at 2 cm wavelength obtained with the 140-foot telescope of NRAO, West Virginia. The single source seen by Westerhout consists of three smaller sources. The one in the middle has been discussed in Chap. 4 (Schraml and Mezger 1969)

next to the radio source. With the revolutionary 'one-mile radio telescope' in Cambridge (England) it was shown in 1971 that the radio nebula was smaller than 14,000 AU and in 1981 the diameter was actually measured: 2000 AU, 100th of the distance of our Sun to its nearest neighbor and a few times the size of the Oort cloud in the solar system. W3(OH) met all the characteristics of a cocoon star (see earlier): a hot star inside a cavity filled with ionized gas and surrounded by an opaque, thick shell. The central star (or perhaps a few of them) ionized the gas through hard ultraviolet radiation. The gas reradiated the energy at somewhat longer wavelengths. The reradiated energy was absorbed again by the opaque shell and converted into infrared radiation: indeed, W3(OH) was a strong infrared source. The ionized gas also radiated some of its energy at radio wavelengths. This radiation passed the shell without hinder and was subsequently detected by radio telescopes. Finally: the OH maser. It was probably located within the shell where the matter was dense and there was strong infrared radiation: two necessary conditions for the presence of a maser.

In around 1970, a new maser was detected in the direction of W3(OH). It had a wavelength of 1.349 cm and was attributed to water molecules. A new radio interferometer from the University of Berkeley was used in 1972 to determine an accurate position of the H_2O maser. It was close to, but well separated from the OH source by 7 seconds of arc. That was enough to consider W3(H_2O) as a separate source, different from W3(OH). In 1984 the same interferometer, but now operating at the even shorter wavelength of 3.4 mm, proved that W3(H_2O) coincided with a strong source of radiation by HCN (hydrocyanic acid) molecules. The presence of the molecules indicated that W3(H_2O) (sometimes called the 'TW object' after the discoverers) was not yet ionizing its surroundings and was thus younger than W3(OH) (Turner and Welch 1984).

New, very precise observation of the H_2O maser a few years later showed that the bright spots of the maser had moved significantly in just a few years, away from W3(H_2O) (Alcolea et al. 1993). This was taken as the presence of a jetstream, fairly common in protostars. This emphasized that W3(H_2O) really was a very young star.

We have now entered the nineties, a time when radio telescopes measured the radiation of the rich spectrum of molecules with wavelengths of a few millimeters or less. Heavy water (HDO), sulphur oxide (SO_2) and methanol (CH_3OH) were found at the site of the radio source W3(H_2O) (Helmich and van Dishoeck 1997). Measurement of the strength of each line showed that the gas was heated by an external source: a star under construction. The methanol radiation was so strong that it was probably generated in the evaporating layers of ice on the dust particles.

In 1999 the source W3(H_2O) was observed again using a new French-German interferometer for millimeter radio astronomy in the Alps near to Grenoble (Wyrowski et al. 1999). The observations yielded more detailed information. W3(H_2O) appeared to have two cores, A and C, separated from one another by a distance of 1.2″ (2400 AU). The observers suggested that A and C were the embryos of two heavy stars (their joint mass was at least 22 times that of the Sun) at such an early stage of its development that no UV radiation escaped the stellar surface. The two embryos were bound to one another by gravity, which meant that the source was an early stage of a binary star. In 2006 measurements were confirmed and supplemented by observations with a new interferometer in California. The result can be seen in Fig. 12.16.

Figure 12.17 dates from 2011 and shows new and old data together in a colorful and informative diagram of both W3(OH) and W3(H_2O)—read the legend.

The two TW protostars together had a mass more than 22 times the mass of the Sun. At least one of the two must be a high-mass star. Because both stars produce jets it is almost certain that the stars are in the center of disks: very young, high-mass stars are thus accreting matter through a disk, just as low-mass stars do. There are now other and similar objects known: forming high-mass stars with jets and disks (e.g. Galván-Madrid et al. 2010). Indeed, high-mass stars may well be forming in a way very similar to that of low-mass stars.

Repetition of the VLBI measurements of the positions of the H_2O maser spots over a period of many years led to a very precise determination of the parallax of the maser and from these one derives the geometric distance: 2.04 ± 0.07 kpc, an

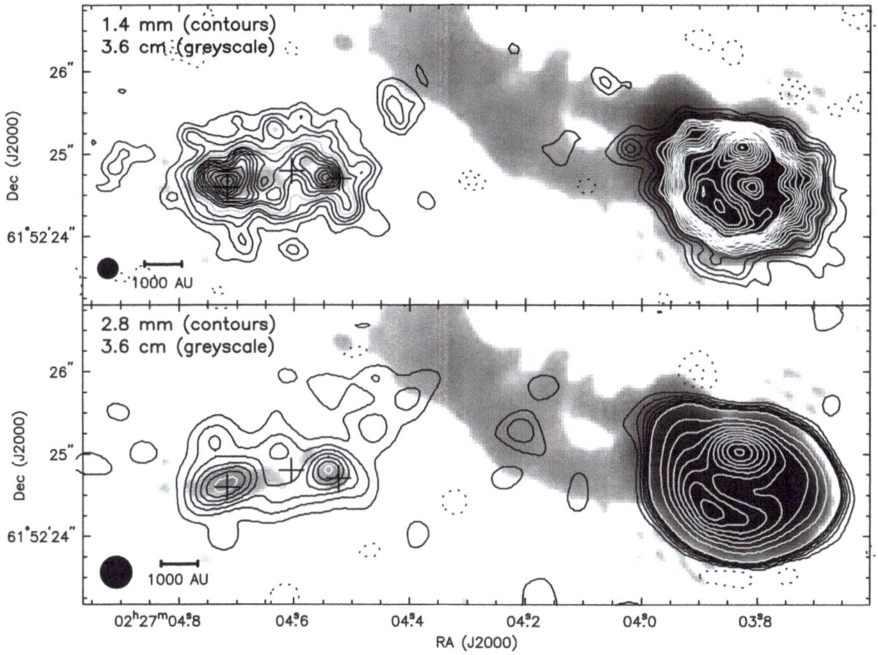

Fig. 12.16 Two images of sources W3(H_2O) (left) and W3(OH) right. The top image shows the contours of radiation by dust particles at 1.4 mm, while radiation at 3.6 cm, from hot ionized gas, is shown as a grey-black shadow. The lower figure is similar, except that here the dust radiation is shown at a wavelength of 2.8 mm. The most important thing in this image is W3(H_2O), the left-hand source it is in two parts: each one is a high-mass star in the making. It can be deduced from velocity measurements that the two are bound to one another by gravity (Chen et al. 2006)

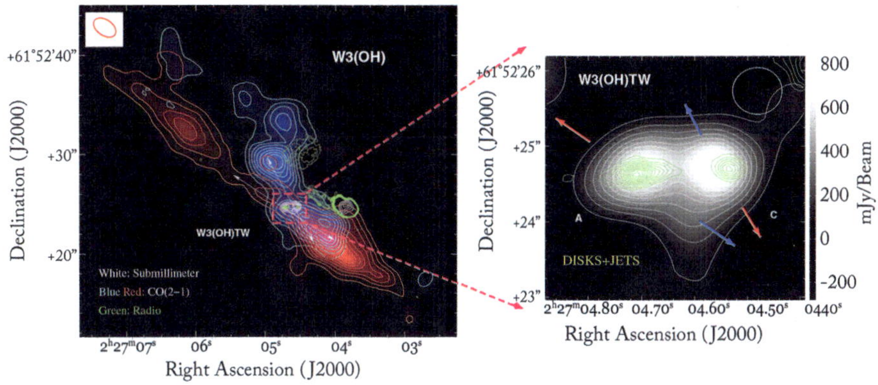

Fig. 12.17 Left image: the region of W3(OH) and of the TW—object with observational results obtained by several different telescopes. Green curves indicate 'bremsstrahlung' from ionized gas in W3(OH) measured at mm-wavelengths. White, red and blue curves indicate measurements of submillimeter radiation of the TW object. Red contours indicate jet streams from one of the two TW stars, and blue the similar streams from the other TW star. The figure on the right is an enlargement of a part of the left figure showing the two stars and the directions of the outflows in blue and in red (Zapata et al. 2011)

accuracy of 3%—far better than what could be achieved optically (Hachisuka et al. 2006; Xu et al. 2006).

Totally different information about the W3 area came from X-ray telescope Chandra. Some 50 X-ray sources were detected in the W3 area and were interpreted as young, very 'active' low-mass stars (Feigelson and Townsley 2008).

In summary, from a beginning in 1958, an unimportant looking radio source led in steps over a period of more than 50 years to more and more detailed and accurate measurements of what in the end was the birth of a high-mass binary star. This was made possible by improved observation techniques (interferometers that could work at ever shorter wavelengths), telescopes high in the mountains with little water vapour above them (e.g. the SMA, the submillimeter array on Mauna Kea). Infrared measurements were essential and an important role was played by the detection of molecules. And, to the surprise of the writer, optical astronomy is not present in this story.

This subchapter was on details of the radio source W3. This same source will be discussed further down with nearby sources W4 and W5 on large-scale star formation.

12.4 The Changing Morphology of Clouds

A Lifecycle

Molecular clouds are gaseous structures subjected to gravity, magnetic fields and pressures. Their behavior is described by well-known equations, but the actors that work upon the cloud are erratic. The calculation of how clouds behaved was a task that demanded much larger and faster computers than were available in 1985. This changed of course in the later years and several groups made extensive calculations. I will use here the conclusions of a review paper of 2007 (McKee and Ostriker 2007).

The basic idea is that molecular clouds represent a phase in a cycle in which large clouds form from a diffuse, rarefied gas that forms a layer in the plane of the Milky Way. The clouds undergo many changes, produce some stars and in the end the cloud is dispersed and again forms a part of the diffuse layer, minus the few percent of the matter that was turned into stellar matter.

Clouds are irregular and unsteady objects. Observation shows that they consist of supersonic, turbulent gas. The turbulent energy is quickly dissipated and has to be replaced by new energy generated by supernovae and by expanding regions of ionized hydrogen that surround high-mass stars. This is an erratic process and clouds never reach a steady state. They come and go, like the clouds painted by the Dutch seventeenth century masters that also form in the real sky, then reform and disappear. Occasionally clumps are formed, even several clumps close together, and these can contain so much mass that collapse will follow. Then an embryonic star will form surrounded by a rotating disk. Accretion may continue, onto the disk and onto the new star. Magnetic forces may force gas into streams outward along the rotation axis of the disk; a jet may also form. In the end the newly formed stars will destroy the cloud and add the molecular gas to the layer of gas in the plane of the Milky Way. End of the cycle.

Figure 12.18 shows a well-known molecular cloud near the Rosetta nebula in the constellation of Monoceros; it consists mostly of clumps of gas.

Denser and Darker Clumps—Snakes

Observations with the ISO infrared satellite in 1996 showed the existence of clouds that were much darker than any previously known examples (Perault et al. 1996). Until then, it had been assumed that all dark clouds would be transparent for infrared radiation: the wavelength was so long that the waves passed through them unhindered. But ISO showed that even at a wavelength of 15 μm some clouds absorbed infrared radiation. This meant that normal light from background stars was weakened at least 25 magnitudes, in other words: more than ten billion times.

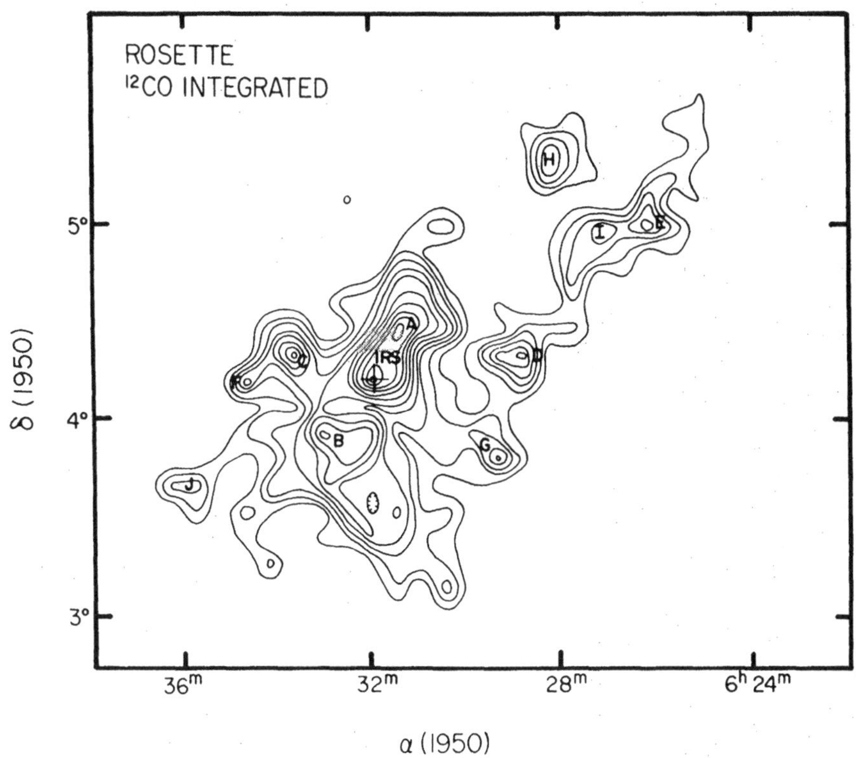

Fig. 12.18 The total density along lines of sight to the cloud near the Rosetta nebula. The maximum length of the cloud is about 100 pc (325 ly) (Blitz and Thaddeus 1980)

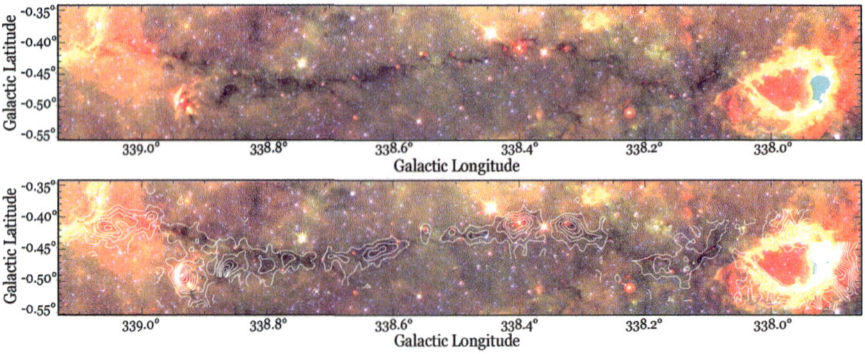

Fig. 12.19 Two images of a part of the Milky Way between the longitudes of 337.9 and 339.1. In the upper image the strength of the infrared radiation is indicated at wavelengths between 3.6 and 24 μm. A long black snake is recognizable that slithers along largely as wide as the Milky Way, This snake, Nessie', is also easy to recognize in the lower image, where the contours show the strength of the millimeter wavelength radiation of the HCN molecule (Jackson et al. 2010)

This discovery was confirmed a few years later by the MSX survey (see Chap. 9) (Egan et al. 1998) and this generated a catalog of more than 10,000 such IRDCs (Infrared Dark Clouds) (Simon et al. 2006).

Then, these very dark clouds often proved to be long, thin filaments. The longest with a snake-like structure, measured 80 pc × 0.5 pc (1 pc = 3.26 light-years) and was fondly named 'Nessie' in memory of the much loved, but never actually seen Loch Ness monster (Jackson et al. 2010) (see Fig. 12.19). These very dark clouds were regarded as the spawning ground of a number of high-mass stars simultaneously (Rathborne et al. 2006). A similar cloud is seen in Chap. 9 in Fig. 9.11.

Long filaments were nothing new in themselves. The filament structure of dark clouds had already been shown in 1979 (Schneider and Elmegreen 1979). The Taurus nebulae, also exhibited a clear filament structure (Heyer and Dame 2015). What was new that these very dark filaments contained so many YSOs. A convincing example can be seen in Fig. 12.20, in which the YSOs are shown in different colors; the measurements were made using NASA's Spitzer satellite. The area richest in YSOs is near to the star called ρ Ophiuchi. The surprising conclusion was: star formation took place in one-dimensional streamers.

The Importance of Filaments

The Herschel satellite came into operation in 2009 to observe the sky in infrared at wavelengths between approximately 55 and 672 μm (0.055 and 0.672 mm). This range of wavelengths had been more or less ignored by astronomers, because the atmosphere of the Earth prevented radiation from reaching its surface and because high-quality detectors, such as would be found in an expensive satellite, had not been available previously. For the research on star formation Herschel's success

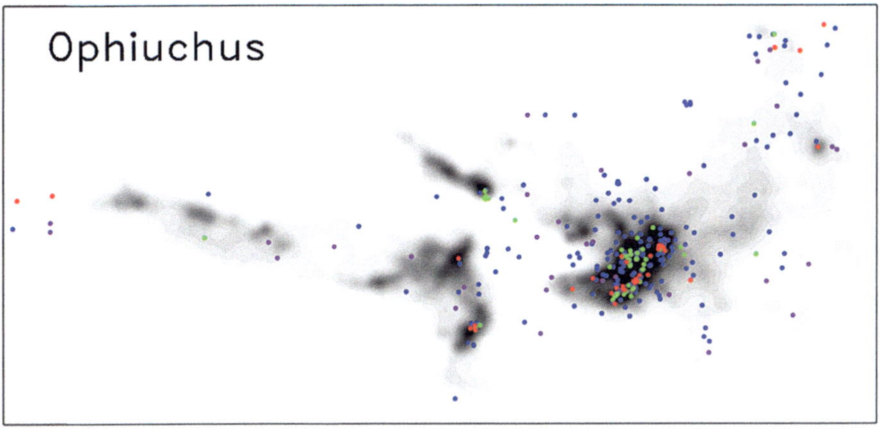

Fig. 12.20 A mozaic of fields in the Ophiuchus complex of molecular clouds, observed with NASA's Spitzer infrared satellite telescope. In grey tints the distribution of dust; the data are based on measurements of the weakening of the starlight. The dots are YSOs: the red ones are class I stars, the green objects between class I and class II, and purple YSOs are class III (Evans et al. 2009)

Fig. 12.21 An image of the radiation of dust particles in a part of the sky measured along the Milky Way using the Herschel space telescope. The longitude is shown horizontally and the latitude vertically. The wavelengths of the radiation are between 70 and 350 μm (0.07 and 0.35 mm) (Molinari et al. 2016)

was enormous and immediate. In the past, some attention had been paid to filament-like structures, but now it appeared that filaments were the basic structure of molecular clouds. The very first publications were jubilant (André et al. 2010; Molinari et al. 2010). The observations showed also convincingly that these filaments were the locations where new stars were formed. Molecular clouds contained many 'Nessies'. Figure 12.21 shows a five degree long piece of the Milky Way as measured at wavelengths between 70 and 350 μm (0.07 and 0.35 mm): the dust condensations radiate like a string of pearls.

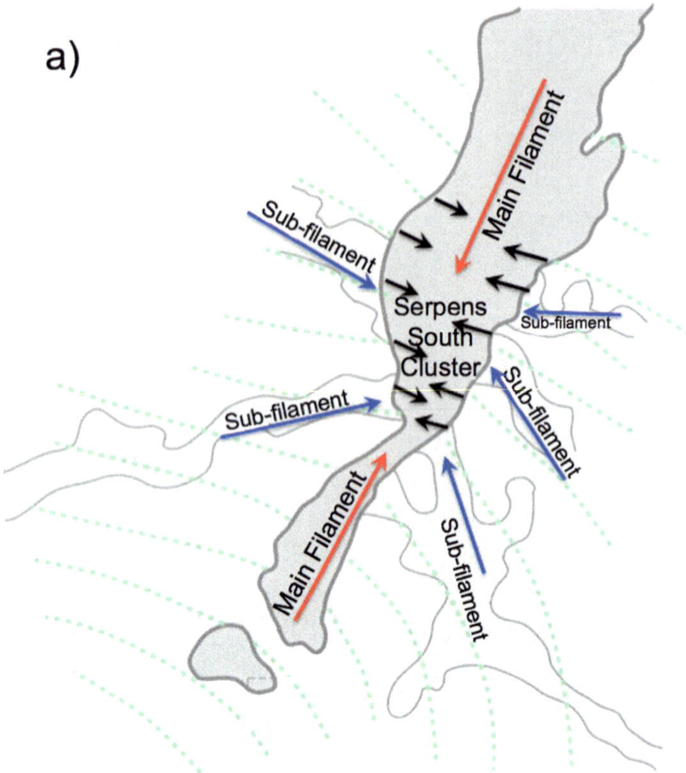

Fig. 12.22 The convergence of several filaments at a junction where a cluster of high-mass and light stars arises (André et al. 2014)

There was more. Various analyses of the filaments and the connection with YSOs showed that where filaments crossed one another, groups (clusters) of high-mass stars appeared to exist (see Fig. 12.22). It was another major discovery: deeply hidden filaments played the key role in the start of the formation of high-mass stars and of clusters. Suddenly this key aspect of star formation had been discovered.

Ultimately it was this anisotropic structure that explained why the formation of high-mass stars did not differ greatly from that of light stars: high-mass stars radiated mainly in the empty space outside the filaments and high-mass accretion took place in a rather thin layer.

12.5 The Formation of Star Clusters, illustrated by the W3/W4/W5 Region

So far we have discussed the details of star and planet formation. But this is a process that takes place on a large scale. The Orion complex is such a region where many high-mass stars are formed. Another region is the W3/W4/W5 complex that

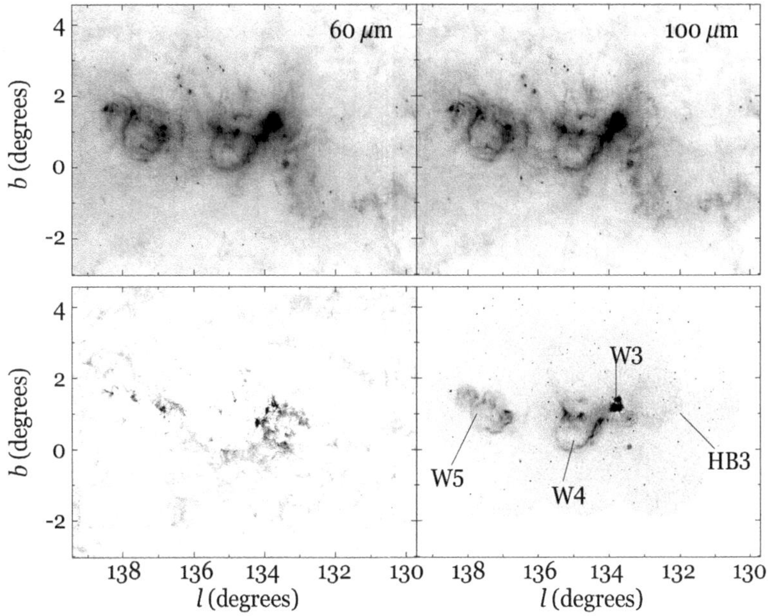

Fig. 12.23 From the upper left, clockwise. The upper two images show radiation by hot dust particles measured at a wavelength 60 μm (left) and 100 μm (right) as measured by the IRAS satellite. Lower right: image of the strength of the 22 cm radio radiation and thus of ionized hydrogen gas (Canadian survey, Normandeau et al. 1997). The names of the three regions are given; HB3 indicates a remnant of a supernova. The fourth diagram, lower left, shows concentrations of CO molecules (Heyer and Terebey 1998)

is located in the Perseus spiral arm and in the sky at high declination (+61°). For observers it is important that the region is far away (130°) from the center of the Milky Way, so that it is possible to observe the region when the center of the Milky Way is below the horizon, traditionally the region that has most interested observers. On the evidence of photographic plates alone, the region was not very interesting; there were just a few bright nebulae that had no striking characteristics. The interest in the region came from the discovery of radio radiation at 22 cm from three sources that were named W3, W4 and W5 (see the discussion earlier in this chapter).

An overview of the three regions W3/W4/W5 of radio radiation, infrared radiation and the radiation from CO molecules dates from 1998; see Fig. 12.23.

The third image in Fig. 12.23 shows that at a wavelength of 22 cm the highest intensity is in W3. A molecular cloud stretches to the right, of which W3 is the left flank. HB3 is a well-known supernova remnant. It has been claimed with good arguments that hot radiation in W4 broke through the ceiling and left the gas layer in our Milky Way, in the so-called Milky Way halo. This earned W4 the name of 'chimney'; more chimneys have been found in our Milky Way and in other galaxies.

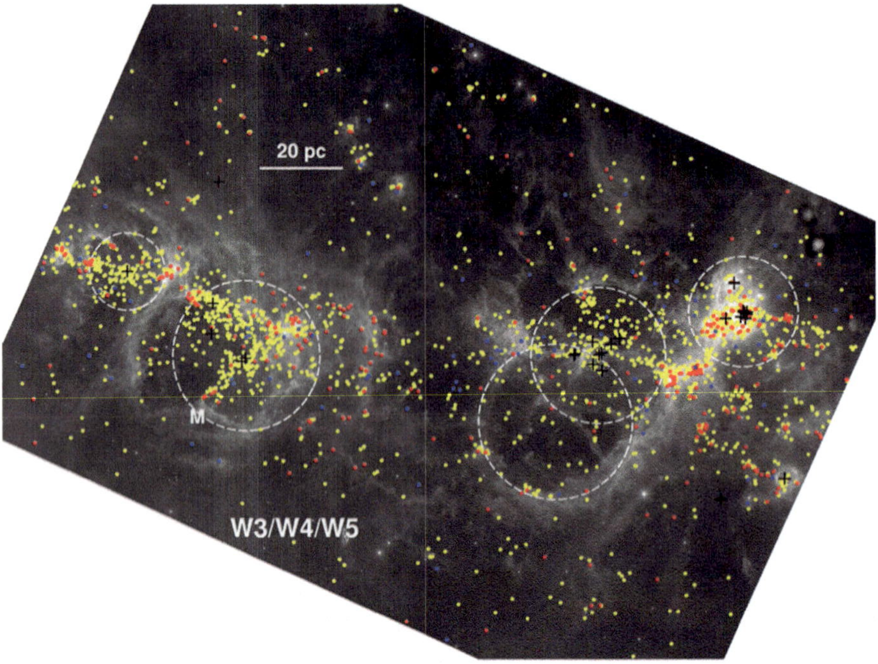

Fig. 12.24 Young stars in the region of W3/W4/W5. Red dots are YSOs of class I, yellow class II and blue are objects with disks from which planets are in the process of being formed. Black crosses indicate the location of an O star, hot stars heavier than around 20 times the mass of the Sun (Koenig et al. 2012)

The expansion of the W4 nebula compressed the gas and dust in the W3 area, which is how star formation started in W3. W5 splits into two adjoining regions, each of which has a cluster of hot stars. New stars are still being formed in W5. Finally, Fig. 12.24 shows that the whole W3/W4/W5 region is full of young stellar objects. W3 shows the highest concentration of new stars and of stars in the making.

Studies of the stars found in these regions show that there are as many as 15 star clusters, of which four are very luminous and probably contain very high-mass O stars (Carpenter et al. 2000). All this shows that the region has produced a great many very high-mass stars and that there are in addition a number of heavy stars under construction, particularly in W3.

Finally, an image obtained recently with the Herschel space telescope (Fig. 12.25). Again, the three regions are easily recognizable, but the image particularly gives the impression of a very turbulent region, as can be expected on the grounds of gas-dynamic calculations. High-mass stars create unrest around themselves.

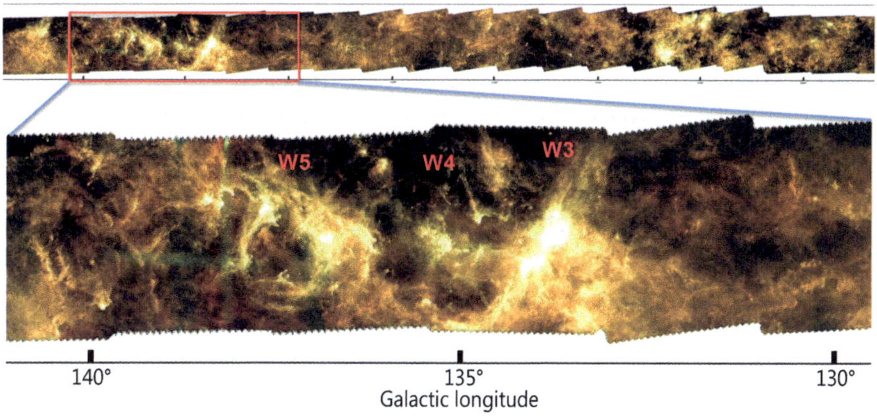

Fig. 12.25 Radiation between 70 and 350 μm of the W3/W4/W5 regions. The strongest radiation is concentrated in W3. W4 resembles a horseshoe pointing upwards: could this be the effect of a major expulsion of hot gas? W5 shows two cavities. In the middle of each is a cluster of hot stars (Molinari et al. 2014)

References

Alcolea, J., Menten, K.M., et al.: Lect. Notes Phys. **412**, 225 (1993)
André, P., Ward-Thompson, D., Barsony, M.: Astrophys. J. **406**, 122 (1993)
André, P., Menshchikov, A., et al.: Astron. Astrophys. **518**, L102 (2010)
André, P., Di Francesco, J., et al.: Protostars & planets, University of Arizona Press, Tucson (Az). VI, 27 (2014)
Bachiller, R.: Annu. Rev. Astron. Astrophys. **34**, 111 (1996)
Beltrán, M.T., de Wit, W.J.: Astron. Astrophys. Rev. **24**, 6 (2016)
Bertout, C.: Annu. Rev. Astron. Astrophys. **27**, 351 (1989)
Blitz, L., Thaddeus, P.: Astrophys. J. **241**, 676 (1980)
Carpenter, J.M., Heyer, M.H., Snell, R.L.: Astrophys. J. Suppl. **130**, 381 (2000)
Chen, H.-R., Welch, W.J., et al.: Astrophys. J. **639**, 975 (2006)
van Dishoeck, E.F.: Faraday Discuss. **168**, 9 (2014)
van Dishoeck, E.F., Blake, G.: Annu. Rev. Astron. Astrophys. **36**, 317 (1998)
Egan, M.P., Shipman, R.F., et al.: Astrophysical Journal. **494**, L199 (1998)
Evans II, N.J., Dunham, M.M., et al.: Astrophys. J. Suppl. **181**, 321 (2009)
Feigelson, E.D., Montmerle, T.: Annu. Rev. Astron. Astrophys. **37**, 363 (1999)
Feigelson, E.D., Townsley, L.K.: Astrophys. J. **673**, 354 (2008)
Galván-Madrid, R., Zhang, Q., et al.: Astrophys. J. **725**, 17 (2010)
Hachisuka, K., Brunthaler, A., et al.: Astrophys. J. **645**, 337 (2006)
Helmich, F., van Dishoeck, E.F.: Astron. Astrophys. Suppl. **124**, 205 (1997)
Heyer, M., Dame, T.M.: Annu. Rev. Astron. Astrophys. **53**, 583 (2015)
Heyer, M.H., Terebey, S.: Astrophys. J. **502**, 265 (1998)
Jackson, J.M., Finn, S.C., et al.: Astrophys. J. **719**, L185 (2010)
Johansson, L.E.B., Andersson, C., et al.: Astron. Astrophys. **130**, 227 (1984)
Kahn, F.D.: Astron. Astrophys. **37**, 149 (1974)
Koenig, X.P., Leisawitz, D.T., et al.: Astrophys. J. **744**, 130 (2012)
Lada, C.: IAU Symp. **115**, 1 (1987)

McKee, C.F., Ostriker, E.C.: Annu. Rev. Astron. Astrophys. **45**, 565 (2007)
Molinari, S., Swinyard, B., et al.: Astron. Astrophys. **518**, L100 (2010)
Molinari, S., Bally, J., et al.: Protostars & planets VI, 125 (2014)
Molinari, S., Schisano, E., et al.: Astron. Astrophys. **591**, A149 (2016)
Moran, J.M., Burke, B.F., et al.: Astrophys. J. **152**, L97 (1968)
Mundt, R., Ray, T.P.: Astron. Astrophys. **232**, 37 (1990)
Normandeau, M., Taylor, A.R., Dewdney, P.E.: Astrophys. J. Suppl. **108**, 279 (1997)
O'Dell, C.R., Wen, Z., Hu, X.: Astrophys. J. **410**, 696 (1993)
Perault, M., Omont, A., et al.: Astron. Astrophys. **315**, L165 (1996)
Raimond, E., Eliasson, B.: Astrophys. J. **155**, 817 (1969)
Rathborne, J.M., Jackson, J.M., Simon, R.: Astrophys. J. **641**, 389 (2006)
Schneider, S., Elmegreen, B.G.: Astrophys. J. Suppl. **41**, 87 (1979)
Schraml, J., Mezger, P.G.: Astrophys. J. **156**, 269 (1969)
Shu, F.H., Adams, F.C., Lizano, S.: Annu. Rev. Astron. Astrophys. **25**, 23 (1987)
Simon, R., Jackson, J.M., et al.: Astrophys. J. **639**, 227 (2006)
Sutton, E.C., Blake, G.A., et al.: Astrophys. J. Suppl. **58**, 341 (1985)
Turner, J.L., Welch, W.J.: Astrophys. J. **287**, L81 (1984)
Walter, F.M., Brown, A., et al.: Astron. J. **96**, 297 (1988)
Ward-Thompson, D., Robson, E.I., et al.: Mon. Not. R. Astron. Soc. **241**, 119 (1989)
Weaver, H., Williams, D.R., et al.: Nature. **208**, 29 (1965)
Westerhout, G.: Bullet. Astron. Instit. Netherlands. **14**, 215 (1958)
Wyrowski, F., Schilke, P., et al.: Astrophys. J. **514**, L43 (1999)
Xu, Y., Reid, M.J., et al.: Science. **311**, 54 (2006)
Zapata, L.A., Rodríguez-Garza, C., et al.: Astrophys. J. **740**, L19 (2011)

Chapter 13
1995–2015: What is left: Compact Objects

A white dwarf, a neutron star or a black hole are left behind after a star dies. These objects are called 'compact' because they are so small compared to the Sun and all other stars. Let us leave the black holes out of the discussion for the moment. White dwarfs and neutron stars have about as much mass as the Sun: a white dwarf is a hundred times smaller, and a neutron star a thousand times smaller still. Thus the density is enormously higher and that has an important consequence: it affects strongly how the gas reacts to the compression caused by the force of gravity inside the compact star. The law of gravity is the same in all normal stars, in white dwarfs or in neutron stars, but the laws of compressibility are totally different. The latter law is called the 'Equation of State' or EoS. In normal stars the EoS is the well-known law of Boyle/Gay-Luzac. In white dwarfs the EoS is known from quantum physics and from the properties of electrons and protons. As a consequence, one can calculate reliably how large a white dwarf will be once one knows its mass. In neutron stars the Equation of State is uncertain, mainly because very little is known about the behaviour of neutrons in a very dense gas. Models of neutron stars are therefore very uncertain, and consequently the size of a neutron star cannot be predicted reliably. Measuring this size is therefore an important research goal in the study of neutron stars.

Black holes are a subject of their own; they may have all kinds of masses. Their size is determined solely by their mass as Schwarzschild and Droste proved in 1916 and 1917, independently of one another (see Chap. 4): $r = 2GM/c^2$ with G being the gravitational constant and c the speed of light. The Sun would be a black hole if all its matter were compressed into a sphere with a radius of $r = 2.9$ km. To imagine a black hole, think of a sphere in space separated from the Universe outside by an invisible membrane that allows everything to pass from the outside to the inside and nothing from inside to outside. What happens inside a black hole we will never see.

Compact objects are studied out of curiosity: what properties do these unusual things have and how does nature make them? There is another argument. Gravity is so strong in these objects that by studying them you test Einstein's General Theory

H. J. Habing, *The Birth of Modern Astronomy*, Historical & Cultural Astronomy,
https://doi.org/10.1007/978-3-319-99082-8_13

of Relativity in a way that cannot be imitated in the lab. There is, for example, the prediction that a light wave that leaves a white dwarf or neutron star will have a longer wavelength because of gravitational attraction, an effect that was seen in the light from white dwarfs (see Chap. 4). The theory also predicts some outlandish phenomena, such as the Lense-Thirring effect, which can be detected only through observation of matter very close to the surface of the compact star. Compact objects are exciting objects.

White dwarfs can be observed by optical telescopes, as has been discussed before (Chap. 5). The recent Sloan Digital Sky Survey (see Chaps. 9 and 16) has discovered thousands of them and ESA's GAIA satellite has determined an accurate distance also for thousands of white dwarfs. A result from GAIA will be shown (Fig. 13.4) at the end of the next section. Neutron stars are found in very different guises. Most have been detected as pulsars or in X-ray binaries. Stellar black holes have only been found in X-ray binaries.

The kind of compact star that is left behind depends in the first instance on the mass of the star at birth. It has not yet been proven, but there is consensus that a star with a mass less than about nine times that of the Sun will leave behind a white dwarf by a rather peaceful process. A star with a mass between 9 and 30 times that of the Sun explodes as a supernova and its remains are a neutron star. Even more massive stars ultimately produce a black hole. Whether the most massive stars (100 times as massive as the Sun or more) leave behind anything tangible is completely uncertain.

Finally, I should mention the existence of close binary stars where both objects are compact objects (WD and WD; NS and NS, BH and BH). If the orbits of these stars once they are formed are sufficiently small, they start to lose energy during their rotation: they will emit gravitational energy, and this will slowly increase as the two circle faster and faster in smaller and smaller circles. The gravitational radiation will be strongest just before the two objects collide. Indeed, in 2016 the gravitational radiation from two merging black holes was detected and briefly thereafter the merging of two neutron stars; more about this in Chap. 14.

13.1 White Dwarfs

White dwarfs remain behind when red giants shed their external mantle, leaving behind the core of the star. Standard theory prescribes that this only happens when the star has become a red giant for the second time, a so-called AGB star. The expectation is that most white dwarfs consist mainly of carbon and oxygen, while other white dwarfs may be made of oxygen, neon and magnesium. There might also be some very light white dwarfs that consist only of helium. To date there are no observations that distinguish between these three kinds.

Because white dwarfs have a hot surface (>30,000 K), the light they emit is strong in the far ultraviolet. The research has benefited greatly from the availability of space telescopes for that wavelength range, such as IUE (see Chap. 9).

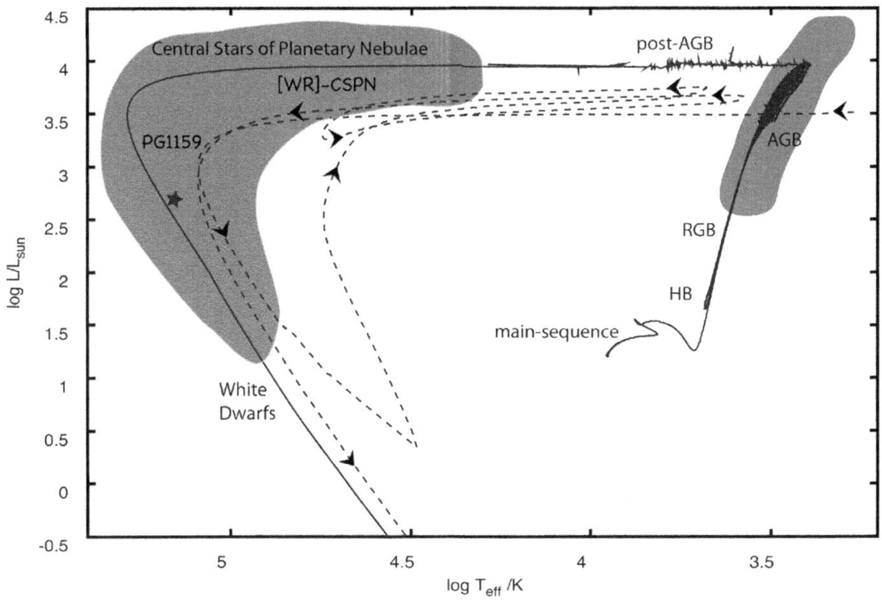

Fig. 13.1 The journey a star twice as massive as the Sun will make through the Hertzsprung-Russell diagram from the main sequence to the white dwarf stage. The different abbreviations mean the following: HB: horizontal branch; RGB: Red Giant Branch; AGB: Asymptotic Giant Branch; [WR]-CSPN: the 'Wolf-Rayet' stage of the central star of a planetary nebula; PG1159: a known hot star that is a white dwarf in the making. The dotted line shows a different path (Werner and Herwig 2006)

Figure 13.1 shows the (calculated) path that a star, twice as massive as the Sun, will follow through the HR diagram from the main-sequence phase up to becoming a red giant and then down to the white dwarf stage. The dotted line indicates a rather more complex path that is also possible. This alternative possibility shows the uncertainties that are inherent in such calculations.

Since 1985, tens of thousands of white dwarfs have been found, mainly thanks to the magnificent Sloan Digital Sky Survey (SDSS) (see Chap. 9) (Kepler et al. 2016). Within a distance of 25 parsec (80 light-years) from the Sun, 232 white dwarfs are now known, which is probably two-thirds of the actual number (Holberg et al. 2016). As a comparison: in our environment there are far fewer white dwarfs than 'standard stars', or main-sequence stars.

Very good recordings have been made and analyzed of the spectra of thousands of white dwarfs (see e.g. Gianninas et al. 2011). Much better models have also been calculated for the atmospheres of white dwarfs, and there is now a close correlation between the observations and calculations. The shape of an absorption line is a good means of determining the gravity at the surface. Figure 13.2 shows how precisely the model and the observation align with one another.

The radius R of the white dwarf and the mass M can also be derived. Figure 13.3 shows a histogram of the mass of a large number of white dwarfs. The diagram is in line with earlier diagrams, but is more detailed.

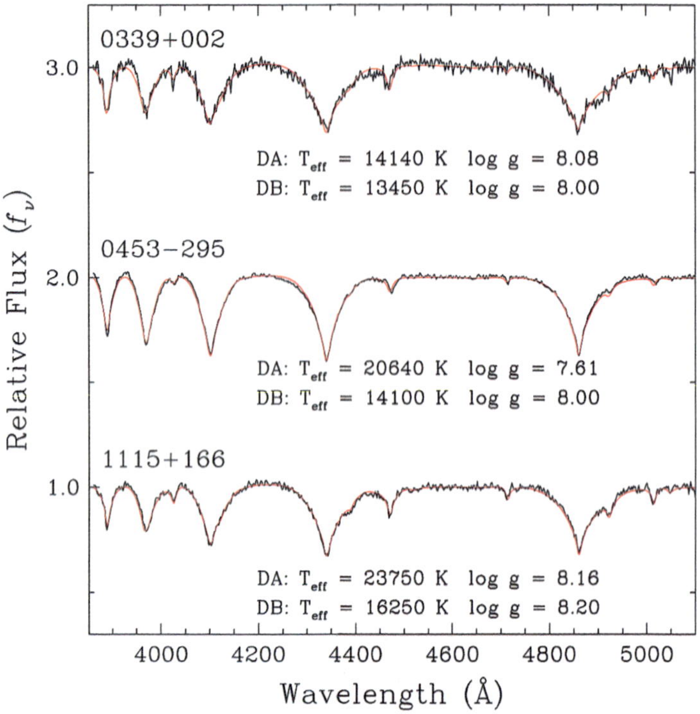

Fig. 13.2 A comparison between part of a spectrum as it has been observed (black) and the same part of a model spectrum (red). The broadening of the lines is in part due to the gravitational attraction, g, on the surface of the white dwarf. Log g is given in each diagram (Gianninas et al. 2011)

Fig. 13.3 A histogram of the mass of white dwarfs. The average mass is easy to determine, with a relatively small spread. In the upper corner two, sets of figures are given, where S/N measures the quality of the individual measurement; the higher the S/N, the higher the quality (Tremblay et al. 2011)

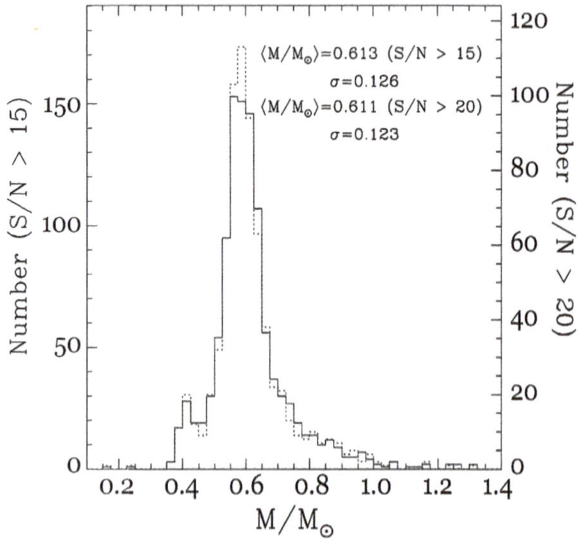

Hundreds of white dwarfs are now known with a strong magnetic field (between 2 and 1,000 million Gauss) (Ferrario et al. 2015). The shape of the magnetic field in many cases resembles that of a bar magnet (a 'dipole field') although multipoles have also been found. Progress has thus been made, but there are still a number of important questions that have to be answered, such as how these magnetic fields originated, how long they continue to exist, and how they influence the further development of the white dwarfs.

Asteroseismology, the study of stellar oscillations or vibrations, was introduced in an earlier chapter and it has been said to hold important promise for the future. This promise materialised in 1991 when the observations were published of the proto-white dwarf, PG1159-035, using the Whole Earth Telescope, a network of telescopes spread over the globe. The luminosity of the star was recorded continuously over a period of more than 10 days without interruption. More than 110 different oscillations were identified, which, when analysed, showed that the star was 0.586 ± 0.003 times as massive as the Sun: this level of precision had previously been impossible to achieve. There were also indications that directly below the surface, the matter that made up the star was divided into layers each of a different chemical composition. A subsequent analysis in 2008 with more observations by the same network of telescopes confirmed the existence of such layers, delivered the speed at which the star rotates (1.39200 ± 0.0008 days) and established that, if there was a magnetic field, it was less than 2,000 G at the surface of the star. Studies of the star over a period of more than 10 years indicate that the oscillations are gradually slowing down (Winget and Kepler 2008). Seismology continues to be an important source of information on white dwarfs, a source that has so far barely been exploited.

White dwarfs cannot produce energy; the heat that they initially contain is radiated away in the course of billions of years, a process that takes so long because the surface area of the white dwarf is so small. The oldest white dwarfs are the faintest radiators. There is an old expectation that by finding the oldest and thus the faintest white dwarf within the Milky Way, one can derive its age. The faintest white dwarf is, of course, the hardest to find and so far it has not been possible to successfully calculate the age of our Milky Way. The best estimate is an age of the disk of our Milky Way of between eight and ten billions of years. If confirmed, this tells us that the disk was born several billion years after the Big Bang (Fontaine et al. 2001; García-Berro and Oswalt 2016).

The theory of white dwarfs predicts a relation between the mass of a white dwarf and its radius. This prediction is based on the expected behavior of the matter that makes up the white dwarf, and is not directly verifiable with laboratory measurements. It is therefore important to try to obtain independent and precise measurements. One example where such measurements have generated precise information can be found in the observations made of a binary star of a white dwarf and a very small red main-sequence star. The mass of both stars can be determined precisely from the measurements, and the radius can be determined from the eclipse of one star by the other (Parsons et al. 2012). The result gives values that are in line with the theoretical predictions. One swallow may not make a summer, but the result is nonetheless encouraging. Here, too, the GAIA satellite can make an important contribution to the discussions by providing very precise distance measurements.

The white dwarfs that occur in a binary star system are fewer in number, but are still numerous. A vast range of companion stars can occur, from main-sequence stars comparable with the Sun to small red stars. Close binary systems have also been found with two white dwarfs or with a neutron star (for example a millisecond pulsar; see later). Interestingly enough, white dwarfs with low mass (less than 0.4 Mzon) only occur in such close binary dwarfs.

A further category of close binary stars with a white dwarf as a receiver of matter was found in 1991 in the measurement data of German X-ray satellite ROSAT. Sources appeared with very low-level X-radiation (Greiner et al. 1991). This new type was named 'super soft' ('SSS: super soft sources'). The explanation (van den Heuvel and Bhattacharya 1992) was that so much hydrogen was dumped on the surface of the dwarf that it burned to form helium: the star behaved like a red giant. That this is at all possible could be seen from Fig. 7.27 in Chap. 7.

Briefly, there are two more peculiarities. Since the 1970s the suggestion has arisen that once the carbon/oxygen within a massive white dwarf has cooled down sufficiently it could adopt a crystal structure: a diamond star! In itself, such an internal structure could be demonstrated using seismological analyses, but this lucky discovery has not yet been made in spite of several efforts. Further news is the presence of disks of dust around white dwarfs and the explanation that massive atoms are found that recently made their way to the surface.

While this book was being finalised (spring 2018), an impressive paper was published based on the second data release of ESA's GAIA satellite. It contains hundreds of thousands of stars with accurately determined parallaxes. Figure 13.4

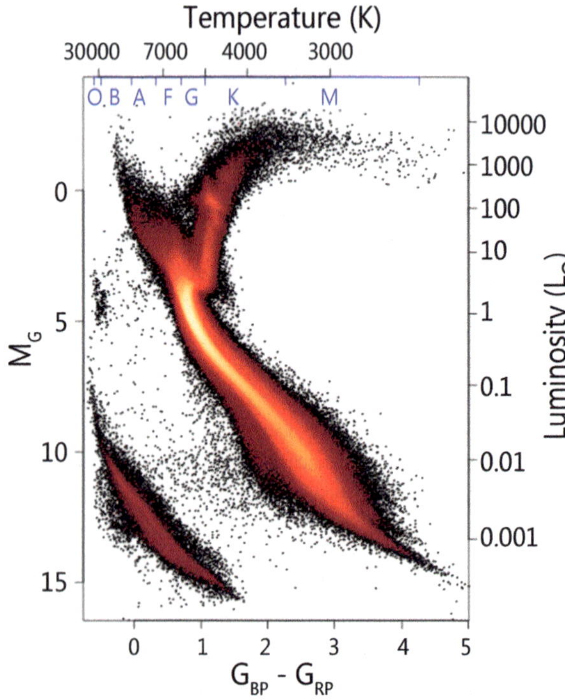

Fig. 13.4 HR diagram showing magnitudes and colors of stars within 100 pc, based on data obtained with the GAIA satellite. Below the main sequence there is a second sequence: that of the white dwarfs. The sequence represents the 'cooling curve' of white dwarfs (Babusiaux and van Leeuwen 2018)

shows an HR diagram of over 200,000 stars within a distance of 100 pc (326 light years). Two sequences are particularly prominent: the upper sequence is the main sequence and below that there is a sequence of white dwarfs. This diagram shows that a major step forward has been made in the identification of white dwarfs and in determining their basic properties, distance, color and luminosity.

13.2 Neutron Stars

Pulsars

The first neutron stars to be discovered were pulsating radio sources (see Chap. 4). Thereafter the search for more of these objects was conducted without interruption, and by 2013 more than 2000 had been discovered (Harding 2013). The fastest pulsar spins around its axis 716 times a second (Hessels et al. 2006).

In October 1968 a visible pulsar was found in the remnants of the Vela supernova and in December 1968 another was found in the Crab nebula. In 1971 the first X-ray pulsar was found and the number of X-ray pulsars rose in the following years, albeit much more slowly than the number of radio pulsars. The German ROSAT X-ray satellite and its Japanese equivalent ASCA identified dozens of pulsars in and around 2000 (Becker and Truemper 1997). A very hot and small area close to the (magnetic) pole of the neutron star emits all the X-radiation. In 1983 observations by ESA's COS-B satellite showed that the pulsars in the Crab and in Vela not only produced radio and light pulses but also gamma radiation pulses (Hermsen 1983).

With such a past history it was no surprise when after 2005 the Crab nebula was also proven to be a strong source of TeV photons, that is, photons with an energy of 10^{12} eV or 10^3 GeV. Figure 13.5 shows the whole magnetic spectrum of the Crab, nebula plus pulsar. The Crab nebula plus Crab pulsar is one of the few objects that has been detected at all possible wavelengths of the electromagnetic spectrum.

Just less than 10 years after the discovery of the gamma pulses from the Crab and Vela, evidence was obtained that one of the strongest sources of gamma radiation in the sky, Geminga, was a pulsar with a period of 0.237 seconds that nonetheless coincided with a very weak, 26th magnitude star, confirming how little light was radiated in relation to gamma radiation. Geminga got its name from the mainly Milan astronomers who, over a period of 10 years, had spent a great deal of energy and intelligence on identifying this source—without success; frustrated, they called the object 'It's not there', which in Milan dialect sounded like 'gh'è minga'. Eventually, in 1991 Geminga was proven to be a pulsar with a period of 0.237 sec (Halpern and Holt 1992). In spite of the faint magnitude of the stellar counterpart, its parallax was measured with the Hubble Space Telescope. This showed Geminga to be at a distance of about 150 parsec (370 light-years), which is close by (Caraveo et al. 1996). This proximity also explained why Geminga was the first identified gamma (radiation) pulsar. All these radio, optical, X-ray and gamma pulsars fall into the category of 'rotation-powered pulsars' (Kaspi et al. 2006).

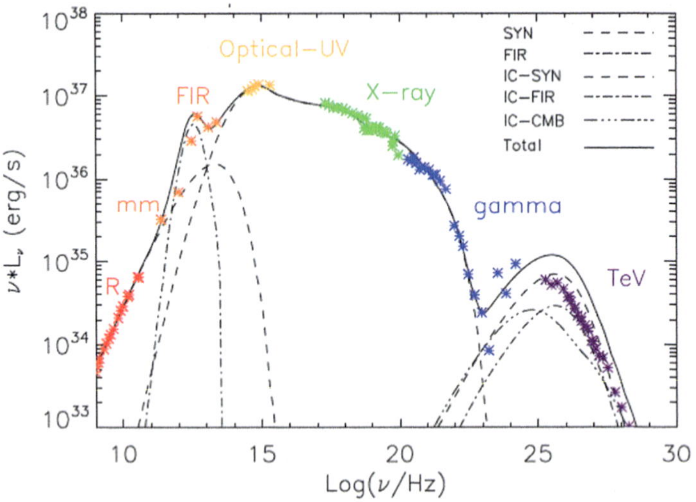

Fig. 13.5 In color the measured spectrum of the Crab nebula, including the neutron star at its center. The combination of nebula and pulsar emits electromagnetic radiation from radio to beyond gamma photon energies. The black-striped curves describe the contributions of different components. TeV stands for 'tera electron volt', photons with an energy that is hundreds of thousands times those of the gamma radiation (Kargaltsev et al. 2015)

After 2009, NASA's FERMI gamma ray telescope identified over 100 more such gamma radiation pulsars—see the paper that had over 200 authors (Abdo et al. 2013): the products of large space projects are no longer collected by a few individuals, but require the collaboration of large groups of experts with different areas of expertise.

Pulsar Winds

As early as the 1950s it was clear that the Crab nebula was made up of two structures: the first of these was formed by the filaments that lay like a network around the second structure, which showed its presence by blue synchrotron light. Relativistically fast electrons produced this light, but they needed to be replenished continuously, which was what the Crab pulsar did during the 900 years after the explosion in 1054 (see Chap. 1). Thus the Crab nebula was also a synchrotron nebula. From the discovery of the pulsars, slowly more and more comparable, relativistic nebulae were discovered—nebulae that were generally referred to as 'pulsar winds'.

Not every pulsar produced a wind: 'empty' supernova shells were discovered with a pulsar at their center and no relativistic gas. Such an empty shell was the supernova remnant Cas A, one of the strongest radio sources in the heavens and the remains of a supernova from around 1680. Cas A was made up of filaments of gas

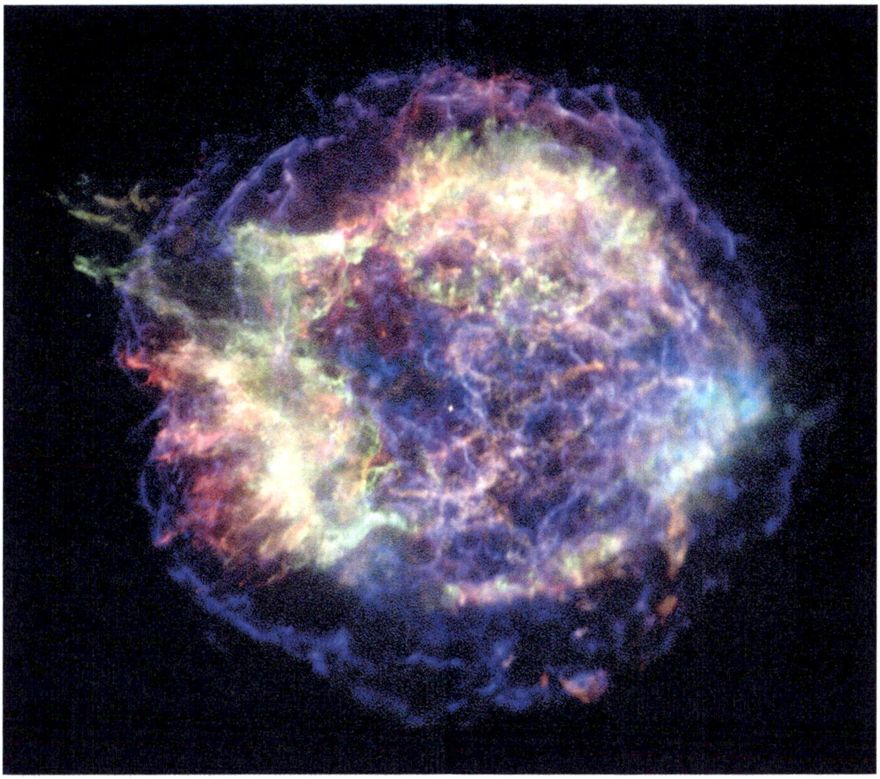

Fig. 13.6 An X-ray recording from Cassiopeia A. The colors represent radiation in several wavelength bands. The neutron star at the center is easily visible. Source: NASA/CXC/MIT/UMass Amherst/M.D.Stage et al.

that formed a skin around an empty space, as was shown by optical, radio and X-ray observations (Helder and Vink 2008). When the Chandra satellite was commissioned in 1999, one of the first discoveries was that of a small, star-like X-ray source in the center of the shell (Tananbaum 1999). As this object emitted X-radiation, it could not be a black hole; neither was there any sign of an accretion disk. The conclusion was that it must be a new kind of neutron star. Pulsations were searched for in vain. More of this kind of neutron star were found in a few other remnants of supernovae (Pavlov et al. 2004). See Fig. 13.6.

In recent years it has become apparent that in a pulsar wind photons are produced with a higher energy than was previously thought: photons with TeV energies (10^{12} eV). These photons never reach the surface of the Earth but are halted high in its atmosphere, an event that can be detected from the Earth's surface, as described in Chap. 9. It is quite possible that cosmic radiation also originates at the same places where these photons are produced.

Black Widows

This is the story of a curious class of pulsars that got the nickname 'Black Widows'. In 1988, the PSR1957+20 pulsar proved to have a very light companion that weighed less than 1% of the Sun: a brown dwarf (Fruchter et al. 1988). It soon became clear that this brown dwarf was losing matter to the neutron star. A few more similar pulsars were identified in the following years.

The almost fatal loss of matter for the poor brown dwarf reminded the discoverers of an American black spider with a very venominous stab; the females are particularly dangerous.

Magnetars, APX and SGR

Chapter 7 discussed the flash of gamma radiation that caused a shock on 5 March 1979 in many space telescopes. Through a lucky circumstance, the direction to the source could be determined very precisely–in the remnants of a supernova in the Great Magellan Cloud. But it was still not known what mechanism or object had caused this enormous explosion. That is, until 1992, when researchers Robert Duncan, living in Canada, and Christopher Thompson from Texas speculated on what could happen in a supernova in the few seconds when a neutron star was created (Duncan and Thompson 1992). The proto-neutron star would be strongly convective, which means that more or less chaotic flows of gas would occur deep inside. And if the neutron star rotated rapidly, a very strong magnetic field could be built up, much stronger than any that had to date been encountered in stars: the neutron star could be an object in which magnetic fields dominated all other forces, including gravity. Duncan and Thompson estimated its strength at 10^{15} Gauss, almost 1,000 times stronger than the magnetic field at the surface of a pulsar in the Crab nebula. (As a comparison: the magnetic field of the Sun has a strength of around 1 Gauss at its surface.) The authors called the hypothetical object a 'magnetar' and speculated that the gamma explosion in March 1979 could have been caused by their magnetar. The article had a major impact. This was a totally new kind of object and, who knows, such objects might exist. It would take many years, but eventually convincing examples of magnetars were identified.

Soft gamma-ray repeaters (SGRs) were introduced in Chap. 7: they were also discovered because of their large explosions. The biggest explosion to date, on 27 December 2004, was so strong that the Earth's atmosphere, in particular the ionosphere, was damaged. Radio amateurs lost their communication with fellow amateurs on other continents, because the ionosphere no longer acted as the mirror that transmitted their signals around the Earth.

A characteristic of the three SGR explosions was that the source brightened very quickly and then slowly faded away. During this latter period, the signal underwent strong oscillations at an interval that brought to mind a pulsar (see Fig. 13.7).

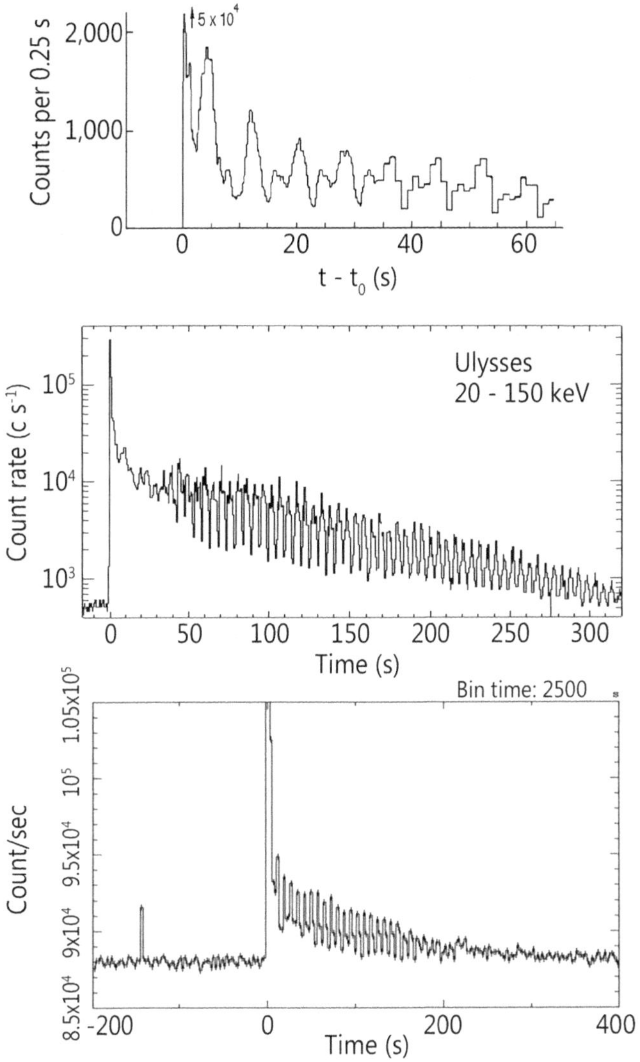

Fig. 13.7 Three SGRs have experienced a large explosion in the past 25 years. In each of the three, the signal arose very rapidly, before fading away much more slowly. Here are shown X-ray detections at photon energies around 100 keV. During the period of fading, there were strong oscillations in the signal that are attributed to pulsar (Mereghetti 2008)

In 1998 and 1999 evidence was found that in at least two of the SGRs a pulsar was present with the same pulse interval as exhibited by the SGR in its 'dying moments' (Kouveliotou et al. 1998, 1999). This evidence and other factors confirmed that an SGR was a pulsar with a magnetic field 100 times stronger than that of an average pulsar. The explosions were attributed to the crust of a magnetar bursting open as a result of an accumulation of magnetic energy.

Anomalous X-ray pulsars (AXPs) were another rare type of object. Originally there was a group of seven X-ray pulsars, all with roughly the same period (between 5 and 12 seconds). Some optical counterparts were found, but they were so weak that there was little to be said about them, except that an accretion disk was visible. AXPs were similar to SGRs without the enormous explosions, but in 2003 two AXPs were discovered that had SGR-like outbursts. Subsequently, AXPs and SGRs were considered to be the same type of object, namely magnetars (Kaspi et al. 2003).

A catalogue from 2014 (Olausen and Kaspi 2014) shows 5 probable and 12 certain magnetars. A considerable proportion of these magnetars coincided with a very young cluster of stars or lay within the remnant of a supernova. This indicated that they must have originated from massive stars, stars so massive that you would expect them to have produced a black hole and not a neutron star. This was something of an enigma.

A Summary of the Different Types of Neutron Stars

Figure 13.8 gives a summary of the different types of neutron stars.

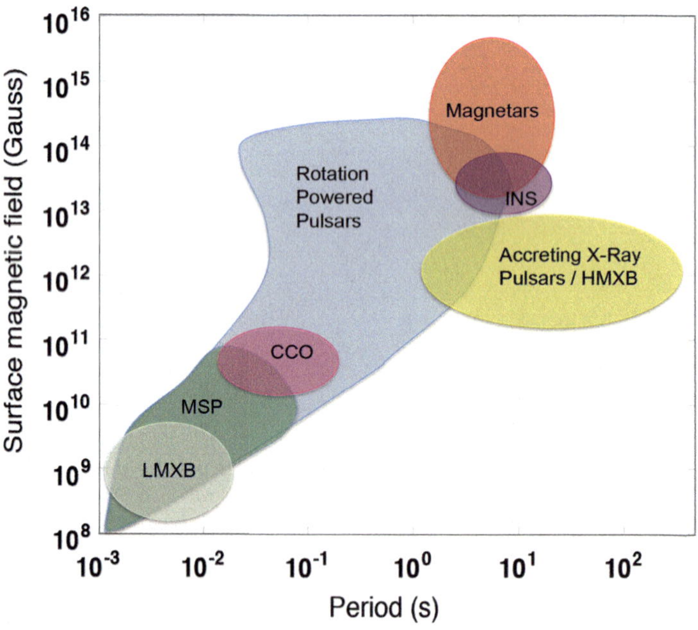

Fig. 13.8 The most important characteristics of different types of neutron stars summarised in a diagram with the period P of the pulsar indicated horizontally and the strength of the magnetic field on the surface of the neutron star vertically. The strength of the magnetic field is calculated from the pulse period and its slow decrease. The acronyms have the following meanings: INS = isolated neutron star; HMXB = high-mass X-ray binary; CCO = Compact Center Object; MSP = millisecond pulsar; LMXB = low-mass X-ray binary (Harding 2013)

13.3 X-Ray Binary Stars: The Hosts for Neutron Stars and Black Holes

Common Properties of X-Ray Binary Stars

The first X-ray source to be identified (Sco X-1; 1966) was a neutron star and a donor star with a mass of only 0.4 times that of the Sun. In 1972 and in another X-ray binary, Cygnus X-1, the donor star had a high mass as did the receiving object: it could not be a neutron star, but had to be a black hole—see Chap. 7. X-ray binary stars thus offered another opportunity for studying neutron stars than via the pulsars but black holes could only be found via binary stars as they themselves did not radiate. Binary stars also provided the valuable opportunity to measure the mass of both stars. On the other hand, the exchange of matter from donor to recipient (generally referred to as the 'accretor') made the analysis of close binary stars much more complex.

One comment before we continue. X-ray binaries are very small; the image you obtain is a blotch of all parts together. Thus all the structural information is obtained from the spectrum (continuum, lines) and its variation is time. That makes the job difficult and it has taken decades before observers agreed on the same overall picture. Some binaries make jets that emit radio radiation; these are large enough to be mapped by VLBI (see Chap. 3).

The classification of X-ray binary stars into 'High Mass X-ray Binaries' and 'Low Mass X-ray binaries' has already been discussed. This classification distinguished between two very different accretion streams: in an HMXB, the massive star loses matter isotropically (= equally strong in all directions), which results in an accretion of the compact from all sides. But in an LMXB, the matter flows through the Roche point, which causes the built up of an accretion disk.

Catalogues of HMXB and LMXB objects comprise several hundreds of objects including a number in the Magellanic Clouds, the most important satellites of our Milky Way (Liu et al. 2005; Haberl and Sturm 2016). Numerous other X-ray binary stars have been found in other galaxies, starting with M31, especially after the large telescopes Chandra and XMM-Newton came into operation in 2000.

Low-mass X-ray binaries are very small. Many would fit inside the Sun and almost all are smaller than the distance from the Sun to the planet Mercury. This smallness is a problem, because binaries of main sequence stars are much farther apart, and those must be the systems that evolved into an X-ray binary. Thus, during this evolution the system had to shrink significantly. But in a shrinking binary the two objects will rotate faster and faster around each other because of the fundamental law of physics that angular momentum is conserved. The Polish astronomer Paczynski proposed a theoretical solution in 1976 (Paczynski 1976). He argued that due to the loss of matter of one of the two stars, the binary system became immersed in a cloud surrounding both stars. He called this cloud a 'common envelope'. If the density of the gas were sufficiently high, the rotation of the stars would slow down by friction; they lost their angular momentum to the gas of the common envelope. In due time the envelope would be dispersed and the double system would reappear to the outside world considerably shrunk in size and still with a not-too-large period of rotation. The 'common envelope' was embraced as a concept by the observers, but definite examples have not yet been found (Ivanova et al. 2013).

Figure 13.9 shows in graphic form a scenario by which a low-mass X-ray binary could have been formed. Notice the influence of the common envelope: after this phase the binary is much smaller.

When a newly detected X-ray binary was studied in more detail, the first step was to take optical spectra of the object, to identify the donor star and to measure the variation in wavelength of spectral lines and to interpret this as a variation in radial velocity through the Doppler effect. As explained in earlier chapters, these measurements lead to an estimate of the minimum mass of the compact object. If this minimum has a value more than 2–3 times the mass of the Sun, the compact object has to be a black hole. In most cases, however, the minimum is below this value and the compact object is most probably a neutron star.

Most X-ray binaries are highly variable, not only in intensity of the X-ray flux, but also in the spectral variations that accompanied the intensity variations. In the beginning, the variations impressed the observers and the enthusiastic description of these variations created a large number of acronyms difficult to understand to the interested outsider; it soon became almost impossible to separate the wood from the trees. Different 'states' were recognized: a given X-ray binary could be in one state for some time and then unexpectedly transform into another state. Slowly, the

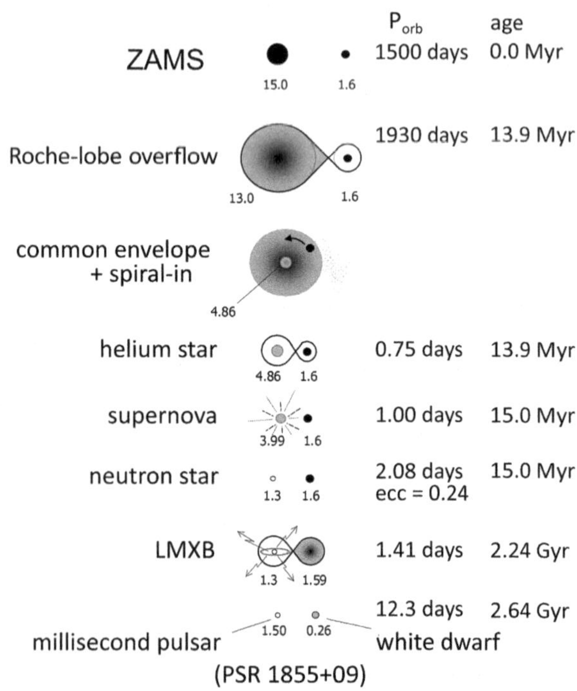

Fig. 13.9 Subsequent stages that lead to the formation of a close X-ray binary. The initial system is at the top (ZAMS = zero-age main-sequence star) and the ultimate binary is at the bottom. In each stage a small picture of the system is shown. Also given is the rotation period of the two objects and the years that passed by. Notice the importance of the common envelope stage; after this (hypothetical) stage the binary is much smaller (Tauris and van den Heuvel 2006)

observers and the theoreticians began to see relations between various manifestations of the X-ray spectrum and flux. A fundamental property of the disk was that it grew from the outside due to the influx of new material. This matter circled around the central object (neutron star, black hole) while moving very slowly in the direction of the center, where it eventually ended up. This slow inward movement was the result of the 'viscosity' of the material. For a long time it was not known what could cause such a viscosity until the discovery that a weak but chaotic magnetic field might do the trick (Balbus and Hawley 1991).

A generally agreed picture appeared of significant 'states' of an X-ray binary. Figure 13.10 shows how far insights had progressed in 1997.

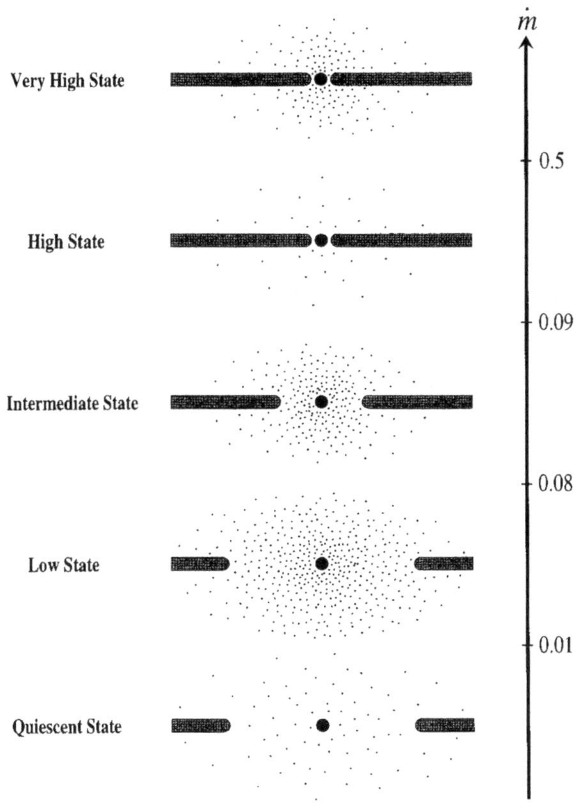

Fig. 13.10 A graphic description of five states of a X-ray binary that contains a black hole. The black dot represents the black hole. The 'accretion disk' is seen edge-on. The disk has a gap near to the black hole. The dots represent the 'corona', a population of relativistic electrons and protons produced by the accretion disk. These electrons collide with soft photons emitted by the disk, and during the collision a photon extracts energy from the electrons and becomes a hard X-ray photon (through the 'inverse Compton effect'). The amount of matter that streams from the donor star (not in the figure) to the accretion disk determines the state. A low stream from donor to accretion disk puts the system onto the 'quiescent state', a very high stream puts it into the very-high state (Esin et al. 1997)

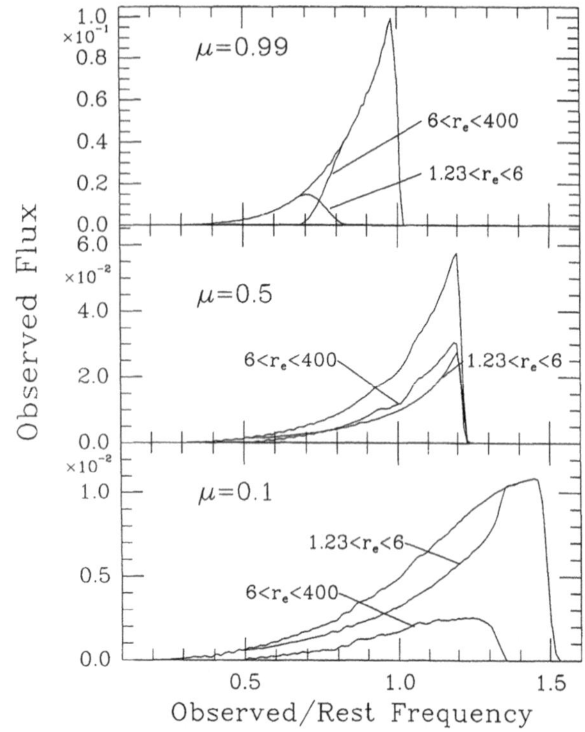

Fig. 13.11 The profile of a spectral line reflected by a disk rotating around a black hole. The shape depends on the angle under which the disk is seen. Here μ = cos θ, where θ equals the angle between the line of sight and the disk; in the top diagram the disk lies almost in the line-of-sight; in the lowest diagram the disk and the line of sight have an angle of 87° (Laor 1991)

Over the years it was realised that there had to be one more component: some of the hard X-ray photons produced by the corona were reflected by the accretion disk, and the arrival of these hard photons on the much cooler disk caused it to emit fluorescent radiation: spectral lines were predicted especially those by iron ions that had lost practically all their electrons except one or two. The profile of the lines would be strongly affected by the Doppler effect, because the reflecting disk was rotating rapidly. Model calculations showed that the line profile would contain information on the inner parts of the disk, those parts closest to the neutron star or to the black hole (Fabian et al. 1989). Figure 13.11 shows such a profile that followed from a calculation. Close examination shows that under favorable conditions (seeing the disk at right angle to the line of sight) one can distinguish the radiation from close to the gap from the radiation of the outer parts.

In some X-ray binaries a jetstream was discovered—see Fig. 13.13 further down. It could be recognised in the X-ray spectra but also at radio wavelengths. The radio spectrum and radio polarisation showed that radiation produced was synchrotron emission—see Chap. 1. In a given source the jet could be switched off and turned on later. Slowly, data were accumulated that in the end showed that the jet was only present when the X-ray binary was in its 'hard state', in which the spectrum is dominated by photons with an energy above 10 keV (Fender 2006).

The curious case of SS433 was discussed in Chap. 7: a star that produced jets with a velocity of 26% of the speed of light. For a decade or so it remained a unique object. But from 1992 on, a few more similar objects were found, e.g. GR1915+105 en GRO J1655-40 = Nova Scorpii 1994, in which the velocity of the jets reached 92% of the speed of light (Mirabel and Rodríguez 1999). In all cases, the object producing the jets was an X-ray binary. The objects were called 'microquasars' because they contained three essential sub-structures that are also found in true quasars: a black hole surrounded by an accretion disk and two jets in opposite directions. The name 'microquasar' suggested, see Mirabel and Rodríguez (1999), that a microquasar was a scaled-down version of a 'macroquasar'. It was a bold thought: a quasar contained a black hole that was sometimes a billion of times heavier than the Sun, and the black hole in microquasars had a mass of at best ten times that of the Sun.

The suggestion had great potential. Microquasars varied in time; jets appeared, but could also disappear. In quasars this might happen as well, but the time scale of the changes could be much too long for us to detect. If the production of jets in nearby X-ray binaries could be traced to some mechanism in a rapidly evolving object, the same mechanism might operate in quasars as well. A review paper in 2017 expresses some optimism that such scaling is allowed. Let me quote from a 2017 review paper (Romero et al. 2017): 'Over the last decades, it has become clear that many aspects of accretion physics are mirrored across the black hole mass scale, irregardless of the size difference and accretion environment'. 'Mass-scaling' is probably possible (McHardy et al. 2006).

The analysis of X-ray spectra and of their variability led over the years to a comprehensive picture. A schematic model dating from 2010 is shown in Fig. 13.12.

The overall spectrum is, as Fig. 13.12 shows, a composite of various substructures (accretion disk; halo of relativistic electrons) plus reflected radiation. Each of these will vary in time. The variability of the spectrum is explained by assuming that the disk and the corona may vary separately in strength.

Figure 13.13 gives another, basically similar but more colorful sketch of an X-ray binary. Note that the compact object could be either a neutron star or a black hole.

An early and unexpected discovery was the existence of 'Quasi-Periodic Oscillations' or QPOs: the presence of broad lines in the X-ray spectra of some sources. The signal was pulsing at a frequency that varied slightly, somewhat like a violinist producing a not-quite pure tone but 'whines' around the correct tone; hence the name 'quasi-periodic oscillations' (van der Klis et al. 1985). The frequencies were close to that of the rotation in the inner parts of the accretion disks. It was therefore generally assumed that QPOs were generated near the inner edge of the accretion disk. This would place their source very close to the surface of the neutron star or to the Schwarzschild radius of a black hole, and that was an area where some peculiar effects might be found predicted by the theory of general relativity (van der Klis 2005).

Although several mechanisms have been put forward to explain QPOs, there appears to be no agreement on any special model and, consequently, the QPOs may have given less insight than they originally promised–but look at the end of this chapter.

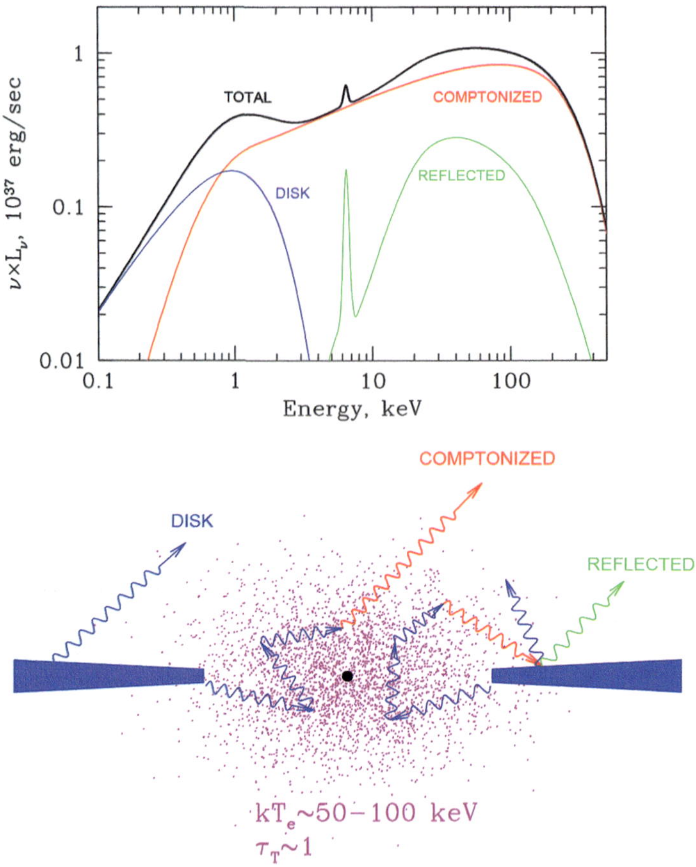

Fig. 13.12 (Lower figure) A sketch of a X-ray binary seen edge-on. The central object (a point) is either a neutron star or a black hole. The photons are shown as wave-like curves. In blue the accretion disk. In red the corona of relativistic electrons that transforms (relatively) cool, blue photons (emitted by the disk) into hard X-ray photons (in red). The disk reflects some of the hard photons as is indicated in green. (Upper figure) The overall X-ray spectrum of the X-ray binary and its decomposition in three parts: radiation emitted by the accretion disk (blue), by the corona (red) and by the reflected photons (green). The green spectrum contains a spike: these are reflected spectral lines; see text (Gilfanov 2010)

As has been said before, it is sometimes possible to calculate the mass of both objects within a binary star. Figure 13.14 shows the mass of a large number of neutron stars. The masses practically all lie between 1.1 and 2.0 M_{zon}. The measurements of the neutron stars with a higher mass are uncertain because the mass of the donor star cannot be determined accurately.

A noteworthy measurement is that of the gravity of pulsar PR J1614-2230 where coincidentally the donor star periodically obscures the pulsar. During the eclipse, the pulsar radiation passes close to the donor; during this transit, the gravity of the

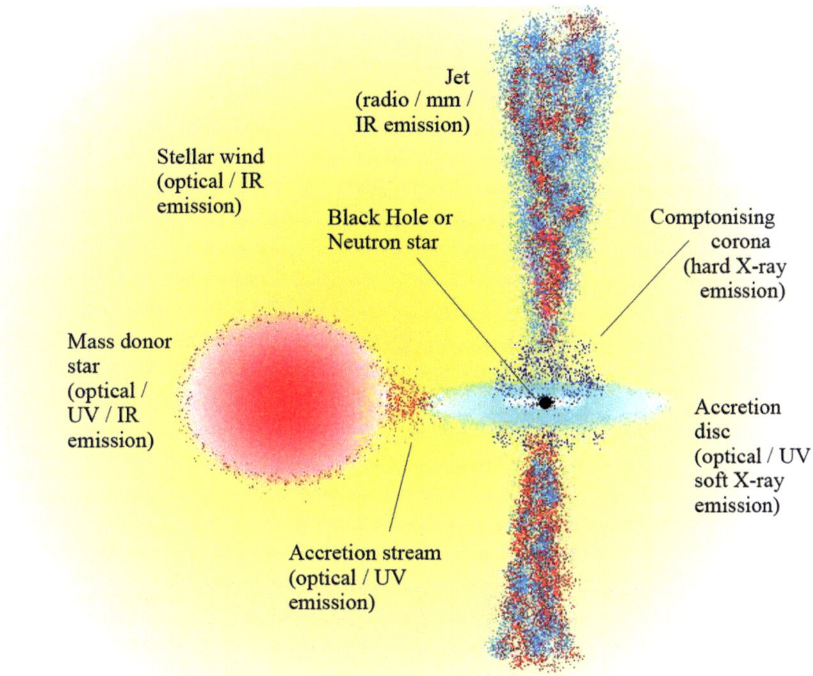

Fig. 13.13 'Water color impression' of a gamma source. The main players are the red donor star, the light blue accretion disk, the neutron star/black hole and the two jetstreams. A 'corona' of dark purple dots has been drawn around the neutron star/black hole that indicate that photons emitted by the disk gain a great deal of energy through collisions with high-energy electrons (known as the 'inverse Compton effect') that have accelerated in the vicinity of a neutron star or black hole. Aside from this, the jets are notable. These are not always present. Remarkable detail: the central object might be a neutron star but it could equally be a black hole: the overall structure is not sensitive to which of the two is present (Fender 1999)

donor star slows the radiation from the pulsar (see Fig. 13.15). Measuring this delay, known as the Shapiro effect, makes it possible to accurately measure the mass of the donor star; because the mass of the total can also be measured (from the circulation period and the pulsar-donor distance), the mass of the pulsar can also be determined. In this case, the analysis yielded a neutron star mass of $1.97 \pm 0.04\ M_{zon}$.

Black Holes

A review in 2006 (Remillard and McClintock 2006) presents a list of 17 X-ray binaries for which the mass of the compact object was determined with enough accuracy to assure that the compact object is a black hole. The lowest mass was in the range of 3.7–5.0 Msun, the highest mass between 10 and 18 Msun. A graphical description of the confirmed black holes is given in Fig. 13.16. The list of confirmed black

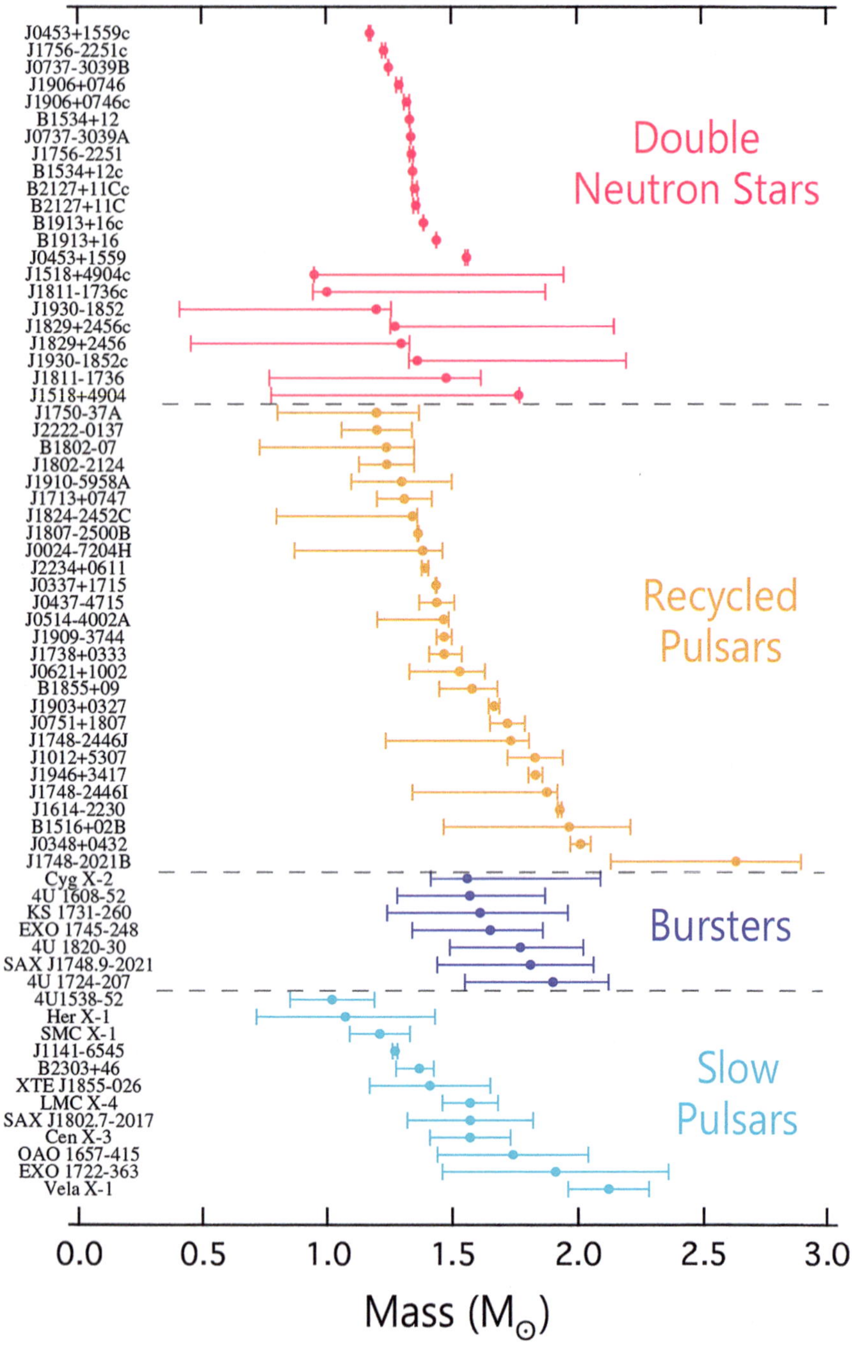

Fig. 13.14 The mass of the neutron star in a number of X-ray binaries. The stars are divided into several groups, but the differences between them do not manifest themselves in systematic differences in mass (Özel and Freire 2016)

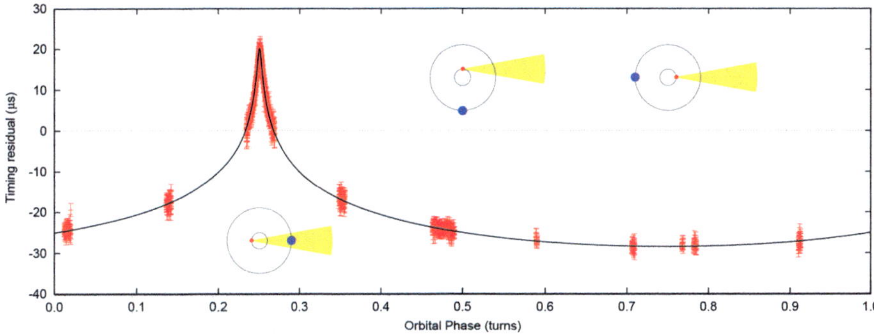

Fig. 13.15 During the rotation of the pulsar around another star the pulsar signal will be delayed by gravity of the other star; the diagram shows the delay time in milliseconds as measured during one revolution (Demorest et al. 2010)

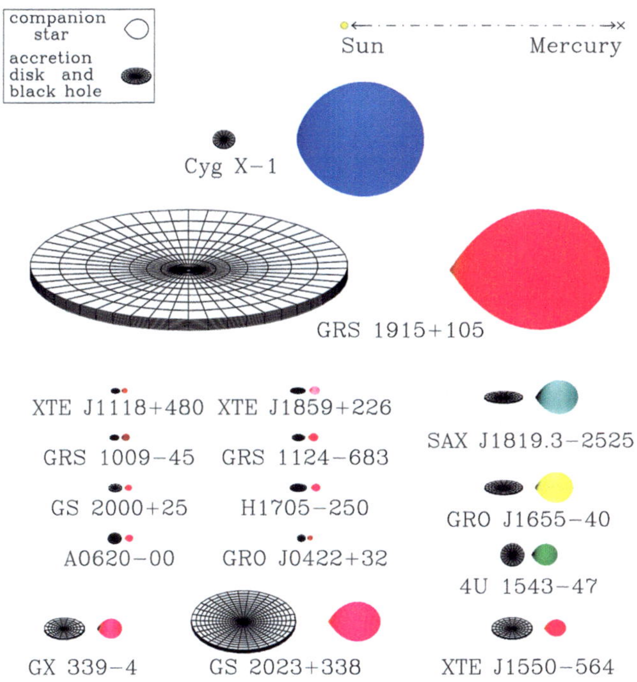

Fig. 13.16 X-ray binaries containing a black hole, all drawn to scale; the distance Sun-Mercury, is shown top-right. In color the donor star and in black the accretion disk with the black hole at its center. Three large objects and 14 small ones are shown. GSR1915+105 is also known as a micro-quasar because of its strong jet- see below (Remillard and McClintock 2006)

holes is certainly not representative for all black holes—the selection of the sources is not random and is probably highly biased.

Schwarzschild's analysis showed that a black hole has a radius indicated by $r_g = 2GM/c^2$. According to the GRT, there are no stable circular orbits around the black hole with a radius smaller than $3r_g$. Objects that arrive within this distance will always move into the black hole. The circular orbit with radius $3r_g$ is the smallest stable orbit and is called the 'Innermost Stable Circular Orbit' or ISCO. The existence of an ISCO has become of major interest for the observations of black holes. Neutron stars may also have an ISCO just outside the stellar surface but it does not play a significant role in the observations.

In 1963, Kerr proved that the GRT allows the existence of rotating black holes—see Chap. 4. The rotating black body is no longer spherical but somewhat ellipsoidal. The spin J (angular momentum) must have value below the maximum $J_{max} = GM^2/c$. For the further discussion it is important that the ISCO of a rotating black body is smaller than for a non-rotating, or Schwarzschild black body. This property is used to measure the spin of a black hole as was successfully done from about the year 2000.

There are two ways of measuring r_{ISCO}: via the shape/form of the iron Kα line (see the discussion above on fluorescence effects) and via spectral fits, where the spectrum depends on the size of the hole in the disk. It is assumed that the radius of this hole equals r_{ISCO}. Measurements through both methods lead to values that agree with one another; this gives confidence in the individual results. In a list of 12 successful measurements from 2014 (Reynolds 2014), there are three where the spin is more than 95% of the maximum. Again, one cannot draw statistical conclusions from this sample.

In 1972 the star HDE 226868 was discovered to be one component of the X-ray binary Cygnus-X1 with a companion that had a mass too high to be a neutron star. It was called a 'candidate black hole', because some doubts remained about its mass. Almost 40 years later and with better measurements at all wavelengths, these hesitations can be cast aside. Radio and X-ray astronomers, mainly from Harvard, determined in 2011 the distance accurately using VLBI techniques, collected optical spectra that gave an accurate determination of the mass of the donor star and the accreting object. They collected X-ray measurements made during the few and short times when the spectrum was dominated by the accretion disk. These measurements gave them the radius of the central hole in the accretion disk, which they equated with the radius of the ISCO. From all these observations they determined the distance (1.86 ± 0.1 kpc), the mass of star (19.2 ± 1.9 M_{sun}) and of the compact object (14.8 ± 1.0 M_{sun}) and the angle between the line of sight and the accretion disk (27.1 ± 0.1 deg) (Reid et al. 2011; Orosz et al. 2011; Gou et al. 2011). The radius of the ISCO was so small that the black hole had to spin at maximal speed. The last conclusion was confirmed by an analysis of the profile of the iron lines as explained above; see Fig. 13.17.

In 1918 the Austrian physicists Lense and Thirring studied the equations of the GRT and concluded that a spinning black hole will drag time and space around, an effect since called the 'Lense-Thirring effect'. It was a theoretical concept that

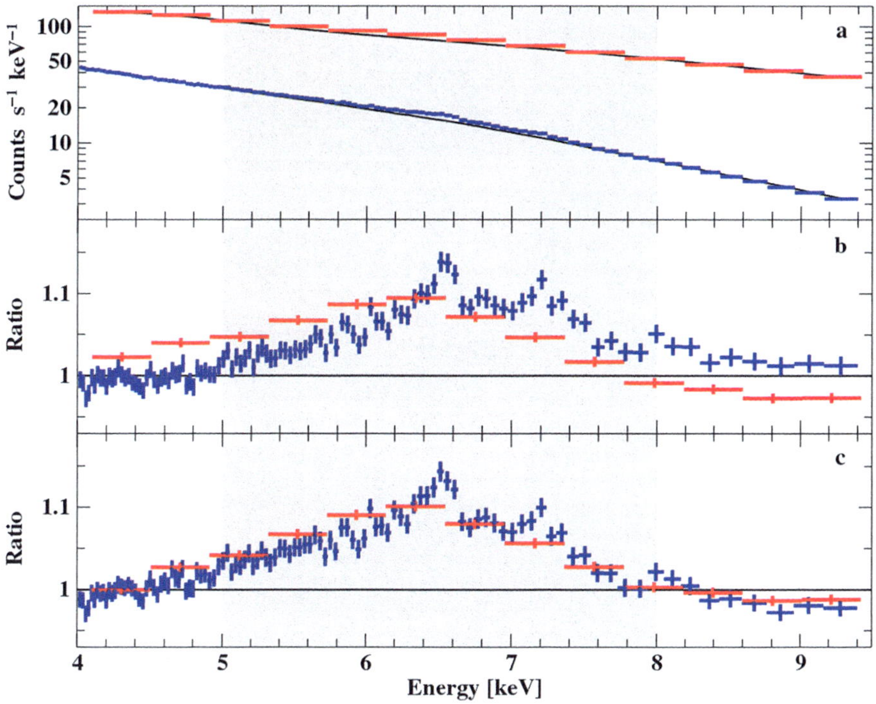

Fig. 13.17 The profile of the iron Kα-line in Cygnus X-1 measured with ESA's XMM-Newton observatory. In the upper diagram (a) the spectrum as measured directly; in the two lower diagrams (named b and c) background emission has been subtracted and the line profiles appeared. The line profile is interpreted by assuming that the line is due to the reflection of hard photons on the accretion disk and the subsequent florescence of the disk by emitting X-ray photons of iron atoms (the Kα-line). The line profile is asymmetric. The long wing to lower photon energies (in spectroscopic language a 'redshift') shows the high velocities in the inner part of the disk (Duro et al. 2011)

could never be checked by measurements. The effect predicts that the rotation axis of the disk should precede around the rotation axis of the black hole. This should make the fluorescent Kα-line somewhat variable in time. In 2016 (almost a century after the prediction!) there is a claim that the effect has been detected—see Fig. 13.18 and read the legend.

Ultraluminous X-Ray Sources (ULXs)

There is a small class of very strong X-ray sources that emit up to a million times more radiation in X-rays than the Sun at all wavelengths together. The highest luminosities exceed the Eddington limit: the maximum amount of light that a star can emit before it is blown apart by light pressure. Ultraluminous X-ray sources (ULXs in short) have so far been found only in other galaxies and usually only a

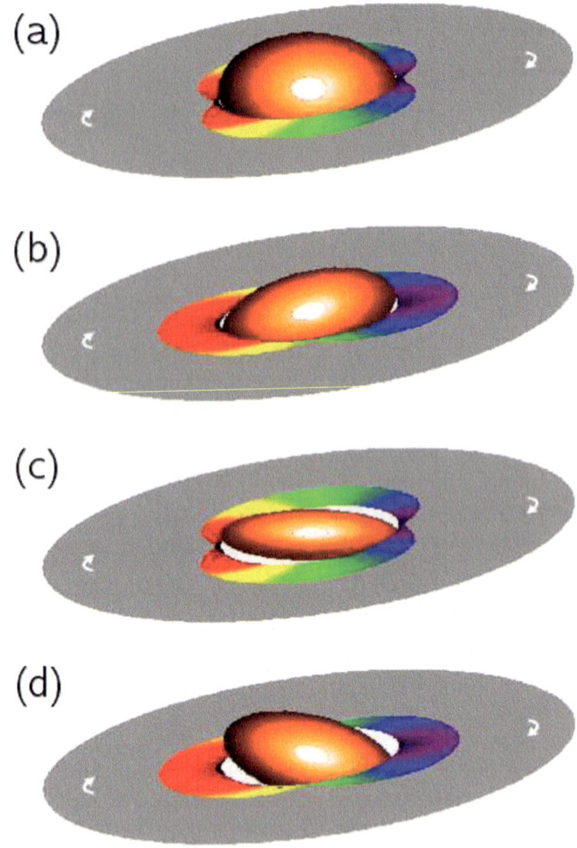

Fig. 13.18 The Lense-Thirring effect predicts that the axis of rotation of the accretion disk will make a precession around the rotation axis of the black hole. The precession is faster in the inner parts of the disk and this will show up as variability in the Kα-line of iron. In this highly schematic figure four different moments show how the inner part of the disk wobbles. This causes a variable heating of more inward parts of the accretion disk that may have been seen in a (minute) change in the spectrum of the Kα-line of iron (Ingram et al. 2016)

very few. The nature of these sources is unclear at this time. Even this basic question does not yet have an answer: do these sources have a neutron star at their center or a black hole? (Colbert and Mushotzky 1999; Swartz et al. 2004; Bachetti et al. 2014; King and Lasota 2016).

References

Abdo, A.A., Ajello, M., et al.: Astrophys. J. Suppl. **208**, 17 (2013)
Bachetti, M., Harrison, F.A., et al.: Nature. **514**, 202 (2014)
Balbus, S.A., Hawley, J.F.: Astrophys. J. **376**, 214 (1991)
Gaia Collaboration; Babusiaux, C.; van Leeuwen, F.; et al.; A&A v616 A10 (2018)
Becker, W., Truemper, J.: Astron. Astrophys. **326**, 682 (1997)
Caraveo, P.A., Bignami, G.F., et al.: Astrophys. J. **461**, L91 (1996)
Colbert, E.J.M., Mushotzky, R.F.: Astrophys. J. **519**, 89 (1999)
Demorest, P.B., Pennucci, T., et al.: Nature. **467**, 1081 (2010)

Duncan, R.C., Thompson, C.: Astrophys. J. **392**, L9 (1992)
Duro, R., Dauser, T., et al.: Astron. Astrophys. **533**, L3 (2011)
Esin, A.A., McClintock, J.E., Narayan, R.: Astrophys. J. **489**, 865 (1997)
Fabian, A.C., Rees, M.J., et al.: Mon. Not. R. Astron. Soc. **238**, 729 (1989)
Fender, R.: In: Proc. ESO workshop 'Black holes in binaries and galactic nuclei' (1999)
Fender, R.: In: Lewin, W., van der Klis, M. (eds.). Compact stellar X-ray sources. Cambridge Astrophysics Series, No. 39. Cambridge University Press, Cambridge, UK (2006)
Ferrario, L., di Martino, D., Gänsicke, B.T.: SSRv. **191**, 111 (2015)
Fontaine, G., Brassard, P., Bergeron, P.: Publ. Astron. Soc. Pac. **113**, 409 (2001)
Fruchter, A.S., Stinebring, D.R., Taylor, J.H.: Nature. **333**, 237 (1988)
García-Berro, E., Oswalt, T.D.: New Astron. Rev. **72**, 1 (2016)
Gianninas, A., Bergeron, P., Ruiz, M.T.: Astrophys. J. **743**, 138 (2011)
Gilfanov, M.: Lect. Notes Phys. **794**, 17 (2010)
Gou, L., McClintock, J.E., et al.: Astrophys. J. **741**, 85 (2011)
Greiner, J., Hasinger, G., Kahabka, P.: Astron. Astrophys. **246**, L17 (1991)
Haberl, F., Sturm, R.: Astron. Astrophys. **586**, A81 (2016)
Halpern, J.P., Holt, S.S.: Nature. **357**, 222 (1992)
Harding, A.K.: Fr. Phys. **8**, 679 (2013)
Helder, E.A., Vink, J.: Astrophys. J. **686**, 1094 (2008)
Hermsen, W.: SSRv. **36**, 61 (1983)
Hessels, J.W.T., Ransom, S.M., et al.: Science. **311**, 1901 (2006)
Holberg, J.B., Oswalt, T.D., et al.: Mon. Not. R. Astron. Soc. **462**, 2295 (2016)
Ingram, A., van der Klis, M., et al.: Mon. Not. R. Astron. Soc. **461**, 1967 (2016)
Ivanova, N., Justham, S., Chen, X., et al.: Astron. Astrophys. Rev. **21**, 59 (2013)
Kargaltsev, O., Cerutti, B., et al.: SSRv. **191**, 391 (2015)
Kaspi, V.M., Gavriil, F.P., et al.: Astrophys. J. **588**, L93 (2003)
Kaspi, V.M., Roberts, M.S.E., Harding, A.K.: In: Lewin, W., van der Klis, M. (eds.). Compact stellar X-ray sources. Cambridge Astrophysics Series, No. 39. Cambridge University Press, Cambridge, UK (2006)
Kepler, S.O., Pelisoli, I., et al.: Mon. Not. R. Astron. Soc. **455**, 3413 (2016)
King, A., Lasota, J.-P.: Mon. Not. R. Astron. Soc. **458**, L10 (2016)
Kouveliotou, C., Dieters, S., et al.: Nature. **393**, 235 (1998)
Kouveliotou, C., Strohmayer, T., et al.: Astrophys. J. **510**, L115 (1999)
Laor, A.: Astrophys. J. **376**, 90 (1991)
Liu, Q.Z., van Paradijs, J., van den Heuvel, E.P.J.: Astron. Astrophys. **442**, 1135 (2005)
Markoff, S., Falcke, H., Fender, R.: Astron. Astrophys. **372**, L25 (2001)
McHardy, I.M., Koerding, E., et al.: Nature. **444**, 730 (2006)
Mereghetti, S.: Astron. Astrophys. Rev. **15**, 225 (2008)
Mirabel, I.F., Rodríguez, L.F.: Annu. Rev. Astron. Astrophys. **37**, 409 (1999)
Olausen, S.A., Kaspi, V.M.: Astrophys. J. Suppl. **212**, 6 (2014)
Orosz, J.A., McClintock, J.E., et al.: Astrophys. J. **741**, 84 (2011)
Özel, F., Freire, P.: Annu. Rev. Astron. Astrophys. **54**, 401 (2016)
Paczynski, B.: IAU Symp. **63**, 75 (1976)
Parsons, S.G., Gänsicke, B.T., et al.: Mon. Not. R. Astron. Soc. **426**, 1950 (2012)
Pavlov, G.G., Sanwal, D., Teter, M.A.: IAU Symp. **218**, 239 (2004)
Reid, M.J., McClintock, J.E., et al.: Astrophysical Journal. **741**, 83 (2011)
Remillard, R.A., McClintock, J.E.: Annu. Rev. Astron. Astrophys. **44**, 49 (2006)
Reynolds, C.S.: SSRv. **183**, 277 (2014)
Romero, G.E., Boettcher, M., et al.: SSRv. **207,** 5 (2017)
Swartz, D.A., Ghosh, K.K., et al.: Astrophys. J. Suppl. **154**, 519 (2004)
Tananbaum, H.: IAU Circular 7246 (1999)
Tremblay, P.-E., Bergeron, P., Gianninas, A.: Astrophys. J. **730**, 128 (2011)

Tauris, T.M., van den Heuvel, E.P.J.: In: Lewin, W., van der Klis, M. (eds.). Compact stellar X-ray sources. Cambridge Astrophysics Series, No. 39. Cambridge University Press, Cambridge, UK (2006)
van den Heuvel, E.P.J., Bhattacharya, D.: Astron. Astrophys. **262**, 97 (1992)
van der Klis, M.: Astron. Nachr. **326**, 798 (2005)
van der Klis, M.; Jansen, F.; et al.: Nature **316**, 225 (1985)
Werner, K., Herwig, F.: Publ. Astron. Soc. Pac., vol. 118, p. 183 (2006)
Winget, D.A., Kepler, S.O.: Annu. Rev. Astron. Astrophys. **46**, 157 (2008)

Chapter 14
1995–2015: Transient Phenomena, Mergers and Explosions

On a cloudless night, just go outside and enjoy the majesty of the Heavens. The moon is up and bathes everything in a faint, pale color, a hue that so many artists have tried with more or less success to capture. Moon and stars shift silently, peacefully and slowly from East to West. All is silent and you are caught up in a sense of infinity and eternity.

This is also what the ancient Greek philosophers experienced in their times, the last four centuries BC: the sky was beautiful and eternal. They called it the 'kosmos', a word that in translation means 'well ordered'. The sky was infinite and unpolluted by the variability, the finiteness and the detritus of life on Earth. This philosophy held until the end of the sixteenth century. It was a great shock to Danish astronomer Tycho Brahe when in 1572 he saw a bright star (we now know that it was a supernova) where one had never been seen before. To confirm that this went against everything he knew, he tested whether the object was within the atmosphere of the Earth: it was not. The conclusion was clear: the sky was not eternal; it changed! A fundament of Brahe's thinking (and that of his colleagues) was destroyed. He had had an inkling of this when he had concluded earlier that comets were not phenomena high in the Earth's atmosphere (as Aristotle had said—and his work was more or less sacred) but farther away than the Moon.

Let us go inside and leave the beautiful and serene night. Yes, the sky is majestic, but changes do occur. X-ray astronomers discovered very early that their X-ray sky was hardly constant and mostly variable. The more energetic the photons were, the more variable the sources: think of the many Gamma Ray Bursts, for example. But in fact, sudden and short-lasting events, 'transients', have been detected at all other wavelengths; most transient objects were totally new. The last two decades have been particularly fruitful. In this chapter, I will discuss some of these transient phenomena in the sky even though most of them are not yet understood.

The first groups dedicated to finding transients responded in around 1990 to a very specific suggestion for how one might find free-roaming planets and black holes by detecting the sudden magnification of a star. Then, in around 1996, came the programs to detect and track supernovae and to use them as a yardstick to measure

H. J. Habing, *The Birth of Modern Astronomy*, Historical & Cultural Astronomy,
https://doi.org/10.1007/978-3-319-99082-8_14

the distance to distant galaxies. After 2000, a general hunt for 'optical transients' was begun when more new groups tried their luck. New types of 'novae' were detected that reached an absolute magnitude M_v of about −12, in between novae (M_V between −5 and −8) and supernovae (brighter than −14). The successes are somewhat bewildering and it was clearly too early in 2015 for a review in a book like this. I will, however, give a few examples of new phenomena that it was possible to explain. I will start with the first and furthest developed program, the search for transients caused by microlensing.

14.1 Microlensing: Searching for Invisible Objects

After 'dark matter' had been discovered in the 1970s, theoreticians immediately speculated that this was an unknown kind of matter: elementary particles that emitted no light, nor any other kind of electromagnetic radiation. The common astronomer was more down to earth and looked for an explanation in the form of dark objects that did not emit electromagnetic radiation (or only very weakly): black holes, neutron stars, very old and cool brown dwarfs or free-floating planets. There could be plenty of them and they would never have been seen. If one wanted to detect them, it could only be done through the gravity that they exerted. How could such a silent and probably large majority be found?

In 1986 Polish astrophysicist Bodo Paczynski (Paczynski 1986) put forward a plan, based on the existence of gravitational lenses. These have been introduced in Chap. 8. The only ones known were enormous concentrations of matter such as galaxies and clusters of galaxies. But a single star could also act as a gravitational lens; after all the Sun had been proven to be a gravitational lens in the famous 1919 test of Einstein's General Theory of Relativity. Paczynski proposed the following: Suppose that invisible object A passed closely in front of star B, then it might happen that for a short while A became a gravitational lens and cause briefly object B to get brighter. Einstein had already conceived of this eventuality in the 1920s and had rejected it as impractical because it would be too rare an occurrence to be identified.

In 1986 things were different: computers had been invented. Paczynski proposed making a photograph of a field with a great many stars and comparing this photo the following day with an earlier photograph. If one of the stars were brighter than in the earlier photo, this star should be kept under observation: it could be that a dark object passed the star and made the star brighter. Paczynski called such an event 'microlensing'. He warned that when such an event had been detected, you could draw very few conclusions, mainly because it could be a relatively light dark object passing the star at a short distance, or it could be a heavy black hole passing at a larger distance. But if you had identified many 'transits', you could use statistical methods to draw a conclusion on the average mass of the dark object and how many of these there were. It would be only a small snippet of information, but it was infinitely better than 'nothing'.

A lot of effort was called for to put Paczynski's plan into action. New instruments, hardware and software needed to be developed but the results could be spectacular. In a short time, there were three consortia of researchers testing Paczynski's proposal, each with its own strategy. There was an American-Australian team headed by Charles Alcock from Berkeley; they wanted to look for MACHOs: 'massive astrophysical compact halo objects'. A team in Poland headed by A. Udalski carried out the 'Optical Gravitational Lensing Experiment' (OGLE'), mainly in Chile; they happily added their fellow countryman Paczynski to their club. In France, a team of 35 participants worked together on the 'Expérience de Recherche d'Objets Sombres' or EROS. I believe I remember that the French group could not agree who was in charge; it was too democratic. I also fear that the choice of these three rather humorous acronyms will confirm the outside world's opinion that all astronomers are nerds ('me too'?). But scientifically the groups were successful: Fig. 14.1 shows a recording of a gravitational lens passing in front of a star.

Years later several more consortia came forward and—what was especially important—worldwide networks emerged that, after an alert, could observe the source continuously for many days, 24 hours a day.

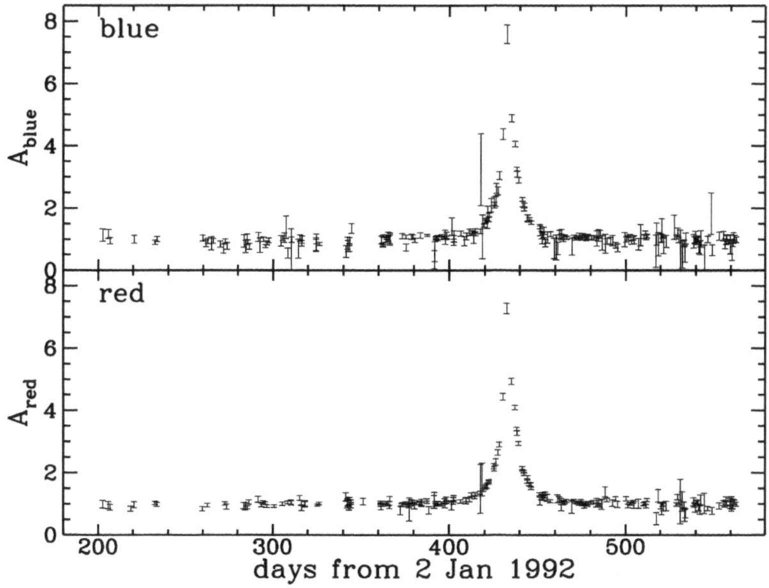

Fig. 14.1 A star brightened because a material (but invisible) object passed in front of it. The event is observed twice at the same time through two differently colored filters and is equally strong in both cases—as it should be when the brightening is due to a gravitational lens. The date of the observation is indicated horizontally, and the number of magnitudes that the star has risen in brightness is indicated vertically. The maximum is a magnitude increase of 7.8, and this implies that at its maximum the light of the stars has increased more than 1,300 times (Alcock et al. 1993)

In a retrospective in 2012 (Mao 2012) about what microlensing had achieved over many years, we can read that, in spite of the at times jubilant tone, the result was very modest—although not unimportant. Hundreds of passages had been recorded and analysed. And these analyses slowly but surely went deeper than was initially foreseen. The original objective had been achieved: determining whether or not dark matter consisted of 'massive astrophysical compact objects'. The answer was 'No'; at most 2% of the dark matter might be made up of such objects. It confirmed a conclusion that had already been drawn on very different grounds. A dozen or so 'candidate black holes' had been identified that moved singly through space, but this number was probably less than the number of black holes that had already been found in close binary stars (see Chap. 13).

14.2 Pseudo Supernovae or Supernova Impostors

In 1961 a supernova, called SN1961V, was seen that exploded in a nearby galaxy, NGC 1058. Zwicky inspected its light curve and found it sufficiently different from all other light curves to allocate it a special position in his classification system of light curves: type V. It was lonely there: SN1961V remained the only supernova of this type V. Another remarkable point was the exceptionally bright star that continued to be present at the location of the supernova after the explosion. Had there originally been two, one of which produced the supernova? In 1989 a new evaluation of all the data, and some new observations of the environment, led to quite a different conclusion: this was not a supernova but an explosion of an LBV star. Similar stars had been found in the Milky Way: η Carinae and P Cyg (see Chap. 4) (Goodrich et al. 1989). Both stars had experienced a heavy explosion, one in 1837 and the other in about 1603 and both had survived the disaster. In a true supernova, the authors said, the star is completely destroyed; at best an extremely faint compact star remains or a black hole. This not being the case, the authors said, shows that these three objects should not be called a 'supernova' but a 'supernova impostor'. Figure 14.2 shows how the authors explain the light curve of SN1961V. The authors also commented on the decline of the light curve at the right end of the horizontal line: this decline, they said, is only apparent; the object forms a dense and dark cloud of small particles that absorb all the light. The implication is that the energy of this absorbed light should then reappear as infrared radiation, as happened in the case of η Carinae (Westphal and Neugebauer 1969).

After the fully automated telescopes became available, more and more supernovae were detected, and the number of 'supernova impostors' grew also. In 2011 a list was compiled of 14 SN impostors detected during the years 1999–2010 (Smith et al. 2011). The general opinion has become that these are all high-mass stars, LBVs, that experience explosions as the consequence of an inherent instability, the cause of which is unknown. The explosion is severe, but the star is not destroyed. In between explosions, the stars remain more or less equally bright.

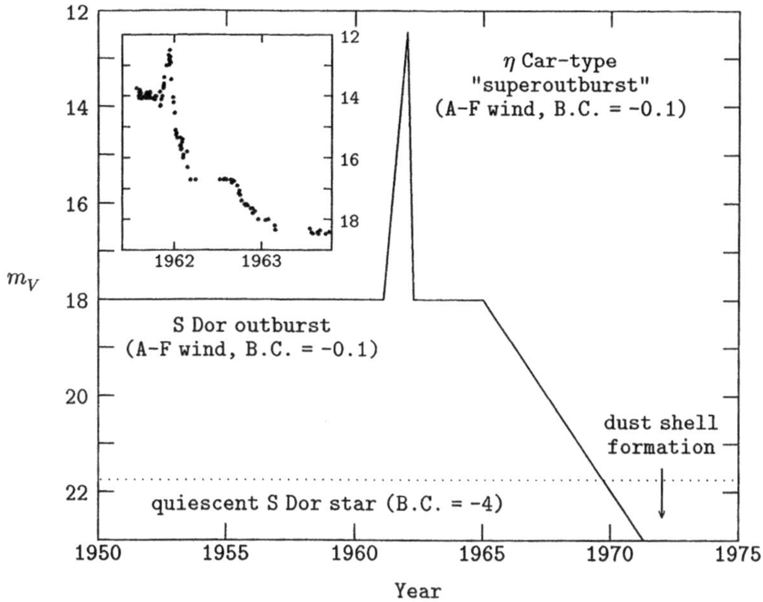

Fig. 14.2 Light curve of 'Supernova 1969V', interpreted as the light curve of an LBV with major outburst (Goodrich et al. 1989)

14.3 Red Novae: Are we seeing a 'Common Envelope Event?'

In 2002 IAU Circular 7785 announced the discovery of a new bright star, a nova, in the constellation of Monoceros (Unicorn). This 'Nova Monocerotis' coincided with a faint variable star V838 Mon that suddenly brightened by 4,000 times (9 magnitudes). The announcement created quite some activity: the object was observed and detected at many observatories—23 circulars appeared before the first scientific analysis was published. The aftermath of this 'nova' was quite unusual: after some decline, the object remained rather stable at a high luminosity whereas the spectrum of the star cooled to what for a star was a very low temperature; in the final stage the spectrum resembled that of the coolest stars known, a star of spectral type M and a surface temperature of a few thousands degrees kelvin (see Fig. 14.3). This was contrary to the behaviour of other novae: the atmosphere of novae had always become hotter during the decline. V838 Mon was not the first object of its kind: two earlier such novae had already been found exhibiting similar behavior, one of them in the Andromeda galaxy, M31.

A welcome surprise was the presence of a 'light echo': different parts of the object become brighter at slightly different times. This time delay between the different illuminations is caused by the light from the central source being reflected on different parts of the object. From these time delays one can derive the size of the

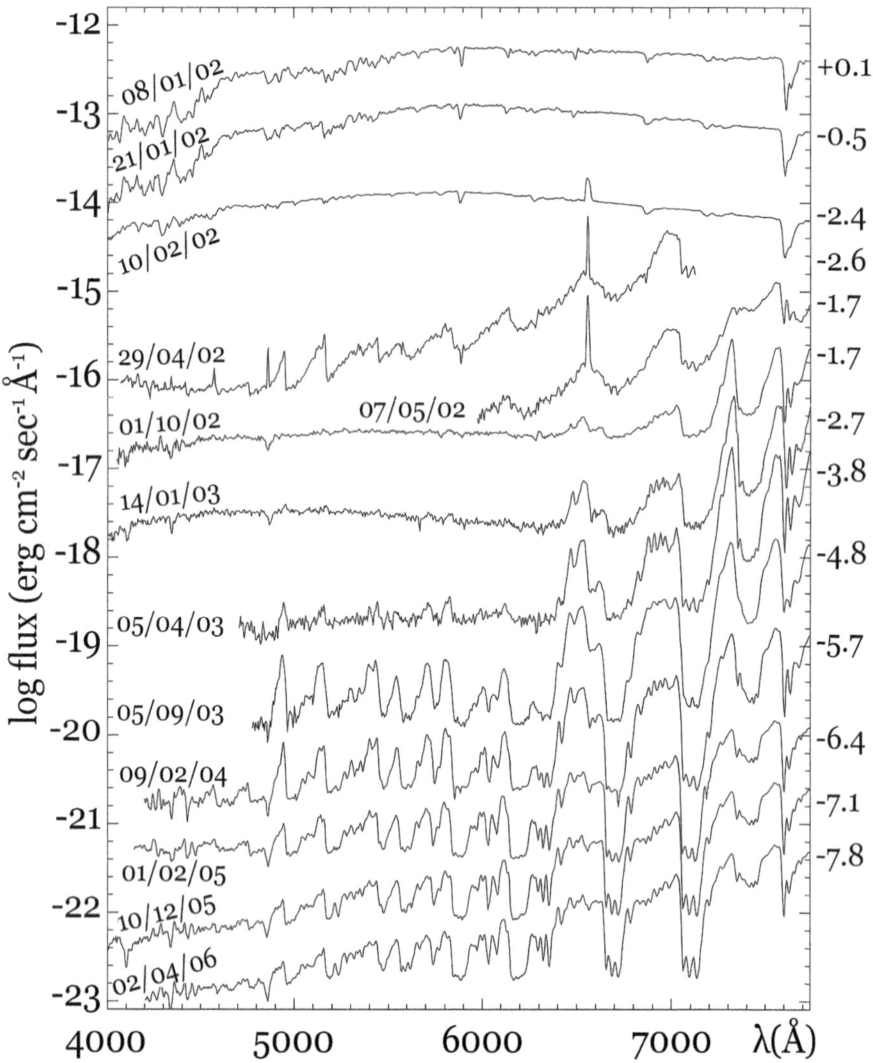

Fig. 14.3 From above to below, optical spectra of V838 Mon at successive points in time. The first spectra show the image of an ordinary star, but progressing downwards, spectra appear with a narrow line of hydrogen, after which broad bands can be seen, all of which are caused by molecules. The star surface has become so cold that molecules continue to exist and are not rapidly broken down. The lowest spectra are those of the coldest red giants (Munari et al. 2007)

objects in light seconds, and thus in kilometers, and from this measurement one can derive the distance (if a ball 1 kilometer in size is seen under an angle of 1 arc second, then simple geometry tells you that the distance is 204,000 km). The first time-delay analysis was not successful, but after a set of Hubble Space Telescope measurements had been taken, the distance was given as at least 6 kpc (20,000 light

Fig. 14.4 Recordings made with the Hubble Space Telescope of V838 Mon on 30 April, 20 May, 2 September, 28 October 2002. The exploding star is in a rapidly expanding, hollowed out cavity. The light reflecting on different parts of the shell of the nova takes different times to reach us. These different arrival times thus contain information about the size of the object expressed in light seconds. Knowing the size allows one to calculate the distance to the object (Bond et al. 2003)

years). See Fig. 14.4. This distance meant that Nova V838 Mon had reached a luminosity of a few thousand times more than that of other novae, but still far below the luminosity of a supernovae. The object was given the nickname 'kilonova' and was sometimes called a 'red nova'.

The major differences between the properties of V838 Mon and other, well-studied novae caused quite some discussion and resulted in different explanations about the nature of this object. Ultimately one explanation was generally accepted (Tylenda and Soker 2006): a small star, the remnant of a much heavier main-sequence star, dumped matter on its companion: a main sequence star. This star grew in size and its surface became cooler—precisely as had been calculated in various theoretical studies. It was predicted that ultimately the small star would be completely absorbed by its bigger companion: a merger event. V838 Mon could be the first example of the merging of two stars.

And then, in 2008, another nova was detected, Nova Scorpii 2008 (Nakano et al. 2008) that was given the name V1309 Sco. Its behavior looked quite similar to that of nova V838 Mon (Mason et al. 2010). It was soon realized that this object was in a field in the sky that the OGLE project had observed on an almost daily basis over a period of 7 years and with a photometric accuracy of 0.1 magnitude. Hundreds of accurate photometric measurements were available before the object exploded. The light curve of the object is shown in Fig. 14.5.

The analysis of the OGLE data showed that the object had been an eclipsing binary: two stars rotated around each other at close distance and twice during one revolution one star blocked off the other. It is a well-known class of objects. In the case of V1309 Sco, the two stars began to rotate faster around each other in the month before the explosion. After that event there was only one star left: the former primary star. Two stars had merged into one. After many years of searching and waiting, the merger of two stars had been witnessed (Tylenda et al. 2011; Ivanova et al. 2013).

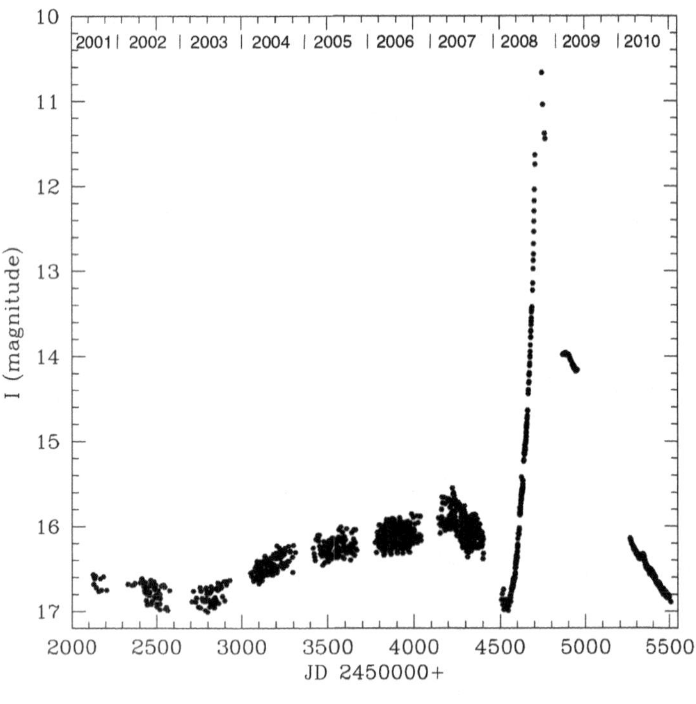

Fig. 14.5 Measurements of the stellar magnitude in the very red I-band. The regularly placed interruptions in the light curve show the times when the object was behind the Sun, as seen from the Earth (Tylenda et al. 2011)

14.4 Radio and Infrared Transients

Fast Radio Bursts

Observations with radio telescopes were made electronically and the measurements were stored. When optical astronomers started to search for transients, radio astronomers realized that they, too, had a lot of archive material and that it would probably be worthwhile systematically searching these archives for 'transients'. Success did not come quickly, but eventually a bona fide 'Fast Radio Burst' was found in the archive of measurements made with the Parkes radio telescope in Australia. The burst in question came from another galaxy (see the legend of the figure), which made it a sensational find (Lorimer et al. 2007).

Figure 14.6 shows how the arrival time of the signal was longer at lower frequencies: the lower the frequency (the longer the wavelength), the later the signal arrived in the receiver. This behavior was well known from pulsar measurements and had been explained conclusively by the delay caused in interstellar space by free electrons.

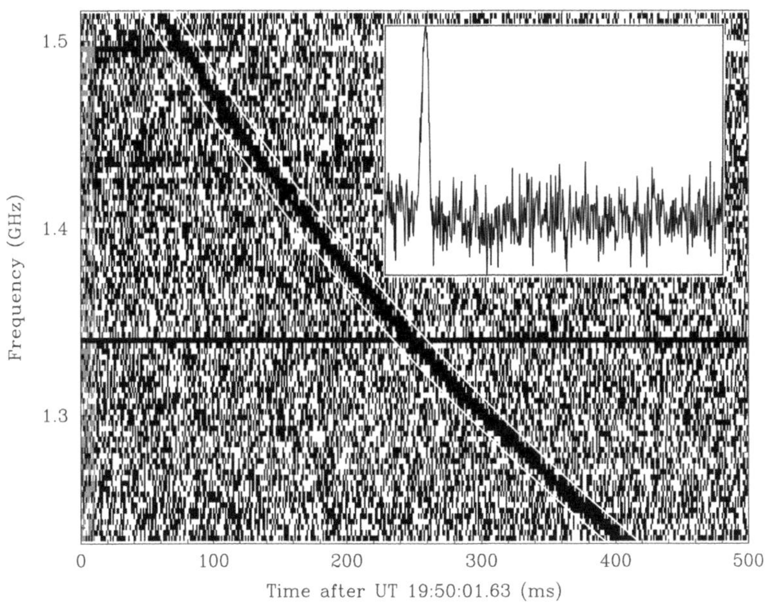

Fig. 14.6 The recording of an isolated radio burst by a radio telescope in Parkes (Australia). The arrival time in milliseconds is shown horizontally. The vertical measurements show the frequency of the receiver, which reduces from top to bottom, or in other words, the wavelength of the radiation recorded increases from top to bottom. The radio burst first arrives at the top of the diagram at the shortest wavelength, thereafter moving downwards. The longer the wavelength, the later the arrival time. This time delay is caused by free electrons between the source and us; how oblique this line is depends on the number of free electrons between the source and us. In this case the number is so large that the signal must come from outside the Milky Way (Lorimer et al. 2007)

In the case of Fig. 14.6, this delay was larger than had been found in pulsars. The conclusion was that the source of this burst must have been extragalactic.

More bursts were found in the archives in Parkes, all of which had an extragalactic origin (Petroff et al. 2015). The only—minor—doubt that remained was that all bursts had been found in the Parkes archives. Could this point to a subtle technical error of the receiver? There was some relief when a burst was also found in the archives of the American Arecibo telescope in Puerto Rico (Spitler et al. 2014), another at the American VLA (Bower et al. 2007), a Japanese radio telescope (Aoki et al. 2014), and the Dutch LOFAR telescope (Stewart et al. 2016). All this removed the last remaining doubts about the reality of FRBs, or Fast Radio Bursts. To allay the concerns of some critical observers, a check was made that a burst was produced when the door of a microwave oven was opened prematurely on the site of the radio telescope. Indeed, it caused a burst in the receiver, but because the arrival time of microwave oven burst was the same at all frequencies, it gave the researchers the confidence that the 'Fast Radio Bursts' measured were a cosmic phenomenon and not a kitchen effect.

The cause of these bursts is completely unknown. It is, however, clear that the observations of FRBs show how many free electrons there are in the space between the galaxy of the burst and us. In themselves, the electrons are not so important because they weigh so little, but for every free electron there is also a free proton: the electrical attraction between protons and electrons is so strong that it is impossible to separate the electrons in a cloud from the protons. Through the electrons, one can find the number of hydrogen atoms along the line of sight and that information is important for cosmology, as will become clear in Chaps. 15 and 16.

14.5 Core-Collapse Supernovae

Many more Supernovae found, and the Complexity increases

The use of robotic telescopes to detect supernovae was very successful, as will be discussed in the following chapters. Hundreds of supernovae were found besides the optical transients discussed above. This produced a wealth of new information, including the realization that there were many more types of supernova than the two in Minkowski's proposition from 1941. For example, a class of exceptionally bright supernovae was found ('superluminous supernovae'). Figure 14.7 comprises a histogram of the maximum luminosities of supernovae. It shows the absence of supernovae with a maximum magnitude between −19 and −22, which suggests that the still brighter supernovae constitute a separate class (Iwamoto et al. 1998). This is also apparent from the appearance of the supernovae shortly after a 'Gamma Ray Burst' (GRB) (see later).

Figure 14.8 shows the light curves of all supernovae, including the superluminous ones.

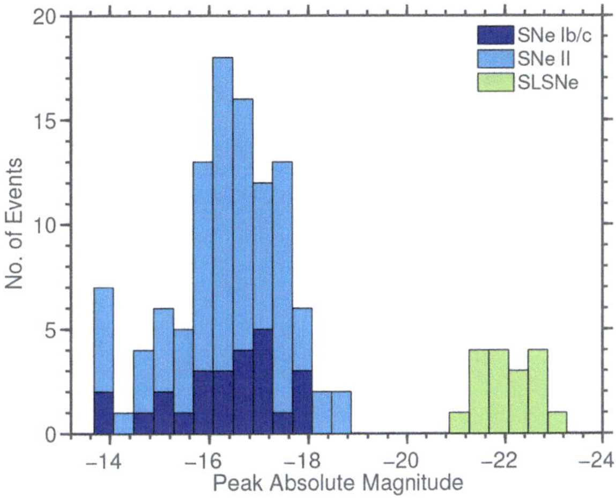

Fig. 14.7 A histogram of the maximum luminosity achieved by different types of core-collapse supernovae. The division shows a gap between −19 and −21, which indicates that the superluminous supernovae, 'SLSNe', are a separate class (Arcavi et al. 2016)

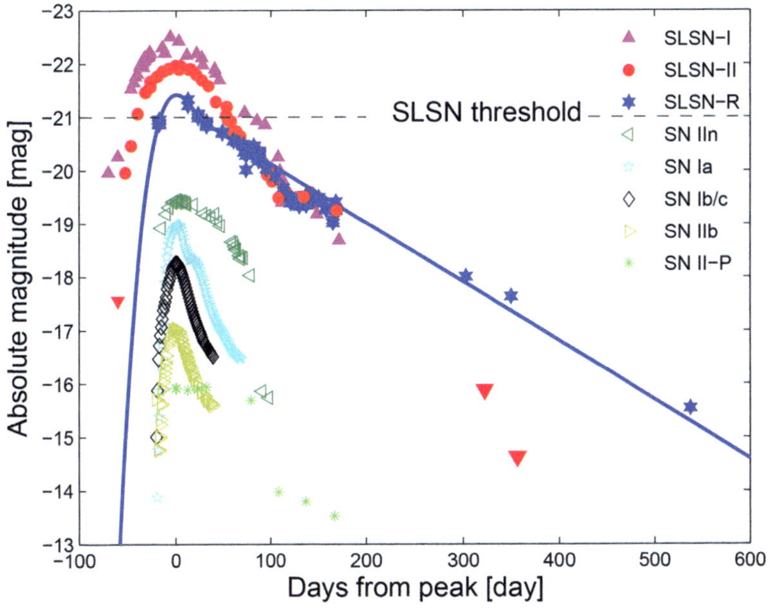

Fig. 14.8 The light curves of supernovae of different types. The maximum luminosity of any supernova is between −22.8 and −16.0. This difference of 6.8 magnitudes means that the most luminous supernova is about 500 times brighter than the least luminous (Gal-Yam 2012)

There is also a close relationship between the maximum luminosity of a supernova and the amount of radioactive nickel that is produced during the explosion (see Fig. 14.9).

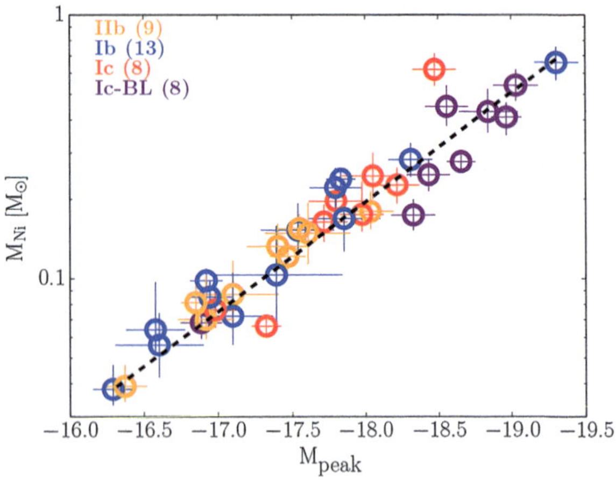

Fig. 14.9 A diagram about supernovae. The greatest luminosity that any supernova has achieved is shown horizontally, and vertically the mass of the nickel atoms produced during the supernova explosion. This mass can be measured from the radio-active energy production that is converted to light (Lyman et al. 2016)

The Star before the Supernova Explosion

We would probably understand a lot more about supernovae if we knew which stars met their end via such an explosion. Unfortunately, supernovae were only discovered after the star had already exploded, which is highly regrettable. If a supernova had appeared, we could try to find images of it in the archives. But that was difficult: most supernovae were in distant galaxies, and existing recordings gave precious little detailed information. In the case of supernovae 1987A, however, it was possible (see below). A second success was SN1993J that exploded in M81, the large galaxy in the Great Bear. The Hubble Space Telescope archive provided observational data of this: the exploding star was a red supergiant with a blue, hot star as companion.

A table from 2015 (Smartt 2015) with 13 successful identifications of stars prior to the explosion, details red—and thus cool—supergiants. The hottest supergiant was only slightly cooler than the Sun. The brightest of these stars, prior to its explosion, emitted a thousand times as much light as the Sun.

The converse question was also interesting: are there massive stars that do not explode as supernovae, but disappear without much ado? The theoretical ideas about supernovae left that option open. Theorists toyed with the thought of ‘failed supernovae’: a hefty explosion that somehow stalled in the interior without the rejection of the outer parts. Were there stars that turned out to have just disappeared, ‘gone without a bang’? That was a difficult question, for which an observational answer seemed impossible: you no longer see a star that was there at first? (Reynolds et al. 2015). A lot of searching has been done, but still no completely convincing examples have been found, although some stars have been detected that are in the process of considerably decreasing in luminosity without a large explosion (Smartt 2015).

SN1987A: Finally a Supernova close by

During the night of 24 February 1987, Ian Shelton, a Canadian astronomer on an observation trip in Chile, saw a bright star in the Large Magellanic Cloud, a star that was not there previously: a supernova. It quickly became brighter and was soon visible to the naked eye. Albert Jones, an amateur astronomer in New Zealand, also detected the supernova at almost the same time. And during that same night recordings were made in Siding Springs (Australia) and in the Karoo in South Africa, as can be read in telegram IAUC 4316. It was the first supernova to be spotted in 1987 and was therefore named 'SN1987a' (Fig. 14.10).

Before Shelton noticed the new star, the supernova had already been reported elsewhere, far below the Earth's surface, but no one had noticed it. In the deep Kamioka mine in Japan in a large water basin, a number of neutrinos had caused flashes of light. In Japan, the Magellanic Clouds are always below the horizon and, if these neutrinos did indeed come from there, then they must have passed through the Earth. That was quite possible, as neutrinos almost never collide with anything. Theoreticians welcomed this flow of neutrinos, because neutrinos could have a

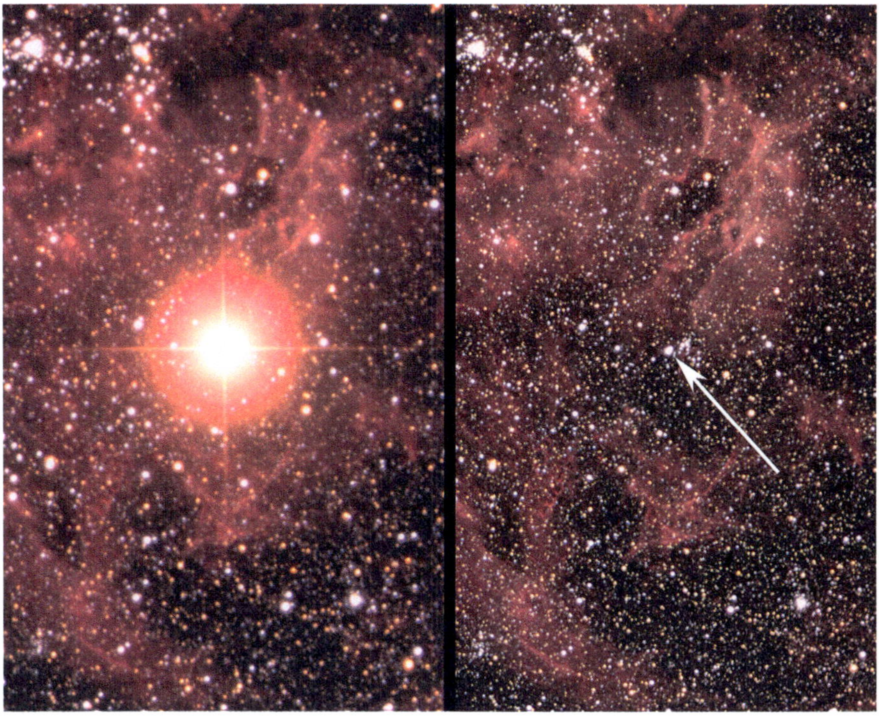

Fig. 14.10 Two pictures of the same area in the Large Magellan Cloud. The image to the right is before the blazing supernova; left is immediately afterwards. Source: David Malin/Australian Astronomical Observatory

major effect on where and how a supernova explosion began. One possible mechanism was that many neutrinos were produced and then escaped; in doing so they would drain a catastrophic amount of energy from the heart of the star. The discovery that this supernova had expelled many neutrinos was undoubtedly one of the most important findings. In 2002 Masatoshi Koshiba, the initiator and leader of the Kamioka project, received the Nobel Prize for physics for his achievements.

That the neutrinos arrived here six hours before one saw the star light up, must be due to the fact that neutrinos were expelled from the center of the star at the speed of light, while the light of the explosion came from the outer areas of the star only after a shock wave had transported the explosion from the center to the outer star layers at a considerably lower speed.

Astronomy was 'ripe' for this occurrence. Computers were powerful enough to calculate sophisticated explosion models. And by now there were a number of large observatories in the southern hemisphere (Australia, Chile, South Africa) that possessed up-to-date instruments (Fig. 14.11). The excitement was felt everywhere and not only among astronomers: the explosion appeared on the cover of several magazines, in fantastic colors.

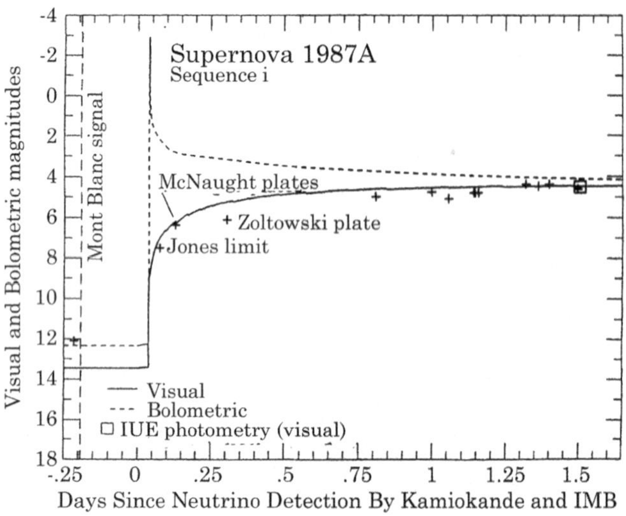

Fig. 14.11 The brightness of the 1987a supernova in the first day and a half. Horizontally, the time is plotted in days; vertically, crosses indicate the visible brightness in magnitudes. The solid line is an average through the observation points indicated by the crosses. The dashed curve shows the 'bolometric' brightness—the brightness of the star when all of the radiation is added up, from the ultraviolet to the infrared. On the horizontal axis, the 0 point is the time at which the neutrinos passed through the Kamiokonde basin. On day −0.25 there is a vertical line with the text 'Mont Blanc Signal': in this mountain there is also a basin filled with water that serves as a neutrino detector. A slight increase could be seen there in the number of neutrinos, a quarter of a day earlier than the Japanese, but the number of detections was too small for a statistically reliable conclusion (Arnett et al. 1989)

The spectrum of the supernova was recorded at frequent intervals, with numerous telescopes. There were very strong hydrogen lines, which was a convincing indication that this was a type II supernova, resulting from the collapse of the innermost part of a massive star. The supernova was in the exact spot in which a blue supergiant had previously been spotted. It was known that supergiants exploded, but the blue color of the star was unexpected: theorists had predicted that before the explosion the star was a red supergiant. This was food for thought (Fig. 14.12).

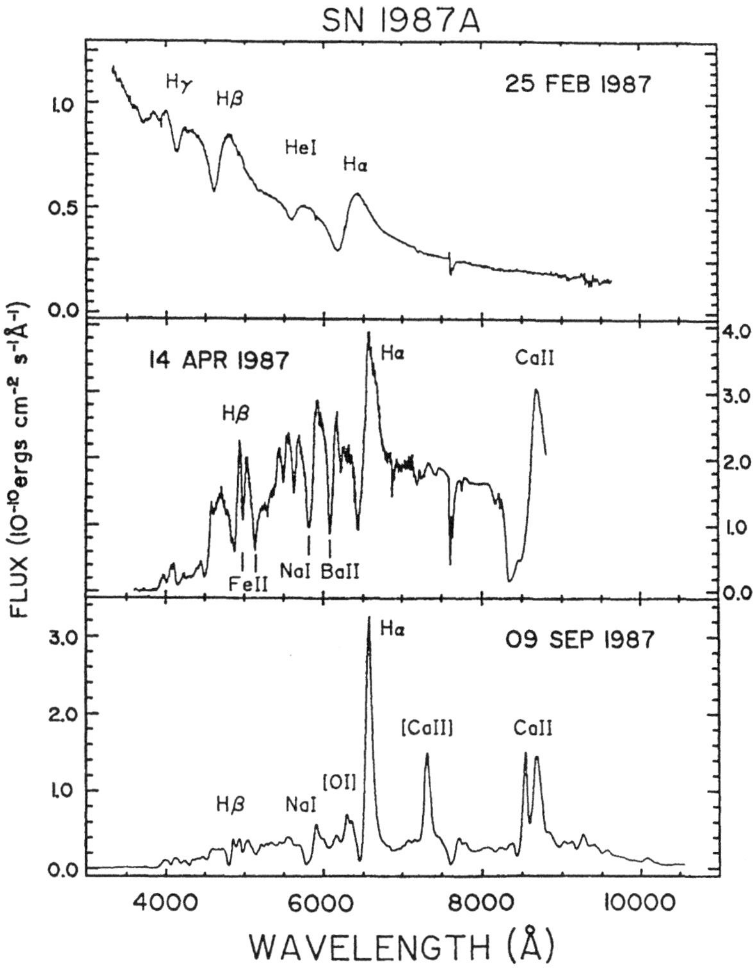

Fig. 14.12 SN1987a spectra recorded on consecutive nights. At the start there are hydrogen absorption lines (the lines form troughs in the background radiation) but in the following days these become strong emission lines that protrude above the underlying continuum (Arnett et al. 1989)

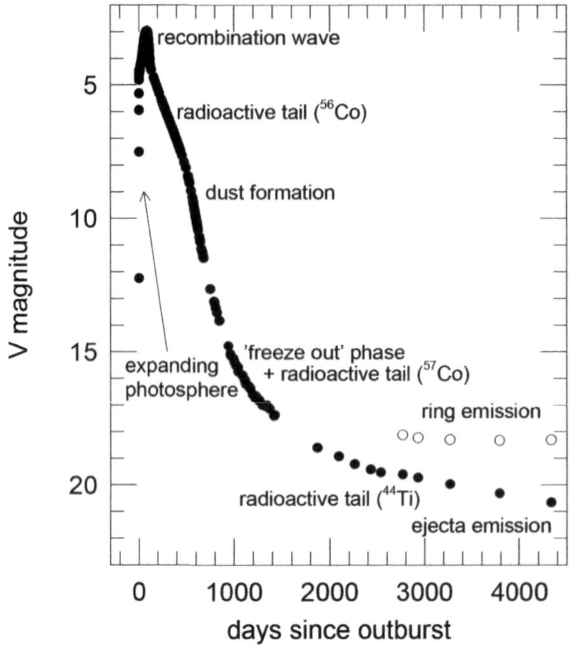

Fig. 14.13 The brightness of supernova SN1987a measured with a yellow filter. In the first days, the photosphere of the original star can be seen that is pushed upwards from below. The recombination of protons and electrons causes a decrease after about day 10. The remnant of the supernova contains a large amount of radioactive nickel that decays to cobalt and functions as a nuclear power plant for a while, being the most important energy source. After some time, the gas has cooled so far that grains of interstellar dust have formed. A period follows in which radioactive cobalt decays, and finally a period in which the decay of radioactive titanium is the most important energy source (Leibundgut and Suntzeff 2003)

Figure 14.13 shows the course of SN1987A's brightness over a period of 4,000 days (11 years) after the explosion. The image indicates in which specific phase the supernova was at the time (see text with the image).

By integrating the light curve, one could determine how much energy was released from the radioactive nickel, cobalt and titanium, and from this one could calculate how much of that material had been released. This information became key for comparing future theoretical models of supernovae.

It is possible that supernova SN1987a ultimately left behind a neutron star in the shape of a pulsar or X-ray source, but neither object type has been observed to date.

SN 1987a was not only in time for the optical and radio observers but also for observers of X-radiation and gamma radiation, especially for the Japanese astronomers. Their large X-ray satellite GINGA (Japanese for 'galaxy') was launched on 5 February 1987. The satellite was directed frequently towards SN 1987A, but recorded only noise until weak signals were picked up halfway through August, a signal that was confirmed in the first days of September. A short while later, American solar physicists directed the NASA satellite SMM towards SN 1987A

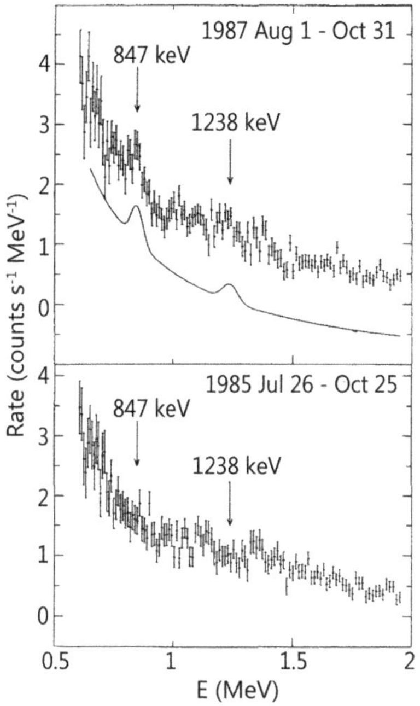

Fig. 14.14 Two spectra of the gamma rays from supernova 1987A, obtained with NASA's SMM satellite. Just as with an ordinary spectrum, the energy from the gamma photons is plotted horizontally and the number of photons received is plotted vertically. The bottom window indicates the number of photons received between 26 July and 31 October 1985; the top window shows the number of photons received in a later, slightly shorter, period. As can be seen, the radiation intensity peaks at 847 keV and again at 1,238 keV (0.847 and 1.238 MeV). Photons with these energies are emitted when a nucleus of the ^{56}Co atom (a nucleus with atomic weight of 56) radioactively decays to a core of iron (^{56}Fe). These spectral lines were recorded in the period between 26 July and 1 August 1987 (Matz et al. 1988)

and detected two spectral lines, one of 847 keV and one with an energy of 1,238 keV; photons with this amount of energy were released through the radioactive decay of cobalt (^{56}Co) (see Fig. 14.14).

About 400 days after the explosion, a ring appeared around the supernova. The material of this ring was too far away from the nova to be there because of the supernova, but must have been produced earlier. This was to be expected: supergiants sometimes produced such rings. The ring was elliptical in shape, but that was an optical illusion: a circle viewed from above at an angle looks like an ellipse. The most interesting thing about the ring was not its presence, but the fact that the supernova light had taken approximately 400 days to reach the ring. In conclusion, the ring had a radius of 400 light-days. By measuring the radius of the circle and equating it to 400 light days, it was determined that the ring, and thus the Large Magellan Cloud, was at a distance of 51.2 kpc (170,000 light-years). This largely agreed with what had been found previously based on quite different measurements (Fig. 14.15) (Panagia et al. 1991).

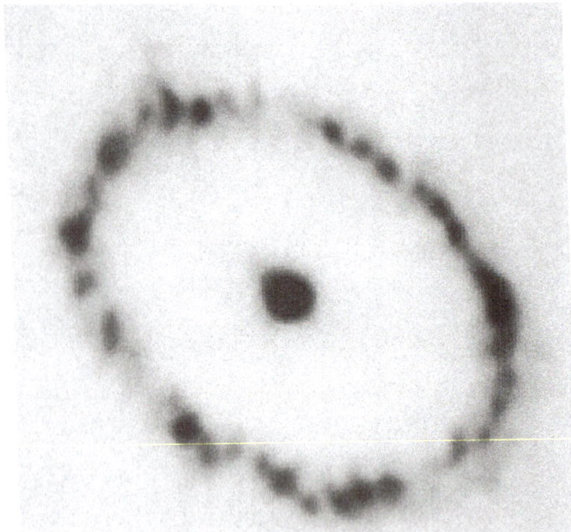

Fig. 14.15 A ring of gas expelled by a star before it exploded as a supernova (Panagia et al. 1991)

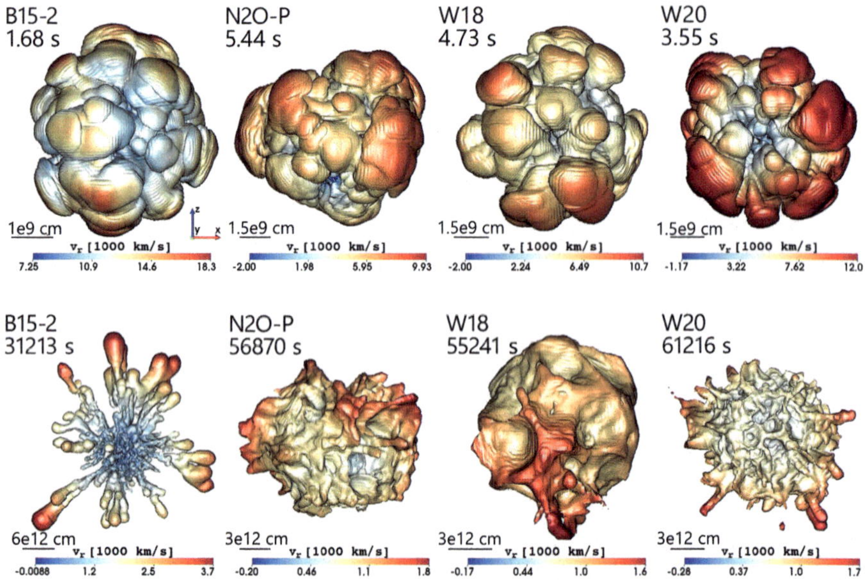

Fig. 14.16 Three dimensional models of an exploding star. Horizontally four different models (B15-2, N20-P, W18 and W20) are shown; the upper image is from shortly after the start of the explosion, the bottom image from a while later. The expansion velocity of the different parts of the explosion is indicated in color (Utrobin et al. 2016)

I will conclude this discussion of SN1987A with a model from 2016, calculated in three dimensions, of an originally spherical star that explodes. The model is 'neutrino driven': neutrinos are the cause of the explosion of the star (Utrobin et al. 2016).

In terms of computer technology, these models are impressive because the calculations are three-dimensional. Figure 14.16 shows the development of four

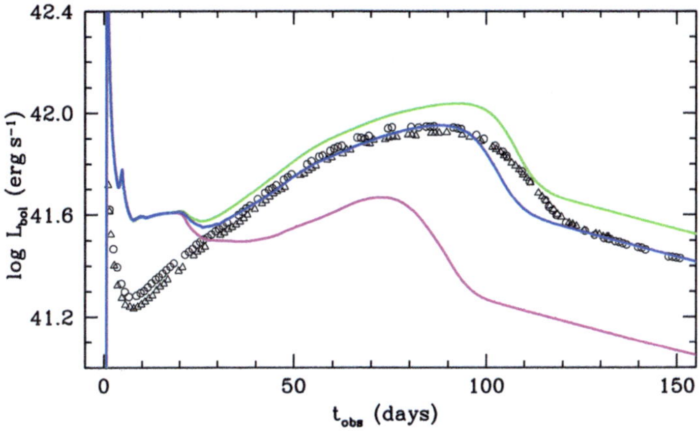

Fig. 14.17 The light curve of SN1987a, observed in black and modeled in color (Utrobin et al. 2016)

different models and Fig. 14.17 shows different predicted light curves in color and two different series of observations in black. The light curves show that theory and reality are quite close, but that there is still room—as well as need—for improvement, a conclusion also drawn by the makers of the models.

Theoretical Considerations for the Collapse of the Core of a Supernova

Chapter 4 discussed the 'onion-skin' model: how the evolution of a high-mass star may lead to a model in which successive layers consist of the nuclear waste of previous fusion processes—see Fig. 4.21 in that chapter.

In the build up of new onion skin, a catastrophe may occur when the star loses a large part of the pressure at its inner core and subsequently collapses. There are three possible causes for this loss of pressure (Janka 2012) and I will list them below for the reader with some knowledge of physics.

1. In the lightest of the massive stars (weighing 9–10 times the mass of the Sun), electrons can be captured by atomic nuclei of neon and oxygen. The loss of electrons means loss of pressure, and is followed by a collapse. These types of supernovae are also called 'electron capture supernovae'.
2. In heavier stars, an iron core is ultimately formed. Upon contraction, the temperature and density are so great that the atomic nuclei of iron are broken down into much lighter helium cores, the so-called α-particles. The energy for this is taken from the gas. The gas pressure decreases rapidly and an implosion follows.

3. In the heaviest stars, the temperature at the core can be so high that the photons have more energy than the rest energy of electrons. Pair production then takes place: the photons are converted into e^+ and e^- particles. In this process, a lot of photon energy is converted into matter. Here, too, the pressure drops and an implosion follows: 'Pair instability Supernovae'. In this last model it can happen that no end product remains: the star disintegrates completely during the supernova. 'Pair production' could be the start of a supernova type Ib or Ic, supernovae that occur shortly after a gamma ray outburst (GRB). This will be discussed later.

As a result of the collapse, a neutron star initially forms at the center, but if the collapse occurs over a sufficiently long period of time the neutron star is compressed into a black hole.

The simple question: 'Which supernovae produce a neutron star and which a black hole?' is not easy to answer. The simplest answer is: lighter massive stars produce a neutron star, and more massive ones result in a black hole. However, this is a too simplistic answer: the result appears to depend on the composition of the stellar gas. If there are few or no heavy atomic nuclei in the gas, then the outcome is different from when the gas has as many atomic cores as, for example, the Sun. Figure 14.18 gives an example. The figure is open to debate; the aim of showing it

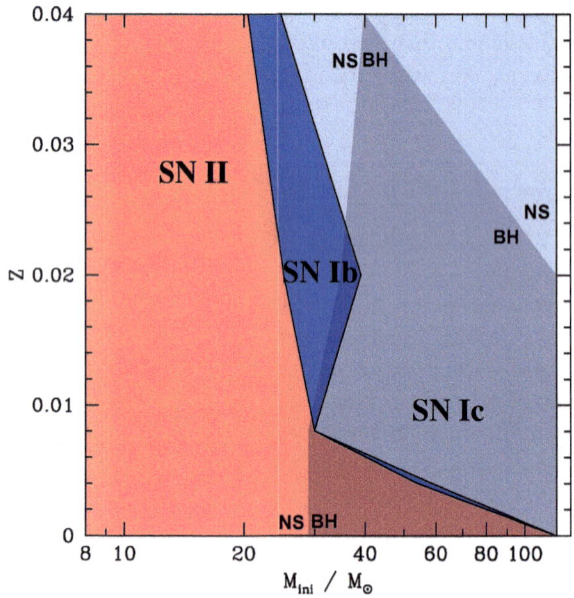

Fig. 14.18 When a core-collapse supernova has finished raging, a neutron star or a black hole is left behind. Which of the two it will be depends on the mass of the star, but also on the number of atoms heavier than helium. In this diagram the mass of a star is plotted along the horizontal axis. Along the vertical axis is the content of all atoms other than hydrogen and helium, compared to the mass of helium and hydrogen together. On this scale, the composition of the Sun is 0.02. Many different types of core-collapse supernovae occur. The letters NS and BH respectively mean 'neutron star' and 'black hole' (Meynet et al. 2011)

here is to give an impression of what the current thinking is. In this figure, for example, there is no place for supernovae that leave nothing behind.

The explosion of a star that leads to a supernova have in common that the core of the star produces many neutrinos, which escape from the core but then meet an impenetrable layer of gas, causing a shockwave which forces these layers outwards: the 'neutrino-driven supernova', which currently seems to be the paradigm for modelers.

Figure 14.16 has already shown how sophisticated modeling has become. The models calculated now also show that a supernova explosion is not necessarily symmetrical.

14.6 Gamma Ray Bursts (or GRBs)

The discovery of bursts of gamma photons has been described in Chap. 4. In 1973 the paper was published that reported what were previously military secrets. Twenty years later there was hardly any solid evidence of GRBs, except that they could appear anywhere in the sky, without any preference for the Milky Way. This meant that the bursts were located either in the halo of our Milky Way, and thus nearby, or at a very large distance. In the latter case, several theorists realized, they potentially constituted a great problem: the explosion must emit 10^{51} erg (10^{44} joules) into space within a fraction of a second (Rees and Meszaros 1992). This was extremely fast: it took a sizeable supernova several days to produce this amount of energy.

Publications on GRBs detected by several Venera satellites came from the Soviet Union (Mazets et al. 1981a, b), including the conclusion that the short-lived gamma bursts (shorter than 2 seconds) were another type of GRB than those with longer bursts.

In 1991 NASA launched the Compton Gamma Ray Observatory, one of its major astronomical satellites. There were three instruments on board, one of which was BATSE, the 'Burst and Transient Source Experiment', intended to detect GRBs. It was quite successful (as were the other two CGRO instruments). BATSE eventually discovered 2704 GRBs and Fig. 14.19 distribution shows the even, uniform distribution of the GRBs across the sky. The Milky Way is along the equator of the image and it is clear that the Milky Way is not the preferred home for the GRBs (Figs. 14.20 and 14.21).

It was the instantaneous emission of so much energy that aroused astonishment. A source that could emit so much energy in milliseconds was at most a few milli-light-seconds in size and therefore far smaller than the Sun. Thus, if GRBs came from very far away, the energy emitted was enormous and they were a new and very impressive phenomenon. But were they actually far away? As long as this was not confirmed, all of the theory was pure speculation.

Why was it so difficult to measure the distance? Gamma photons may be rich in energy, and thus easy to detect, but the detection devices had very little directional

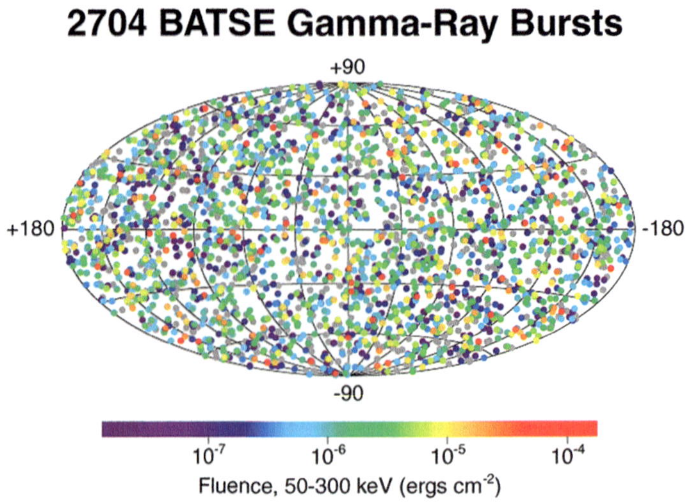

Fig. 14.19 The location in the sky of more than 2700 Gamma Ray Bursts detected by BATSE. The different colors indicate the strength of the GRB. The sources are randomly distributed across the sky; there is no preference for the sources to be located in the Milky Way. Source: Astronomy Picture of the Day, 28 June 2000, G. Fishman et al., BATSE, CGRO, NASA

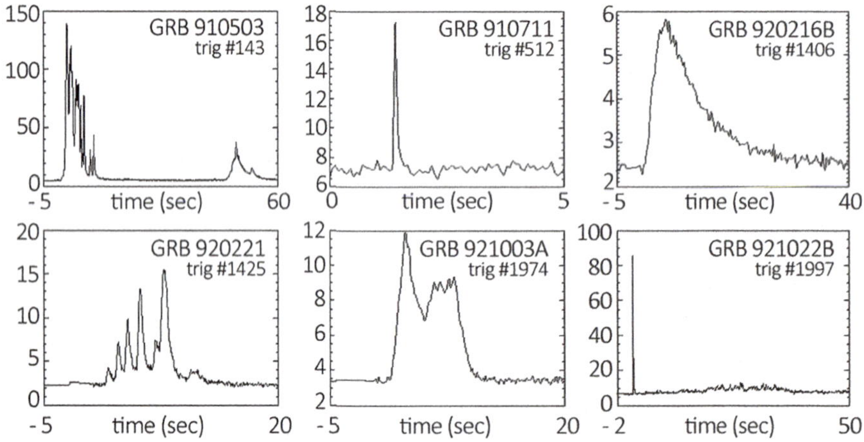

Fig. 14.20 The figure shows a dozen 'light' curves reflecting the evolution of a GRB. The curves have nothing in common. Notice how fast a burst can rise to its peak, much faster than in a supernova. Source: Wikimedia Commons

sensitivity. BATSE discovered more than 2700 GRBs but none could be identified with an object of a known type. The situation was a reminder of the early days of radio and X-ray astronomy, when sources were discovered but, due to poor positioning, it was impossible to identify them with a definite optical object.

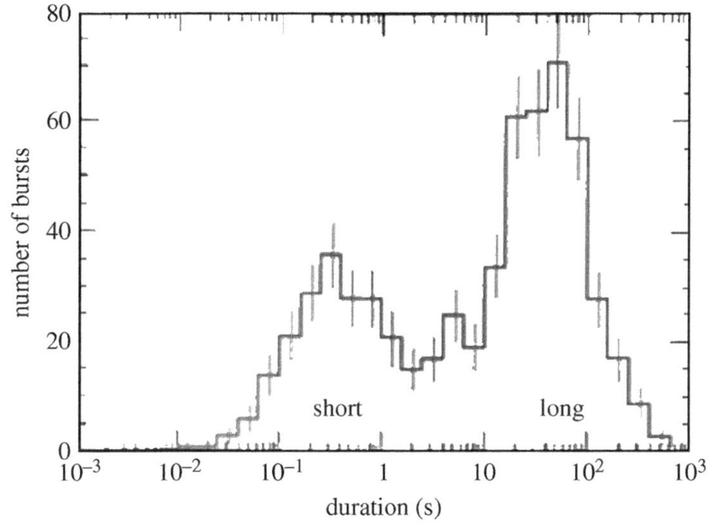

Fig. 14.21 A histogram of the duration of a GRB, measured as the time in which 90% of the total energy is emitted. There are clearly two groups of GRBs, which will be referred to as 'long' and 'short', confirming the earlier Russian conclusion. Source: Wikimedia Commons

On 10 February 1997, a theoretical article was published that again pointed out that such a powerful and rapid explosion as a GRB must produce an 'afterglow'; the object would definitely manifest itself at different wavelengths (Meszaros and Rees 1997). Within several weeks, the confirmation was received. On 2 March, message IAUC 6572 from Harvard, the center for such emergency communications, was disseminated throughout the entire world. Another new GRB had been discovered that had lasted 80 seconds. The detection was made on 27 February with the Italian-Dutch X-ray satellite Beppo-Sax and was called GRB970227: the date gave it its name. Four days later the message was received that Beppo-Sax had observed a weak X-ray source at the position of GRB970227; the source became rapidly fainter and disappeared a week later (Costa et al. 1997). Could the X-ray source be part of an afterglow?

Astronomy groups in the US and Europe sprang into action, directed their telescopes and saw optical and radio sources, but nothing that could be associated with this GRB. That is, until two PhD students from Amsterdam sent a telegram from the observatory in Gran Canaria on 12 March, to say that the light source that had been seen on 1 March at the position of the X-ray source had disappeared on 8 March. This light source also coincided with a small galaxy. That was enough information for an article stating that this was evidence for the GRBs being a cosmological phenomenon and thus explosion in which enormous amounts of energy were released in a split second (van Paradijs et al. 1997). It was the breakthrough that had been waited for so long (Fig. 14.22).

The way was now open to identify detections of other GRBs. The GRB of 27 May 1997 was identified with a galaxy with a redshift of z = 0.835, which showed

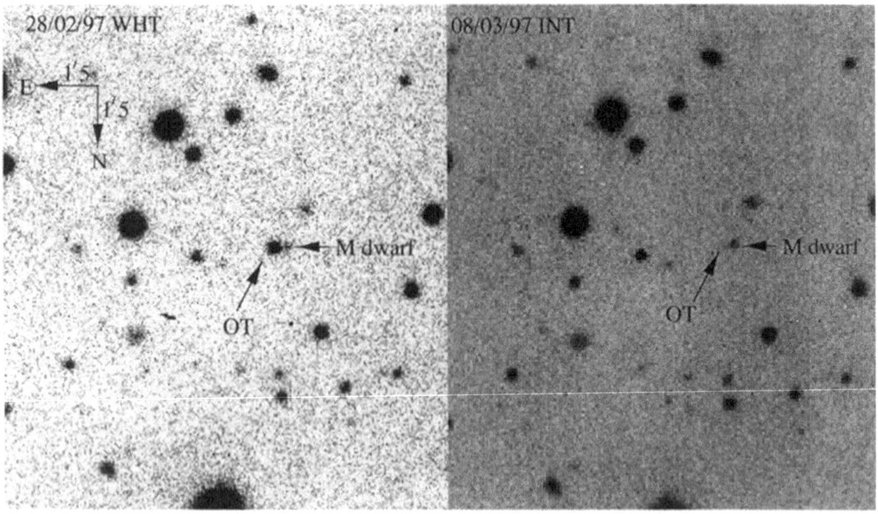

Fig. 14.22 A GRB is arriving from the location indicated by an arrow. The recording to the left was made on 28 February 1997, the one to the right on 8 March. In between the source has disappeared (van Paradijs et al. 1997)

that this galaxy was moving away from us at more than half the speed of light (Metzger et al. 1997). Shortly after this, another GRB, 980425, was identified with a supernova (Galama et al. 1998). The supernova was unusually strong and led to the suggestion that this was a superluminous supernova that emitted up to a hundred times more light than the other core-collapse supernovae (see earlier discussion) (Iwamoto et al. 1998). In 2011 a GRB was found in the galaxy with the largest redshift ever measured: $z = 9.4$ (Cucchiara et al. 2011). This galaxy was moving away from us at 98% of the speed of light. It followed that you could detect GRBs at distances greater than supernovae could be seen. GRBs promised the possibility to investigate the Universe at distances so far away that we were looking at the time in which the first galaxies were formed.

The detection of an afterglow and of a connection with those very bright supernovae made it necessary to seriously consider what type of object could cause a GRB. Theorists had speculated about this for years, long before GRB970228 and GRB890425 were found. The search was for an object at most a light-second large that could emit 10^{46} joules within one second. To give an indication of the scale of the event: 10^{46} joules is as much energy as when the matter of half a neutron star is converted into photons. Trapped energy that could suddenly escape from a star was considered, for example, due to a hole in the outer layers of a star: the brief instant of escape resulted in the GRB (Wijers et al. 1997).

What was happening inside the star was guesswork for the time being. Often, magnetars were seen as a possible source of energy of this type of 'supernova'. 'Pair-instability', supernovae that have been mentioned above, was also considered.

In 2015 there was as yet no certainty. A third possibility was thought to be the fusion between two neutron stars that rotated closely around one another. They emitted gravitational waves due to which rotation energy would be lost, and the two stars would come closer together and eventually merge into one object. Finally, it was also possible that the explosion meant that a neutron star disappeared into a black hole. The 'long' GRBs would mainly originate during supernova-type explosions and the 'shorter' GRB's would originate from mergers of binary stars into a single star. The capture of a neutron star by a black hole should then explain the 'ultra long' GRBs, a few of which had been observed. But it was very much all speculation.

On 11 February 2016 it was announced that LIGO, an American instrument that could record gravitational waves, had produced the first success: in September 2015 a gravitational wave had passed the detector (Abbott et al. 2016). Gravitational waves such as these had been predicted by theorists who were engaged in two neutron stars that rotate around one another and by emitting gravitational waves lose so much energy that they eventually merge. Whether the detected waves actually originate from such a 'merger' remains to be seen, but it is evident that this is exciting research (Levan et al. 2016).

14.7 Type Ia Supernovae

The existence of two major types supernovae, those of type I and those of type II, has been discussed in Chap. 7. The generally accepted hypothesis was that the type I supernovae were white dwarfs that burned out in a very short time. Most white dwarfs were made up of carbon and oxygen with only a thin layer of helium and/or hydrogen around it. This explained the absence of spectral lines of those atoms in the spectrum.

Two different mechanisms were suggested for the explosion, both of which involved a close binary star. According to the first mechanism, matter flowed from the other star to the white dwarf until the Chandrasekhar limit (1.44 × the weight of the Sun) was reached and the white dwarf exploded. In the second case, two dwarfs moved slowly towards one another and eventually fused. Both options were still open.

In the 1970s a few type I supernovae were found that differed from the large majority. There were two anomalies: a strong dark (absorption) line of silicon was missing and the light curves were abnormal. In 1985 it turned out (Wheeler and Levreault 1985) that these few different cases were core-collapse supernovae after all. They were classified as Ib or Ic; these coincided with gamma ray bursts. The old-fashioned, 'classic' type I objects were now called type Ia.

The type Ia supernovae formed a homogeneous group, as noted by Minkowski and Zwicky already in 1941—see the prologue. This became even more apparent in a diagram from 1968 in which the maximum brightness of these supernovae was plotted against the radial velocity of the corresponding galaxy (see Fig. 14.23).

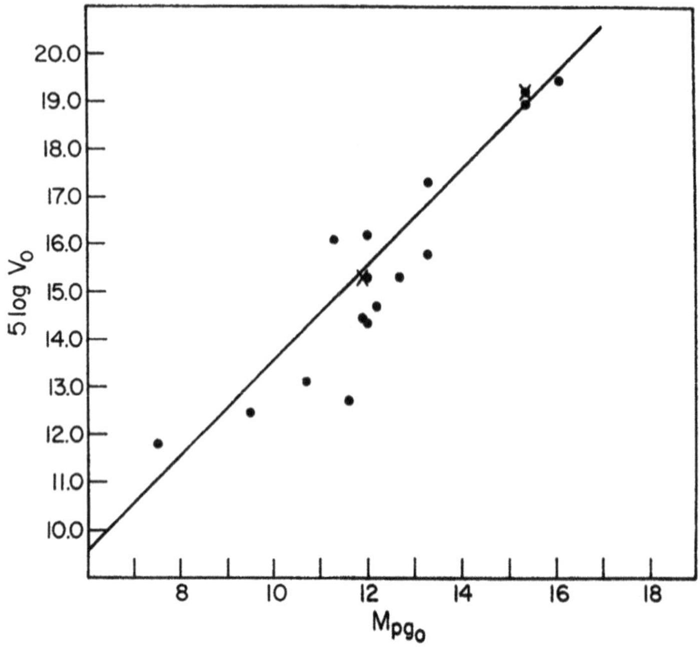

Fig. 14.23 The correlation between two different properties of supernovae of type Ia. The maximum apparent magnitude M_{pg} is plotted horizontally, and vertically the logarithm of its radial velocity V_0, multiplied by 5. The suffix 'pg' to the maximum magnitude indicates that the magnitude was measured on photographic plates. There is a well-defined correlation between M_{pg} and 5 log V_0. If all supernovae had the same absolute maximum magnitude, the points would all lie on the straight line. The spread around the line could then be explained by measurement errors (Kowal 1968)

The straight line that followed the data points so closely was where all of the supernovae would appear if they all had the same luminosity at their brightest instance. The image suggested that this could well be the case.

This was an important suggestion. All stars with known luminosity could be used as 'standard candles': if you knew beforehand how much light a particular object emitted, then you could calculate the distance from the amount of light that you received. Cepheid stars (a particular type of variable star) had been used for this purpose. The supernovae of type Ia (SNIa) were 10 magnitudes brighter than the Cepheids and they could be used to measure 100 times greater distances. Cosmologists were desperate for such measurements.

As we will see later, the type Ia supernovae indeed turned out to be a standard candle. The use of these objects led to the fundamental discovery that the Universe expands at an accelerated rate, a discovery that yielded a Nobel prize. However, I will delay the discussion of how this developed until Chap. 16 on cosmology where it is more appropriate.

Let us return to the supernovae of type Ia and to how they were formed. When a companion star transferred matter to a white dwarf, the white dwarf contracted because of the remarkable effect that, the greater mass of the white dwarf, the smaller it is. When the dwarf's mass reached the Chandrasekhar limit, the size of the object should become zero, and the pressure infinite. However, things would never get that far because before that point was reached carbon and oxygen atoms would begin to fuse, forming atoms of iron, nickel and cobalt. The resulting conflagration would spread rapidly and the star would explode: this was the start of the supernova. Most likely, little or nothing of the fire remained: neither neutron star nor black hole.

The newly formed cobalt and nickel atoms were thought to be radioactive and to disintegrate spontaneously to form iron atoms. This process released a large amount of energy, particularly in the form of gamma rays. A nuclear power plant! This energy was gradually released over a period of a hundred days (the decay time of cobalt into iron) and because so much radioactive material had been created, the supernova's 'afterglow' could be seen for a long time afterwards. In 2014 the INTEGRAL satellites detected the same two spectral lines that had been seen earlier in the SN1987a explosion (see above) (Churazov et al. 2014; Diehl et al. 2014). The detection of these lines confirmed the radioactive decay of cobalt (Fig. 14.24).

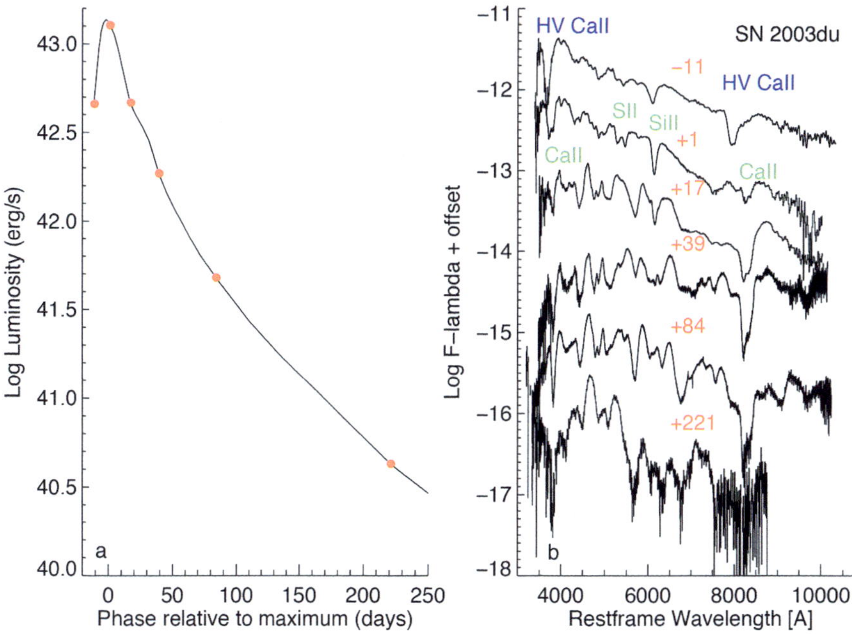

Fig. 14.24 The left-hand figure depicts the light curve of prototype supernovae 1a. To the right are the spectra of the supernova measured on the days indicated in orange, counting from the day of maximum luminosity; the horizontal scale is given in Ångstrom; 1 Å = 10 nm. For some of the absorption lines the associated atoms are indicated. In particular, the Si II line (of Si^{+}) is typical of supernova Ia (Howell 2011)

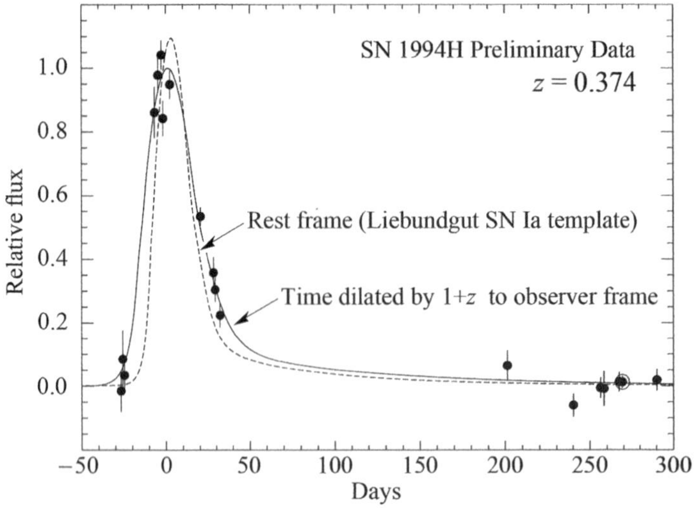

Fig. 14.25 Two light curves. The dashed line is the standard light curve of type Ia supernovae, and the solid line is the expected distortion of the light curve, as predicted by the theory of relativity, given the large difference in velocity (90,000 km/s; z = 0.374) between the sender (SN 1994H) and the receiver (us) (Goldhaber et al. 1999)

When a supernova has a large redshift, effects of relativity theory are noticeable: the time interval at the source differs from the time interval in our Milky Way (see Fig. 14.25).

An important question you could ask about supernovae, type Ia and all others, is: What did the star look like immediately before it exploded? As mentioned earlier, that is a difficult question: you would have to be lucky enough to have a recording in your archive of the exact location, and the recording would have to be of good quality. Type Ia supernovae had very faint white dwarfs as progenitor. No 'progenitors' were found, although there was some indirect evidence. However, this has so far not been conclusive.

Model calculations of all types of supernova have continuously been made over several decades. The recent, very fast computers with large memories finally made it possible to make 3-D calculations. Success may well be within reach, but to an outside observer it still appears as 'work in progress'. The last word has not yet been said.

14.8 Gravity Waves

When any object is placed somewhere else, the distribution of matter in the Universe changes. The force of gravity is consequently changed everywhere in the Universe. Einstein showed in his General Theory of Relativity that the remainder of the Universe is informed about the change in gravity by gravity waves that travel with the speed of light. When the information arrives objects will briefly get smaller and bigger.

The effect is very, very small, unless a dramatic change occurs in the distribution of matter. The existence of gravity waves was proven indirectly by Joseph Taylor who studied the pulse rate of two neutron stars rotating around their common center of gravity. He noticed that the stars slowly got closer to each other, an approach that was incomprehensible except when the system lost energy through gravity radiation, and that explanation predicted very accurately what happened—see Chap. 7.

Another event that should produce very strong gravity waves would be the merging of two neutron stars or of two stellar back holes. The gravity waves associated with these events could be calculated from Einstein's theory and although the arrival of the gravity waves on Earth would have a very small effect, it might be observed. Over the last 40 years there have been several attempts to measure the arrival of gravity waves, but no signal has been seen. Initially one looked for changes in the length of a metal bar: when a gravity wave passed, it should quickly oscillate between becoming a bit longer and smaller. No effect was found—signals were seen, but these did not correspond to gravity wave.

Finally a large-scale apparatus was built called LIGO, Laser Interferometer Gravity-Wave Observatory. The central idea was that a gravity wave would cause a wave-like change in the distance between two mirrors. Laser light should monitor this variation in distance with the required accuracy. Caltech in Pasadena (Cal.) and MIT in Boston undertook the initial construction of LIGO, This first attempt failed to detect the waves. A larger project was started under the same name. Support came from the National Science Foundation in the USA with participation from Australia, the UK and Germany. This time it was successful. On 14 September 2015, a signal was received of a gravity wave with the predicted shape of a wave caused by two colliding and merging black holes with a mass of about 30 times that of the Sun. On 16 June 2016, the signal of a second wave of the merging of two black holes was measured, the first with a mass of 14.2 times the Sun, the second with a mass of 7.5 times that mass. Since then, other gravitational wave events have been recorded.

In August 2017 a wave was detected with the signature of the merging of two neutron stars. The signature arrived at the same time as a Gamma Ray Burst detected by two gamma ray space telescopes, Fermi and Integral. A short optical 'transient' was recorded in the NGC4933 galaxy and was clearly related to the gravity wave and then GRB. The proof of coincidence between a GRB, a gravity wave detection and an optical transient is a promising beginning of a new field for study.

References

Abbott, B.P., Abbott, R., et al.: Phys. Rev. Lett. **116**, 1102 (2016)
Alcock, C., Akerlof, C.W., et al.: Nature. **365**, 621 (1993)
Aoki, T., Tanaka, T., et al.: Astrophys. J. **781**, 10 (2014)
Arcavi, I., Wolf, W.M., et al.: Astrophys. J. **819**, 35 (2016)
Arnett, W.D., Bahcall, J.N., et al.: Annu. Rev. Astron. Astrophys. **27**, 629 (1989)
Bond, H.E., Henden, A., et al.: Nature. **422**, 405 (2003)

Bower, G.C., Saul, D., et al.: Astrophys. J. **666**, 346 (2007)
Churazov, E., Sunyaev, R., et al.: Nature. **512**, 406 (2014)
Costa, E., Frontera, F., et al.: Nature. **387**, 783 (1997)
Cucchiara, A., Levan, A.J., et al.: Astrophys. J. **736**, 7 (2011)
Diehl, R., Siegert, T., et al.: Science. **345**, 1162 (2014)
Galama, T.J., Vreeswijk, P.M., et al.: Nature. **395**, 670 (1998)
Gal-Yam, A.: Science. **337**, 927 (2012)
Goldhaber, G., Deustua, S., et al.: ASIC. **486**, 777 (1999)
Goodrich, R.W., Stringfellow, G.S., et al.: Astrophys. J. **342**, 908 (1989)
Howell, D.A.: Nat. Co. **2**, E350 (2011)
Ivanova, N., Justham, S., et al.: Astron. Astrophys. Rev. **21**, 59 (2013)
Iwamoto, K., Mazzali, P.A., et al.: Nature. **395**, 672 (1998)
Janka, H.-T.: Anuual Review of Nuclear and Particle Science. **62**, 407 (2012)
Kowal, C.T.: Astron. J. **73**, 1021 (1968)
Leibundgut, B., Suntzeff, N.B.: Lect. Notes Phys. **598**, 77 (2003)
Levan, A., Crowther, P., et al.: Space Science Reviews. **202**, 33 (2016)
Lorimer, D.R., Bailes, M., et al.: Science. **318**, 777 (2007)
Lyman, J.D., Bersier, D., et al.: Mon. Not. R. Astron. Soc. **457**, 328 (2016)
Mao, S.: Reviews in Astronomy and Astrophysics. **12**, 947 (2012)
Mason, E., Diaz, N., et al.: Astron. Astrophys. **516**, A108 (2010)
Matz, S.M., Share, G.H., et al.: Nature. **331**, 416 (1988)
Mazets, E.P., Golenetskii, S.V., et al.: Astrophys. Space Sci. **80**, 3 (1981a)
Mazets, E.P., Golenetskii, S.V., et al.: Nature. **290**, 378 (1981b)
Metzger, M.R., Djorgovski, S.G., et al.: Nature. **387**, 878 (1997)
Meynet, G., Georgy, C., et al.: Bulletin de la Société Royale des Sciences de Liège. **80**, 266 (2011)
Munari, U., Navasardyan, H., Villanova, S.: Astron. Soc. Pac. Conf. Ser. **363**, 13 (2007)
Nakano, S., Nishiyama, K., et al.: IAU Circ. 8972 (2008)
Paczynski, B.: Astrophys. J. **304**, 1 (1986)
Panagia, N., Gilmozzi, R., et al.: Astrophys. J. **380**, L23 (1991)
van Paradijs, J., Groot, P.J., et al.: Nature. **386**, 686 (1997)
Petroff, E., Bailes, M., et al.: Mon. Not. R. Astron. Soc. **447**, 246 (2015)
Reynolds, T.M., Fraser, M., Gilmore, G.: Mon. Not. R. Astron. Soc. **453**, 2885 (2015)
Smartt, S.J.: Proc. Astron. Soc. Aust., **32**, 16 (2015)
Smith, N., Li, W., et al.: Mon. Not. R. Astron. Soc. **415**, 773 (2011)
Spitler, L.G., Cordes, J.M., et al.: Astrophys. J. **790**, 101 (2014)
Stewart, A.J., Fender, R.P., et al.: Mon. Not. R. Astron. Soc. **456**, 2321 (2016)
Tylenda, R., Soker, N.: Astron. Astrophys. **451**, 223 (2006)
Tylenda, R., Hajduk, M., et al.: Astron. Astrophys. **528**, A114 (2011)
Utrobin, V.P., Wongwathanarat, A., Janka, H.-T., Müller, E.: Astron. Astrophys. **581**, 40 (2016)
Westphal, J.A., Neugebauer, G.: Astrophys. J. **156**, L45 (1969)
Wheeler, J.C., Levreault, R.: Astrophys. J. **294**, 17 (1985)
Wijers, R.A.M.J., Rees, M.J., Meszaros, P.: Mon. Not. R. Astron. Soc. **288**, L51 (1997)

Chapter 15
1985–1995: Galaxies

Discovering how galaxies originated has been a mission for astronomers ever since they first became aware of the existence of these objects. But the desire for this discovery was even stronger in the period from 1985 to 1995, a desire that was fueled on the one hand by an increasing understanding of the structure of galaxies and on the other by the promising attempts to see into the past through distant objects. But researchers were also motivated by another, very different, goal: the joy of measuring something new, seeing something that nobody had ever seen before, not 'because it's necessary, but because it's possible.'

15.1 Elliptical Galaxies

As described in Chap. 8 the elliptical galaxies were a lot more complex (and interesting) than had been assumed. The objects looked simple and the isophotes were almost elliptical. Nevertheless three-dimensional models with three different axes were needed to explain both the distribution of stellar light and the velocities as measured. Models were no longer easily calculated. The difficulties did not go away in this period.

A Number of very different Types

Thanks to CCD cameras, the light distribution of a large number of elliptical galaxies could now be measured precisely. The lines of equal brightness, the isophotes, looked at first sight elliptical in shape, but closer examination show pronounced differences (Jedrzejewski 1987). Two types stood out. In one of these the ellipse was more 'boxy', and in the other more 'disky' (see Fig. 15.1).

H. J. Habing, *The Birth of Modern Astronomy*, Historical & Cultural Astronomy,
https://doi.org/10.1007/978-3-319-99082-8_15

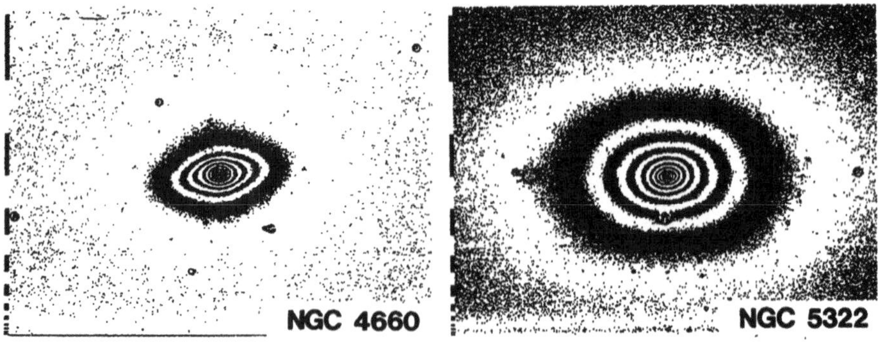

Fig. 15.1 A disky galaxy (NGC 4660) and a boxy galaxy (NGC5322) (Bender et al. 1988)

This difference seemed small, but it was in fact major: in Chap. 8 it has already been discussed that intrinsically bright elliptical galaxies rotated slowly—or not rotate at all—around their axis, while their fainter counterparts did rotate at the expected speeds. The first group looked 'boxy' while the second group was 'disky'. There were two fundamentally different types of galaxies underlying the boxy/disky distinction (Davies et al. 1983; Bender et al. 1992). And there were more differences. The bright galaxies were more massive and they sometimes had strong radio sources at their core: a—rather old—example was the large galaxy M87 that coincided with the Virgo A radio source. The massive galaxies also proved to be surrounded by a halo of X-radiation (Forman et al. 1985). Both characteristics—radio source and X-ray halo—were missing in the fainter galaxies (Bender et al. 1989).

Another aspect was that a part of an elliptical galaxy could have a different rotation axis than the rest of the galaxy. Probably the most sensational example was the discovery that the core of the galaxy IC1459 rotated in one direction while the remainder revolved in the other direction (Franx and Illingworth 1988). The only conceivable explanation was that the core was the main galaxy that had swallowed another smaller galaxy. This theory was referred to as 'hierarchical', a concept that can better be described as a form of 'land-grabbing on a cosmic scale'.

Characteristics of the Stellar Population

The quality of the spectra of elliptical galaxies improved so much that they could be used for more than simply information about stellar motions inside the galaxy. An elliptical galaxy was transparent and the spectrum of the galaxy would be the sum of very many individual star spectra. In the literature from the 1940s (Stebbins and Whitford 1948), researchers toyed with the idea that you could figure out the composition of a galaxy from its spectrum. It was a nice idea but for a long time it did not generate any lasting results. In earlier times, this could have been due to the very limited quality of the spectra, but now the spectra were of such good quality that the plan to discover the population of such stars could be implemented.

It would be a pointless to begin from scratch and mix a random set of stars in order to explain the spectrum: there were too many free parameters to play with. Some advance planning was called for. In analogy with what was known about our Milky Way, modelers started with a population of newly formed stars, both light and massive. They then calculated how this population changed as the stars became older and what kind of composite spectrum this resulted in. Only then could the spectrum of this population be calculated and compared with the spectrum observed.

An unexpected problem also occurred. The spectrum of a star depended not only on its age but also on its chemical composition, in particular the 1% of atoms other than hydrogen and helium had a big effect on the spectrum. Different stellar populations could explain the same galactic spectrum: a young population with more heavy atoms and an old population with few heavy atoms. The solution had to lie in finding details in the spectrum that differentiated between the two. The search for such distinguishing spectroscopic details took many years, but in 1994 they were found (Worthey 1994): using spectroscopic analysis one could detect what kinds of stars populated the elliptical galaxies.

X-Rays prove the Presence of Dark Matter in and around Elliptical Galaxies

Measurements with the Einstein X-ray satellite showed that elliptical galaxies produced X-radiation. To the surprise of the observers, much of the radiation came from outside the visible part of the galaxies and it clearly surrounded them (Forman et al. 1985). Signs of matter outside the optical galaxies had been looked for, by radiotelescopes, for example, but no signs had ever been found. After some discussion, the conclusion was drawn that the radiation came from hot gas ejected by all the dying stars inside the galaxy during its lifetime. This gas had not escaped because of the large amounts of dark matter that filled and surrounded the galaxy. That explanation demanded that the galaxy was many times as massive might be derived from the amount of starlight. Dark matter had to be a dominant ingredient of elliptical galaxies.

15.2 Disk Galaxies

Disk galaxies were hybrid objects: there was a disk that contained not only stars but also interstellar gas, and there was a bulge that surrounded the center. In the disk the stars moved in circles around the nucleus, with the force of gravity being compensated for by the centrifugal force. New stars were formed from giant molecular clouds and orbited around the center, just like the rest of the disk.

Three topics will be discussed in this section: the formation of new stars in the disk of a galaxy, the chemical composition of the stars, and the area around the center of our Milky Way.

Disk Galaxies: The Formation of New Stars

Massive young stars were hot and they radiated a lot of ultraviolet light. This could be captured by surrounding gas clouds and converted into infrared radiation and in radiation of molecular spectral lines. By measuring all the radiation one could gain a good idea of the number of recently formed stars, and because these stars only existed for a short while, it was possible to calculate how many stars were formed per year (or rather per million years). In 1959, Dutch-American Maarten Schmidt (Schmidt 1959) used this method to demonstrate that the number of stars produced by a disk galaxy depended on the amount of hydrogen in the system, divided by the area of the disk. This 'Schmidt law' has since been used extensively in research on galaxies.

However, there were instances where the law did not seem to apply. In 1989, American Rob Kennicutt discovered that there was a limit to the validity of Schmidt's law: if the local hydrogen density was below a particular level, no stars were produced. This modified 'Kennicutt-Schmidt law' has since been used with great success for modeling disk galaxies. The law has proved very important in interpreting very distant and difficult to observe galaxies that we see in a distant past. Figure 15.2 shows that star formation ceased when the density of the hydrogen was too low.

The Center of our Milky Way and the Andromeda Nebula: Very Massive Black Holes?

In the 1930s, the very first radio observations showed that a lot of radiation was emitted by the Milky Way and that this radiation was strongest in the direction of the constellation of Sagittarius; that is, the direction of the center of the Milky Way. Radio observations after 1945 confirmed that the strongest radiation very probably indicated the direction to the center, quite accurately but not exactly. The charts made of the radiation emitted by 'Sagittarius A' (the strongest radio source in the constellation of Sagittarius) contained a complex collection of different types of sources. In the 1960s a strong infrared source was detected radiating from the radio source, followed much later by X-ray detections. These discoveries led to the suspicion that our Milky Way had an 'active nucleus', although the activity was much less than the nucleus in radio galaxies. Second-rate nucleus or not, the Milky Way

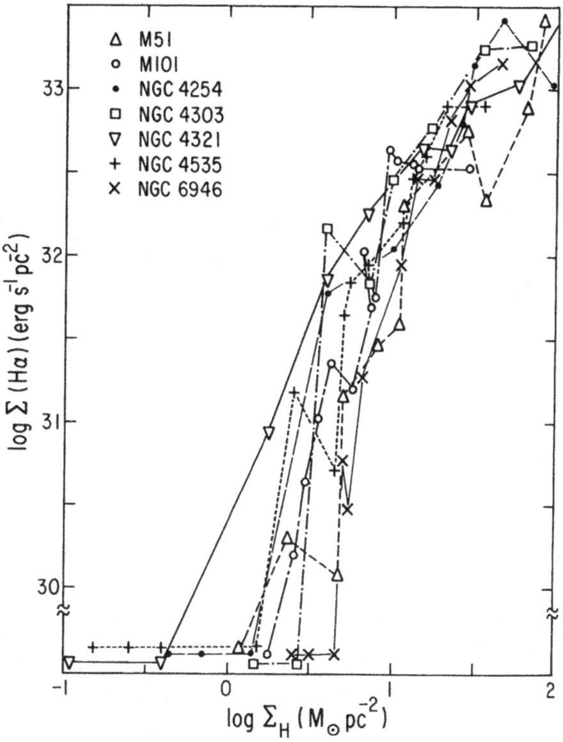

Fig. 15.2 Star formation in spiral galaxies reduces as the gas density becomes lower, until a threshold level is reached below which stars are no longer formed. The density of the interstellar gas is shown horizontally and a measure for the numbers of newly formed stars is shown vertically (Kennicutt Jr. 1989)

center was always one of the first objects to be observed whenever there were new observation opportunities (a new instrument, or an improved technique).

An important juncture was reached in 1974, when Balick and Brown in the US demonstrated by interferometric means that Sagittarius A contained a very small source, smaller than 0.1 arcsec (Balick and Brown 1974). Over the years, this small source has been given a range of different names, but it eventually became known as Sgr A*. At our distance from the Milky Way center, 0.1 arc sec is equivalent to 800 Astronomical Units or 800 times the distance of the Earth to the Sun. Finding such a small object was an impressive technical achievement, but it was not enough to determine whether SgrA* was a black hole: a hypothetical black hole, a billion times as massive as the Sun, had a Schwarzschild orbit that was another 40 times smaller. Even though everyone expected that Sgr A* was a 'super-massive black hole', it had not yet been proven.

In 1983 an interesting question arose: did SgrA* move perceptibly across the sky? If it did, it could be a random object from the Milky Way, but if it remained static, a good argument had been found that SgrA* was the center of our Milky Way. Measurements gave the answer: SgrA* was stationary when its position on the sky was compared with the positions of very distant quasars (Backer and Sramek 1982). The measurements were repeated in 1999 with greater precision and confirmed the earlier result.

The center of the Milky Way was also an important object for observations with new infrared techniques. The infrared radiation was mapped, although the images were not as sharp as those produced by radio astronomers. A bonus from the infrared measurements were bright spectral lines, such as neon, that provided information about gas movements in the vicinity of SgrA*, although nothing about the source itself.

In 1993, with better equipment and an interferometer with a longer baseline radio astronomers made measurements at a wavelength of 7 mm showing that SgrA* was smaller than 3.3 astronomical units (Backer et al. 1993). The measurement was another wonderful example of craftsmanship, but once again it gave no new direct evidence that the object had to be a black hole—SgrA* was just a very weak radio source, an unpretentious, negligible object compared with, for example, the centers of radio galaxies such as Cygnus A or Virgo A.

Let us take a look at our large neighbor, the Andromeda nebula M31. Balloon observations 30 years previously had already shown that M31 had a very small core: see Chap. 5. In 1993 the Hubble Space Telescope offered the possibility of repeating the measurements with a superior instrument. The result can be seen in Fig. 15.3. Two centers are present in the shape or two peaks on a contour map. Top left is the

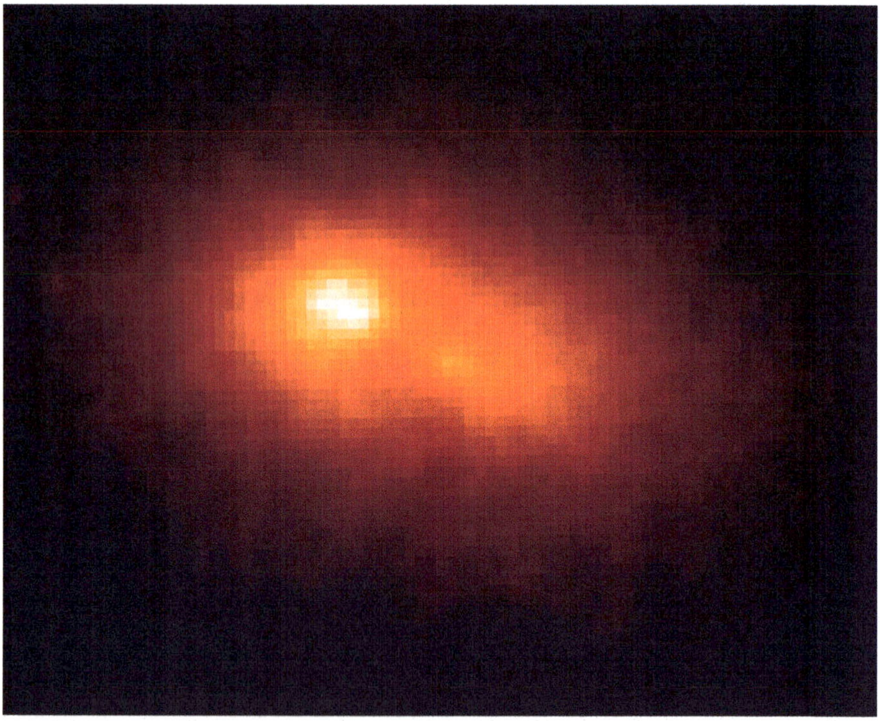

Fig. 15.3 The center of the Andromeda nebula as 'planetary camera' for the Hubble Space Telescope. The distance between the two brightest points is less than 1 arc second (Lauer et al. 1993)

strongest. Analyses of the image and of spectra recorded previously led to the rather unexpected conclusion that the fainter of the two was a black hole with a mass of between 30 and 70 million times the mass of the Sun. The other peak could be a cluster of stars; it might even be the core of a small galaxy that had been swallowed by M31. The authors (Lauer et al. 1993) were by no means certain of their conclusions because the two sources were less than 1 arc second apart. In measurements from the ground they would always be seen as one source and not as two separate sources: the Hubble Space Telescope generated new knowledge that was impossible to discover from the surface of the Earth.

Stars from Prehistory

It has already been said several times: the chemical composition of stellar atmospheres was always the same: 90% hydrogen atoms, 9% helium atoms and all other atoms together made up the last 1%. This very uneven distribution was explained in the 1950s when calculations were made of what happened to the gas that filled the Universe immediately after the Big Bang and before the stars and galaxies started to form (see Chap. 5). From this it followed that the matter could not contain any atoms heavier than helium[1] and so these other atoms must have been made in stars and those were formed much later. The more mass a star had, the more new atoms it created. Supernovae were special objects because during their explosion the temperature and the pressure were many times higher than in a standard star and many new atoms were made that could not have formed in ordinary stars.

To measure the atomic composition of a star you look at the stellar spectrum. In a main sequence star, the stellar spectrum reflects the atomic composition of the gas from which the star was created. Measuring the composition of the atmosphere required spectra of high quality that had been stretched out over a large number of detector pixels. Figure 15.4 explains this. In order to get enough light on each pixel, a large telescope was needed. Such telescopes came into use between 1985 and 1995.

In the 1950s, stars were discovered in which the percentage of 'other' atoms (those heavier than helium atoms) was lower than, for example, in the Sun (see Chap. 2). This gave rise to the idea that the percentage of 'other' atoms had grown gradually over the long existence of our galaxy: when stars disintegrated due to their old age, they returned 'enriched' gas to the interstellar environment and subsequent generations of stars contained a higher percentage of 'other' atoms. People started to suspect that the history of our Milky Way could be deduced on the basis of the increasing percentage of 'other' atoms.

[1] I am ignoring the small amount of lithium that also came from the Big Bang. It does not play any meaningful role here.

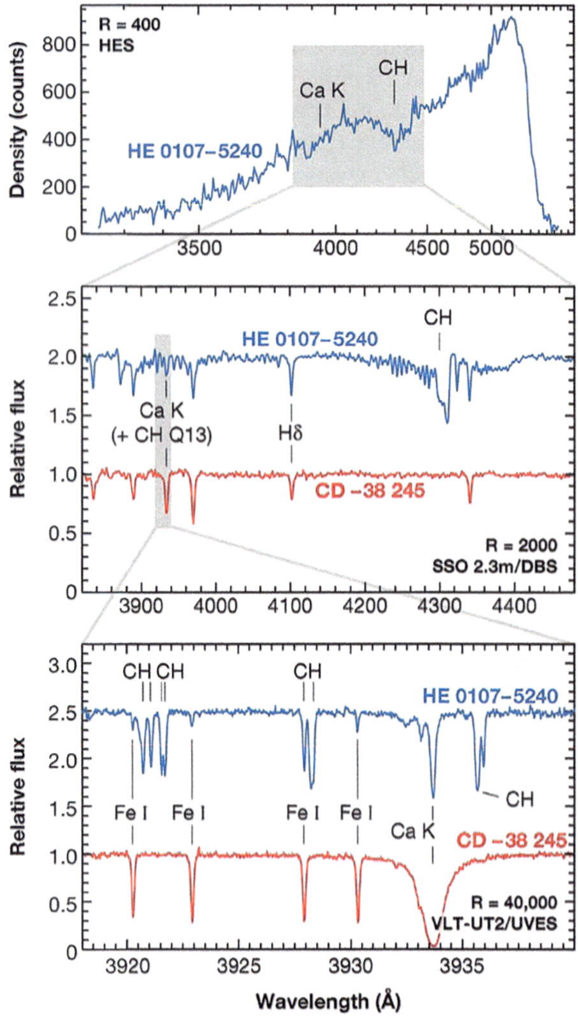

Fig. 15.4 Three versions of a partial spectrum of the same star. The second image from the top is a part of the upper image, in which details can now be seen that could not be measured in the uppermost image. Red and blue curves represent different stars. The lowest spectrum is detailed enough to measure how many atoms (and molecules) of different types were present. Measuring detailed and reliable spectra generally required a large telescope (Beers and Christlieb 2005)

Theoretical studies, for example the B^{2}FH article from 1957 mentioned in Chap. 2, provided a number of guidelines. Three helium atoms ('alpha particles') could be compressed to one carbon atom, a process called the 'triple-alpha process'. Carbon atoms gave rise to consecutively heavier atoms by repeatedly capturing an alpha particle: ^{12}C, ^{16}O, ^{20}Ne, ^{24}Mg, ^{28}Si, ^{32}S, ^{36}Ar, ^{40}Ca, ^{44}Ti, ^{48}Cr, ^{52}Fe, ^{56}Ni; these atoms were often referred to as alpha atoms. Atoms heavier than nickel were formed by the capture of protons (^{1}H), which could happen in two different processes: the rapid r-process and slow s-process.

This was all basic theory and there was no way of being sure how it occurred in practice. But it inspired the observers to search for recurrent patterns in the abundances of the atoms and so investigate the history of the galaxies.

The idea that the next generation of stars contains more heavy atoms led to various model calculations of the history of our Milky Way. Since 1957, numerous such models have been made that describe the birth of stars, their decay and the new stars that were born from the matter left behind; gradually more and more heavy atoms were added to the interstellar matter and the new stars showed the growth of the number of heavy atoms. Later generations of stars were described as 'enriched' by more heavy atoms. But all the models failed. The failure was most convincingly seen in the so-called 'G-dwarf problem': all models predicted that in our local neighborhood there should be many of these stars but with low 'metallicity'. But there were no such 'low-metallicity' G-dwarfs present and there were no good explanations why the models failed.

As has been said, the quality of the spectra gradually increased. The content of all kinds of atoms was measured more precisely. Spectral lines were also found of elements heavier than nickel, such as those of the europium atom (Z = 63), an element that must have been formed via the so-called r- or s-process. In spite of the progress with observations, the theoreticians had insufficient observational data to make successful models. In 1992 a well-known British astronomer, Bernhard Pagel, ended a review with the words: 'Playing with all these ideas is fun, but we do need better data' (Pagel 1992).

A long-term observational study by Danish and Swedish astronomers was published in 1993 (Edvardsson et al. 1993). Their programme had been very extensive and the analysis had taken several years. Some 190 stars were chosen that were comparable with the Sun and their spectra were measured precisely and analysed carefully; the age of each star was determined and the orbit was deduced that the star had followed in our Milky Way. This amounted to first-class work, extremely painstaking, and it generated some impressive observational material.

Figure 15.5 shows how the abundance of iron (Fe), α-elements and barium atoms was on average lower in older stars than in younger ones. But the figure also showed that at each age there was a considerable variation in abundances, for example, in the percentages of iron atoms. These differences, the authors reported, were not due to measurement errors, but must be an intrinsic effect. They concluded that at any time in the past stars were born with very different compositions. There was no simple, straightforward relation between age and atomic composition. This was a sobering thought.

Besides the research described above, there was also another related, but different study: a few observers searched for 'primordial' stars, the stars that belonged to the first generation of stars formed after the Big Bang. In the 1970s a major programme was started on Mount Palomar that aimed to select candidate primordial stars. This, too, was painstaking work because it was like searching for a needle in a haystack: there were hundreds if not thousands of stars on each photographic plate, of which only one or two were potentially primordial. In spite of the time-consuming nature of the work, it took until 1992 before a list of some 440 stars had been compiled in which the percentage of iron was at least 100 times less than in the Sun. Among these stars were three of which the iron percentage was even 10,000 times less than that of the Sun. This collection, referred to as the 'HK catalogue', would provide the material for further, more detailed research for many years (Beers et al. 1985).

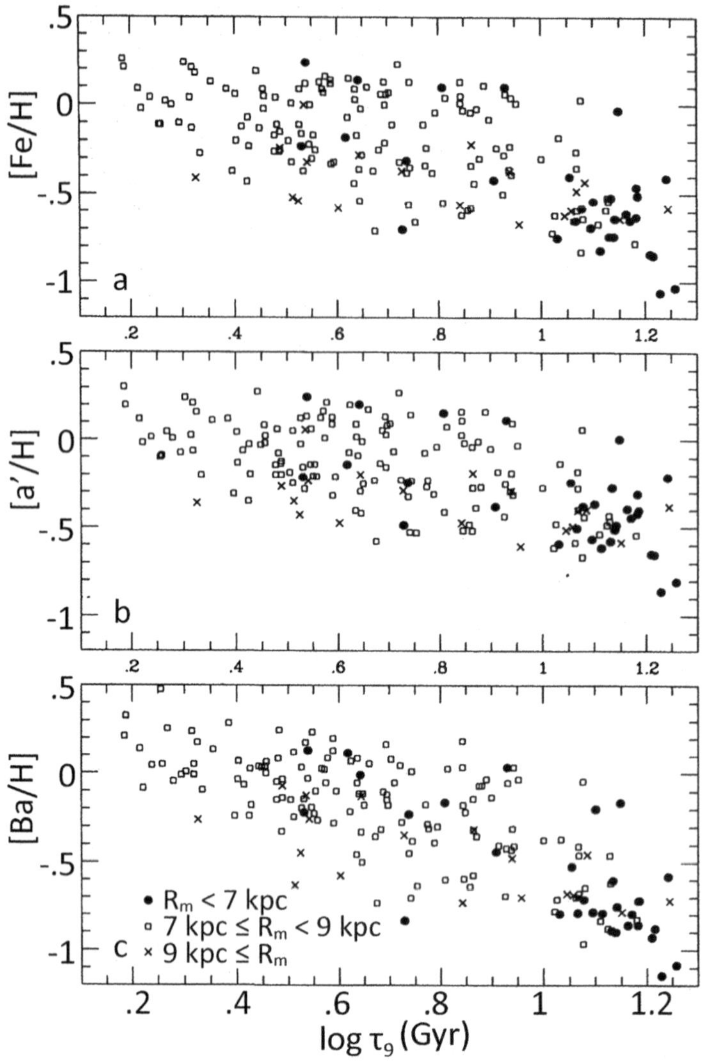

Fig. 15.5 Three diagrams in which the abundance of an element in a star is plotted against the age of the star. The age is given horizontally, and the abundance vertically. The abundance of iron is indicated as '[Fe/H]' which means: the amount of Fe atomic nuclei divided by the number of hydrogen atoms, and this ratio compared with the value of Fe/H in the Sun; all in logarithmic numbers. The center diagram indicates the abundance of the above-mentioned α-atoms (Edvardsson et al. 1993)

One such follow-up study revealed some surprises (Sneden et al. 1994). A star that had the least amount of iron appeared to have ten times too much carbon. This result did not fit in any conceptual scheme. There was no basic understanding of the enrichment of matter; there was more work to be done.

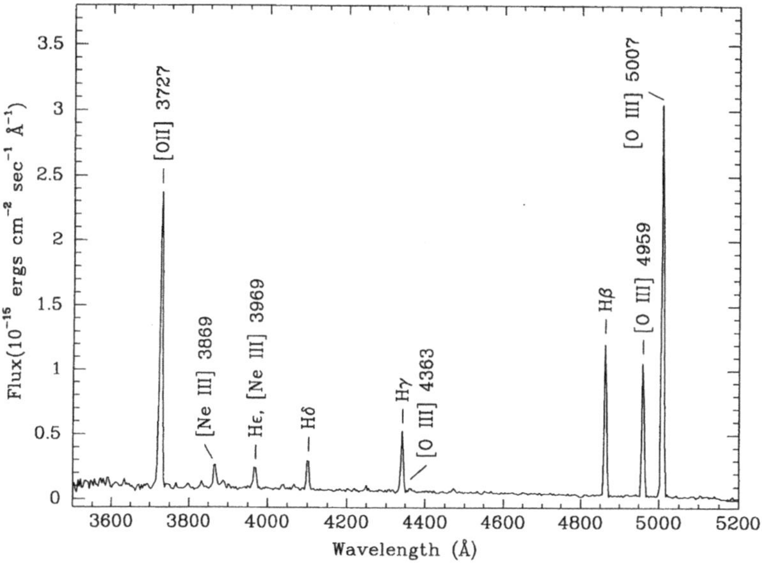

Fig. 15.6 Emission line of the spectrum of galaxy NGC 2541. The strongest lines have been identified (Zaritsky et al. 1994)

Chemical Abundances in other Spiral Galaxies

The atomic abundances in other spiral galaxies could be measured by taking spectra of light emitting nebulae. Figure 15.6 shows the spectrum of such a nebula. The measurement of the oxygen lines showed that the abundance of oxygen was greater when the spiral galaxy itself contained more stars: in other words, larger galaxies contained relatively more oxygen atoms: when more stars were born, the production of new atoms was enhanced (Zaritsky et al. 1994). This result was understood qualitatively, but there was no quantitative explanation.

15.3 Dwarf Galaxies

If all the known galaxies were put in a row, depending on the amount of light that they produce, what one would be left with is a long row of faint galaxies. These were the dwarf galaxies with up to a hundred million stars. The two Magellanic Clouds were examples of relatively large dwarf galaxies. The Magellanic Clouds contained lots of interstellar matter and they produced many new stars. In Chap. 5 the small galaxy I Zw 18 was mentioned, in which so many stars were formed that the gas would have been exhausted very rapidly. The impression was that for a long time this galaxy had formed hardly any stars and then, for some unknown reason, it had started to form stars very rapidly.

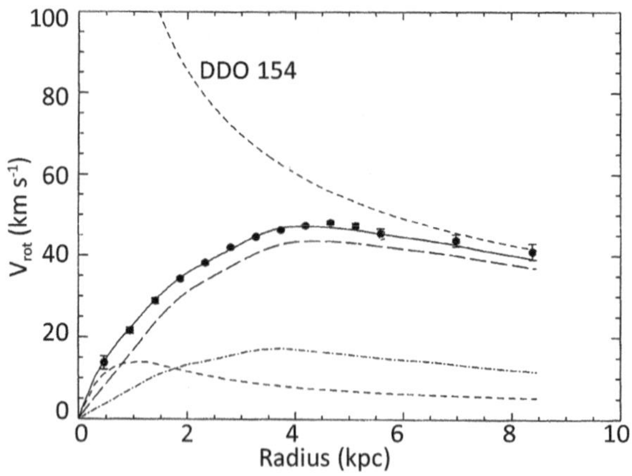

Fig. 15.7 A diagram showing the rotation velocity in the DDO 154 dwarf galaxy. The distance to the center is shown horizontally, and the rotation velocity vertically. The dots are measurement points. The curve of long dashes is the contribution by the halo. The lowest arc of dashes is the contribution of the stellar disk. The upper curve of narrow dashes is the rotation velocity if all matter were concentrated in the center instead of in a big envelope around the galaxy. (Analysis by C. Carignan, cited in Ashman, 1992)

A related object, DDO 154, contained so much hydrogen gas that the rotation of the galaxy could be measured using a radio telescope. It was the same old story: the rotation velocity of the gas continued to be high even when astronomers looked outside the visible galaxy: there was a lot of dark matter: see Fig. 15.7. This result added strength to the idea that dwarf galaxies are also embedded in a larger halo of dark matter.

A significant number of the faintest dwarf galaxies appeared to be completely transparent: you could look straight through them. They probably contained almost no interstellar matter. Figure 15.8 shows an example.

15.4 Mergers and Starbursts

Earlier chapters discussed matter that enters a galaxy from outside. Mergers of galaxies were also discussed. It is an issue that will increasingly become an important subject because, as we will see in later chapters, a successful new theory will predict that larger galaxies arise from mergers of smaller galaxies.

In the search for mergers in the distant and earlier Universe, a difficult problem cropped up: from a great distance it was sometimes difficult to see the difference between two merging galaxies and a galaxy with an active core, an AGN. Both types had a core that was very bright both in infrared and at radio wavelengths. The only way to distinguish between the two was through detailed study (Condon et al. 1991). An example is object Arp 220 which will be discussed shortly. Let us first look at a few examples of merging galaxies.

Fig. 15.8 'SagDig', one of the dwarf galaxies in Sagittarius. It is transparent (you can see galaxies behind it) and thus contains little interstellar matter. Most individual stars in the image are foreground stars and are located within our Milky Way. Source: NASA, ESA, and the Hubble Heritage Team STScI/AURA

Too close to Big Brother

An exciting find in 1994 was the discovery of a small galaxy that was being consumed by our own large Milky Way. The discovery was made by chance: the discoverers were engaged in a study of older stars close to the bulge of our Milky Way. They recorded spectra and derived the velocities of the stars. The velocity of the stars fitted quite well within an expected pattern, but there was a group that was

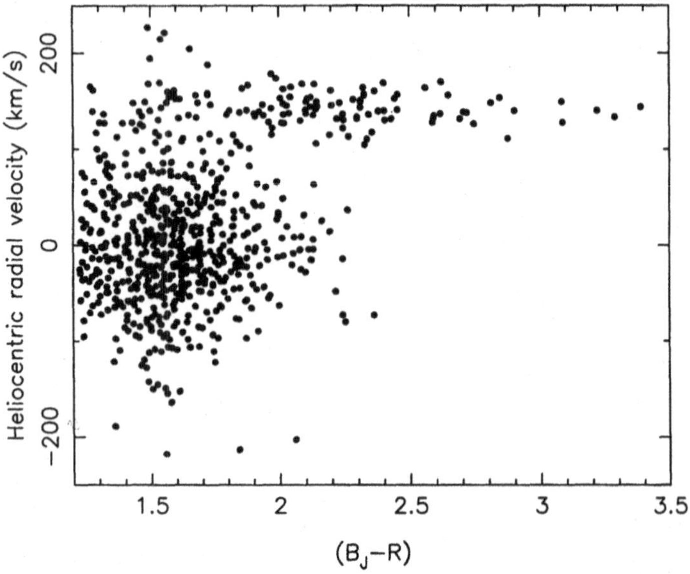

Fig. 15.9 In an area of the sky just above the center of our Milky Way, the radial velocity was derived for a large number of stars from the redshift of spectral lines. The color of the stars is indicated horizontally (the difference between the magnitude in blue and in red) and the velocity in km/second is indicated vertically. To the surprise of the authors, there was a horizontal series of stars traveling at a velocity of roughly 140 km/second. This was a stream of stars moving towards the plane of the Milky Way (Ibata et al. 1994)

clearly different. When the observers had recovered from their surprise, they decided they were seeing a stream of stars falling into the Milky Way (Fig. 15.9).

Figure 15.10 is a montage that shows where the small dwarf is ingested by our Milky Way.

The Magellanic Stream and the Structure of the Small Magellanic Cloud

The Magellanic Stream has already been discussed in Chap. 5: it is a stream of gas from the Large Magellanic Cloud into the halo of our Milky Way. The subject continued to occupy researchers, even though they had not managed to create a definitive model. In 1986, Australian Don Mathewson reported the interesting fact that the Small Magellanic Cloud was made up of two parts, one of which, seen from our perspective, was found some way behind the other. The evidence was clear. Mathewson explained the tearing apart of the Small Cloud as the result of a collision with the Large Cloud around 400–500 million years ago. Japanese researchers Murai and Fujimoto had already reached the same conclusion in 1980 (Murai and Fujimoto 1980). In 1987 Mathewson and several colleagues used this breaking apart of the

Fig. 15.10 The area around the center of the Milky Way and the small galaxy that is falling towards the Milky Way surface. Below the Milky Way, one part of the small galaxy can be seen and above the Milky Way a few stars are found that seem to belong to the small galaxy. They are on their way out of the Milky Way. Source: Astronomy Picture of the Day, 16 February 1998, R. Ibata (UBC), R. Wyse (JHU), R. Sword (IoA)

Small Cloud as the basis for a model to explain the Magellanic Stream as shown in Fig. 15.11. There still is no certainty that this was the real course of events, but it encourages the research on the local history in our Universe.

M82

If a large number of stars are formed in a short time and if these include massive stars, then after an astronomically brief time (a few millions of years), the massive stars will explode as supernovae. This is what happened in galaxy M82; the galaxy became the prototype of a starburst galaxy. Figure 15.12 shows a whole series of small radio sources that coincide with the galaxy: it is the cigar-shaped object in Fig. 5.10 in Chap. 5. Figure 15.13 shows the location of these radio sources on a photo of M82. The radio radiation of these sources was stronger at longer wavelengths, which was a reliable indication that the radio radiation was synchrotron radiation, and therefore from high-energy relativistic electrons in a weak magnetic field. If these sources were the result of supernova explosions, then with time the radio radiation of the sources would become fainter; measurements confirmed that this was the case. All in all, these were strong indications that the galaxy M82 had recently experienced a period of increased star formation.

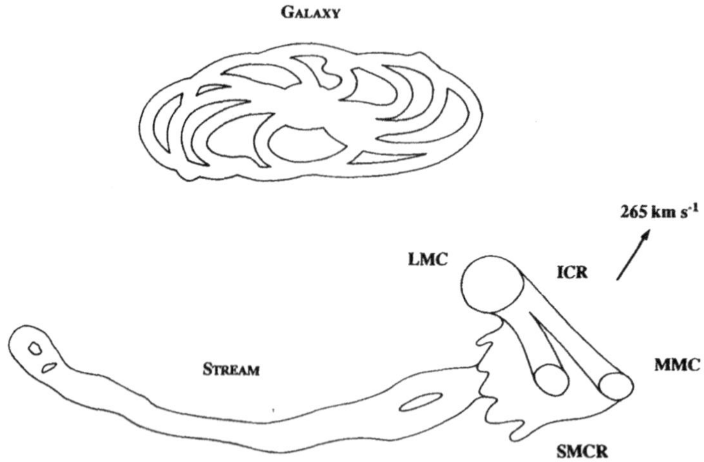

Fig. 15.11 A sketch of the position of our Milky Way, the Great and Small Magellanic Cloud and the Magellanic Stream. The whole complex outside our Milky Way revolves like a satellite around our Milky Way. An explanation of the acronyms: LMC, Large Magellanic Cloud; SMCR, Small Magellanic Cloud Remnant; MMC, Mini Magellanic Cloud; ICR, Inter-Cloud Region (Mathewson et al. 1987)

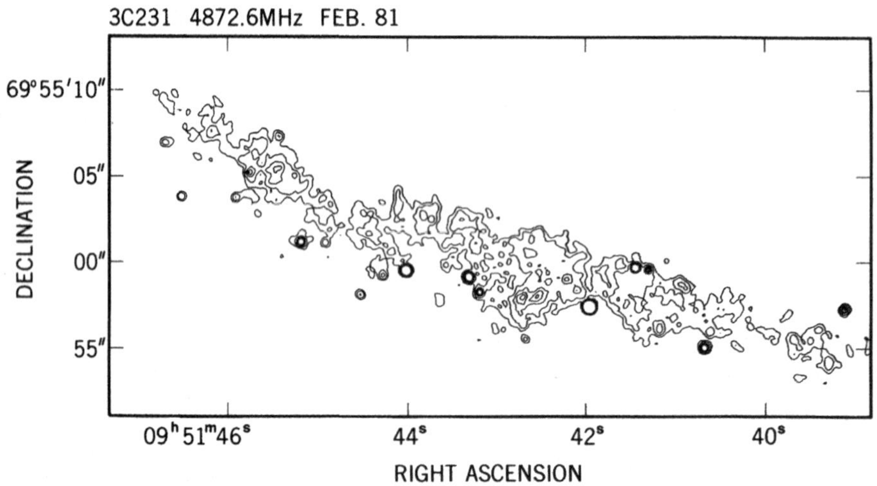

Fig. 15.12 A map of the radiation of M82 at a wavelength of 6 centimeters. All small sources are a remnant of a supernova; new supernovae explode about once every 10 years. For comparison: in our own, much larger Milky Way, there is one explosion every 50 years (Kronberg et al. 1985)

Super Winds

Imagine a disk galaxy that experiences a starburst. Soon numerous supernovae will explode in quick succession. Figure 15.14 shows a photo of a galaxy with strong infrared radiation and a suspected 'starburst'. The photo shows gas flows that are

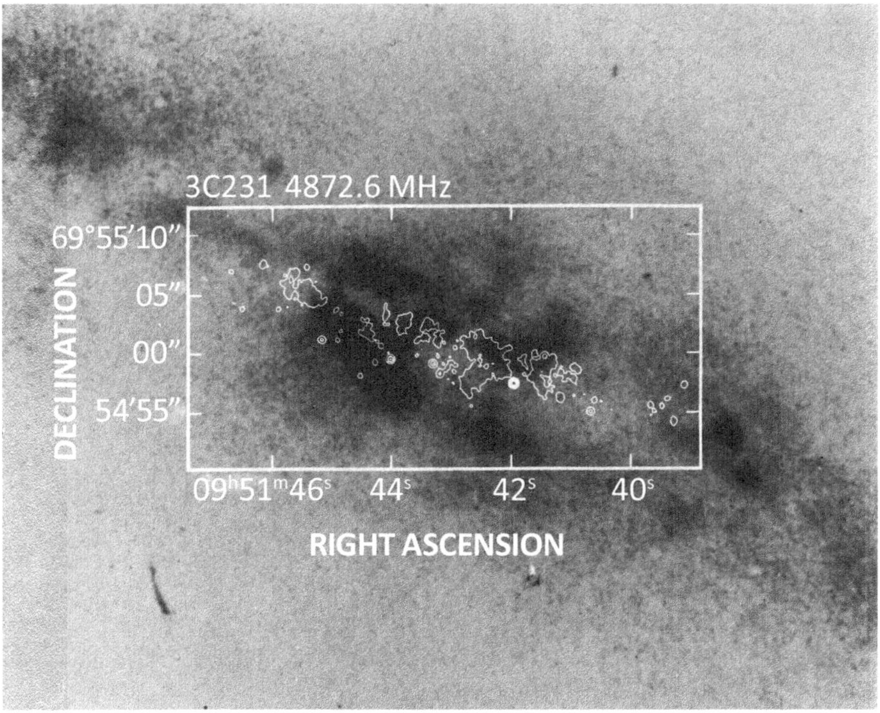

Fig. 15.13 The radio chart of Kronberg_Biermann_85a projected on a photo of M82 (Kronberg et al. 1985)

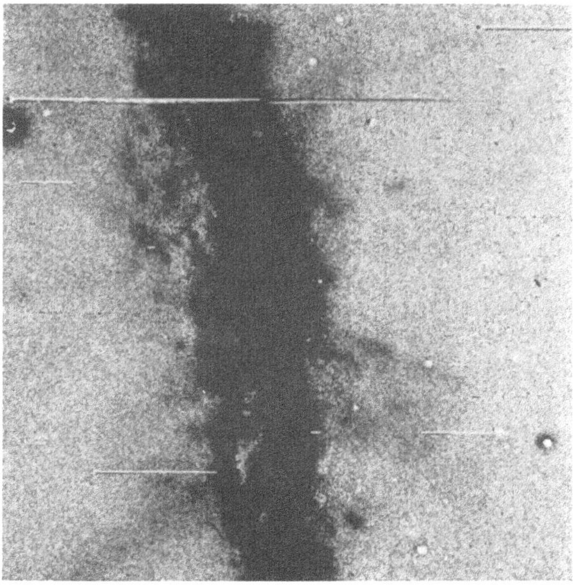

Fig. 15.14 A part of the Seyfert galaxy NGC 3079, photographed edge-on and through a filter that only allows light to pass through from a spectral line of hydrogen (Hα) and of ionized nitrogen. We are looking at the disk galaxy from the side and see projections from the surface that are attributed to explosions in the central plane of the disk (Heckman et al. 1990)

perpendicular to the surface of the galaxy. This outflow is so much stronger than what is normally found in old stars and has been referred to as a 'superwind'.

ARP 220: an unusual Case

In 1971 it was discovered (see Chap. 5) that M82 had experienced a starburst and produced a lot of stars in a very short space of time. The discovery caused a sensation: a starburst was at that time something completely new. The thought was entertained that M82 could be an active galaxy, a galaxy with a monster at its core. This proved not to be the case: what happened in M82 was not directed from the core. We will look now at Arp 220, object number 220 in Arp's 'Atlas of Peculiar Galaxies'. The image on the photographic plate was difficult to interpret. Although Arp 220 is very distant, a billion light years away, the IRAS survey showed that it was an exceptionally strong infrared source. The galaxy even became the prototype of a new class of galaxies, the 'Ultra Luminous Infrared Galaxies' or ULIRGs.

Figure 15.15 shows a detailed map of Arp 220. Two peaks can be seen on a large plateau. Surprisingly, the valley between the two hills was only an illusion: right above the top there was an obscuring cloud producing an illusionary dip. The observation of the radio radiation of CO showed that the dip was filled with a molecular cloud.

Other observations were made that showed amazingly minute details (see, for example, Fig. 15.16). Compare the scale of this image with the one above. Again

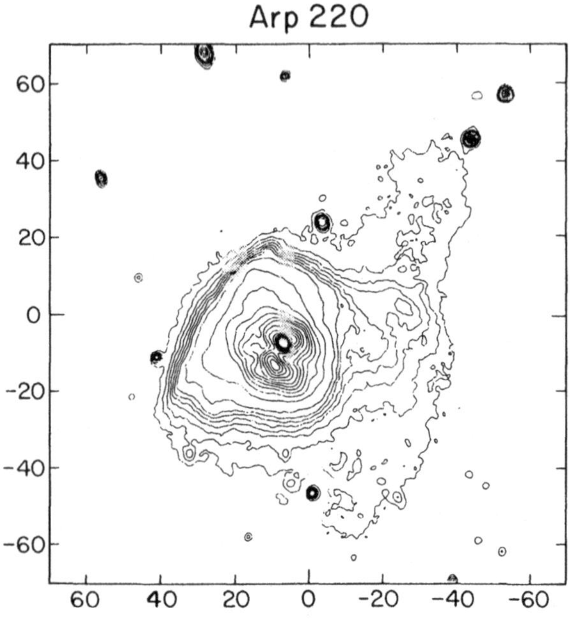

Fig. 15.15 A contour chart of Arp 220 with lines of equal luminosity in red light. An almost triangular plateau can be seen on a hill with two peaks and a valley between. X- and Y-axis are measured in arcseconds (Sanders et al. 1988)

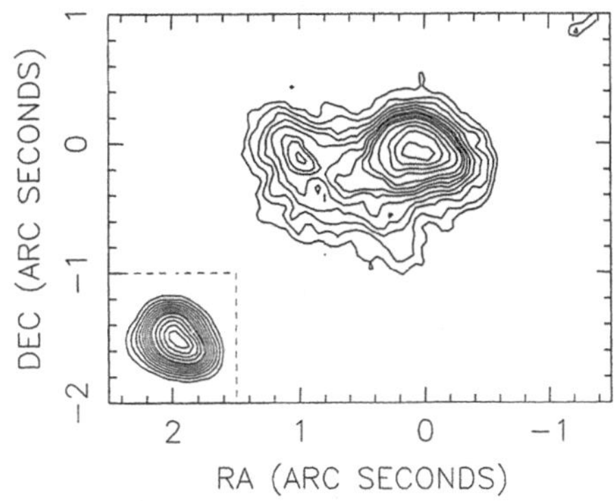

Fig. 15.16 Chart of the 2.2 μm emission of Arp 220. The innermost 3 × 3 arcsec are pictured with a resolution of 0.4 × 0.5 arcsec (Graham et al. 1990)

there were two sources of radiation that were now closer together. The authors suggested that we were seeing here the cores of two galaxies. In the near future they will probably merge: Arp 220 is en route to a merger of two cores to one massive black hole. Arp 220 could (astronomically) soon become a quasar.

15.5 Active Nuclei of Galaxies (AGNs)

Let us look now at galaxies with an active nucleus and at the research in the period from 1985 to 1995. Galaxies with an active nucleus are quite different from objects that appear to be two merging galaxies. Nonetheless, and this was certainly so in the case of Arp 220 (above), seen as a very small object at large distances, the difference between the two types of objects was not always easy to detect and observers had to search hard for signals that pointed to one or the other. The difficulty lay in the fact that a galaxy with an active core and a merger of two galaxies often had strongly distorted images. Historically, the problem had already been pointed out in 1954 when Baade and Minkowski took a photograph of the galaxy at the center of the radio source Cygnus A and, after some hesitation, decided that the central object consisted of two colliding galaxies; see Chap. 1. After this, Cygnus A became the prototype of a galaxy with an active core.

Quasars

In 1978, gamma radiation was measured coming from the 3C273 quasar (Swanenburg et al. 1978) and for a long time this appeared to be the only quasar that emitted gamma radiation. This changed in 1991 when NASA launched its 'gamma satellite', CGRO (see Chap. 9), after which more quasars were discovered that emitted gamma radiation. The first was 3C279 and it was a strong source (Hartman et al. 1992). 3C279 was also a 'blazar': a quasar oriented such that the jet pointed in our

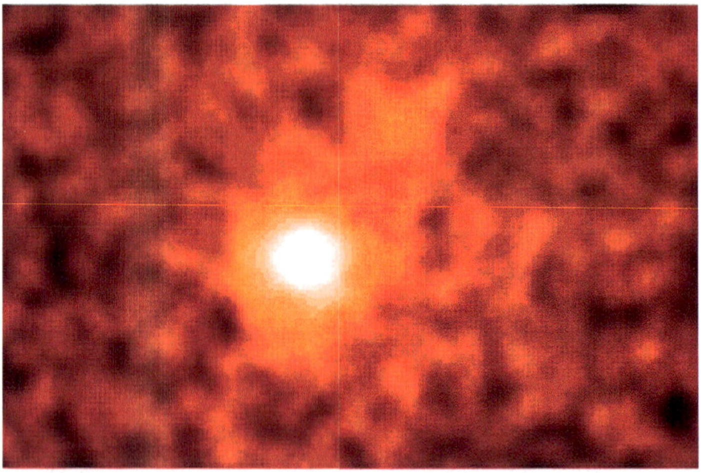

Fig. 15.17 The gamma radiation of 3C279: a peek into the maw of the monster, the entrance to hell. Astronomy Picture of the Day, 26 December 1998, EGRET team, Compton Observatory, NASA

direction and we are looking directly into the jet. The photons of the gamma radiation had energies of up 5 GeV, more than a billion times the energy of for example, blue light. The authors found something else: 3C273 was discovered in 1978 and then the observers did not detect 3C279 although both were close to each other in the sky. This indicates that the gamma radiation from 3C279 must be highly variable. The association of variable gamma radiation when you look down into a jet inspired a number of theoreticians to put forward an explanation (Maraschi et al. 1992; Sikora et al. 1993). This explanation convinced them that the GRBs, gamma ray bursters, were of the same kind (see Chap. 14). Figure 15.17 shows what you see in the maw of the monster.

Another important discovery arose from further studies of ULIRGs, the Ultra-Luminous InfraREd Galaxies. Researchers examined the plates of the good old Palomar Sky Survey (see Chap. 3) for the images of the ULIRGs and found that in all cases the image was distorted: the objects looked like merging galaxies. The conclusion was drawn that the ULIRGs were objects in the stage just prior to the flaring up of a quasar (Sanders et al. 1988).

In this period, large numbers of new quasars were discovered. In 1980 there were 1550 known quasars, in 1986 3681 and in 1992 7315 (Hewitt and Burbidge 1993). This seemed to represent enormous progress, but it was child's play compared with the numbers that would be discovered within a decade (see Chap. 17). Figure 15.18 shows in a histogram the number of known quasars and their redshift. The decrease in the number of quasars with a redshift larger than 2.5 might have been caused by the difficulty of detecting quasars that far away, but later studies showed that the quasar density reached a maximum at the time associated with the redshift 2.5 (Schmidt et al. 1995).

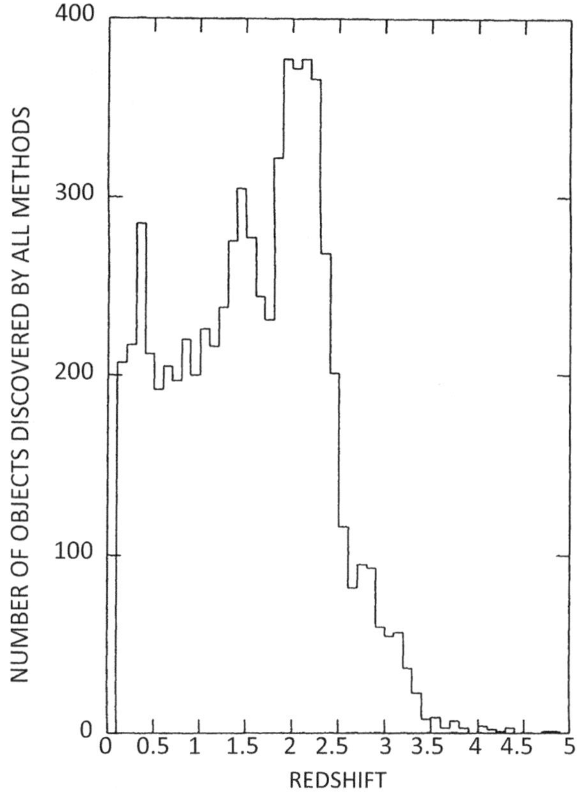

Fig. 15.18 A histogram of the numbers of quasars with a redshift of '*z*'. The total number is more than 7,000. The redshift parameter '*z*' is indicated horizontally, and the number of quasars with the value '*z*', is shown vertically. The number of quasars with *z* greater than 2.5 is much smaller and indeed fewer quasars existed in earlier periods than later (Hewitt and Burbidge 1993)

Unification

In the period from 1985 to 1995, quasars, Seyfert galaxies, radio galaxies, and blazars were often observed with all the available instruments, telescopes on Earth or in space. More and more studies were made and each added complexity rather than simplifying the matter. As a reaction the word 'unification' cropped up. Could all the different active galaxies be regarded as a manifestation of one and the same mechanism, a monster, a black hole millions of times more massive than the Sun? Black holes with the mass of a star had been discovered in close X-ray binary stars (such as Cygnus X-1, for example) and in theory it was quite possible that such objects existed with a mass billions of times greater than that of the Sun.

An attempt at unification has already been discussed in Chap. 8 (see Fig. 8.15). In fact, this unification theory was mentioned even earlier, in Chap. 5 (see Fig. 5.20). This figure suggests at least that there is a continuous series of standard galaxies transitioning to radio galaxies and quasars: it is simply a question of increasing energy. In Chap. 8 it was mentioned that in normal, non-active, galaxies there could be an 'AGN-light' nucleus, a more modest type of 'monster'.

In 1985 a discovery was made that drastically simplified the mental picture that people had of AGNs and that promoted unification. This is how it went. It has been indicated earlier that among the AGNs there were two types of 'Seyfert' galaxies. These had very different spectra although both types were disk galaxies. In 1985, Antonucci and Miller, two American astronomers, measured the spectrum of *polarised* radiation from a Seyfert 2 galaxy and saw to their surprise that the spectrum of this radiation was that of a Seyfert 1 galaxy. How was that possible? After thinking about it for some time, they came to the conclusion that the polarisation was caused by the reflection of the radiation by a cloud of free electrons that floated above the nucleus (Antonucci and Miller 1985). (Reflection produces polarisation: the light reflected from a water surface, a pond, a swimming pool, is significantly polarised; 'Polaroid' sunglasses remove the polarised light and thus the reflected light.)

Antonucci and Miller decided that by looking at the polarized light they were looking into the core of the galaxy, just as it is possible to look from one's living room at who is standing at the front door by using a strategically placed mirror on the outside of the window. In Seyfert 1 galaxies you were looking directly into the AGN, whereas with Seyfert 2 galaxies you were looking from the side, and the polarized light came to you after being reflected.

The two types of Seyfert galaxies therefore had very different spectra, but the same hellish machine was present at the core: what you saw depended on the direction from which you were looking. Blandford's reasoning was confirmed that all the 'activity' of galaxies could be explained by two factors: how massive the black hole was and the amount of gas that disappeared in the back hole, or in other words the amount of fuel that the black hole received. The discovery by Antonucci and Miller gave a third parameter: the angle from which you looked at the object.

Unification was in 1993 the subject of a widely discussed review article (Antonucci 1993). The AGNs were divided into either 'radio-loud' or 'radio-silent'. All the other differences could be explained by the difference in the angle from which the galaxy was observed. The author concluded that this explanation was probably somewhat too simple but it was nonetheless a useful start: in the jargon, 'a zeroth order of approximation'. In another article about unification there was an image of how the monster, the hellish machine, looked in a 'radio-loud' galaxy (see Fig. 15.19).

Radio-silent AGNs and radio-loud AGNs had the same monster at their center, but with radio-silent AGNs the jets were missing. How could this difference be explained? As yet no definitive answer has been found, but there was speculation that in radio-loud quasars there is a rotating black hole and in the other type a non-rotating black hole: it is the rotation of the black hole that causes the jet and the radio radiation (Wilson and Colbert 1995).

Figure 15.19 gained the status of the paradigm of AGN research. In the above-mentioned article about this unification (Antonucci 1993), there is a list of some 15 acronyms, each of which refers to an object of a particular type. Unification meant that all these 15 types were very similar objects.

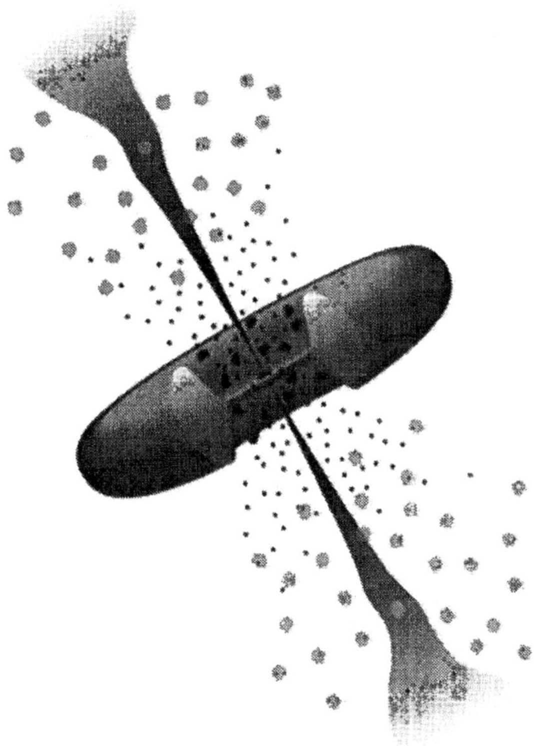

Fig. 15.19 The direct environment of a massive black hole. The black hole is too small to be seen, but it is at the center of the object. Around it there is a disk and around this there is a donut of absorbent matter that prevents the black hole from being observed obliquely. The small dots represent an electron cloud. The large circles represents clouds of ionised gas. Two jets are emitted from the center, from top left to bottom right. The radio radiation comes primarily from the jets (Urry and Padovani 1995)

There is much that is not understood about the AGN. The old question still remains: do all galaxies have a massive black hole in their center, that can be inactive for a long time, a kind of dormant volcano? In our Milky Way, SgrA* *could* be an example of such a sleeping danger.

I will in conclusion give an estimate from 1993: 1% of all spiral galaxies were Seyferts; radio galaxies were 100 times less common and quasars were a further 10 times less common than that. Active galaxies were uncommon (Osterbrock 1993).

Massive Black Holes in the Center of Galaxies

So the conclusion was that a black hole is at the center of all active galaxies and that it is the cause of all high-energy processes. How reliable was the evidence that such a black holes actually existed?

An influential review article appeared in 1995 about the research on massive black holes (Kormendy and Richstone 1995). SgrA* was mentioned in this article,

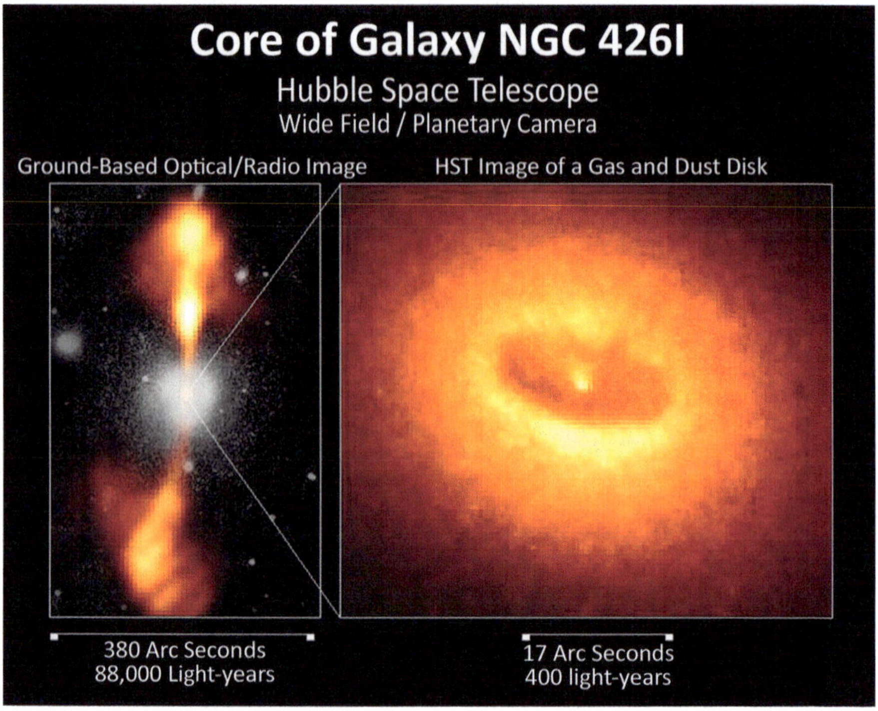

Fig. 15.20 Image taken with the Hubble Space Telescope of the galaxy NGC 4261. In the diagram on the left, the (optical) galaxy is in white and the radio radiation of the two jet streams are in color. To the right is a detailed recording of the center of the galaxy. It shows a ring or donut around a central source. Because a lot of energy is emitted as radiation by such a small source, it is assumed that this small sources is a massive black hole (Jaffe et al. 1993)

as was the center of the Andromeda nebula. The authors added their supposition that such objects also occurred in other well-known and nearby galaxies, such as NGC 3115, NGC 4258 (M106) and NGC4594 (M101, Sombrero). M87, the radio source Virgo A, could well contain a very massive black hole (possibly a billion times the mass of the Sun).

A surprising discovery was that of a massive back hole in M32, an elliptical-shaped galaxy and a satellite of M31. A strong absorption line of calcium could be seen. Its width indicates the average velocity of all stars in that direction. If you looked in the direction of the center, the line was much broader. This meant that here the stars sped through space much more rapidly than elsewhere in M32 and this was attributed to a concentrated ball of matter at the center, around eight million times the mass of the Sun. At the same time the area in which this broadening of the line could be seen was so small that the matter must have contracted to a black hole.

Disks around Massive Black Holes

Radio galaxies emitted a lot of energy in the form of jet streams. This could be produced in only one conceivable way: gas that revolved in a disk around a black hole and that while revolving moved very slowly towards the whole and eventually fell into the hole. The energy was in origin gravitational and it was sufficient to energize the jet. Such a rotating disk had never been seen but it had to be there: the disk was seen close to the stellar black holes within our Milky Way, such as Cygnus X-1 (see Chap. 4).

Whether they had been observed or not, there should be such a disk in every AGN. Finding them was difficult because they were just a small detail in a chaotic environment. Success came in with the Hubble Space Telescope when astronomers looked toward the center of the radio galaxy NGC 4261 (see Fig. 15.20).

A short while later images and spectra of the center of the radio galaxy M87 were obtained using the Hubble Space Telescope. The measurements made it possible to derive the rotational velocity of gas around the center. From the velocity the centrifugal force was calculated and thus the gravitational force of attraction. From this it followed that within a distance of 50 light-years (around 5×10^{19} cm) material was amassed that was at least two billion times the mass of the Sun. Was this a black hole? The answer would be affirmative if all this mass were within a sphere with a diameter of 10^{15} cm. This was much smaller than the observations showed and the conclusion was that the center of M87 might be a black hole although this was not yet proven.

15.6 Clusters of Galaxies

The Virgo cluster of galaxies is the nearest large cluster. In the 1980s, Allen Sandage, Hubble's successor on Mount Wilson and Palomar, undertook a very extensive and precise study of the cluster along with two Swiss colleagues, Gustav Tammann and Bruno Binggeli. Clusters are among the largest structures in the Universe and when studying very distant clusters it is important to have a good inventory of a nearby cluster. Each of the hundreds of galaxies of the Virgo cluster was carefully photographed and analyzed and an extensive inventory was drawn up. Figures 15.21 and 15.22 show some of the results (see legends with the figures).

It occurs to me that these images represent the end of a somewhat outdated scientific vision. The images embody the spirit of Hubble's tuning fork vision of the world of galaxies. This vision was largely phenomenological rather than quantitative. In the future the quantitative aspects will increasingly determine the discussions and the role of Hubble's tuning fork will decrease.

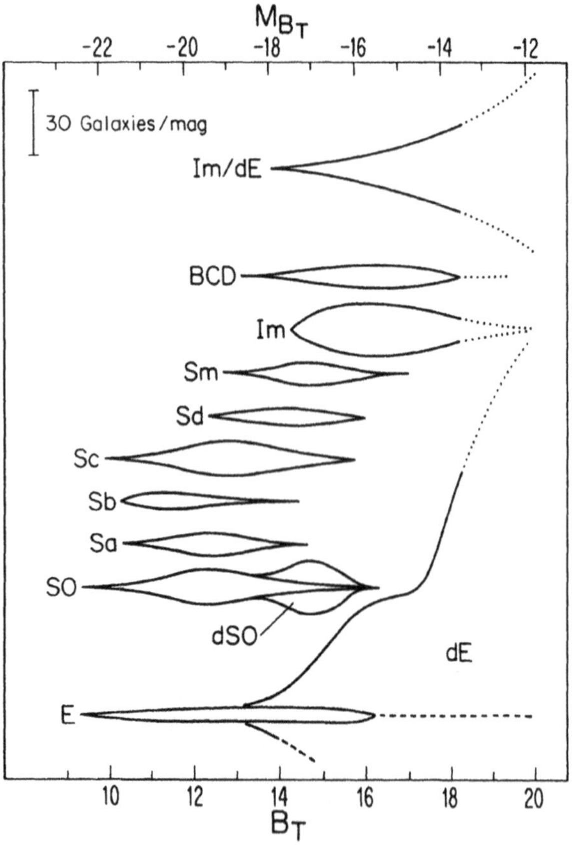

Fig. 15.21 The frequency of galaxies of different types in the Virgo cluster. Some of these types do not occur in Hubble's tuning fork. Along the horizontal upper edge is the absolute magnitude, the magnitude the object would have if it were at a distance of 10 pc. The magnitude increases toward the right, which means that less and less light is received. The numbers of galaxies of different types are indicated vertically. The dotted extensions are estimated because these dwarf galaxies are so weak that only a few of them had been detected. The apparent magnitude of the galaxies is indicated along the horizontal lower edge. There is a constant difference of around 32 between apparent and absolute magnitude (Sandage et al. 1985)

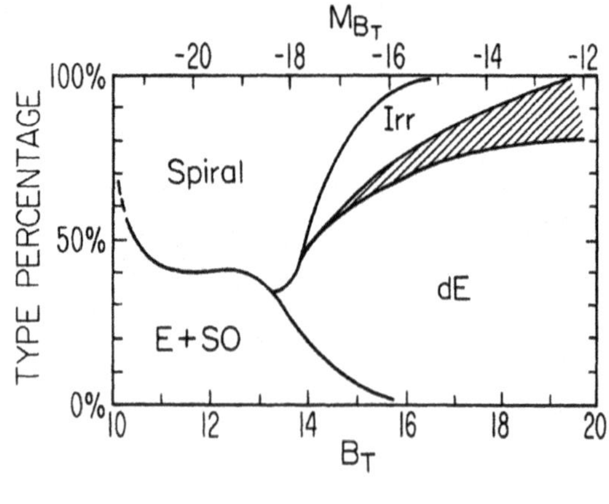

Fig. 15.22 The numbers of diverse types of standard galaxies in the Virgo cluster. The luminosity, expressed in absolute magnitudes M_{BT}, is indicated horizontally along the lower edge. The percentages of the different types are given vertically. See further the inscription for Fig. 15.21 (Sandage et al. 1985)

Gravitational Lenses

Chapter 8 discussed the discovery of the first gravitational lens. In 1985, only five of such lenses were known: the number increased slowly. In December 1989, Roger Blandford gave a lecture in England in which he expressed a very clear message: 'Gravitational lenses are welcome as a means of finding out more about dark matter, but so far that promise has not been met. However, that's now about to change.' He was wrong as far as the past was concerned (see immediately below) but in his vision for the future he was right: something did happen.

Two years before this lecture, in 1987, an object was discovered by chance that ultimately proved to be a new kind of image made by a gravitational lens (Soucail et al. 1987). The discoverers were French astronomers from Toulouse (not from Paris; how could that be possible?). They looked at photos of Abell 370, a cluster of stellar galaxies with a redshift of $z = 0.374$ and they saw something unusual: an 'arc'. Figure 15.23 shows a more recent photo. The arc is very prominently visible. In the first instance the astronomers involved thought that the object was part of the cluster and that it was a completely new type of physical object. But very soon their Polish colleague Paczynski came up with the suggestion that the arc was a distorted image of the galaxy far behind

Fig. 15.23 A colorful recording of a cluster of galaxies called A370. In the upper right part, an arc, new type of object, can be seen. It is, however, the strongly distorted image of a galaxy far behind the cluster. Source: NASA

the cluster. He was quickly found to be right. The French recorded a spectrum of the arc and this showed clearly that the redshift was much larger than the redshift of the cluster. It now appeared from the photo archives that similar arcs could also be found in other clusters, but that they had never previously been noticed.

This surprising result led to the idea of looking in clusters of galaxies for other distorted images. Because the distortion is produced by the gravitational forces in the cluster, one could perhaps chart the force of gravity and thus determine all the matter in the cluster, the visible and the dark (Fig. 15.24) (Schneider et al. 1986; Tyson et al. 1990; Kaiser and Squires 1993).

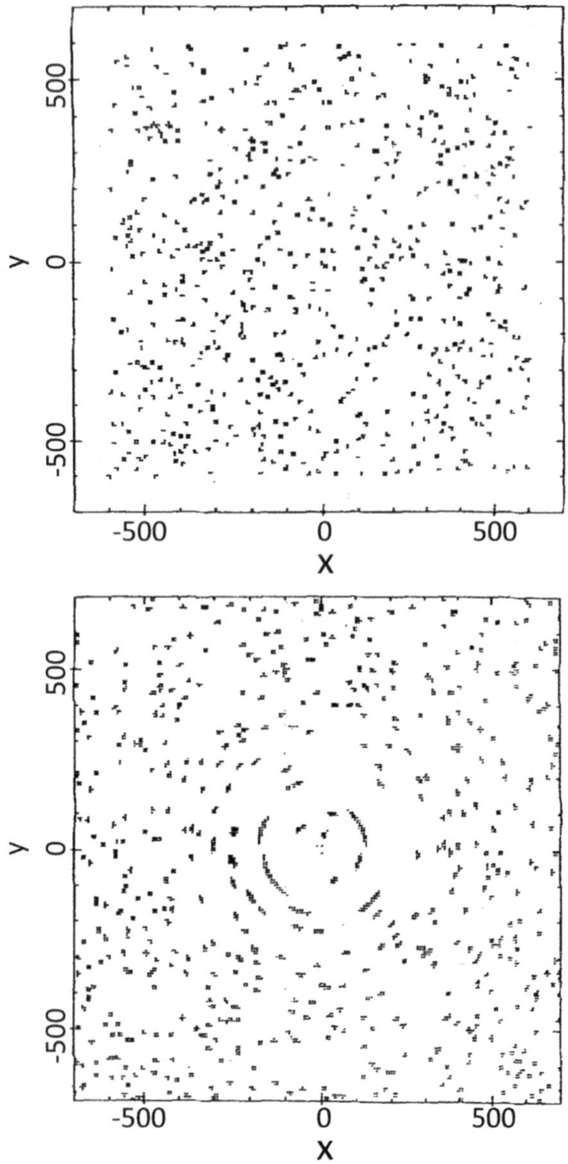

Fig. 15.24 A computer simulation of the same field with galaxies observed without a gravitational lens (above) and (below) seen through a lens somewhere between our line of sight to the cluster (Fort and Mellier 1994)

It was these arcs and small arcs that made it possible to trace the dark matter in clusters of galaxies. In Fig. 15.25 the dark matter is indicated in blue. The distribution of the dark matter has the same shape as the distribution of the galaxies: dark and light-emitting matter have the same distribution.

Besides charting the dark matter in clusters of galaxies another development started. When a galaxy was located behind a cluster, the image of that galaxy was made brighter by the cluster lens: put more simply, more rays of light from the galaxy were curved towards us than if there were no cluster. This meant that it would be possible to observe galaxies that by themselves were too faint to detect, but that had by accident been made brighter and observable a by gravitational lens.

Fig. 15.25 A recording of the Abell 1689 cluster made using the Hubble Space Telescope with the distribution of the dark matter projected in blue. Source: Astronomy Picture of the Day, 24 August 2010, NASA, ESA, E. Jullo, P. Natarajan, and J.-P. Kneib

15.7 Large Scale and Even Larger: Cosmology

Voids and Walls

The remarkable distribution of galaxies in space has already been discussed in Chap. 8: the existence of 'voids' with thin walls between them. A major survey of redshifts of galaxies was carried out at Harvard. The results confirmed convincingly (if that was necessary) the existence of 'voids' and 'walls' (see Fig. 15.26).

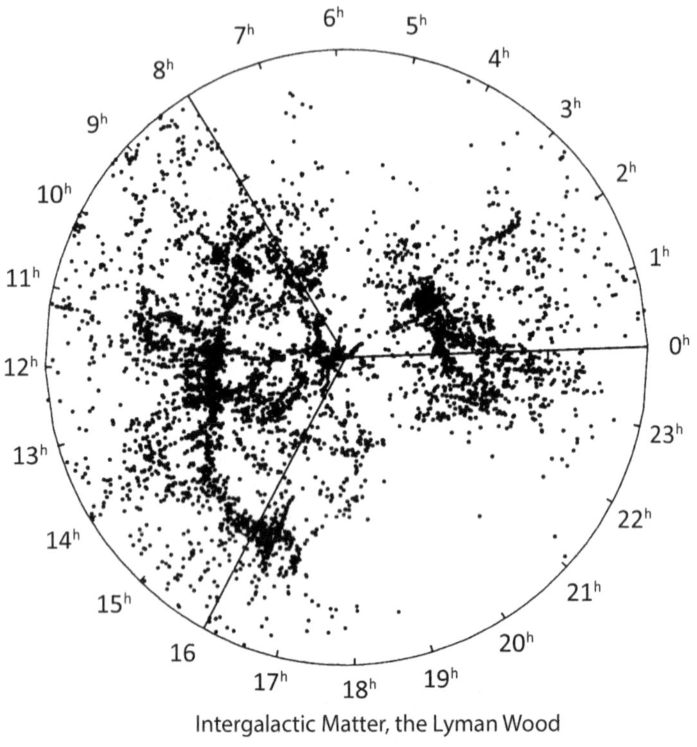

Fig. 15.26 Draw a band around the heavens between 20° and 40° of the equator and make a circle with our Milky Way as the center. Take all 6,220 of the brightest galaxies in this band and measure the redshift of each of them; convert this shift into velocity. Then draw a circle; 0 km/s is the center and the outer edge of the circle at 15,000 km/s. Place each galaxy as a point at the right distance from the center and in the right direction, as indicated at the outside of the circle. This is a flattened map of the distribution of the galaxies around our Milky Way. There are two white regions in the circle: this is where the Milky Way prevents us seeing any galaxy. A 'Great Wall' can be seen in the left half of the circle at about 8,000 km/s from the center (Geller and Huchra 1989)

Intergalactic Matter, the Lyman Wood

Chapter 8 also discussed the absorption lines in the spectra of quasars. A distinction was made between the many thin and faint lines and the few broad lines, which according to an old-fashioned nomenclature were called 'damped lines'. The conclusion was that the thin and faint lines were clouds of hydrogen along the line of sight from us to the quasar. The broad lines were ascribed to galaxies that intersected the line of sight.

This conclusion was proven in 1986 (Bergeron 1986). A galaxy with a redshift of $z = 0.43$ was detected right next to a quasar with redshift of $z = 0.50$. The quasar is thus farther away than the galaxy. In the quasar spectrum a broad absorption line was seen with redshift of $z = 0.43$. This left no doubt: the galaxy caused this broad absorption line in the quasar spectrum. More cases appeared and a generalisation could be made: the broad and strong absorption lines in quasar spectra were caused by galaxies along the line of sight; the thin lines were made by much smaller clouds of hydrogen.

Research on the Lyman absorption lines in quasars was work for large telescopes and long exposure times. A successful combination was that of the 5 m telescope on Mount Palomar with an electronic light detector built in London. Superior observations were obtained that led to a number of important conclusions. The large number of thin lines, often referred to as 'the Lyman Forest', were hydrogen gas clouds that must have been distributed randomly in space. Not only were hydrogen lines present: a close pair of lines was also seen at many wavelengths. These were explained as a 'doublet' made by carbon ions. A remarkable conclusion was that at larger redshifts the carbon 'doublet' became fainter with respect to the hydrogen lines. Within a larger redshift we are looking deeper into the past: thus there was less carbon around in the distant past than presently. This fits in with the earlier discussion of the 'enrichment' in younger stars, that is the increase in the number of 'other' atoms, no hydrogen atoms nor those of helium (Fig. 15.27).

In later years the 'damped' lines were used to find galaxies with large redshifts, as will now be discussed.

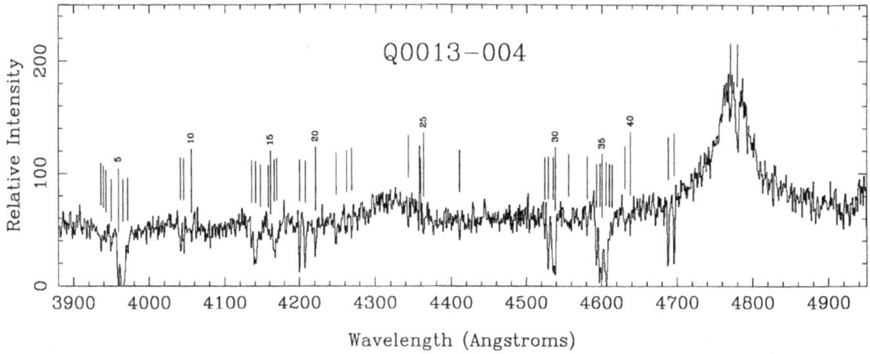

Fig. 15.27 The spectrum of quasar Q0013-004. The broad, high emission line to the right is a line of C^{+++}-ions with a rest wavelength of 1549 Ångström (154.9 nm) and a redshift of $z = 2.083$. Each vertical line indicates a spectral line that could be traced back to absorption by atoms somewhere between the quasar and us (Sargent et al. 1988)

How far can we look back into the Past?

The desire to look far into the Universe is very old. Add to this the consideration that the further we stare, the deeper we look into the past. A simple example is the supernova 1987A that caused some consternation and enthusiasm after its discovery in the Large Magellanic Cloud (see Chap. 14). Because the cloud is 180,000 light years away, that explosion had taken place 180,000 years ago, but the news only reached us now, by looking far away we look into the past. 180,000 years is very little compared to the age of the Universe, 13.5 billion years, and to go that deep, we need objects that billions of years ago emitted the light that we receive now. How and where can we find these objects, and, once we have found them, how can we work out the time it took the light to reach us? Let us call the time the light needs to reach us the 'look-back time', t_{lb}. The look-back time of the Large Magellanic Cloud is thus 180,000 years.

The redshift, z, is what you can measure directly from the spectrum of a galaxy. If z is small, you can calculate the look-back time quite easily. A galaxy with a redshift of $z = 0.15$ is moving away from us with a speed of $z \times c$ and in this case it is $0.15\ c = 45{,}000$ km/s. Because the Universe expands, the speed tells us the distance '*d*' according to the Hubble relation (see the Prologue) that $v = H \times d$. Use for *H*, the Hubble constant, a value of 75 km/s/Mpc and the distance is 45,000/75 Mpc = 600 Mpc. This is about two billion light-years and the look-back time t_{lb} is thus two billion years.

This is only true if z is much smaller than 1. If the redshift z is larger than, say 0.25, the conversion of z into velocity v is more complex; see the discussion in Chap. 5. The next conversion of v into distance and thus in look-back time is also more complex. The outcome depends on what is assumed about the expansion of the Universe, or to phrase it differently, about our adopted model for the Universe. As explained in the Prologue, cosmological theory implies that there are many possible models and that the observers have to choose which model is right. Unfortunately the observers had never been able to produce the data needed to make the choice. Consequently there was no accepted model in the period we are now discussing. Quite surprisingly the correct model will be found around the year 2000, as we will see in Chap. 16.

For the moment, the conclusion is that for a given redshift we can calculate the look-back time as long as z is small. But if z is larger than, say, 0.25, we do not know enough about the expansion of the Universe to transform z into t_{lb}. This has to wait until we finally will know the proper model for the expanding Universe. Our only consolation is that there is a one-to-one relation between z and t_{lb}; when z increases t_{lb} increases; z is a useful proxy for time to describe the evolution. We therefore describe evolution in terms of z and not in years (through t_{lb}).

Quasars are the most luminous objects and they are your best friend if you want to look deep into the past of the Universe. Maarten Schmidt (Schmidt 1968), who discovered the quasars in 1963, was the first to derive in 1968 how many quasars were around at redshift $z = 1$: between 80 and 150 more that at the present time. The

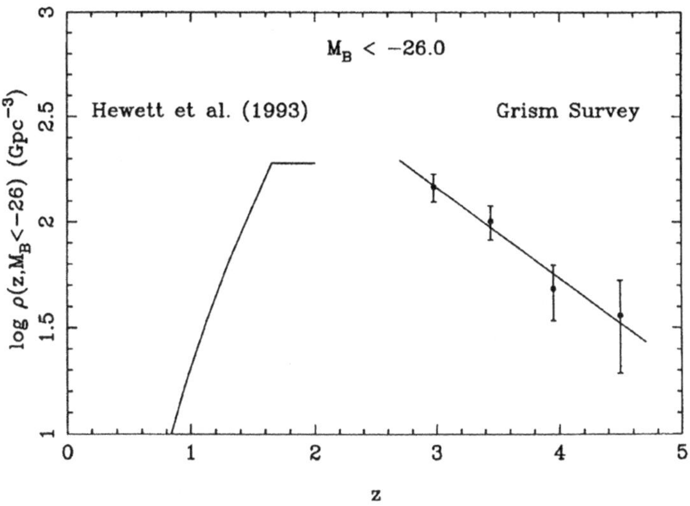

Fig. 15.28 The space density of quasars through cosmic time, as indicated by the redshift z (Schmidt et al. 1995)

Universe had evolved! Schmidt's result was not questioned but it took about 25 years before a deeper study was published. In 1993 a new publication stated that if z was larger than 3, there were less quasars: the population of the quasars had reached its maximum at between $z = 2$ and $z = 3$ (Warren et al. 1994). In 1995 a larger, better sampled collection of quasars led to Fig. 15.28, that shows that the population of quasars reached a maximum at the distance corresponding to $z = 2$ or 3—see Fig. 15.28.

Quasars were particularly powerful light sources, and thus easier to find at large distances than regular galaxies. The desire to see whether normal galaxies had the same history as quasars was much more difficult to satisfy. For example, Butcher and Oemler had discovered in 1978 that surprisingly a number of blue galaxies had been found in two distant clusters. It was not at all clear what that meant: why were these galaxies more blue? The clusters were so far away that the galaxies were too weak to properly depict. It was not until 16 years later, and after the Hubble Space Telescope was refurbished (see Chap. 9), that images were obtained that showed that the blue systems had a seriously distorted image. This morphology indicated the merging of galaxies. The Butcher-Oemler discovery had been explained: in distant clusters the blue galaxies were indeed merging objects.

Finding galaxies with a high 'z' was very difficult. To understand what you saw, it was necessary to record a spectrum, and only when you knew its redshift you could be certain that the galaxy was far away. But recording a spectrum took a long time, even with a large telescope. You could not simply randomly select galaxies; you first had to pre-select what you were going to look at. And therein lay the difficulty. Radio systems were the first support.

A radio source gave a direction in which candidates were to be found for the label 'distant galaxy'. See the following list of consecutive discoveries of the farthest galaxies. The story begins in 1960 when Minkowski found the furthest galaxy ever with the (at that time) spectacular redshift of $z = 0.45$; here a lookback period of $t_{lb} = 4$ Gyr applies; remember that the look-back time up to the Big Bang is $t_{lb} = 13.5$ Gyr. In 1982 a first threshold was passed with the discovery of a galaxy with $z = 1.13$, $t_{lb} = 8.5$ Gyr (Spinrad 1982); the second threshold was passed in 1988 with a galaxy with $z = 2.27$, $t_{lb} = 10.8$ Gyr (Chambers et al. 1988) and in the same year a third threshold with a galaxy with $z = 3.395$ was recorded, $t_{lb} = 10.8$ Gyr (Lilly 1988). The story ended for the time being with a system in which $z = 3.8$, $t_{lb} = 12.2$ Gyr (Chambers et al. 1990). In the meantime, quasars had already been found with $z > 4.0$.

Records were well and good, but if you wanted real results then a larger number of galaxies were required with a whole range of redshifts. In 1995 a new and successful strategy was devised to find distant galaxies (see Chap. 16).

15.8 Background Radiation

The discovery in 1964 of 'background radiation' was one of the most important findings since 1945 and was just as significant as the discovery of the expansion of the universe by Hubble in 1929. Yet, after 1964, only very little more was discovered about this radiation, even though many astronomers were working in this field. Chapter 8 discussed that with radio telescopes on the ground, balloons and even with the U2 spy plane that could fly so high, it was discovered that the radiation in one half of the sky was just slightly stronger than in the other half. It was suspected that our Milky Way moved through space and with a very simple mathematical equation the difference between the radiation of both halves could be described.

Research into more details about the background radiation was raised to a much higher plane by measurements with a space telescope: COBE, the Cosmic Background Explorer, launched by NASA in November 1989 and decommissioned in December 1993. The telescope was designed to measure the background radiation as accurately as possible. The aims of the satellite have been achieved to everyone's satisfaction. The results are shown in Figs. 15.29 and 15.30. COBE was one of NASA's most successful missions. The brightness of the radiation in these images is given by a temperature T instead of the brightness B. The relation between B and T is given by Planck's equation and that reflects the fact that we are in a Universe that is filled with radiation in thermal equilibrium.

The first result to be mentioned (though not the main one) was a confirmation, but now in great detail, of the 'dipole' radiation mentioned in Chap. 8. The strength of the dipole mounted to 3.365 mK ± 0.027 mK, and it was mathematically accurately

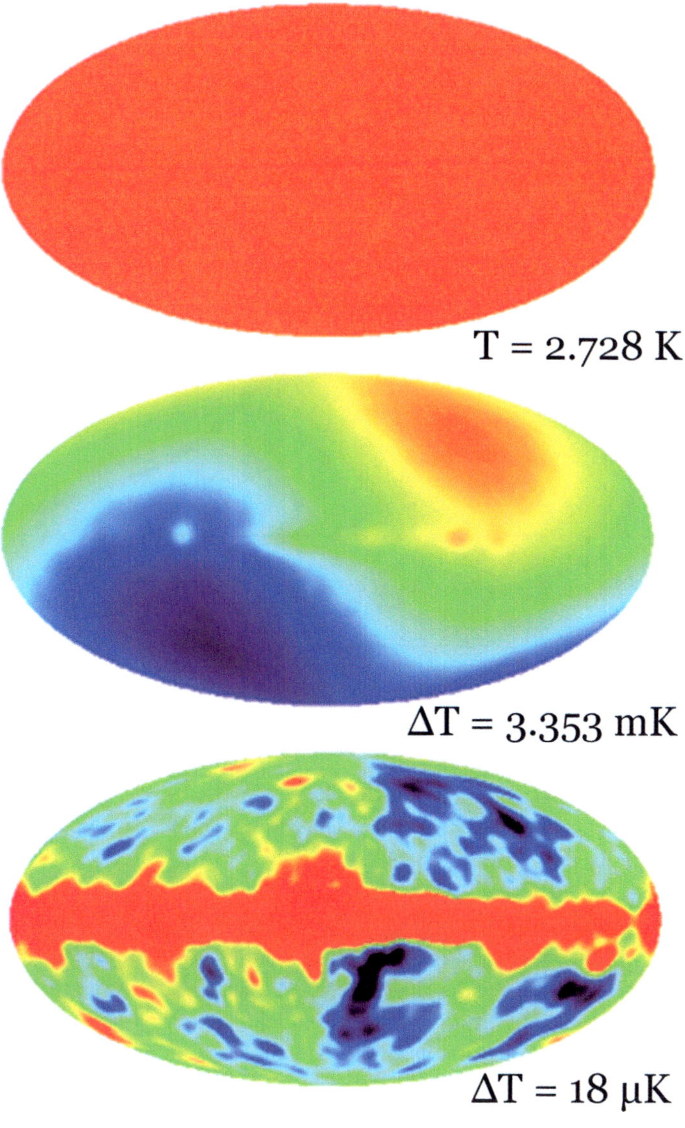

Fig. 15.29 Three times the submillimeter radiation from the full sky as measured by the COBE satellite. The intensity of the radiation is given in degrees kelvin using Planck's equation to convert intensity into temperature. The upper image shows the cosmic background radiation, after the elimination of imperfections of less than 1 in 1,000. The second image arises if we look separately at the systematic deviations associated with a dipole field. The bottom image is created if one removes the dipole and colors the imperfections of 1 in 100,000. Source: NASA

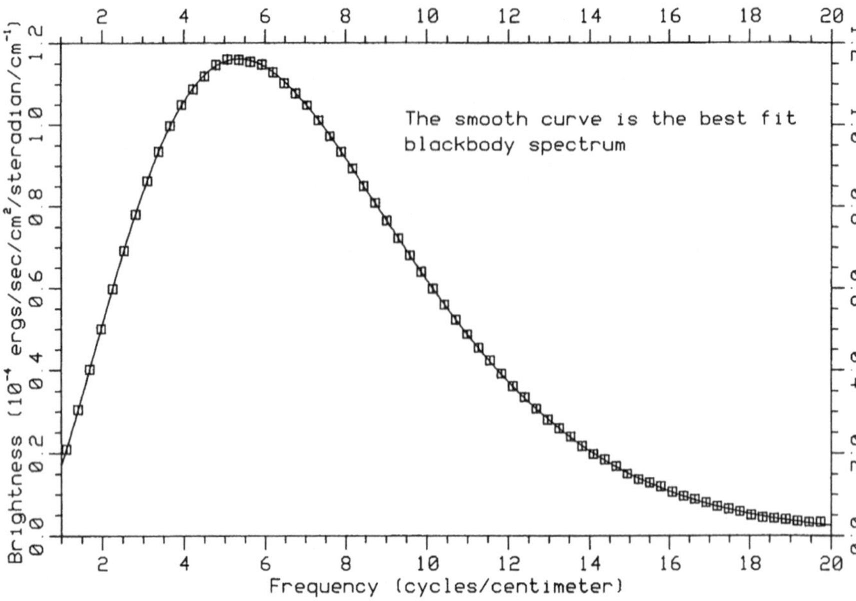

Fig. 15.30 Diagram with (vertically), the strength of the cosmic background radiation at different 'frequencies' (horizontal). The squares represent the measured values and the solid curve is the Planck curve at a temperature of 2.735 K. The 'frequencies' are indicated in 1/λ (Mather et al. 1990)

described by a simple equation; see the box. This led to the conclusion that the Local Group rushed through space at a velocity of about 630 km/s in a precisely determined direction ($l = 276°$, $b = 30°$).

In the middle image of Fig. 15.29 the radiation is excellently described by the equation $T_{dp} = 0.00353 \times \cos\theta$. Here θ is the angle between the direction in which you look and the direction given by the galactic coordinates $l = 264°$, and $b = 48°$. This comparison emerges if you assume that our Milky Way moves in the given direction at a velocity of $v/c = 0.00353/T_{bg}$, where $T_{bg} = 2.73$ K; c is the velocity of light; the velocity of the Sun relative to the background radiation is 0.00353 × 300,000/2.73 = 370 km/s in the direction indicated above (Kogut et al. 1993).

There was something in this result that dumbfounded every astronomer. The strength of the dipole radiation (3365 mK) was measured with a precision of 0.027 mK, within 1%! In cosmology such great accuracy had never previously been achieved. And, as we shall see, that precision would be bettered twice in the subsequent 20 years.

If the dipole radiation was removed, radiation was left over that was very small (the brightness in the order of one-millionth degree kelvin, μK) and it had a messy pattern: see the lower ellipse. In it, the remnants of the radiation from the Milky Way are prominent, even though most of that radiation had been removed. The conclusion was that the background radiation was isotropic: equally strong in all directions in the sky, a ruling that applied to the level of one in a hundred thousand.

Now look at Fig. 15.30. This shows the spectrum, the variation of the strength of the radiation over all wavelengths. The spectrum is very accurately described by a rather complicated formula, Planck's formula, when $T = 2.735 \pm 0{:}06$ K is used for the temperature; again a value that has been measured very accurately (2%). This precise agreement between a measured curve and a theoretically predicted curve always impressed me tremendously. Planck's formula was derived by him in 1900 and was the solution to a very different problem: take a closed furnace; heat the walls of the furnace to an arbitrary temperature; the Planck curve will then predict the intensity of the electromagnetic radiation in that oven at all possible wavelengths. Apparently the same curve also predicts the spectrum of the cosmic background radiation as if the Universe were our oven (a very cold oven, fortunately).

I will repeat what I wrote earlier: COBE turned cosmology into a precise science; this was a field of research field that had been plagued by difficult and inaccurate measurements for as long as it existed. Finally, accurate results had arrived! In 2006, the Nobel Prize for physics was awarded to John Mather en George Smoot for the COBE project. According to the Nobel Prize committee, 'the COBE project can be regarded as the starting point for cosmology as a precision science'.

The theorists, however, were not satisfied and they had good reason. In 1965, when the Princeton physicists explained the measurements of background radiation (see Chap. 5) they argued that after the radiation and light were 'decoupled', the current structure of the Universe began to emerge. And the start had to lie in small irregularities in the hot gas in the manner described in Section 15.8. The background radiation could therefore not be entirely uniform: there *had to be* irregularities (for some reason these were called 'anisotropies'). These anisotropies had to be less than one in a hundred thousand. Could the theorists convince NASA to launch a satellite that measured the radiation at least ten times as accurate as COBE?

Cosmogony, or How did the Structure of the Universe come about?

Chapter 8 discussed the growing role of computer simulations in determining how the distribution of matter in the Universe could develop from small irregularities in the hot, gaseous matter at the time that radiation and matter were 'decoupled'. Computer simulations became important tools to understand how many irregularities in the density could grow by gravitational attraction. The simulations were becoming more comprehensive which was made possible due to the rapidly increasing capacity of calculators. More and more processes could be included in the calculations, for example the development of clusters of galaxies and the consequences of collisions between galaxies.

The developments did not go smoothly. Time and again crises arose due to unforeseen problems, but the practitioners of this profession remained optimistic. One such issue was that it seemed necessary to suppose that greater fluctuations in density resulted in many more stars being formed than were calculated. An additional factor was added to the calculations, which ensured that not too many small galaxies were created. Such 'fudge' factors are often introduced when calculations do not work out, but they are extremely suspicious. It's just like cheating for a while at cards in a game of solitaire: you obtain good results but the success does not give you a feeling of satisfaction.

In Chap. 8 a group of four young astronomers was mentioned, Davis, Efstathiou, Frenk, and White. They also published the most frequently cited models in this period (see Fig. 15.31).

Two years later, the same authors showed another simulation similar to the previously mentioned observations by de Lapparent, Geller, Huchra (1987) (see Fig. 15.32). Although the comparison with the observations was not perfect, it was so good that the authors were clearly pleased with the success of their simulations.

But five years later the group appeared to have come to a standstill. Anyway, in 1992 they published an article in *Nature*, entitled 'The end of cold dark matter?' In it, they reported that their CDM model was inadequate. In 1994 one of these authors, White (1994), wrote an overview of the status of their research. He mentioned a number of established facts: (1) the density of the baryons is at most a few percent of the critical density; (2) there is much more hidden than visible matter; (3) immediately (10^{-34} seconds) after the starting signal of the Big Bang an 'inflation' occurred in which the Universe for as far as we can see was smoothed over; (4) the force of gravity caused the development of small fluctuations into galaxies and clusters. After these strongly positive conclusions, it is somewhat surprising to read the last sentence of White's review: 'No current model appears consistent with a straightforward interpretation of all observational data.' There was, in short, a major crisis in their approach to cosmogony.

'How to tackle this crisis?' was the next question. There were suggestions; competing groups were waiting in the wings. I will mention two: (1) add a little HDM (hot dark matter) to the CDM; (2) restore the cosmological constant Λ.

About that second suggestion: it was explained in the Prologue that Einstein had changed his basic equations for his general theory of relativity by adding a factor which he called Λ. He did this to make certain that the Universe had always existed and would always exist; in jargon: the Universe was static. But when Hubble discovered that the Universe was not static, but rather that it expanded, Einstein concluded that Λ could be removed from his equations, and deemed the introduction of Λ 'his greatest blunder '(at least that is what Gamow says). Einstein might have removed the term, but theorists working on the theory of relativity did not forget Λ. In 1992, an article appeared in a highly rated review journal, regarding the need to

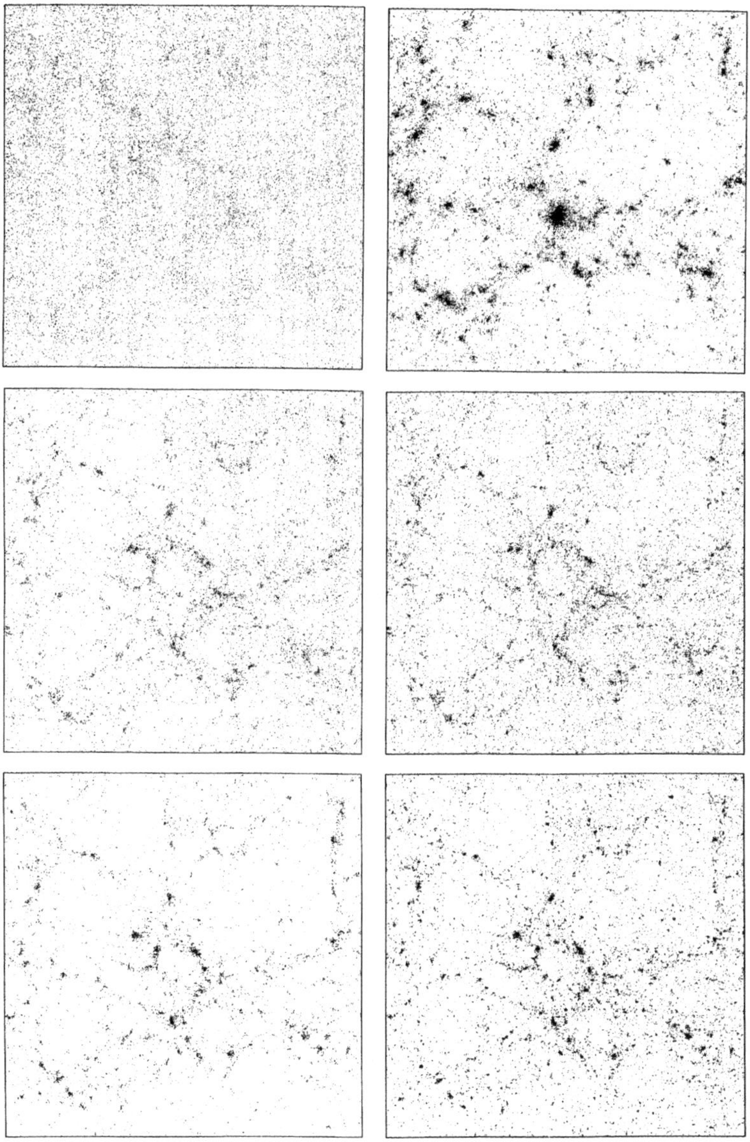

Fig. 15.31 The development of a randomly chosen initial distribution over time: top left is the initial situation, centre row the situation when the scale factor has increased from 1 to 1.8, and in the bottom row the scale factor is 2.4. Top right the scale factor is 4.5. The bottom two in the right-hand column are not relevant here (Davis et al. 1985)

include Λ in the FLRW-equation (Carroll et al. 1992). In concluding their article, the very competent authors posed the question of how one could demonstrate the existence of Λ. A number of tests were discussed but not the research in which a few years later Λ made its comeback.

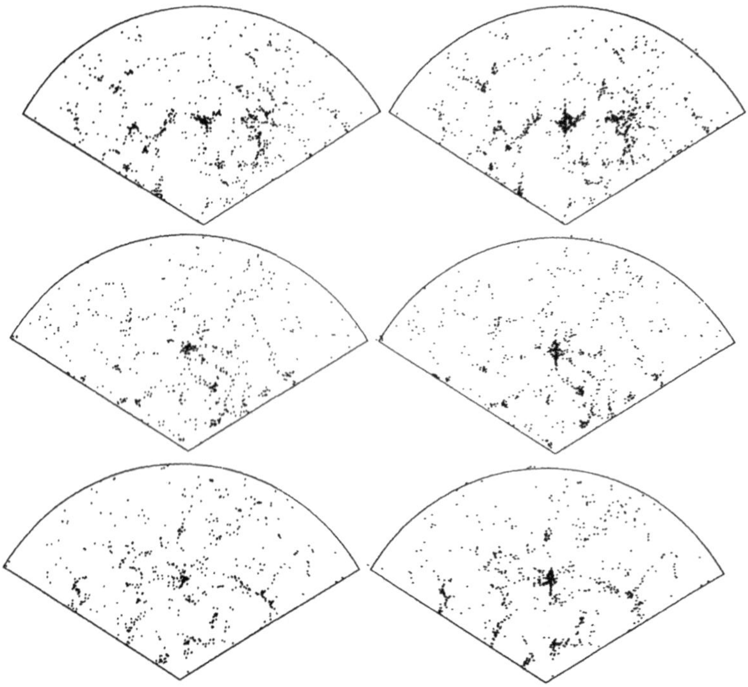

Fig. 15.32 Some simulations of the distribution of galaxies (White et al. 1987)

References

Antonucci, R.: Annu. Rev. Astron. Astrophys. **31**, 473 (1993)
Antonucci, R.R.J., Miller, J.S.: Astrophys. J. **297**, 621 (1985)
Ashman, K.: Publ. Astron. Soc. Pac. **104**, 1109 (1992)
Backer, D.C., Sramek, R.A.: Astrophys. J. **260**, 512 (1982)
Backer, D.C., Zensus, J.A., et al.: Science. **262**, 1414 (1993)
Balick, B., Brown, R.L.: Astrophys. J. **194**, 265 (1974)
Beers, T.C., Christlieb, N.: Annu. Rev. Astron. Astrophys. **43**, 531 (2005)
Beers, T.C., Preston, G.W., Shectman, S.A.: Astron. J. **90**, 2089 (1985)
Bender, R., Doebereiner, S., Moellenhoff, C.: Astron. Astrophys. Suppl. **74**, 385 (1988)
Bender, R., Surma, P., Doebereiner, S., et al.: Astron. Astrophys. **217**, 35 (1989)
Bender, R., Burstein, D., Faber, S.M.: Astrophys. J. **399**, 462 (1992)
Bergeron, J.: Astron. Astrophys. **155**, L8 (1986)
Carroll, S.M., Press, W.H., Turner, E.L.: Annu. Rev. Astron. Astrophys. **30**, 499 (1992)
Chambers, K.C., Miley, G.K., van Breugel, W.J.M.: Astrophys. J. **327**, L47 (1988)
Chambers, K.C., Miley, G.K., van Breugel, W.J.M.: Astrophys. J. **363**, 21 (1990)
Condon, J.J., Huang, Z.-P., et al.: Astrophys. J. **378**, 65 (1991)
Davies, M., Efstathiou, G., et al.: Astrophys. J. **266**, 41 (1983)
Davis, M., Efstathiou, G., et al.: Astrophys. J. **292**, 371 (1985)
de Lapparent, V., Geller, M.J., Huchra, J.P.: Astrophys. J. **302**, L1 (1986)
Edvardsson, B., Andersen, J., et al.: Astron. Astrophys. **275**, 101 (1993)

Forman, W., Jones, C., Tucker, W.: Astrophys. J. **293**, 102 (1985)
Fort, B., Mellier, Y.: Astron. Astrophys. Rev. **5**, 239 (1994)
Franx, M., Illingworth, G.: Astrophys. J. **327**, L55 (1988)
Geller, M.J., Huchra, J.P.: Science. **246**, 897 (1989)
Graham, J.R., Carico, D.P., et al.: Astrophys. J. **354**, L5 (1990)
Hartman, R.C., Bertsch, D.L., et al.: Astrophys. J. **385**, L1 (1992)
Heckman, T.M., Armus, L., Miley, G.K.: Astrophys. J. Suppl. **74**, 833 (1990)
Hewitt, A., Burbidge, G.: Astrophys. J. Suppl. **87**, 451 (1993)
Ibata, R.A., Gilmore, G., Irwin, M.J.: Nature. **370**, 194 (1994)
Jaffe, W., Ford, H.C., et al.: Nature. **364**, 213 (1993)
Jedrzejewski, R.I.: Mon. Not. R. Astron. Soc. **226**, 747 (1987)
Kaiser, N., Squires, G.: Astrophys. J. **404**, 441 (1993)
Kennicutt Jr., R.C.: Astrophys. J. **344**, 685 (1989)
Kogut, A., Lineweaver, C., et al.: Astrophys. J. **419**, 1 (1993)
Kormendy, J., Richstone, D.: Annu. Rev. Astron. Astrophys. **33**, 581 (1995)
Kronberg, P.P., Biermann, P., Schwab, F.R.: Astrophys. J. **291**, 693 (1985)
Lauer, T.R., Faber, S.M., et al.: Astrophys. J. **106**, 1436 (1993)
Lilly, S.J.: Astrophys. J. **333**, 161 (1988)
Maraschi, L., Ghisellini, G., Celotti, A.: Astrophys. J. **397**, L5 (1992)
Mather, J.C., Cheng, E.S., et al.: Astrophys. J. **354**, L37 (1990)
Mathewson, D.S., Wayte, S.R., et al.: Proceedings of the Astronomical Society of Australia. **7**, 19 (1987)
Murai, T., Fujimoto, M.: Publ. Astron. Soc. Jpn. **32**, 581 (1980)
Osterbrock, D.E.: Astrophys. J. **404**, 551 (1993)
Pagel, B.E.J.: IAU Symp. **149**, 133 (1992)
Sandage, A., Binggeli, B., Tammann, G.A.: Astron. J. **90**, 1759 (1985)
Sanders, D.B., Soifer, B.T., et al.: Astrophys. J. **325**, 74 (1988)
Sargent, W.L.W., Boksenberg, A., Steidel, C.C.: Astrophys. J. Suppl. **68**, 539 (1988)
Schmidt, M.: Astrophys. J. **129**, 243 (1959)
Schmidt, M.: Astrophys. J. **151**, 393 (1968)
Schmidt, M., Schneider, D.P., Gunn, J.E.: Astron. J. **110**, 68 (1995)
Schneider, D.P., Gunn, J.E., et al.: Astron. J. **91**, 991 (1986)
Sikora, M., Begelman, M.C., Rees, M.J.: Astrophys. J. **421**, 153 (1993)
Sneden, C., Preston, G.W., et al.: Astrophys. J. **431**, L27 (1994)
Soucail, G., Fort, B., Mellier, Y., Picat, J.P.: Astron. Astrophys. **172**, L14 (1987)
Spinrad, H.: Publ. Astron. Soc. Pac. **94**, 397 (1982)
Stebbins, J., Whitford, A.E.: Astrophys. J. **108**, 413 (1948)
Swanenburg, B.N., Hermsen, W., et al.: Nature. **275**, 298 (1978)
Tyson, J.A., Valdes, F., Wenk, R.A.: Astrophys. J. **349**, L1 (1990)
Urry, C.M., Padovani, P.: Publ. Astron. Soc. Pac. **107**, 803 (1995)
Warren, S.J., Hewett, P.C., Osmer, P.S.: Astrophys. J. **421**, 412 (1994)
White, S.D.M.: Reviews in Modern Astronomy. **7**, 255 (1994)
White, S.D.M., Frenk, C.S., et al.: Astrophys. J. **313**, 505 (1987)
Wilson, A.S., Colbert, E.J.M.: Astrophys. J. **438**, 62 (1995)
Worthey, G.: Astrophys. J. Suppl. **95**, 107 (1994)
Zaritsky, D., Kennicutt Jr., R.C., Huchra, J.P.: Astrophys. J. **420**, 87 (1994)

Chapter 16
1995–2005: Galaxies and another Revolution

The theme of this book is the spectacular development of astronomy research from 1945 to 2015. Earlier chapters have described how the years from 1960 to 1975 were a revolutionary period marked by the discovery of completely unexpected phenomena: hotbeds of new stars, neutron stars, black holes, X-ray binary stars and radio radiation that showed a picture of what the Universe looked like before the first stars and galaxies formed.

A second revolution followed in the years between 1995 and 2002. This was again something that could not have been predicted. It was the result of a number of very different research projects that were started independently of one another, but the results of which demonstrated a staggering synthesis. First, after a stagnation of 60 years, finally a convincing choice could be made of a well-defined mathematical model of the expanding Universe; the success came from new data provided by a survey for distant supernovae (Nobel Prize), by two balloon flights and from two NASA satellites (Nobel Prize). The model was happily called 'the concordance model'. Second, the theory by which galaxies were the results of collisions of smaller gas clouds, a theory of which the beginnings go back to the 1970s, was further developed by large computer simulations and resulted in stunning images of the deep sky that at first and second sight looked quite realistic: the origin of galaxies had finally been explained. Third, the proof was given that black holes exist, as massive as a million to a billion times the mass of the Sun (SuperMassive Black Holes, or SMBHs). It consequently became clear that these SMBHs were the source of the active cores of Active Galactic Nuclei (AGNs). Moreover, the mass of such a black hole appeared to coincide closely with the mass of the bulge of the galaxy in which it was found. Fourth, two large-scale optical surveys were undertaken, one in the northern hemisphere and the other in the southern hemisphere; the emphasis was on obtaining spectra of hundreds of thousands of galaxies.

H. J. Habing, *The Birth of Modern Astronomy*, Historical & Cultural Astronomy,
https://doi.org/10.1007/978-3-319-99082-8_16

16.1 The Concordance Model of the Universe

In the Prologue it was explained how, starting from Einstein's General Theory of Relativity, an equation could be derived, called the Friedmann-Lemaître-Robertson-Walker equation (in short: the FLRW equation) that described how the expansion of the Universe would be slowing down under the influence of the force of gravity that was produced by all the matter in the Universe. If one knew the present expansion velocity (the Hubble constant H) and ρ, the present density of matter in the Universe, the FLRW equation allowed you to calculate density and expansion velocity at any time in the future or in the past. It was also clear that there were three different kinds of solution. If the expansion velocity now was high and the density low, then the Universe would keep expanding forever. But if, on the contrary, the speed of expansion was low and the density was high, the expansion would at some point in the future stop expanding, after which it would reverse and collapse. Between these two extremes there was one special situation when H and ρ had a specific value and when the expansion would go on forever, but slower and slower, ultimately approaching zero velocity; this solution was called the 'critical Universe' and in spite of it being very special it played a major part in the subsequent history.

One of the unpleasant conclusions was that the observers failed to measure the precise and reliable present-day values for H and ρ needed for the choice of the right cosmological model. Such measurements were beyond the observational possibilities in the 1930s and they remained that way until around the year 2000. Things changed somewhat around 1980 when, after the formulation of the inflation theory, the theoretical cosmologists found strong arguments that the density of matter was *exactly* equal to the critical density. The observers, however, insisted that they could find only a small fraction of this matter: they refused to assume that the Universe was critical. When dark matter was discovered, it gave some slight hope that this would be enough to reach the critical density limit, but it soon appeared that even the dark matter was not enough. The discrepancy remained.

If the Universe had the critical density, a very important consequence was that the Universe at large had Euclid's geometry, the geometry that we all learn at high school and that implies such well known theorems as that of Pythagoras. This may surprise some readers who do not know that in the nineteenth century a few mathematicians (Gauss, Riemann) had proven that a perfectly correct mathematical theory could be developed with a geometry different from that of Euclid. In Einstein's General Theory of Relativity the basic geometry was not firm: it was quite possible that the Universe had a non-Euclidean geometry. This uncertainty plagued theoretical cosmologists, who always had to leave more options open than they would have liked.

The Expansion of the Universe is Accelerating

Hubble had discovered in 1920 that the Universe was expanding: all galaxies are moving away from us at a speed (v) that increases as the distance (d) becomes greater: $v = H \times d$. The H is a number that will later be known as the 'Hubble constant'.

Does this same relationship apply to galaxies that were much more distant than Hubble and its successor Sandage could have seen? In theory, this was probably *not* the case: Einstein's General Theory of Relativity, about which there was no doubt, showed that the further you looked—and consequently the further back in time—the greater the Hubble constant must have been. It was therefore important to find galaxies much further away than those that had been found to date and then to measure the velocity at which they were moving away from us (called the 'radial velocity') and their distance.

Determining the velocity was not difficult. You took a photograph of the spectrum and measured the shift z of the spectral line. As stated before, the redshift was specified by a number z. Measuring the speed was thus not the problem; the difficulty was measuring the distance.

Distances were—and still are—measured using 'standard candles': an archaic term for an object of which you already knew how bright it was. If this standard candle was a large distance away and you measured how bright the light was, you could work out the distance by dividing the (known) strength of the 'standard candle' by the value measured; the division gives you the square of the distance. The most frequently used and most reliable standard candles were variable stars of the Cepheid type that had been used by Hubble in his pioneering studies. With newer technology, the maximum distance at which these standard candles could be recognized had been increased somewhat. But to make real progress a new type of candle was needed.

The problem had been known for a long time, and discussions about an alternative standard candle had centered on one specific candidate: a type Ia supernova at its maximum brightness. Different types of supernovae were discussed in Chap. 14 and it was remarked there that the light curves of supernovae differed from one to another and that their maximum brightness differed also. But the light curves of all supernovae of type Ia appeared very similar to one another; see Chap. 14.

Type Ia supernovae reached a maximum magnitude of -19, which was around 11.5 magnitudes brighter than the brightest Cepheid. The longest distance that you could measure with such a supernova would be 200 times greater than with a Cepheid. This offered some interesting opportunities for future measurements. But there was a major practical problem. How could you discover enough supernovae? Assume that in a large galaxy a supernova explodes once in 50 years; at least this seems to apply for our Milky Way. To encounter a supernova in another galaxy you would have to watch the galaxy for 50 years, day in, day out, in order to witness the instant that a supernova might explode. Because such a supernova reaches its peak very shortly after it appears, it was important to detect the explosion quickly. This was the only way of being sure that you were there when the object achieved its highest intensity.

To make such a plan realistic, you would have to observe many galaxies continuously in the hope of occasionally finding a supernova—which would by no means always be of type Ia. It was this need for constant monitoring that was the biggest obstacle to the plan. A group of Scandinavian astronomers started this kind of research, but after a couple of years of solid work, during which they had come across only one supernova, they gave the project up. A similar monitoring problem had, however, previously been resolved: the project of the search for stars that brightened when a gravity lens passed in front of it (see Chap. 14). Automated telescopes and computers used as monitoring instruments had proved to be a useful solution. The conclusion was that it was possible to monitor many galaxies often enough using the same method.

There were two groups researching this problem. They were competitors, but at the same time they maintained a professional relationship with one another, with some astronomers even being members of both groups. The first group (the 'Supernova Cosmology Project') was led by Saul Perlmutter from Berkeley (California) and the second (the 'High-z SN Search Team') by Brian Schmidt from Canberra, Australia, and Adam Riess, also from Berkeley. Both groups made recordings of a field with galaxies and repeated the measurements 3 weeks later. A computer programme was used to compare the two plates with one another, trace differences between the plates and find recently exploded supernovae. As might be expected, this was simple to explain, but difficult to do. There were many problems, and the fact that both projects were highly successful is a tribute to their excellent organisation and professional skills. Both groups found the supernovae they had been looking for and determined the light curves of the explosion. The analysis was painstaking, having to take into account all kinds of difficulties, including such questions as whether the light of a given supernova could have been weakened by interstellar matter.

Eventually, in 1998 and 1999, both groups presented definitive results, which are summarized in Fig. 16.1. The most important feature is that in all three images the measurement points to the furthest right are just above the line. This seems to be a questionable result, but both parties obtained the same result independently of one another, and moreover various statistical tests confirmed the slight deviation from the straight line. Because the points were above the line, this meant that the supernovae were somewhat weaker than had been expected (greater magnitude means weaker light!). These supernovae were therefore further away than Hubble's theory of relativity predicted and that implied that the expansion of the Universe had accelerated very slightly in the time between the supernova transmitting the light and our receiving the signal.

In Fig. 16.1, the measurement points diverge only very little from the straight line but in general they do diverge. It is this minor divergence that had such colossal consequences, because it was related to the General Theory of Relativity. There is no need to go into the complex mathematical equations of the theory to understand this. The theory was based on the idea that in the Universe there was just one force that worked over large distances: gravity. All that gravity could do was to slow down the expansion, by trying to hold everything close together. This excluded the possibility of an acceleration of the expansion. Einstein's theory in its existing form contrasted with the measurements, regardless of how small the divergences from the measurement points were.

This was a shocking idea—for a short while. Those who knew the history of the General Theory of Relativity remembered what was discussed in the Prologue: following criticism by De Sitter, Einstein had modified his theory and introduced a

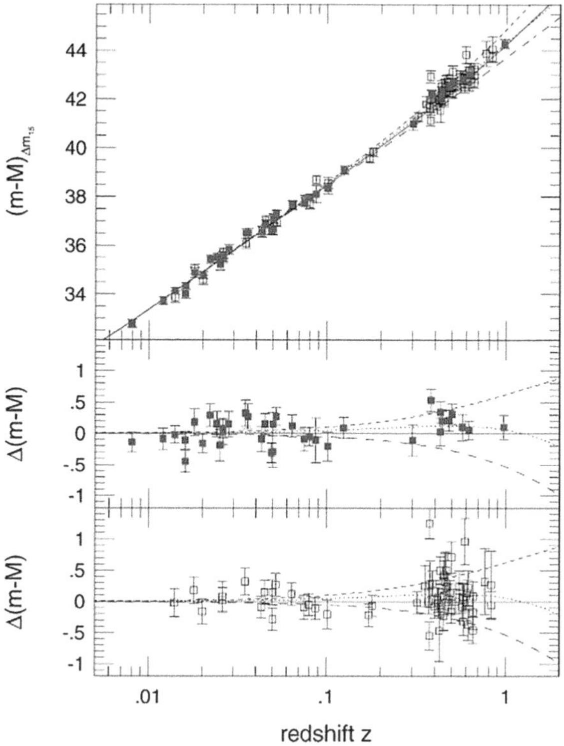

Fig. 16.1 Three diagrams of the relation between the distance and the radial velocity of galaxies in which a supernova of type Ia has been detected. Upper diagram: horizontally: the radial velocity expressed in the (logarithm) of the redshift *z*. Vertically the logarithm of the distance to the galaxy, calculated by comparing the observed maximum brightness of the supernova (*m*) with its intrinsic brightness (*M*). A straight line is drawn through the measurement points. The middle and lower diagrams are enlargements, indicating how much the different measurements of each of the two research teams diverge from the straight line (Leibundgut 2001)

completely new force, symbolised by Λ, the Greek capital L. A few years later, Hubble discovered that the Universe was not static, but was expanding and Einstein happily removed the Λ: De Sitter's objection was no longer valid. But now, in 1998, there was a convincing reason to put Λ back into the FLRW equation, even though it was not understood what kind of force it was (Einstein and De Sitter did not understand it and even in 2016 Λ is still a force that is not fully understood). In the way Λ was introduce in the equation it implied that it acted like a pressure that increased with the volume of the Universe. As long as the Universe had been small, Λ was also small but now it was a significant force.

This new evidence that the Λ-related force actually existed offered a way out of the dilemma facing observers and theoreticians: the value of Λ bridged the gap between the value $\Omega = 1$, that the theoreticians wanted to see and the small value $\Omega = 0.3$ that the observers had actually measured: you could convert Λ to a parameter Ω_Λ which appeared to have roughly a value of 0.7. This was what was necessary to bridge the gap between the critical density demanded by the theoreticians and the density that the observers could provide.

Saul Perlmutter, Brian Schmidt and Adam Riess received the Nobel Prize for Physics in 2011 'for the discovery of the accelerating expansion of the Universe through observations of distant supernovae'. This was the second Nobel Prize for research on the expansion of the Universe.

'Anisotropies' and Fundamental Properties of the Universe

As mentioned above, the COBE satellite had shown the high level of isotropy and homogeneity of the early Universe, but the observations also seemed to show that extremely small 'anisotropies' were present, smaller than 1 in 100,000. This discovery stimulated theoretical physical research on the characteristics of the hot, cosmic plasma of photons, electrons, baryons and particles of heavy matter. The good thing was that the matter was so sparse that the behavior of the particles was 'linear', which made it easy to calculate its properties.[1]

One of the persons playing a leading role especially in the early discussions was Soviet cosmologist Yakov Borisovich Zel'dovich (1914–1987). His studies of the behavior of this cosmic plasma showed his deep insight and his creativity in hypothesizing how nature might behave under cosmological conditions. He was truly a great scientist.

One of the topics under discussion were the 'Baryon Acoustic Oscillations'. To get an idea of what this means, think of the flat surface of a pond. Somewhere a droplet falls on the surface, causing a wave that spreads outwards in a circular shape. The surface of water is a two-dimensional example and a three-dimensional image illustrates the idea even better. In that case, think of a small disturbance in the air that moves away from its source like a sound wave and is heard everywhere in the room. Such disturbances, termed 'baryonical acoustic oscillations', were conceivable in the hot plasma of the Universe. And because these disturbations were very small, many of their characteristics could be calculated reliably- non-linear effects could be ignored. Therefore it was possible to draw conclusions about the fundamental characteristics of the disturbations, the 'anisotropies', in the smooth water of the cosmic pond. The observations by COBE had shown the existence of such disturbations but had not been sufficiently accurate to characterize their properties.

Impressed by these and similar ideas, NASA decided to launch a satellite as a successor to COBE: MAP, the Microwave Anisotropy Probe. The plan was announced at an astronomy conference in 1997. It always takes a number of years to build a satellite and it was much quicker to build an instrument for a balloon. MAP was thus scooped by two balloon projects, BOOMERanG and MAXIMA.

The BOOMERanG and MAXIMA experiments each consisted of a balloon with a millimeter radio telescope suspended beneath it which measured the background radiation. Originally, the two experiments were one, but a difference in estimating the risks of the balloon flight led to a split. MAXIMA was launched in August 1998 and June 1999 for 8-hour flights above the southern United States. BOOMERanG departed in 1998 for a 10-day flight from an Antarctic station (Fig. 16.2). The balloon was carried along by a fixed airflow and made an orbit around the south pole, ultimately returning to base; hence the name of the experiment. Both experiments

[1] In physics, the law of linearity states that an agent that is twice as strong will have an effect that is also twice as great and not, for example, an effect that is 2^2 as great.

Fig. 16.2 The BOOMERanG balloon and telescope are prepared for launch from a station in Antarctica. Source: Wikimedia Commons

were successful; they confirmed one another's results and an important conclusion was drawn from the observations.

The telescopes observed the radiation in a region of the sky at four different wavelengths around 1 mm. For the result, see Fig. 16.3.

At first sight the three maps on the left look somewhat chaotic, just like noise. Mathematical analysis of peaks and troughs in the chart showed, however, that there was a preferred angular distance between the peaks: the positions of the peaks were not purely random. This preferred angle was compared with the distance that the acoustic waves (sound waves) had travelled between Big Bang and decoupling: one knew how long the waves could have travelled and how fast they did that. This distance got the name of Baryon Acoustic Horizon. By comparing the preferred (angular) distance between the peaks with the size of the Baryon Acoustic Horizon and with the distance from us to the oscillations (15 billion light years, the age of the Universe) it followed that the ratio was precisely that of Euclidean geometry and it excluded the possibility that the Universe had another geometry. It was in the end a single, and very direct, argument that after more than 70 years determined which of the three possible models of the Universe from Einstein's General Theory of Relativity was applicable to the cosmos: it was the Euclidic version.

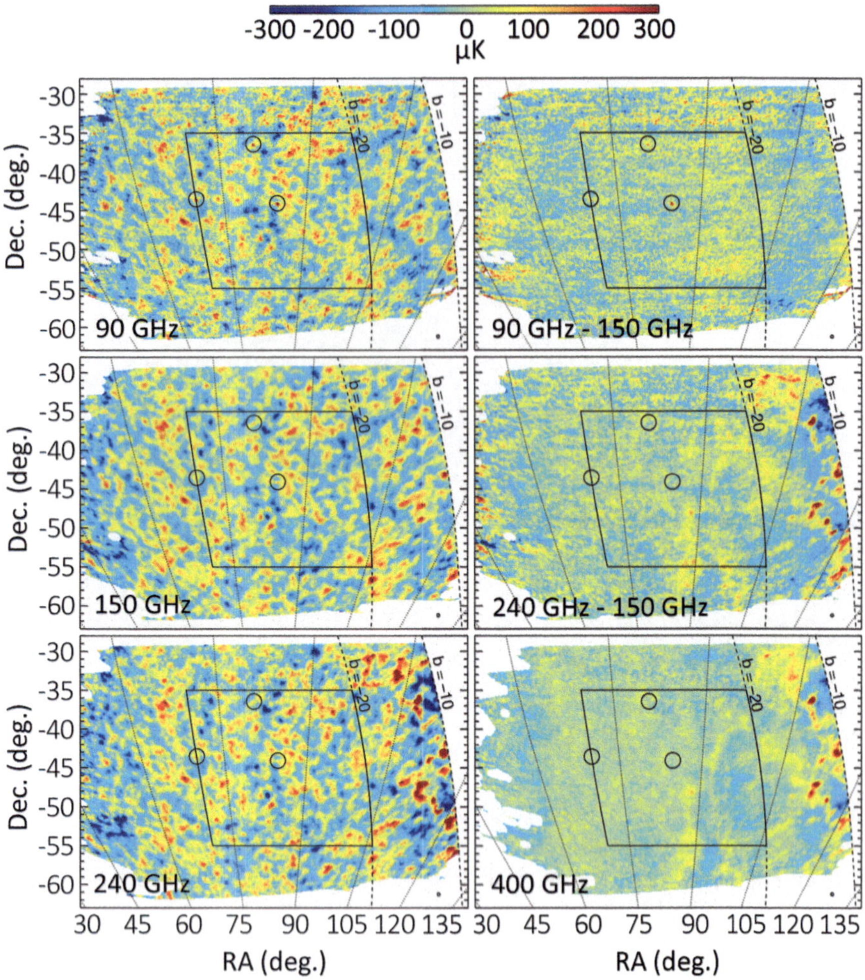

Fig. 16.3 Two times three charts of the millimeter radiation of a part of the Universe observed with the BOOMERanG telescope. Concentrate on the three charts to the left, that from top to bottom were measured at 90, 150 and 240 GHz consecutively (wavelengths 3.3, 2.0 and 1.25 mm). The charts show that the sources of the radiation are the same in all three versions (De Bernardis et al. 2000)

This was followed immediately by another important conclusion: if the Universe was Euclidic, then it must be 'critical', $\Omega_M + \Omega_\Lambda = 1$. This conclusion directly, and independently endorsed the results of the supernova project.

A politically interesting aspect of the two balloon experiments was the large contribution provided by NASA, while the balloons were in direct competition with the MAP telescope that had yet to be launched. The balloons delivered the great news, even while the MAP satellite was being launched. Did this mean that the mission was no longer relevant? By no means, as was ultimately proven. Indeed, MAP (or WMAP, as the satellite was now named[2]) confirmed the results of BOOMERanG and MAXIMA, but the much larger area of the Universe that was observed with WMAP and the greater precision of the measurements provided the definitive basis for the 'concordant' model of the Universe. Since the publication of the WMAP data in 2003 (Spergel et al. 2003), this model has provided cosmology with a very solid, empirical foundation.

It is worth mentioning some of the findings of WMAP: the age of the Universe was 13,779 ± 0.059 billion years (an accuracy of more than 1%); the Hubble constant was 69.32 ± 0.80 km/s/Mpc; the Universe comprises 4.628 ± 0.093% 'normal matter', 24.02 ± 0.87% dark matter and 71.35 ± 0.9% dark energy. I quote the figures to two decimal places in order to show how remarkable it was that both CODE and WAP (and a few years later this would apply to the third satellite, the European PLANCK mission) were able to generate fundamental information about the Universe so accurately; compared with the situation less than 10 years previously, it was miraculous.

A 'Concordant' Model of the Universe

The model of the Universe, with the values provided by WMAP, was referred to as the 'concordant model' but was also known as the 'Standard Cosmological Model' and the 'Lambda CDM model'. From 2003, it directed almost all cosmological research and particularly research on the very distant Universe onto objects with a redshift parameter z greater than, for instance, 2.

To see what was achieved, look at Fig. 16.4, in which the time is indicated horizontally in billions of years and the 'scale' of the Universe is given vertically. This last concept requires an explanation: think of a map of a city that stays the same for many years with all the houses remaining in the same place; in this map they maintain the same coordinates. But the scale of the map changes: at first it is 1:100,000, and a few years later it appears to be 1:110,000, indicating that the distances between all the houses has increased by 10%. In the same way, the scale of the Universe is increasing continuously and very uniformly, while the galaxies maintain the same coordinates on the chart.

[2] Shortly after the launch of MAP, David Wilkinson, one of the Princeton cosmologists involved in the project since its very start, died. In his honour, his name was added to the project title: Wilkinson Microwave Anisotropy Project, WMAP.

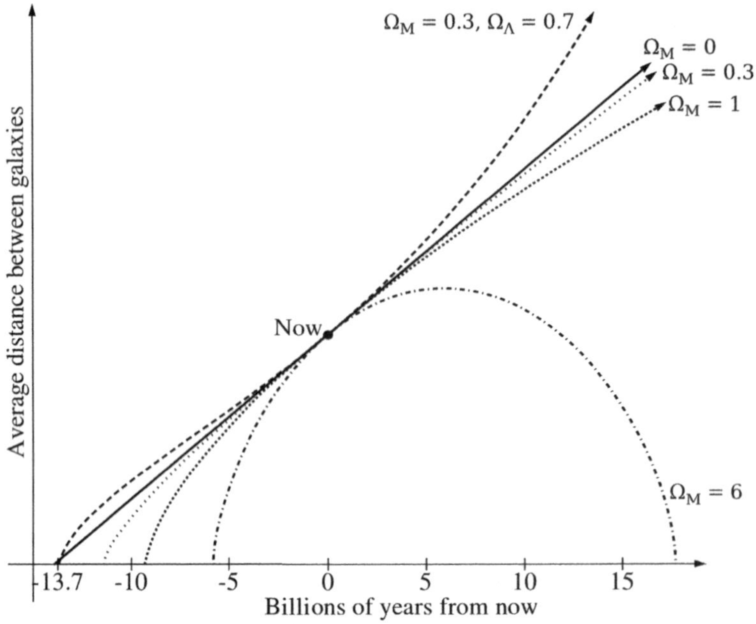

Fig. 16.4 The scale of the Universe is given vertically, and the cosmic time horizontally. Start at the point 'Now': this is where the expected further development of the Universe begins. A number of curves have been drawn, each of which predicts a different model of the Universe. Look first at the curves where Ω_Λ is not mentioned: these are models where the Λ force is absent ($\Omega_\Lambda = 0$). Follow the curve with the label $\Omega_M = 6$. The letter '*M*' stands for matter: in this model of the Universe the density of the matter is six times the critical value. Very shortly after 'Now', the expansion of the Universe stops, after which it will start to contract: the average distance between galaxies will reduce until it reaches zero, after 15 billion years. Now take the curve with $\Omega_M = 1$: this is the critical Universe where the scale factor continues to increase until into the far distant future. Finally, there is the curve $\Omega_M = 0.3$ and $\Omega_\Lambda = 0.7$, which is the current ('concordant') model in which both gravity and the Λ force play a role. Source: Wikimedia Commons

16.2 ΛCDM: A Successful Theory for the Origin of the Cosmic Web and its Stars and Galaxies

Chapter 15 discussed how after 1984 a theory gradually emerged in which all the large objects that we knew at the time—galaxies and clusters of galaxies—were formed from very small irregularities that had been present in the primary gas, the primeval soup of photons and elementary particles, from the beginning. This theory assumed that of these elementary particles a large proportion (80%) were part of the 'dark matter'. High-energy physicists, looking for the basic structure of elementary particles have not yet detected particles that are sensitive to gravity but do not have an electrical charge. The remaining 20% were made up of elementary particles referred to as baryons: electrons, protons and alpha particles, that could be indicated with chemical names: e^-, H^+ and He^{++}. They could and would transmit other types of electromagnetic radiation if they encountered one another. (Strictly speaking one should call electrons 'leptons', not 'baryons', a difference that is usually ignored.) All this matter, the dark and the light-emitting, was attracted in the same way by gravity.

One of the characteristics of gravity, as Newton already knew, was that it caused irregularities to clump together. Large blocks of matter—known as haloes—were thought to have formed from this matter. After this, the dark particles and the baryons separated: whenever two baryons encountered one another, they emitted light or other electromagnetic radiation, which caused them to lose energy and—under the influence of gravity—sink to the center of the halo. Since the particles of dark matter did not lose energy, there was a natural separation between the dark matter that was present everywhere in the halo and the light-emitting matter that was almost wholly at the center of the halo. The assumption was that the light-emitting matter would form galaxies that were either elliptical or in the shape of a disk. After this, the stars were formed in the galaxies. It was also realized that haloes could encounter one another and merge to form a larger halo. The galaxies present in the haloes would then also collide with one another and merge.

This is the story, told in words. What made the biggest impression, in hindsight, was that simulations by computers reproduced the observations except for minor details. The starting point of each calculation was a large nebula containing small droplets of matter that moved through the nebula in line with Newton's theory of gravity. Sometimes two droplets merged to form a new droplet with twice the mass. This continued, and droplets of different mass were created. The first calculations, in the 1980s, started with nebulae containing thousands of droplets, until ultimately, in 2005, ten billion droplets were being monitored.

The merging of initially small and gradually larger galaxies was probably not completely effective: following a merger, smaller fragments would remain that could develop into dwarf galaxies. The first estimates showed that there must have been more dwarf galaxies around our Milky Way than had so far been encountered (Klypin et al. 1999). 'Where are the dwarfs' continued to be an issue for some considerable time.

An effective illustration of how large galaxies were formed from small fragments as a result of merging is given in the 'family tree' diagram in Fig. 16.5. In this diagram the time runs from top to bottom, though it is indicated by the redshift z. All kinds of fragments knit together and form an increasingly dense family tree with a decreasing number of galaxies, until eventually you arrive at the base (the 'Now') and all elements are merged into one galaxy. The winning galaxy is indicated with a star.

In the early 1990s the researchers found themselves in a crisis. This was caused, as previously explained, by uncertainty about the average amount of matter that the Universe contained: what was the value of Ω_M: was it 0.3 or was it =1.0? Observers stuck to the first value; theoreticians demanded the second. In this crisis there were a few brave researchers (or speculators) who thought that if you took Einstein's Λ out of hibernation, you could combine these two very different demands. But, take care! Such a step was typically an ad hoc argument, a suspect class of assertions that

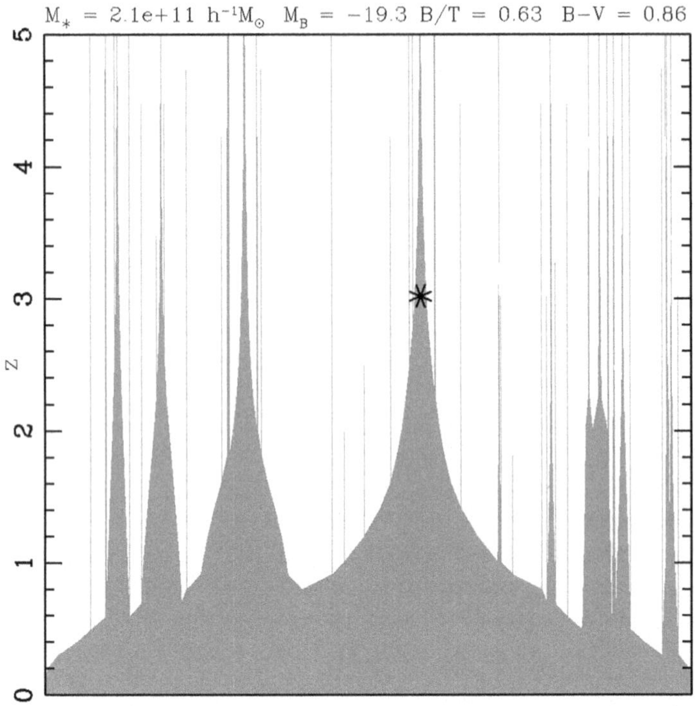

Fig. 16.5 Family trees of galaxies. The time runs from top to bottom; at the top $z = 5$ (distant past), at the bottom $z = 0$ (now). An indication is given of how different fragments of proto galaxies falling downwards merge with one another, making the line thicker. Eventually a galaxy appears with the characteristics indicated in the line at the top. The star indicates the 'winning' fragment (Baugh et al. 1998)

were used to resolve a problem by devising a completely new argument. (Journalists also always want to have two independent sources before publishing their news). Less adventurous (or possibly more realistic) researchers were more cautious. An article from 1992 contained the sentence: 'Most cosmologists regard this solution (i.e. that Λ does not equal 0) as a long shot' (Davis et al. 1992). It was therefore a great relief when in 1998/1999 the solution came from a completely unexpected source, namely from the supernova projects discussed above. And the value of Λ that these observations required was enough for the CDM calculations This success transformed the name of the theory into ΛCDM. The results were convincingly confirmed in 2003 by the analysis of measurements taken with the WMAP satellite. And so, within a few years—with a long run-up time—a new paradigm emerged about the formation of galaxies from small irregularities in the primal soup of the Big Bang.

The culmination of the development of the ΛCDM model was an article in *Nature* in 2005 that has since been cited over 2,500 times (Springel et al. 2005). The 17 authors are from all regions of the world and parts of the calculations have been

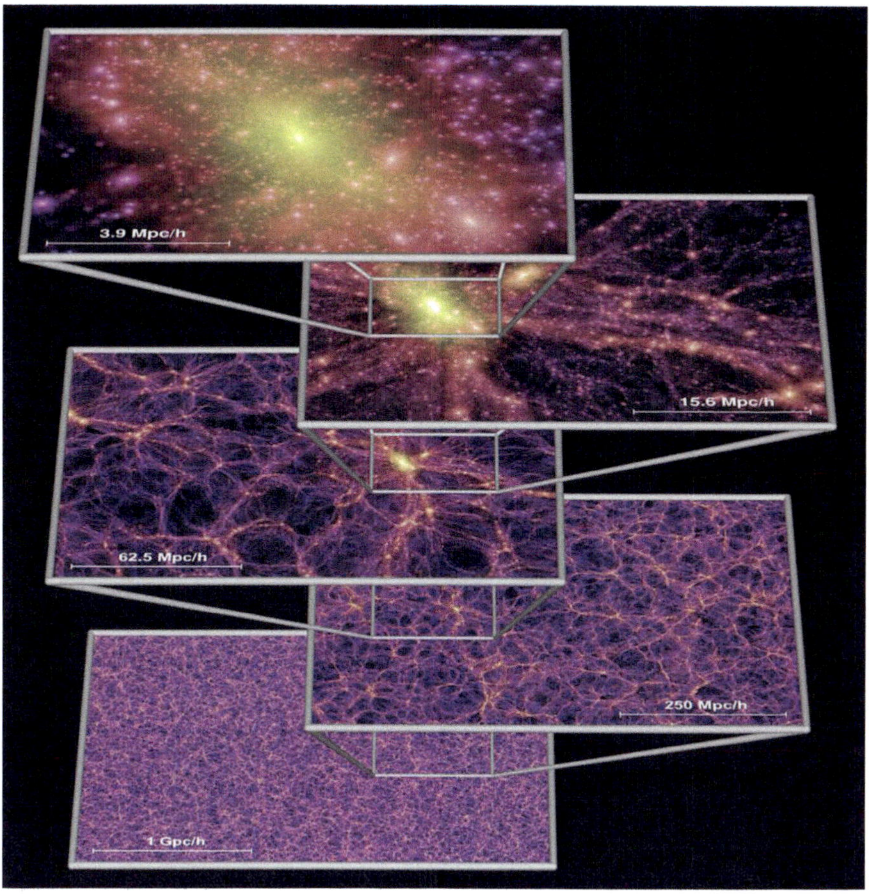

Fig. 16.6 Five images show the result of a simulation in ΛCDM cosmology. The top image has the smallest scale and shows the division of light-emitting matter around a highly elliptical galaxy. The second image shows how the previous figure is embedded in a larger whole; the same applies for each subsequent image. The bottom image shows that on the largest scale the Universe appears very uniform (Springel et al. 2005)

made on independent computers. The article is based on a simulation of the formation of galaxies from small wrinkles in the early Universe. The starting point can be found in ten billion (!!) droplets that were monitored in great detail. The droplets can undergo all kinds of processes (implosion or merger, for example) and they are eventually found in a 'calculated' Universe that matches the observations amazingly well: see Figs. 16.6 and 16.7. The model produced was named the 'millennium model', because the year 2000 was important in the international planning of the calculations.

This figure is a triumph of science: using computers, it is now possible to reproduce how the Universe developed after the Big Bang to become what it is today.

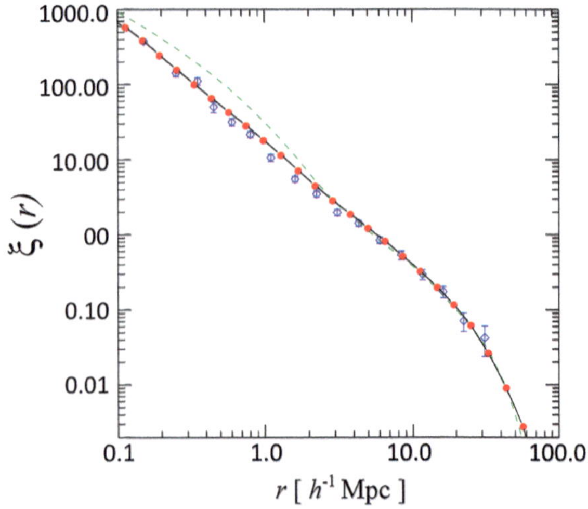

Fig. 16.7 The autocorrelation function of the Millennium simulation (in rood) and as observed by the 2dFGR survey (in blue) (Springel et al. 2005)

Having said that, it is necessary to add some caveats. The calculations are based on the existence of dark matter (of which nobody knows what it is) and dark energy (of which also nobody knows what it is) and starts with a Big Bang and an inflation. These assumptions are at the very least speculative. In all honesty, the giant is standing on feet made of clay and we have to hope that the rainy season doesn't start too soon.

How well did this simulation match the observations? This was a discussion that would last many years; it will be addressed in the coming chapters. The article from 2005 already contains an important test.

The comparison of simulation and observation is the comparison between two large collections. To do this, the autocorrelation function of both sets—a well-known technique from statistics—was chosen, which in this case boils down to the following. Take a random distance and ask how often there are two galaxies with the same distance; the answer is the number ξ(s). Figure 16.6 shows that the distribution of the matter is very irregular; there are holes and threads and thin membranes of material, besides strong concentrations. But Fig. 16.7 shows that the ξ(s) function taken from the Millennium Simulation is very similar to the 2dFGRS survey. This is reassuring: the statistics of distances and open spaces is the same in the simulation and in reality Fig. 16.6 is visually similar to the distribution of the galaxies as described in the big redshift survey, and the first solid statistical test confirms the agreement.

Chapters 8 and 15 discussed the birth and early development of the ΛCDM theory where, as a result of a merger, ever larger objects were formed as a result of merger with small objects. Chapter 15 ended with some signs of doubt among the greatest defenders of the theory. Were they on the right track? Looking back, you can see how the development of the ΛCDM model in the eighties and nineties was hampered by uncertainties about the correct model of the Universe: how much matter did it contain? How old was it? How fast was it expanding and was there a dark energy that constantly speeded up this expansion? All the uncertainties about these characteristics of the Universe disappeared with the arrival of the concordant model, as already discussed earlier in this chapter. The SDSS- and 2dFGRS-surveys that are yet to be addressed also contributed important information. The ΛCDM theory gained the status of a paradigm following the 2005 article in Nature.

The results of these millennium calculations will therefore serve in this and in the next chapter as a guide for the description of the study of the Universe and of the systems it ordered. A short period must be excluded: the period of recombination mentioned above, approximately 300,000 years after the Big Bang, when gas atoms and light had cooled down to such an extent that the electrons, protons and α particles united to form hydrogen or helium atoms. This turned out to be a brief period in which observations had to lead the research because the theory cannot describe this period.

Finally, the millennium project has some successors. Still bigger simulations will come later. Nonetheless, I feel justified in saying that the above result represented an important breakthrough in our knowledge about the Universe.

16.3 The Existence of Supermassive Black Holes and their Relation to the surrounding Galaxy

In Chap. 15, the theory was discussed that the source of the jets and other high-energy phenomena in radio systems and quasars was to be found in matter falling into the extremely deep gravitational well of a supermassive black hole (SMBH). In addition, it became possible to unite the very different manifestations in one model, a process called 'unification'. The most basic issues that the radio systems and quasars had raised seemed to have been resolved.

But despite these successes, an essential element in the story was missing: did SMBHs actually exist? Ever since the first discovery of a radio galaxy, the core of such galaxies has been investigated. Every time new detection possibilities became available, there were new attempts to prove the presence of a SMBH in radio galaxies and quasdars. A well-known example was the research in the 1980s by British-American astronomer Wall Sargent, who, with the help of the 5 m telescope at Mount Palomar (the largest at the time), studied the centers of radio systems, in particular the center of M87, also known as the radio source Vir A. Sargent succeeded in proving that a large amount of matter had accumulated in the center, enough for a SMBH, but he could not prove that the volume was so small that it had to be a black hole. The Earth's atmosphere was the fly in the ointment here, as was mentioned earlier: irregular movements in the atmosphere spread out the image of any small object. Sargent was certainly not the only person to pursue this goal; others tried and also failed.

The first true evidence came in 1995 and 1996: in three different cases and using fundamentally different techniques, the existence of a 'supermassive black hole' was finally demonstrated.

'Supermassive Black Holes' do exist

The first proof of the presence of a supermassive black hole in the center of a galaxy is credited to three Japanese radio astronomers (Nakai et al. 1993). Using a radio telescope, they measured at a wavelength of 1.2 cm details in a spectral line of H_2O in the nearby galaxy M106 (or NGC4258). In their case the telescope was equipped with a spectrograph with a much larger bandwidth than necessary: only lines were expected with small redshifts. But the broad receptor was what was available to them and they set about making their measurements.

They detected the spectral lines with a redshift of around 500 km/s, which was the velocity at which the galaxy flew away from us. But they also, unexpectedly, found peaks at velocities that deviated by 1000 km/s or more. These high-velocity lines would have been missed if the bandwidth of the radio-receptor, for efficiency reasons (and as should be), had been much narrower. Serendipity! With an ad hoc fabricated interferometer (Miyoshi 1998), they were able to prove that the high-velocity peaks originated from the center of the galaxy. They decided to seek help from radio astronomers at Harvard, who had access to the VLBA, the new and large interferometer centering on the VLA. Observations were made which proved highly successful (Miyoshi et al. 1995). The peaks at high velocity corresponded with a very small gas disk (a radius of between 0.13 and 0.25 parsec; about 0.4 lightyears) that orbited the center at high speed. It was easy to show that the matter in this center had a higher density than the average matter of every conceivable star cluster. The conclusion was therefore that they had discovered the first supermassive black hole. The measurements were very reliable and the conclusion was equally so. Excellent craftsmanship (Fig. 16.8).

The discovery of a second, very heavy black hole followed shortly after, also from Japan, but the observation technique was completely different. The Japanese ASCA satellite was used to measure the X-radiation of a galaxy, MCG-6-30-15, that had an active core (AGN). In the spectrum, the strength of the radiation decreased very regularly with increasing photon energy, but there was a deviation, an excess of radiation at a photon energy between 5 and 7 keV (see Fig. 16.9). Taking into account the photon energy, this extra radiation had to be attributed to a spectral line of strongly ionised iron—for specialists, the K-line. A spectral line with this shape had been seen before (see Chap. 13) and the explanation of the shape was given by a reflection on a rotating disk. In this case, though, the rotation speed was about 100,000 km/s. Such a high rotation speed implied the gravity of a large mass of material. The conclusion was inevitable: this was a SMBH, a supermassive black hole.

And then a third and very different sort of evidence of a heavy black hole appeared. This time the evidence was gained from the analysis of the infrared radiation of the SgrA* environment, the small source of synchrotron radiation in the center of our Milky Way and long considered to be a potential black hole. No light was received from this source: the interstellar matter between the center and us absorbed all the light. But infrared radiation did penetrate, although somewhat weakened. In the

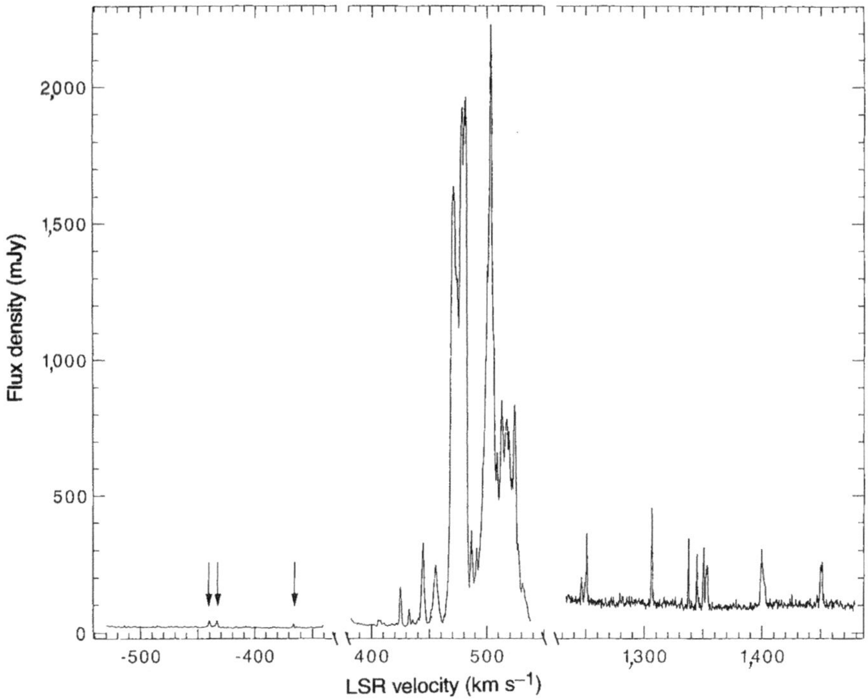

Fig. 16.8 The spectrum of the spectral line of water (H_2O) at 1.2 cm in galaxy M106. Horizontally, the velocity is calculated based on the Doppler effect. The basepoint of the velocities is that of the Sun. The system itself has a speed of around 500 km/s. The presence of lines at −500 km/s (1000 km/s less than that of the system itself) was surprising and unexpected (Miyoshi et al. 1995)

observations, individual stars were found near to SgrA*. If SgrA* was a heavy black hole, and if these stars really were close by, they would probably move at high speeds. This would in turn be reflected in an annual movement of these stars. The latest techniques in image sharpening ('active optics', see Chap. 9) were needed to see very small annual movements.

Two groups of researchers took on the challenge. One group was located in the Max Planck Institute in Munich and the researchers made their observations using the large ESO telescopes in Chile (Genzel et al. 1997). The other group was associated with UCLA in Los Angeles and used observations from the 10 m 'Keck' telescope in Hawaii (Ghez et al. 1998). The competition was fierce but fair, and in the end both groups reached the finish within the same timeframe: they showed that there were a number of stars very close to SgrA* that followed elliptical paths around a center of gravity with a mass of 2 10^6 M_{zon}, a center of gravity that coincided with Sgr A*. Figure 16.10 shows the orbits of many stars all moving in highly elliptical orbits that have one common focal point: the location of the black hole. The mass of this black hole was two million times the mass of the Sun: clearly a supermassive black hole.

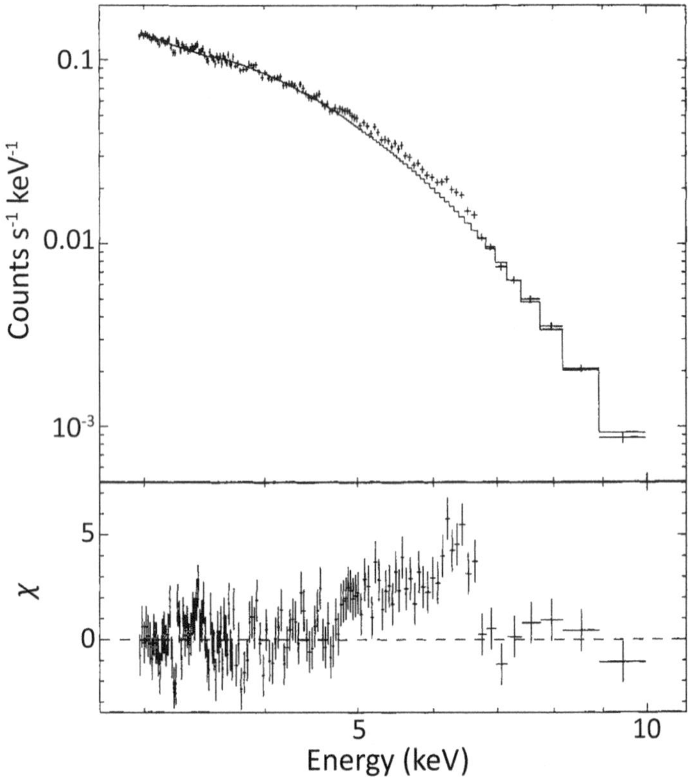

Fig. 16.9 Above: the spectrum of the X-ray of source MCG-6-30-15 measured with the ASCA satellite. The intensity of the radiation is shown vertically, the photon energy horizontally, expressed in keV. Below: after subtraction of the continuum radiation, a faint line remains: 'excess' of radiation around 7 keV. From left to right the excess slowly rises and then very quickly falls back to zero. This shape is explainable by a fast rotating disk around the center of that galaxy (Tanaka et al. 1995)

Sgr A* was host to a supermassive black hole, but it had a very low level of activity. It is a good example of a 'sleeping monster', a type of object first suggested in 1969 by Donald Lynden-Bell.

An unexpected Relationship between Galaxies and Massive Black Holes

These three exceptional and independent pieces of evidence of the existence of a heavy black hole were an enormous moral support to researchers who were still trying to find black holes in the centers of galaxies by means of optical observations, as discussed above. It has been mentioned previously that the fundamental difficulty

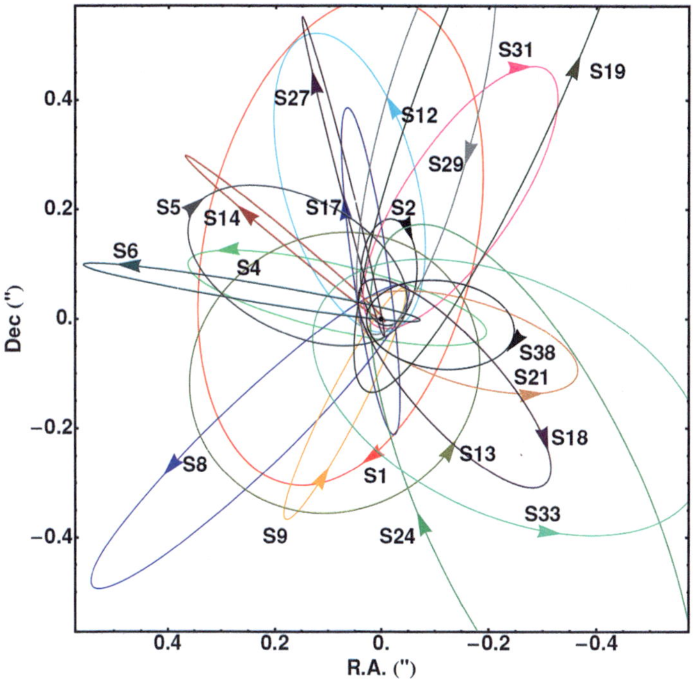

Fig. 16.10 The elliptical orbits of stars near SgrA*: all orbits have one common focal point: that of the black hole (Gillesen et al. 2009)

was in proving that the large mass of matter was concentrated in a very small volume. Optical observations from the Earth were frustrated by irregularities in the Earth's atmosphere during the observations. This kind of research was consequently given high priority once the Hubble Space Telescope was repaired, and very sharp images were produced.

Between 1988 and 2003, a dozen or so much-cited articles appeared in which an SMBH (supermassive black hole) was found in no fewer than 30 out of 36 galaxies that were observed. Very surprisingly there was evidence of a direct relation between the mass of a black hole and that of the bulge of the surrounding system (see Fig. 16.11).

The existence of this relationship meant that the mass of the bulge and of the black hole were directly related to each other, and that black holes were not objects created by accident: their existence was part of the formation of the entire galaxy. Estimates of how many galaxies had dormant and very black holes could never be made with any certainty, but the predication was that it probably applied to all major galaxies (Kauffmann and Haehnelt 2000). Quasars were the first objects in which a SMBH was suspected; now it seemed that many more systems had such an object at their center, but that often it was 'dormant' and not 'fed' with gas clouds and stars as was the case with quasars.

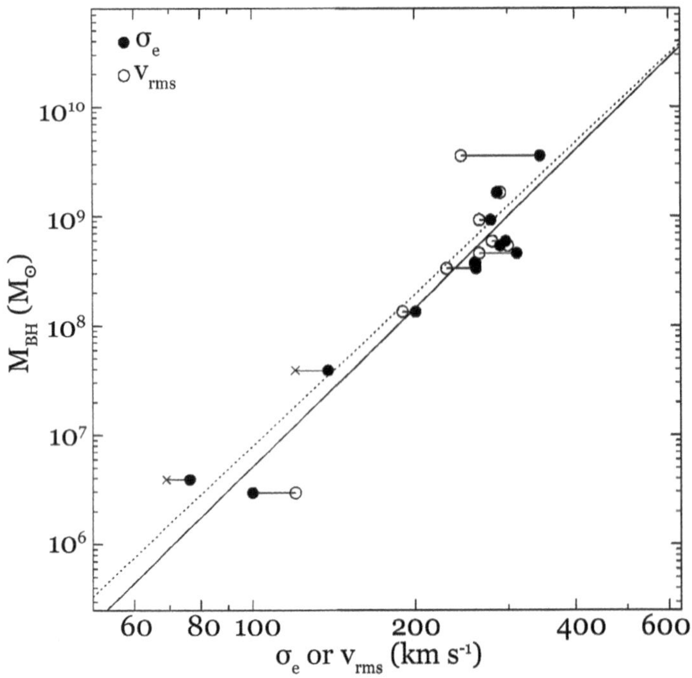

Fig. 16.11 A diagram showing the correlation between the mass of a supermassive black hole and the mass of the bulge of the host galaxy. The mass of the bulge is given in the horizontal direction and vertically the mass of the black hole. Note: The mass of the bulge is indicated by the (square) average speed of the stars in the bulge, which, based on much earlier research, is an excellent measure of the total mass of the bulge (Ferrarese and Merritt 2000)

16.4 Large Surveys

The Sloan Digital Sky Survey

The first major success of radio astronomy was the discovery in 1948 that the strong radio source Taurus A coincided with the small but special Crab Nebula. X-ray astronomy revolutionized astronomy in 1972 by demonstrating that the strong source Cygnus X-1 coincided with a heavy star that formed a binary star with a black hole—the first demonstration that this completely theoretical concept had been realized by nature.

Thirdly, when American espionage satellites observed unpredictable outbursts of gamma radiation, it quickly became clear that they did not come from the Soviet Union, but from space. But that is where all the knowledge stopped. What kinds of objects did these Gamma Ray Bursts (GRBs) produce? The answer came more than 20 years later when an ordinary optical telescope was used to find glowing gas at the site of a GRB and then a major redshift was found in the spectrum: the explosion was deep in the Universe and the magnitude of the explosion was greater than what had been seen until then.

These three examples made it clear that, despite the advent of 'new astronomies' such as radio and X-ray astronomy, the astronomy of visible light had remained of decisive importance. You could philosophize about why it was that way (because so much light is produced by stars and our eyes are optimized for sunlight?) but the fact does not leave room for discussion.

If, however, you compared the state of affairs in optical astronomy in the 1980s with that in the other astronomical disciplines, then there was a backlog. While radio astronomers and X-ray astronomers searched the sky systematically and in as much detail as possible for sources in order to store the data in openly accessible electronic databases, such archives did not exist for optical astronomy. The largest existing archive dates back to around 1950 and consisted of the Palomar Sky Survey's photographic records, supplemented in the southern hemisphere by British and European records. Examination of those records had produced numerous catalogs, for example of galaxies, but no matter how diligent the recording and subsequent analysis had been, the result was of lower quality when compared with what could be achieved with CCD cameras.

The project to make an inventory of the sky even at visible wavelengths was an obvious one and with good reason. What has impressed me, the writer, as outsider and as a professional researcher are the correct assessment of the size of the project, the quality of the data supplied and its delivery in a sufficiently short period of time. The SLOAN Foundation was found willing to provide the major investments, including for a state-of-the-art telescope in the New Mexico desert. The survey is therefore known as the 'Sloan Digital Sky Survey', abbreviated to SDSS. The size of the group of researchers involved is also impressive; all nearly 200 authors of the first publication of results were based in the US, supplemented by a strong Japanese group. Initially the emphasis was on measurements of galaxies, so the Milky Way was not observed.

The Plan and Its Implementation

The project was based on an automated, state-of-the-art telescope with a mirror diameter of the mirror of 2.5 m, which was equipped with a large battery of CCD chips (30 pieces with four million pixels each and an instantaneous field of view of 2.5°). The telescope was facing south and recorded the stars, galaxies and other objects that passed through the rotation of the Earth, everything under the control of a computer. Of course, the telescope could not look below the horizon and therefore most of the southern sky remained unobserved.

Observations started in 2000. Two surveys were carried out at the same time: during the first one all light sources were detected until a very faint magnitude (the 23rd) and in the second survey spectra were obtained from sufficiently bright objects. It was estimated that about one million of the brightest galaxies, and 100,000 quasars would be registered. The computer took the first step: it separated the stars (points) from the galaxies (smudges). The resulting data were made freely available to all interested. You did not have to be a professional astronomer to access the data.

The success of the survey can be measured in different ways. (1) A repetition decision (SDSS II) was taken at a very early stage and it was later decided to add two more surveys (SDSS III and IV). The latter also has an extension to the southern sky with a telescope in Chile and includes parts of the Milky Way. (2) A second measure of success is the high citation scores of a large number of articles that appeared shortly after the first dates were released.

Initially, the survey was intended to finally provide an overview of the galaxies to a much deeper limit than had previously been reached. But because the survey was based on measurements with a telescope with a diameter of only 2.5 m, all galaxies had a small redshift: the SDSS made an inventory of the local, contemporary Universe. This did not apply to the quasars: quasars were discovered with SDSS up to a redshift of $z = 6$.

In 2002 a first publication of the data appeared (Early Data Release) (Stoughton et al. 2002). It proudly listed a large number of studies that had already been carried out with the early data. There was a broad array of subjects: quasars with a large redshift and therefore far away in the Universe; clusters of galaxies; image deformation by gravitational lenses; the division of all galaxies into only two types, red and blue; currents of stars in the halo of our Milky Way; new variable stars; the discovery of a brown dwarf star, so cold that the atmosphere contained methane; moving objects within the solar system. Not all discoveries were new, but it did make a difference whether you showed a correlation of some property in a list of 50 objects or in a list of 150,000! Finally there was an accurate and complete record of the sky to a depth that matched contemporary research.

SDSS and 'Galaxy Zoo': Green Peas; 'Hanny's Voorwerp'

A somewhat special project, which the professional astronomers looked at with some suspicion, was the classification of the shapes of the galaxies with the help of volunteers. Enthusiasts who came forward were given a relatively simple introduction and could then download the celestial images and classify them on the basis of a few clear instructions. The project, called 'Galaxy Zoo', was a great success. In the end, there were more than 100,000 participants who collectively viewed and classified 900,000 galaxies. Because many systems had been viewed by dozens of people, the quality of the classification could be determined statistically. The results were very satisfactory.

Of course, there were also some surprises. One of these was the discovery of the Green Peas galaxies, a new type of galaxy that had never been found before. Further research showed that it was a new manifestation of the existing class of Blue Compact Galaxies or BCGs. A Dutch school teacher, Hanny van Arkel, noticed a truly exceptional object: a green nebula that defied all explanations. Still known by its Dutch name ('Hanny's voorwerp'), it took observations by the Hubble Space Telescope to come up with the explanation that it was the light of a quasar reflected in a gas cloud.

Properties of Galaxies in the SDSS Survey

One of the first analyses of SSDS data led to a still surprising result. Using a database of 150,000 galaxies, the prime conclusion was that it consisted of two distinct samples: red galaxies and blue galaxies (Strateva et al. 2001). This led to what has been named the 'color-magnitude diagram of galaxies'. See Fig. 16.12.

Another image tells the same story, but rather more exactly; see Fig. 16.13.

The blue galaxies proved to correspond to spiral systems and gas-rich, irregular galaxies. The red galaxies were a less homogeneous group, but they all had in common that, unlike the blue galaxies, they contained virtually no cool gas and did not appear to form any stars. The red systems consisted mainly of elliptical systems, from small dwarf systems to gigantic cD systems. In addition, the red group also includes the 'lenticular systems', usually coded as S0.

Not everything was new. The idea that galaxies were red or blue also occurred in a review paper from 1994 (Roberts and Haynes 1994). This dichotomy, red or blue, was an astonishing result. From 1936 onwards, Hubble's tuning fork, in which 12 kinds of systems were recognized, had gradually taken on the scent of holiness. Now, measurements of the color (and not shape) of more than 100,000 galaxies showed that a division into only two types, red and blue, sufficed. And what was more: in further studies of both the SDSS and remote systems, this red/blue dichotomy remained ample. It seemed that Hubble's morphology had only come about very late in the development of galaxies.

First we look at the number of systems. The distance of each system can be calculated thanks to the measured redshift. So the 'absolute' brightness can be deduced from the measured brightness, the brightness that the system would have if it were at a distance of 10 parsecs.

This conversion of the amount of emitted light, L, to the total mass, M, is usually based on the global course of the spectrum, also called the 'spectral energy distribution' or SED. Figure 16.14 shows the numbers of red and blue galaxies for a given mass. The

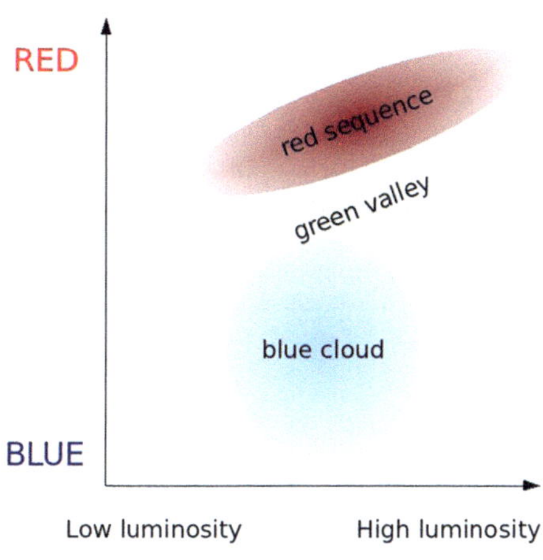

Fig. 16.12 A somewhat schematic representation of the colors and luminosities of galaxies. Source: Wikipedia, Joshua Schroeder

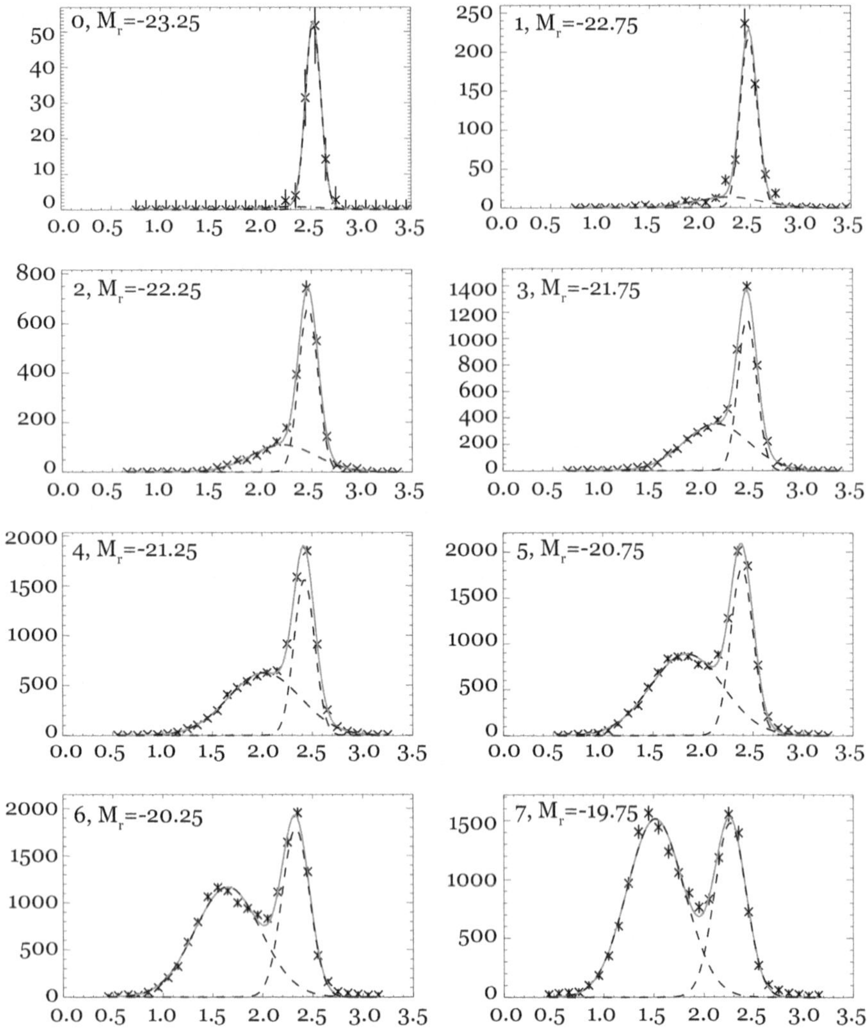

Fig. 16.13 Histograms of the color of galaxies in a diagram of differing luminosity and numbered 0–7. In each diagram the color of the galaxy is given horizontally. A color of 1.5 indicates 'blue', a color of 2.5 indicates 'red'. In each diagram, 1 or 2 clock functions of Gauss are drawn through the data points. Diagram 0 contains the brightest systems, all of which are red. More and more blue systems appear in successive diagrams. In the last diagram, the blue ones predominate in number. The most surprising conclusion to come from this picture is that two of Gauss's clock functions describe the measurements excellently in all diagrams. This explains why one talks of two separate groups of systems, without intermediate objects (Baldry et al. 2004)

heaviest galaxies are in the left part of the diagram: there are many more red galaxies than blue. The reverse is true of the lightest galaxies, which are mainly blue.

A very different result was a spectral line of oxygen found in many galaxies (see Fig. 15.6 in Chap. 15). The strength of this line, which occurs only in the spectra of the blue galaxies, can be compared with the strength of the lines of hydrogen and

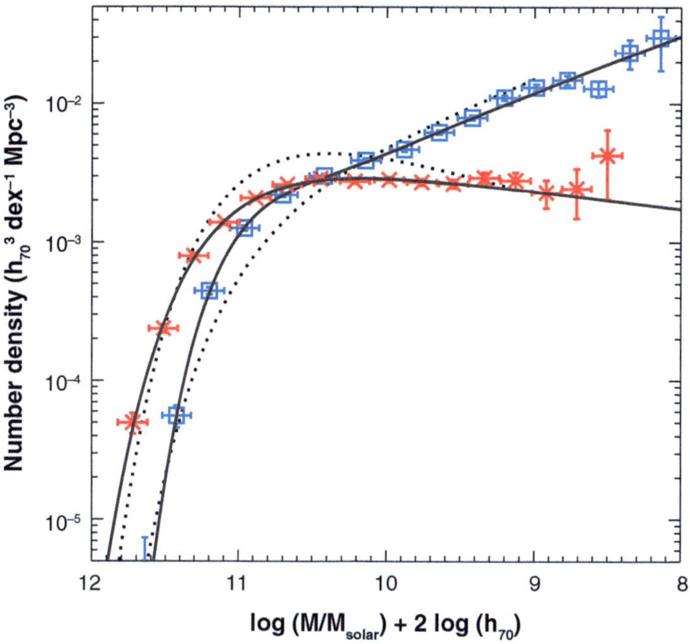

Fig. 16.14 Diagram of the distribution of the star masses of red and blue galaxies (Renzini 2006)

from the relationship between the two the ratio between the total number of oxygen atoms present and that of the hydrogen atoms can be determined; in other words, it was possible to calculate the abundance of oxygen. Surprisingly, there was a clear link between the size of the galaxies and oxygen abundance: the larger and heavier the system, the higher the oxygen abundance. As already mentioned, atoms such as oxygen must have been produced in the nuclei of stars and returned to circulation via supernovae. Massive galaxies apparently do this more efficiently than light galaxies. This was a fundamental fact that said something about the evolution of galaxies, although that information could not yet be read from this curve. More massive galaxies have higher abundances, but the most massive galaxies reach a saturation point that is twice the Sun's abundance. The central curve is called the mass-metallicity relationship (Fig. 16.15).

The Quenching Problem

The separation of all galaxies into blue and red evoked a number of problems that have still not been solved. Star formation is a transient phenomenon, because sufficient gas must be supplied to form stars. But if all the gas has been consumed, or if the production of new stars is otherwise prevented, then this system will automatically become sterile in a short time, in a few millions of years; in particular, the bright blue stars will disappear and the color will be predominantly red. The

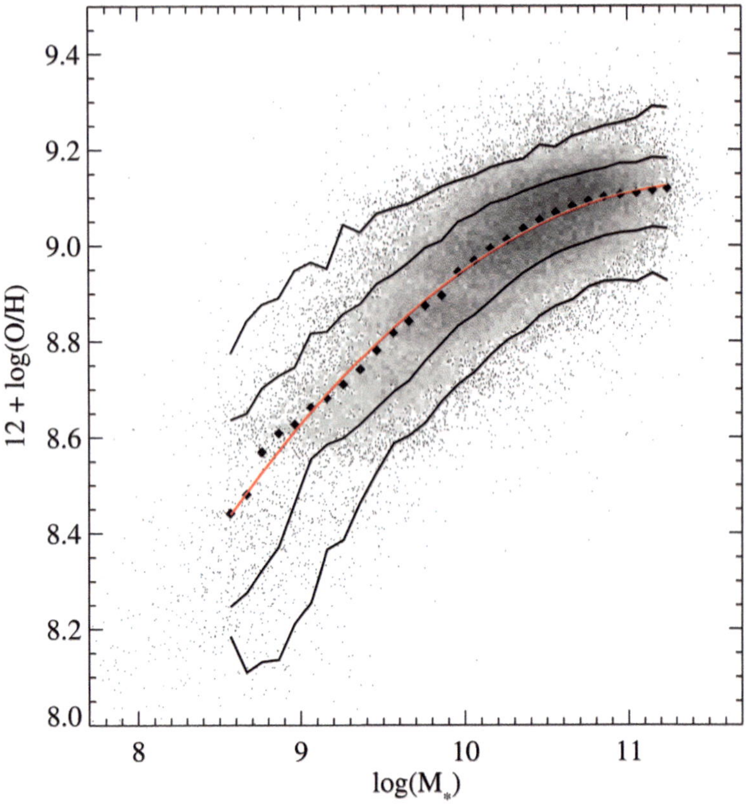

Fig. 16.15 Mass-Metallicity Relation based on a small number of emission lines of oxygen (Tremonti et al. 2004)

generally accepted hypothesis is that all gas in the blue system disappears in a short time, a phenomenon called 'quenching'.

How this quenching takes place is as yet an unsolved problem. An eruption of a super-massive black hole in the center of the system would probably drive all the clouds of gas out of the system, but also a violent phase of star formation could lead to a large number of supernovae occurring in close succession that reach the same level. Another possibility is that a merger takes place between two equally large systems and all the gas is expelled (Peng et al. 2010). In 2016, quenching was still an unexplained event.

2dFRGS: The 'Two-Dimensional Fiber Galaxy Redshift Survey'

The SDSS covered a large part of the sky, but not the southernmost part. The southern hemisphere now had the well-equipped 3.9 m Anglo-Australian Telescope (AAT) and there was a lot of experience in simultaneously measuring a lot of spectra by means of fiber technology. In the British-Australian astronomical community, the plan emerged to use the AAT to measure the spectrum of approximately one million galaxies in the southern sky: for cosmological research, this southern supplement to northern surveys (such as the SDSS) was essential. The plan became a project called 'two-dimensional fiber galaxy redshift survey', where 'two-dimensional' indicates that one simultaneously measured the spectra of many galaxies in the two-dimensional field in the image field. The project was usually referred to by the acronym '2dFGRS'.

The 2dFGRS measurements would also complete another, existing database. This database with data from two million galaxies in the southernmost part of the sky was created by digitizing large survey photos taken on the AAT and on the ESO observatory in Chile (Maddox et al. 1990; Hambly et al. 2001). These data and those of the 2dFRGS together formed a successful completion of the extragalactic measurements of the SDSS in the northern sky.

In the years between 1997 and 2002, spectra were measured on 272 nights. In total, measurements were taken in 1,550 square degrees (7% of the southern sky). The limit magnitude in blue light was 19.5. There were spectra for 221,000 galaxies including 6000 quasars; the highest redshift of a galaxy was 0.3, the highest z-value for the quasars was z = 2.3 (Colless et al. 2001). Like the SDSS, the 2dFGRS was limited to nearby galaxies and thus to the present Universe (Figs. 16.16 and 16.17).

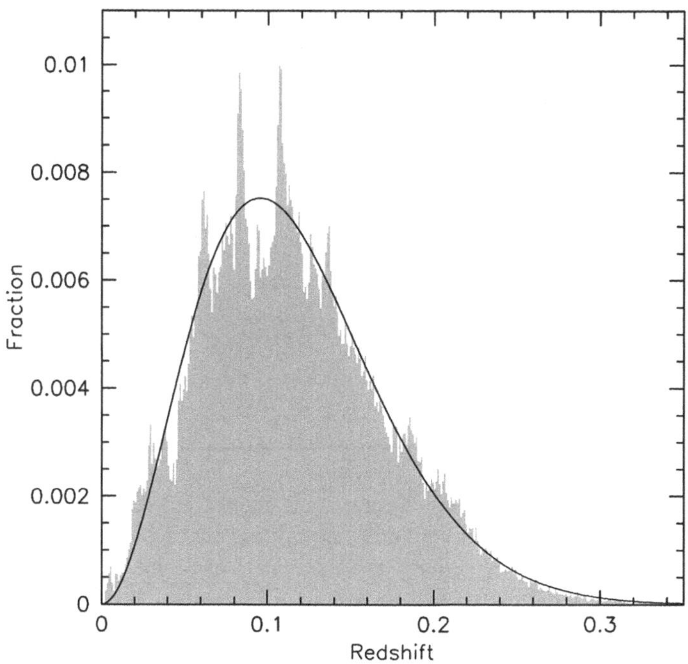

Fig. 16.16 Histogram of the measured redshift in the 2dFRG survey (Colless et al. 2001)

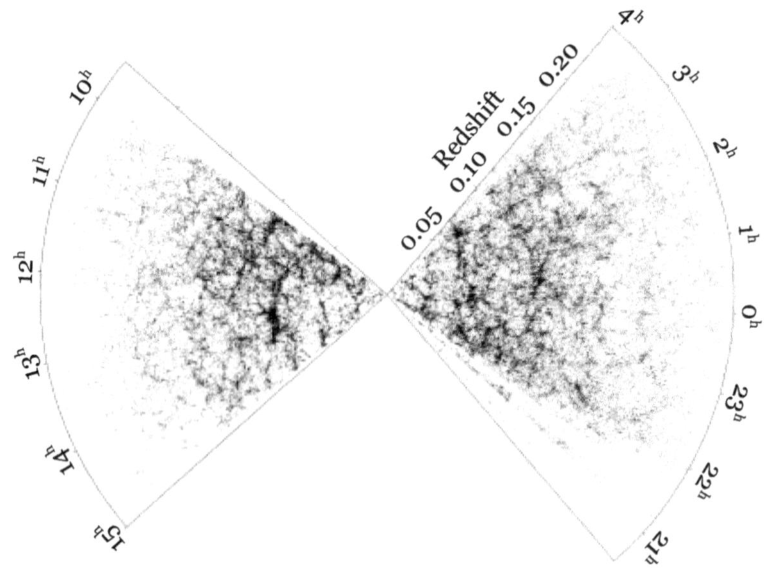

Fig. 16.17 A small part of the 'cosmic web' as can be deduced from the redshifts recorded by 2dFGRS (Colless et al. 2001)

Other Surveys

SDSS and 2dFGRS were both large projects that succeeded in their objectives. Since 2002 they have had a lot of influence on the research of galaxies and of the structure of the Universe. The two complemented each other because the first had mainly explored the northern hemisphere and the second the southern one. Where they overlapped, the measurements gave the same results. There were also significant differences in quantity and quality. SDSS had five color filters covering the entire optical wavelength range (from 355 to 890 nm) and the survey went deeper (up to 22nd size); 2dFGRS had only one color filter (at 445 nm) and went up to a magnitude of 19.5. 2dFGRS had the specific goal of exploring the local structure of the Universe while SDSS has served, in addition to this goal, the stellar research community.

Yet these differences became less important when you look at the treasures they contain. All over the world, astronomers could feel like Ali Baba or the Count of Monte Cristo: after years of poverty or misery, an overwhelming treasure was suddenly at their feet. The comparison was not entirely correct, because the astronomical treasures were real, but unpolished diamonds: only expert hands made them valuable. For the 2dFGRS estimations the diamond polishers were mainly in Great Britain, with the SDSS it is striking that besides American groups there was also a very productive group in a Max Planck Institute in Munich.

Finally, there must have been a major difference in the financial basis, which that was many times greater for SDSS than for 2dFGRS. The second is like a pioneering action: with few resources but with enthusiastic and competent participants, a maximum of results was obtained, the kind of actions the British have always been good at. SDSS, on the other hand, was a great, broadly supported and smooth-running machine, which once again leads me to think that Americans are not English, but English-speaking Germans.

The success of SDSS and of the 2dFGRS led to more surveys being set up—quite similar, but aimed at more and better measurements. For example, there are the 6dF and WiggleZ surveys as a follow-up to the 2dFGRS survey and the BOSS (Baryon Oscillation Spectroscopic Survey) as a follow-up to the three previous SSDS surveys SDSS I, II, III. There is also the dark-energy survey. The results reinforced and specified those of the previous surveys: the determination of cosmological parameters. However, they did not yet provide any important news. The most important conclusion was always that the ΛCDM model with an Euclidean geometry is the right one.

The progress in automation also made it possible to make surveys at other wavelengths. A few radio astronomical surveys are worth mentioning, such as the NRAO-VLA survey at 21 cm wavelength continuum radiation (Condon et al. 1995) and the WENSS survey at 92 cm (Rengelink et al. 1997). There was HIPASS (Australia, 1997–2002), which blindly searched 71% of the sky for 21 cm line radiation (Meyer et al. 2004); in the northern hemisphere the ALFA project was based on the Arecibo 300 m telescope in Puerto Rico (Giovannelli et al. 2005). Both 21 cm line surveys yielded some curious objects, which contain a lot of gas, and almost no stars. The suggestion was made that these were original 'clouds of primeval matter' that had never formed stars. The most striking example, however, a large cloud in the Virgo cluster of galaxies, was probably a gas cloud that had been driven out of a nearby spiral system (Minchin et al. 2005; Bekki et al. 2005). These radio surveys were limited to the local Universe.

A general rule is that radio and X-ray surveys look less deep into space than optical and infrared surveys. Exceptions are galaxies with active cores (AGNs), as was apparent on the discovery of Cygnus A in 1954.

References

Baldry, I.K., Glazebrook, K., et al.: Astrophys. J. **600**, 681 (2004)
Baugh, C.M., Cole, S., et al.: Astrophys. J. **498**, 504 (1998)
Bekki, K., Koribalski, B.S., Killborn, V.A.: Mon. Not. R. Astron. Soc. **363**, L21 (2005)
Colless, M., Dalton, G., et al.: Mon. Not. R. Astron. Soc. **328**, 1039 (2001)
Condon, J.J., Cotton, W.D., et al.: Astron. J. **115**, 1693 (1995)
Davis, M., Efstathiou, G., et al.: Nature. **356**, 490 (1992)
De Bernardis, P., Ade, P.A.R., et al.: Nature. **404**, 955 (2000)
Ferrarese, L., Merritt, D.: Astrophys. J. **539**, L9 (2000)
Genzel, R., Eckart, A., et al.: Mon. Not. R. Astron. Soc. **291**, 219 (1997)

Ghez, A.M., Klein, B.L., et al.: Astrophys. J. **509**, 678 (1998)
Gillesen, S., Eisenhauer, F., et al.: Astrophys. J. **692**, 1075 (2009)
Giovannelli, R., Haynes, M.P., et al.: Astron. J. **130**, 2613 (2005)
Hambly, N.C., MacGillivray, H.T., et al.: Mon. Not. R. Astron. Soc. **326**, 1279 (2001)
Kauffmann, G., Haehnelt, M.: Mon. Not. R. Astron. Soc. **311**, 576 (2000)
Klypin, A., Kravtsov, A.V., et al.: Astrophys. J. **522**, 82 (1999)
Kowal, C.T.: Astron. J. **73**, 1021 (1968)
Leibundgut, B.: Annu. Rev. Astron. Astrophys. **39**, 67 (2001)
Maddox, S.J., Sutherland, W.J., et al.: Mon. Not. R. Astron. Soc. **243**, 692 (1990)
Meyer, M.J., Zwaan, M.A., et al.: Mon. Not. R. Astron. Soc. **350**, 1195 (2004)
Minchin, R., Davies, J., et al.: Astrophys. J. **622**, L21 (2005)
Miyoshi, M.: IAU. **184**, 407 (1998)
Miyoshi, M., Moran, J., et al.: Nature. **373**, 127 (1995)
Nakai, N., Inoue, M., Miyoshi, M.: Nature. **361**, 45 (1993)
Peng, Y.-J., Lilly, S.J., et al.: Astrophys. J. **721**, 193 (2010)
Rengelink, R.B., Tang, Y., et al.: Astron. Astrophys. Suppl. **124**, 259 (1997)
Renzini, A.: Annu. Rev. Astron. Astrophys. **44**, 141 (2006)
Roberts, M.S., Haynes, M.P.: Annu. Rev. Astron. Astrophys. **32**, 115 (1994)
Spergel, D.N., Verde, L., et al.: Astrophys. J. Suppl. **148**, 175 (2003)
Springel, V., White, S.D.M., et al.: Nature. **435**, 629 (2005)
Stoughton, C., Lupton, R.H., et al.: Astron. J. **123**, 485 (2002)
Strateva, I., Ivezič, Z., et al.: Astrophys. J. **122**, 1861 (2001)
Tanaka, Y., Nandra, K., et al.: Nature. **375**, 659 (1995)
Tremonti, C.A., Heckman, T.M., et al.: Astrophys. J. **613**, 898 (2004)

Chapter 17
2005–2015: Harvest Time

'***The Universe is in Transition***' (Kormendy and Kennicutt 2004).

In the years from 1995–2005 extragalactic research made a number of important steps forward, as reported in the previous chapter: a definite and well specified model for the Universe was generally adopted: 'the concordant model'. The ΛCDM theory on how galaxies and stars formed was widely accepted, and large surveys brought a database of galaxies of unheard of size and quality.

So: what next? The first thing that comes to mind is the search for distant galaxies, galaxies with high redshifts that are our messengers from earlier times. Neither the SDSS nor the 2dFGRS survey went very deep into the past. Luckily the exploration of galaxies in deep space was already going on before 1995 and was enthusiastically continued after 2005, with quite some success as will become clear shortly; especially the history of star formation over the last 12 billion years was clarified. Other aspects of the evolution of galaxies (formation of bulges and disks; developments of bars) were described by new ΛCDM simulations but could not be checked thoroughly by observations—an example is the formation history of our own Milky Way. It can be calculated but not observed.

A big unknown that remained was the early episode in the history of the Universe when recombination began, at $z = 1100$, roughly 300,000 years after the Big Bang and ended with the reionization that occurred between $z = 11$ and $z = 6$. This episode is an almost white area on the map of evolution and if Mr Livingstone has already been there, we are unaware of it.

I called the early episodes a big unknown; this word applies even more strongly to the fundamental unknowns, dark matter, dark energy and inflation. The reader will find a few words about them at the end of this chapter.

The research in the years 2005–2015 has been very intensive, and the publication rate has been enormous. Writing present-day history is almost impossible. I describe the results from some impressive papers, but they do not cover all the developments. I apologize to those who wrote what I have missed.

H. J. Habing, *The Birth of Modern Astronomy*, Historical & Cultural Astronomy,
https://doi.org/10.1007/978-3-319-99082-8_17

17.1 Searching for High-z Galaxies

Let me start with a somewhat negative statement: the study of very distant galaxies is very hard. These galaxies are very faint objects and because among the faint objects there is always a majority of intrinsically faint galaxies; the truly distant galaxies are always a minority on a photographic plate—see Fig. 17.1 below. Since taking a spectrum of a faint object takes much on a large telescope, the first step in setting up a successful program is to find a method to separate the minority of interesting objects from large majority of non-interesting, faint objects. Even when the method chosen is successful, its success is limited. Several selection methods have been developed over the last 20 years and some were successful, but they all selected a sample of galaxies within a limited range of properties and therefore of limited importance.

To compare samples of galaxies, the size of the sample has to be as large as possible. In many studies the sample size of very distant galaxies was over 20 and sometimes over 100, but smaller samples often occurred and then statistical discussions were dangerous. A further problem is called 'cosmic variance': distant galaxies are not uniformly distributed over the sky: if you select a small area to study in detail, you may get different results from what you found in another small area. The cosmic web causes quite some non-uniformity on the small scale distribution of galaxies. See for example Fig. 17.2.

The demand for large samples of faint objects led to the generally accepted procedure to constrain the studies of distant galaxies to a limited number of small fields

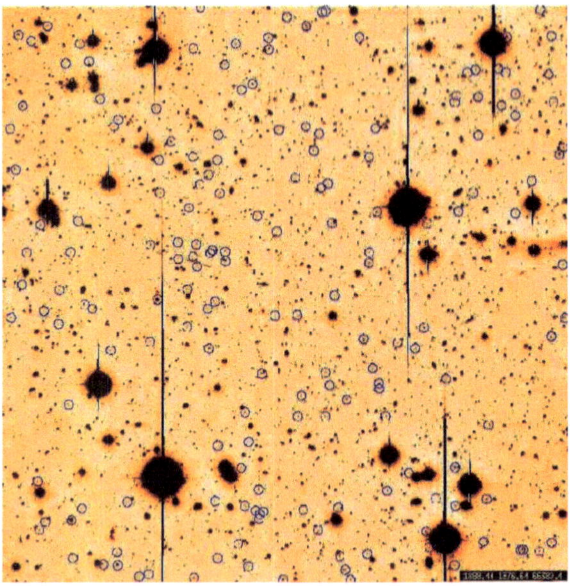

Fig. 17.1 A small field on the sky. The strong, black dots are all stars within our Milky Way. The very small dots are all galaxies. Most of them are faint and nearby. The encircled dots are probably distant galaxies selected by Steidel's technique (Pettini 2003)

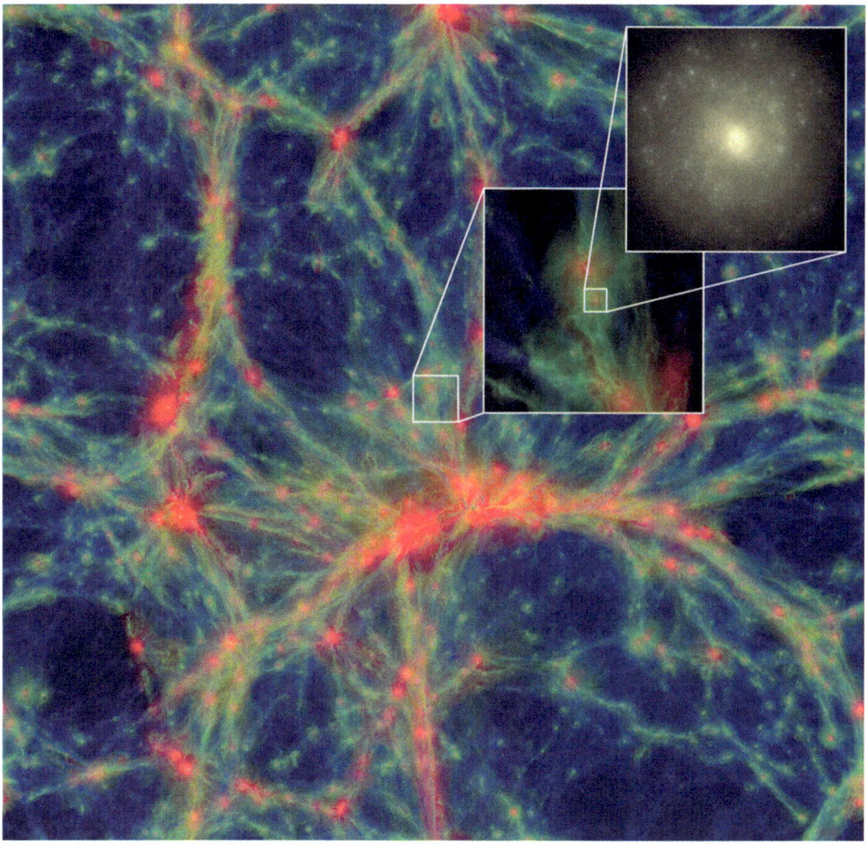

Fig. 17.2 A small piece of sky reproduced in a large simulation. There may be considerable differences between the content of selected areas depending on where they are located—'cosmic variance' (Somerville and Dave 2015)

on the sky (e.g. 10 × 10 arcminutes) and to concentrate on the objects in those fields. The first field was the Hubble Deep Field (see Chap. 16) but others have been added, e.g. the Hubble Ultra Deep Field. Famous were the two fields called GOODS (great Observatories Origins Deep Surveys) on which the largest space telescopes from ESA and NASA have made long exposures from the X-ray to the infrared wavelengths.

Observing very faint objects is time-consuming. Taking a spectrum is strongly preferable over taking measurements through narrow band filters, but you pay for it in telescope time. An example is the infrared spectrum of a galaxy with a near-infrared magnitude of K = 19.5. It was taken with a 8.2 m telescope during an exposure that lasted 29 hours (Kriek et al. 2009). The farther the galaxy is away (and thus: the younger the Universe it lives in), the more difficult it is to assess reliably its properties.

To make the discussion more specific, let us look quantitatively at the faintness problem. The concordant model of the Universe states that 13.6 billion years have passed since the Big Bang. The farthest galaxy is therefore 13.6 billion light-years away (minus a few hundreds of thousands of years). This implies that a galaxy in the very young Universe will be 46 magnitudes fainter than its intrinsic magnitude (this is the distance modulus: 5.0 log 13.6 10^9–5 = 46). Even a very luminous galaxy with an intrinsic magnitude of −20 will appear as an object of the 26th magnitude (−20 + 46 = 26).

The CFHT Survey

A first survey dedicated to finding galaxies as distant as possible was a survey undertaken with the 3.6 m Canada-France-Hawaii telescope, the results of which were published in 1995 and 1996. Some 1,000 galaxies were discovered. Figure 17.3 show that on average the galaxies get brighter when z and the lookback time increase up to 8 Gyr.

The CFHT survey was a success, it achieved what it set out to do. And it made clear that if you really want to penetrate into the Universe beyond $z = 1$ (lookback time 8 Gyr), you need a telescope with a mirror larger than 3.6 m. Luckily these came into operation around the same years that CFHT produced its results.

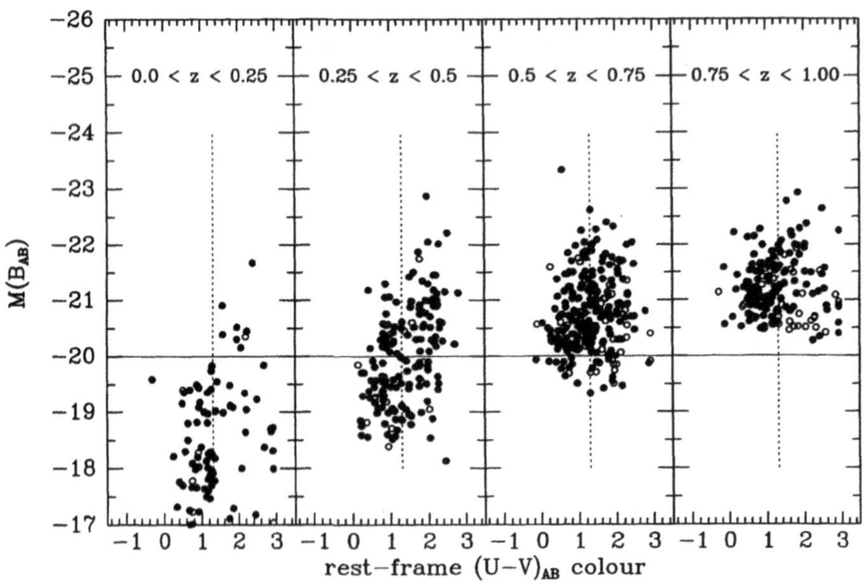

Fig. 17.3 Color-magnitude diagram of galaxies from the CFHT survey. The diagrams are arranged by increasing redshift. At earlier times (larger z) the galaxies were intrinsically brighter. For a conversion of z into lookback time, see figures below (Lilly et al. 1995)

Lyman-Break Galaxies: z = 2–3

In the 1970s and 1980s the search for galaxies with a redshift over $z = 1$ had been very frustrating (except for quasars). This ended when in the early 1990s a young American astronomer, Charles Steidel, came up with a bright idea. First look at the problem: read the legend to Fig. 17.1.

The large black dots are stars and they do not interest us. All the very small dots are for the large part galaxies and among these galaxies there will be some that are truly distant. The question was, how to weed them out? Steidel considered the ultraviolet spectra of young and hot stars. See Fig. 17.4.

The spectrum shows a 'break' just below 900 Å (90 nm); the break is caused by absorption by hydrogen in the atmosphere of the star. Now suppose you want to find galaxies with a redshift $z = 3$. The spectra of galaxies with many young stars will have an ultraviolet spectrum like the one in Fig. 17.4. When the galaxy has a redshift of $z = 3$, the break occurs at 360 nm (4 times 90 nm). This is in the optical part of the spectrum. Thus observe the field through three filters, one in red, one in green and the third in the UV, with a wavelength below 360 nm. The galaxy will show up in the images taken through the red and green filters, but not through the UV filter. Figure 17.5 shows a specific case.

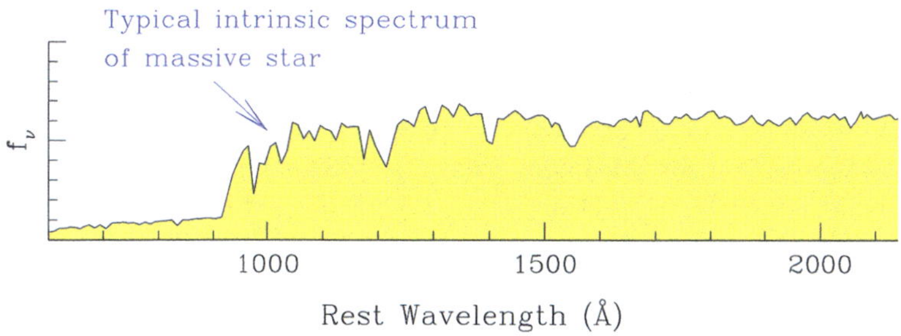

Fig. 17.4 The spectrum of a young and massive star. Wavelength in the horizontal scale; 10 Å = 1 nm (Pettini 2003)

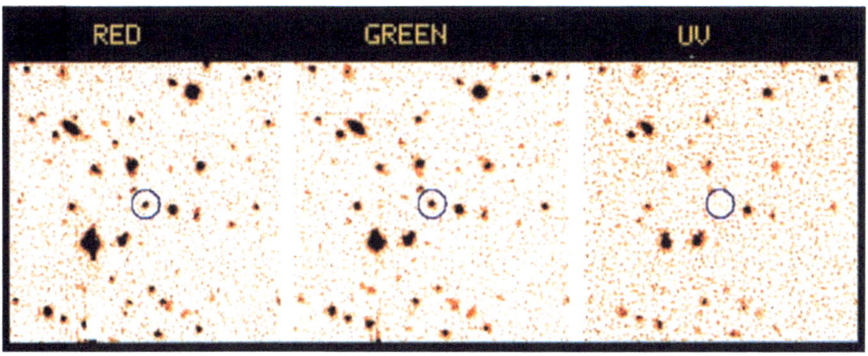

Fig. 17.5 A small piece of sky observed through three filters. When observed through the red and through the green filter the encircled object shows up. But is it not seen in the UV-filter image. This object turned out to be a distant galaxy (Steidel 1999)

The filter observations produced a large number of candidate galaxies and the spectra subsequently obtained with the 10 m Keck telescopes proved that most of the candidates were galaxies with the redshift Steidel wanted to find (Steidel et al. 1996). The method was a real success and broke through the barrier that had been so frustrating. From now on galaxies could be found significantly deep in the Universe to see evolutionary changes.

What other properties did these 'Lyman-break' galaxies have? In how far were they different from the galaxies in the Universe of today? For one thing, these galaxies showed such strong lines of hydrogen that they must be forming new stars at a very high rate - perhaps 500–1,000 times higher. Another question was: into what type of present-day galaxy will these distant galaxies evolve? No solid answer was forthcoming.

Around the millennium, the year 2000, there was abundant proof that the Universe had evolved over the last 11–12 billion years. The Universe then was different from what it is now. The Universe is in transition.

Further than $z = 3$

Steidel's technique was restricted in its application; galaxies with $z = 4$ or 5 did not show up. But variants were later invented and led to the detection of galaxies with larger redshift, galaxies living in a Universe of more than 12 billion years ago, when it was at most 1.5 billion years old. One other detection technique was to use radio sources to point to a possible high-z galaxy. In 1988 a radiogalaxy was found with $z = 3.395$ (Lilly 1988), in 1990 one with $z = 3{,}800$ (Chambers et al. 1990) in 1999 a galaxy with $z = 5.19$ (van Breugel and de Breuck et al. 1999). In more recent years the records were improved: galaxies at $z \sim 8$ were discovered with the HST and in the Hubble Ultra-Deep Field a small smudge that might well be a galaxy with $z \sim 10$ (Bouwens, Illingworth et al. 2011), and a 'remarkable luminous' galaxy at $z = 11.1$ (Oesch, Brammer et al. 2016).

Records are set by exceptions. Strong conclusions about the Universe when it was younger than about one billion years were not obtained from data on galaxies; one swallow did not make a summer.

17.2 Star Formation through the Ages

Gas Clouds at high z

A galaxy that forms stars must contain gas clouds. An unexpected surprise was the detection of a CO emission line in a quasar with a redshift of 4.69 (Omont et al. 1996) (Fig. 17.6). Later the (forbidden) line of ionized carbon with a rest wavelength of 0.156 mm (and thus undetectable from the surface of the Earth) was detected at a wavelength of 1.1 mm, because of its redshift of $z = 7.49$ (Venemans

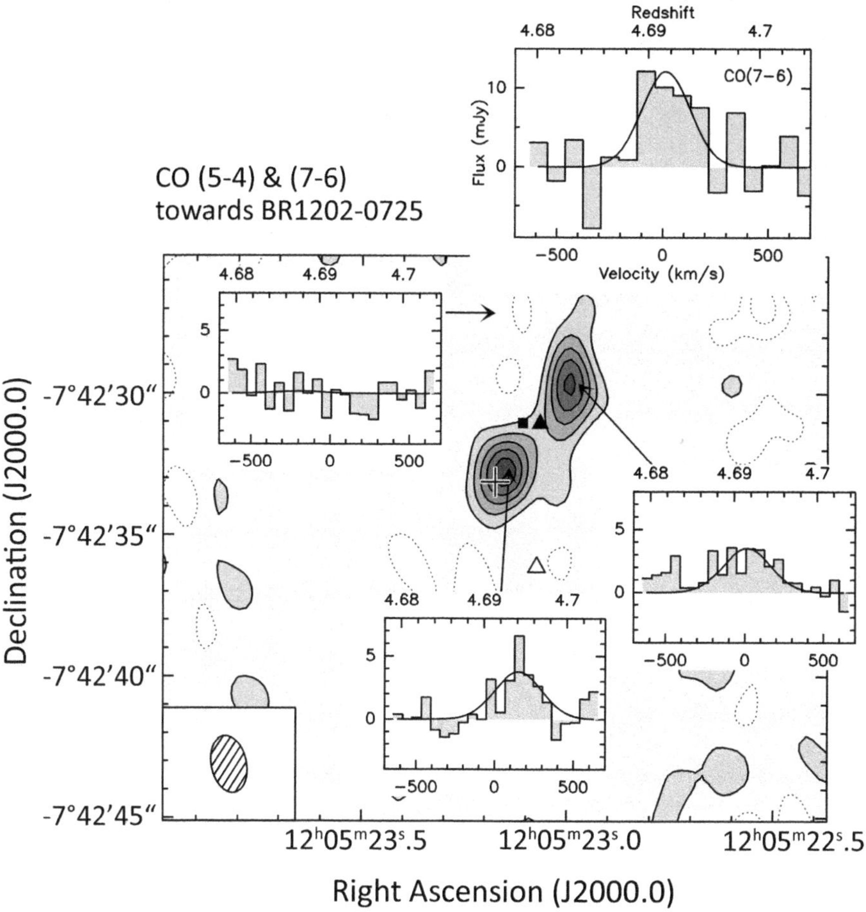

Fig. 17.6 The line profile of the CO J = 5–4 and 7–6 lines as measured with the IRAM 30 m telescope in several directions around the quasar BR1202-0725. The background is a continuum map of the quasar (Omont et al. 1996)

et al. 2017). Figure 17.7 shows a detection of the line of ionized carbon and a line of CO in a galaxy with a redshift of 'only' z = 6.42.

Several other studies have confirmed the presence of gas in such distant galaxies and thus so early in the history of the Universe. The radio interferometer for mm studies in the French Alps produces detailed maps, and, of course, there is now ALMA, whose production will become enormous.

Newly formed Stars

The formation of new stars, and especially young and high-mass stars, contributes strong emission lines in the spectra of a galaxy—see Chap. 15. Consequently the rate at which a galaxy forms new stars can be reliably measured. It explains why the

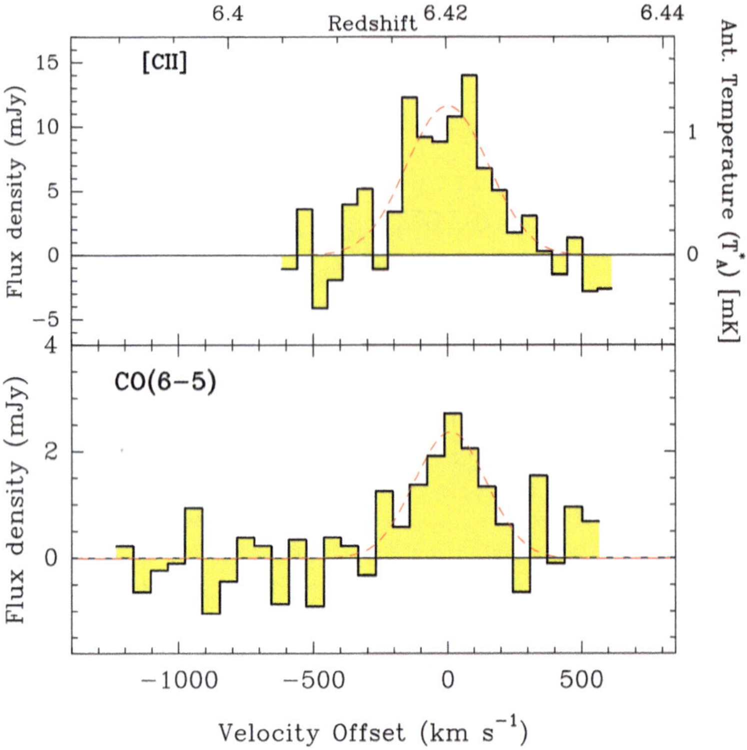

Fig. 17.7 The [CII] line rest wavelength 156 µm and a CO line in a quasar with a redshift of z = 6.42. The CO measurement is from an earlier paper (Maiolino et al. 2005)

'star formation rate', abbreviated to SFR, plays such a large role in the studies of the early Universe. This is undoubtedly the reason why this aspect of the evolution of galaxies has been studied so intensively whereas there are hardly any observational studies of the evolution of other galactic properties such as the formation of bulge and disk or of bars.

One of the first discoveries was the existence of what is known as 'the main sequence of star formation'. When one makes a plot of the mass of a galaxy in the horizontal direction and the SFR in the vertical direction, a broad bar appears: the SFR is systematically larger in galaxies with more stars. This in itself is not surprising as it is also true for human populations: when there are more people, more children are born. What is surprising is that in the immediate past the production of new stars was significantly higher, as Fig. 17.8 shows: even at z = 0.75 (1 Gyr ago) the galaxies formed stars at a higher rate.

The well-received review by Madau and Dickinson (2014) gives me the opportunity to summarize many excellent studies by many researchers, using the HST and other big new telescopes. Figure 17.9 shows how the average rate of star formation increased in the first 3.6 billion years after the Big Bang and then started to go down. The diagram resembles the similar map of the quasars (see Fig. 15.28 in Chap. 15).

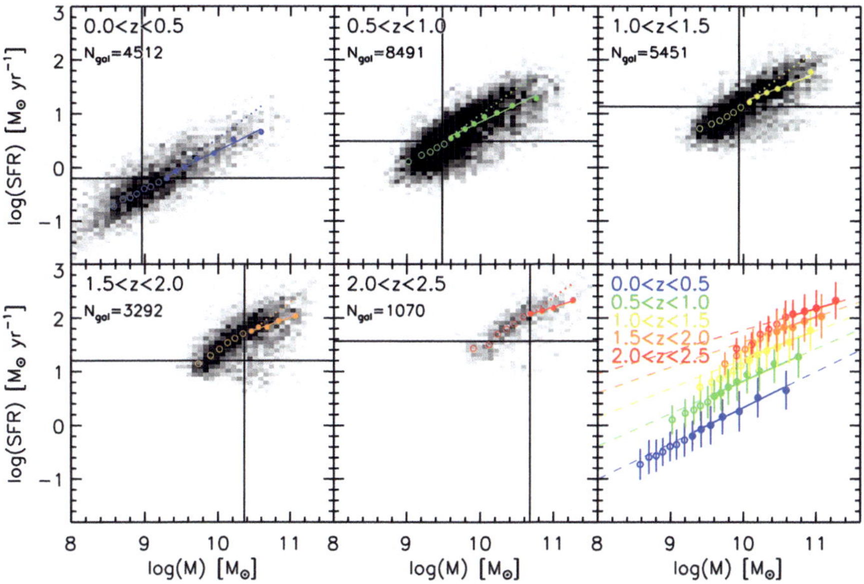

Fig. 17.8 Five diagrams referring to the same properties of star-forming galaxies each diagram for different values of *z*; the sixth, colored diagram contains the previous five. One notices the correlation, called the 'main sequence', between SFR and log M. The SFR was higher in earlier times: star formation has reduced in the last ten gigayears (Whitaker et al. 2012)

Fig. 17.9 The average SFR in the Universe as a function of *z* or in lookback time. 10 Gyr ago there was a peak. The colors refer to different samples of galaxies (Madau and Dickinson 2014)

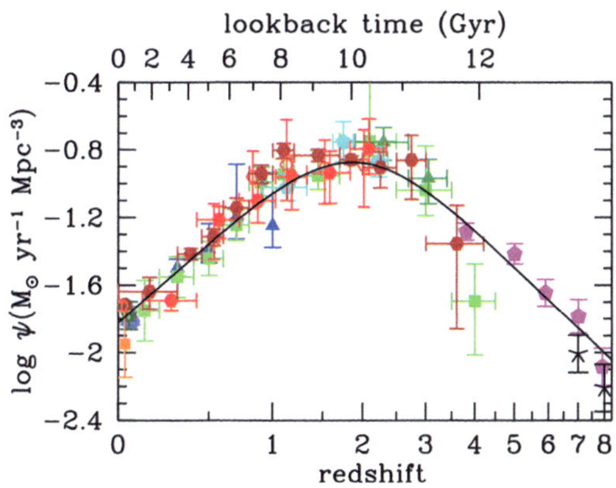

Figure 17.10 shows the consequence of rearing many children and a low death rate: the mass of the matter locked up in stars increases.

Figure 17.11 shows that an older population produces less children: define the efficiency of star formation, sSFR, as the ratio of SFR and the mass of the galaxy; or in other words, how many new stars are born for the same average star. The figure shows that this efficiency is going down all the time: the Universe gets older and becomes less productive.

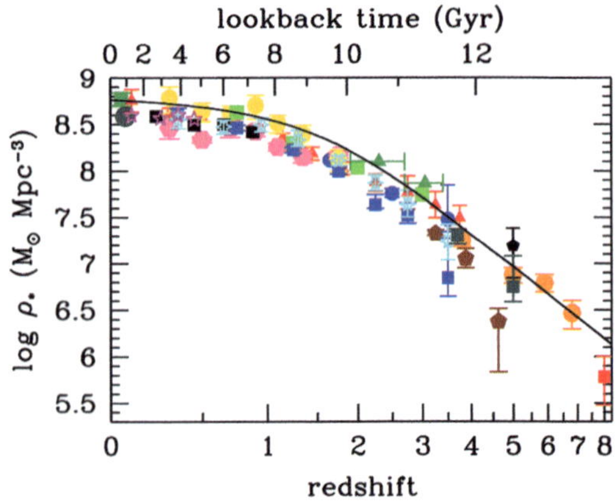

Fig. 17.10 The average density of stellar matter through the ages (Madau and Dickinson 2014)

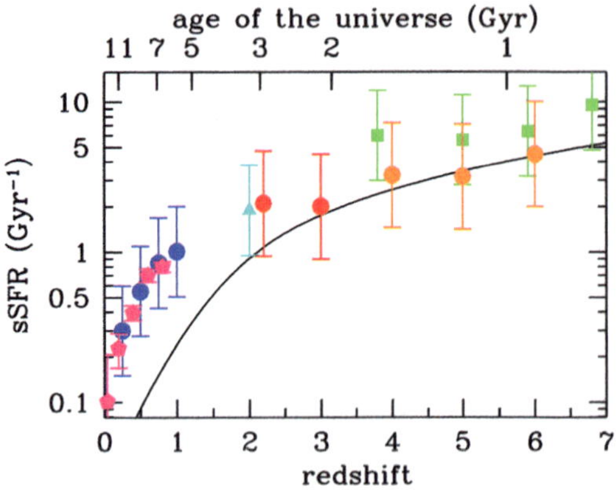

Fig. 17.11 Star formation efficiency through the ages. sSFR is the 'specific star forming rate': the star forming rate SFR divided by the mass of the stars in the galaxy (Madau and Dickinson 2014)

The diagrams shown represent an enormous amount of work: many observations, many analyses. The results are satisfying, yet, they represent correlations between measured quantities. Science must explain how these results come about. Simulations are a useful help to come to some explanation but, when they are complex, they do not clarify all the causes and consequences. Like demography again, it may well be that the study of the stellar populations are affected by many factors and that a clear understanding of what precisely happens escapes any analysis.

17.3 Clusters Galore

Clusters of galaxies are still a major topic of research. The Coma was discovered long ago, before photography made its entry, by Charles Messier and by William Herschel (see review by Biviano 2000). Abell (1958) made a frequently used catalogue of clusters found on plates of the Mount Palomar Photographic Survey. Later clusters appeared to be strong X-ray sources and many appear in for example, the

ROSAT databases. In 1972 Soviet scientists Sunyaev and Zeldovich pointed out that galaxy clusters should leave a mark on the background radiation, a weak imprint. Observations proved them right, but the effect was so slight that it did not lead to a practical method to detect clusters. This has now changed. Two millimeter wavelength telescopes, one at the South Pole, the SPT (Carlstrom et al. 2011), the other, the Atacama Cosmological Telescope, ACT Kosowsky (2003), is located at an altitude of 5,000 m high in the same desert as ALMA. The Earth's atmosphere above ACT and SPT is clear enough to detect the faint signature of the Sunyaev-Zeldovich effect. Even more productive was The Planck survey of the cosmic background radiation was even more effective: it had enough sensitivity to detect over 1,200 clusters (Planck Collaboration XXVII 2016). This sample is now studied and compared with the hundreds of clusters detected by X-radiation.

In 1972 the Americans Gunn and Gott surprised everyone with their thorough study of clusters of galaxies in which they showed that the formation of these clusters is still going on, although the heyday of making quasars and new stars is over. Their conclusion agrees with the fact that there are still so many clusters that are not 'relaxed': statistical physics predicts that in the long run a closed system of bodies will reach an equilibrium in which the distribution of the bodies and their velocities follow well-known mathematical laws. Some clusters of galaxies (e.g. the Coma cluster) have reached such a state, but others not yet (e.g. the Virgo cluster). Simulations confirm that that the formation of very massive units is still going on—see Fig. 17.12.

The Bullet Cluster: Unexpected Evidence of the Existence of Dark Matter

The X-radiation from the cluster 1E0657-56 was detected by the Einstein satellite, one of the first X-ray satellites. Data from ASCA (see Chap. 9) showed that the X-ray emitting gas was exceptionally hot (photon energy 17 keV). Observations with the ROSAT X-ray satellite showed that the cluster actually consisted of two clusters as could also be seen very clearly on photographs. And in 2002 Chandra observations showed that actually one of the clusters had passed *through* the other at the incredibly high speed of 4000 km/s and was now flying away from the place of collision. In its flight it had created a bow shock (think of a ship creating waves at its bow) as was clearly visible in the Chandra images (Markevitch et al. 2002). In the Fig. 17.13 one easily recognizes the bow shock in the contours. This gave the cluster its nickname: the Bullet Cluster. Apparently it is a strong concentration of matter that acts as the bullet.

The picture became even more interesting after the mass of the two clusters had been measured through gravitationally deformed images of galaxies that were far behind the cluster. The mass of each cluster coincided precisely with the visible clusters of galaxies. The hot, X-ray emitting gas did not coincide with the clusters but was predominantly in between. This was to be expected: during the collisions the galaxies passed without much hinder through each other: there is always so much empty space between galaxies that the chance of collision is low. (If the mesh

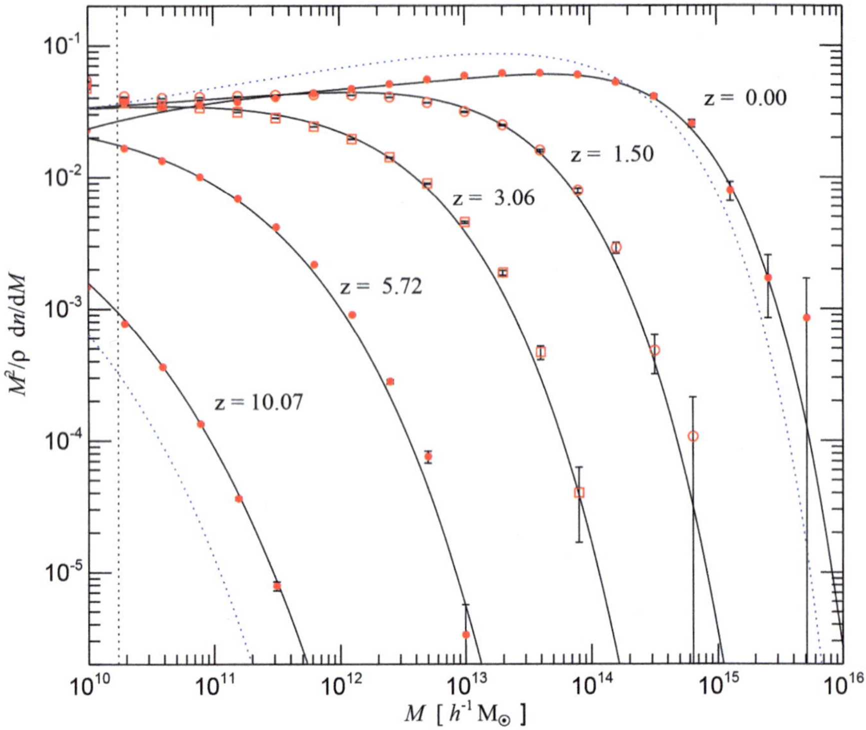

Fig. 17.12 The growth of coherent objects in cosmic time. Each curve presents the relation between the mass of a cluster and the number of that mass. Each curve is labeled with the appropriate redshift z. As z (and time) increases the curves shift to the right. The largest, most massive objects (clusters!) have accumulated only recently at $z = 0$ (Springel et al. 2005)

of the fishing net is too large, the fish will swim through it). But the X-ray gas clouds were too large and thus the gas masses of the two clusters collided. Therefore there was so much very hot gas in between the galaxy clusters. And now the surprise: the dark matter still coincided with each cluster (or vice versa). The dark matter clouds had also passed through one another without collision, showing that dark matter particles did not collide. That had been one of the major assumptions about the nature of dark-matter particles; it was now proven (Clowe et al. 2006).

17.4 Early Times: Re-ionization

Quasars and Re-ionization

About 380,000 years after the Big Bang ($z = 1100$) the matter and radiation had cooled so far that electrons and protons combined permanently to hydrogen atom for the first time. This combination had happened sometimes before, but then the

Fig. 17.13 Image of the bullet cluster. The red color shows the X-radiation with the 'bullet' at the right edge. The galaxies are seen as yellow objects. In blue the dark matter found by studying background galaxies whose distorted shapes were produced by the gravitational force of the cluster. See text for a discussion. Source: Astronomy Picture of the Day, 24 august 2006, NASA/CXC/CfA, STScI, ESO WFI, M. Markevitch et al.; D. Clowe et al.

two were quickly separated again by collisions between the new atoms and other protons. But now the temperature had fallen so far that ionizing collisions did no longer occur. This was also the moment that gravity began to get a hold on the matter and that small existing irregularities began to grow until in the end galaxies and stars were formed. The stars then began to ionize the matter again: reionization. The estimate is that ionization began at around $z = 11$ and was complete at $z = 6$.

Among the earliest objects that formed there have been quasars. How they formed is still unknown, but the fact is that quasars have been detected up to a redshift of $z = 7.5$ (Bañados et al. 2016) and there are over 100 known with $z > 5.8$ (Mortlock 2016).

In Chap. 5 the so-called Gunn-Peterson effect was introduced: if there is neutral hydrogen gas in between the galaxies this will absorb all the radiation with a wavelength below 121.6 nm, the limit of the strong absorption line Ly-alpha of the hydrogen atoms. For quasars with a redshift larger than 3 the hydrogen limit is now in the UV part of the spectrum. From the fact that many quasars with even higher redshifts were perfectly visible, it followed that the intergalactic matter was largely ionized.

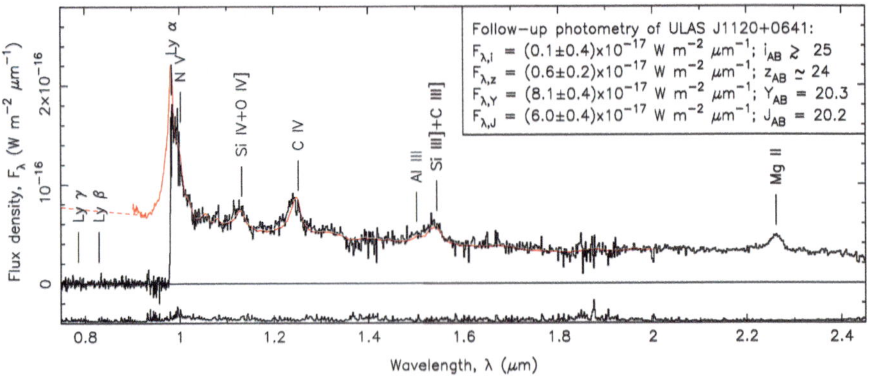

Fig. 17.14 The spectrum of quasar ULAS J1120+0641, with a redshift $z = 7.08$. Notice the sharp cut-off at the Lyman-alpha line that has a rest wavelength of 121.6 nm but is redshifted to the wavelength near 1 μm. The presence of neutral hydrogen atoms causes the cut-off (Mortlock et al. 2011)

The impression is that this is largely due to the quasars themselves: they are sources with an enormous output of ionizing radiation.

Immediately after the recombination the intergalactic medium was filled with neutral hydrogen atoms. The first quasars found themselves in an environment that had not yet been ionized. If so, the radiation below the redshifted Lyman limit must have been absorbed by all these neutral atoms. And indeed—the spectrum of the quasar ULAS J1120+0641[1] with a redshift $z = 7.08$ showed that there is no signal below the predicted edge at the wavelength of $(z + 1) \times 121.6$ nm (Fig. 17.14).

Figure 17.15 shows the transparency of the intergalactic medium below the (redshifted) limit caused by hydrogen atoms: for quasars with a redshift above 4.0 the intergalactic gas becomes increasingly neutral, and at $z = 6$ it is completely neutral.

Re-ionization and the 21 cm Line

If the intergalactic hydrogen is neutral for $z > 6$, will it emit the 21 cm line of hydrogen and can we detect it? This is a current (2016) question. The theoreticians predict that the line will be emitted but it has not yet been detected. Beautifully coloured simulations tempt the observers to detect the radiation, so far without result. These simulations, however, are based on uncertain assumptions about how the UV-light is turned on, that is: on the formation of the first quasars and of the first UV producing stars. See Fig. 17.16 for the results of one such simulation.

[1] ULAS is an abbreviation for 'UKIDSS Large-Area Survey. J indicates the coordinate system and the numbers specify the coordinates.

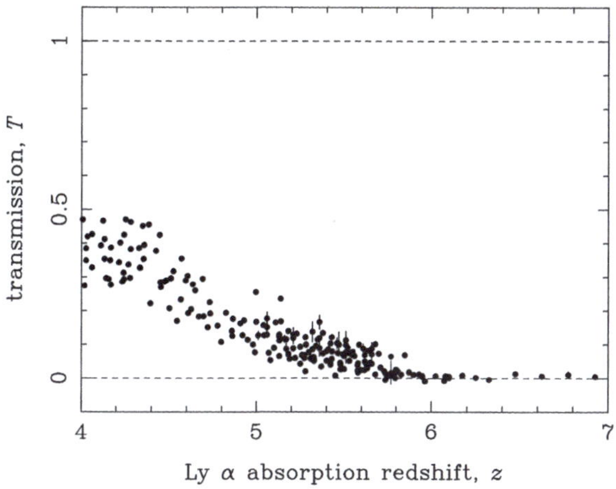

Fig. 17.15 The transparency of the intergalactic medium measured in quasar spectra (Mortlock 2016)

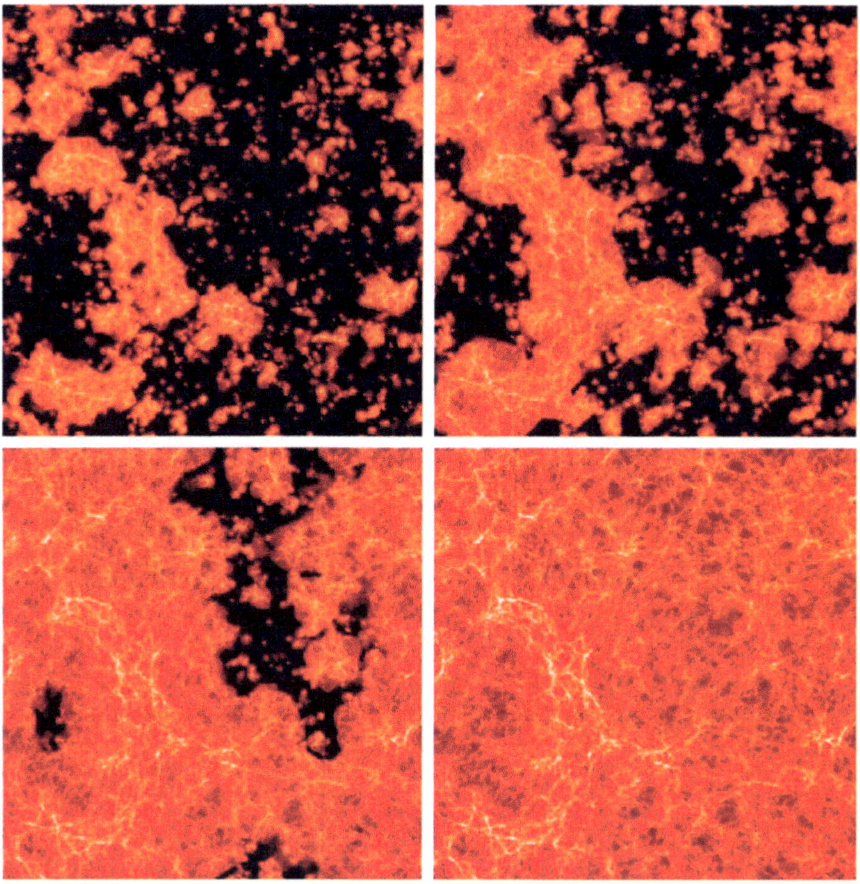

Fig. 17.16 Diagrams from a simulation of the reionization phase at four consecutive times. Shown is a slice of the Universe with a size of 50 × 50 × 10 Mpc (co-moving). Neutral hydrogen is in black and ionized hydrogen in red. Above, left: $z = 9$; above right $z = 8$; below left $z = 7$ and below right $z = 6$ (Morales and Wyithe 2010)

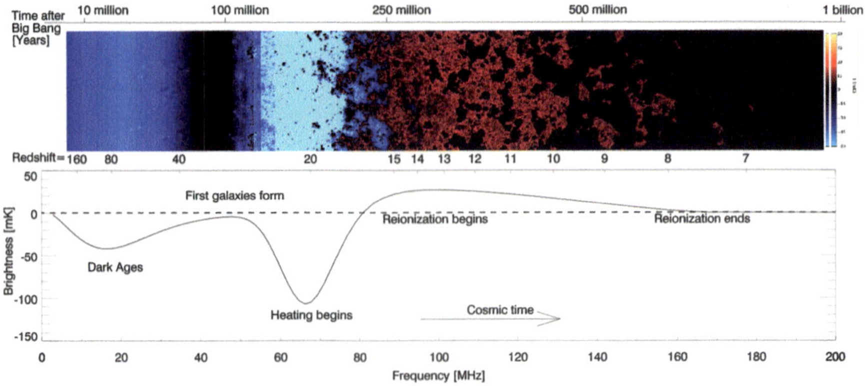

Fig. 17.17 Predictions of the strength of the cosmological 21 cm line in two diagrams and three distance/time wavelength scales. The top scale is in years passed since the Big Bang. In the middle the corresponding value of the redshift z, and below the redshifted frequency of the 21 cm line. The uppermost diagram gives the strength of the (redshifted) 21 cm line in colors. After a blue period in which the hydrogen atoms absorb background radiation there follows a dark period when the gas is completely unobservable. This is followed by a light blue period with strong absorption, before the hydrogen will start to emit the 21 cm line radiation (indicated by the red colour). Finally the hydrogen is ionized, the spectral line disappears and the diagram becomes black. The lower diagram shows the spectral line after it has been drawn out by redshift over this enormous range in wavelengths (Pritchard and Loeb 2012)

Another simulation led to Fig. 17.17. In the lower diagram, it shows the spectral line that is drawn out by redshift to lower and lower frequencies (longer and longer wavelengths). At every frequency below that of the line at rest (1420 MHz) one does no longer recognize the emission profile of a line, but the line has become a very faint continuum. To prove the presence of this continuum very sensitive measurements are needed. The detection of this (pseudo) continuum is an unsolved problem—for the time being.

The 21 cm emission line could be discovered while I am finishing this book. It will be an important discovery because it informs us about the very early years of the Universe.

17.5 Baryonic Acoustic Oscillations

In Chap. 16 the Baryonic Acoustic Oscillations have been discussed: sound-like small oscillations in the plasma that filled the Universe before star and galaxy formation began. It was predicted that the oscillations would behave as outgoing sound waves and that they all would have reached the same distance from there source before they were stopped by the beginning of the recombination. This common distance is called the 'baryonic sound horizon' or simple 'sound horizon'. Indeed, the sound horizon was detected in the analysis of the Cosmic Background Radiation (see Fig. 16.4 in Chap. 16).

It was soon realized that the sound horizon might also show up in the distribution of the galaxies and indeed this was the case; the horizon was seen in the data of the 2dFGR-survey (Cole et al. 2005) and the SDSS survey (Eisenstein et al. 2005); its radius was about 150 Mpc in todays Universe. That conclusion offered an interesting application. If you split the sample of galaxies in subsamples of different intervals in z, you could hopefully detect the horizon in these subsamples and measure its angular size: that gave you the distance. Thus you would obtain at each value of z the distance of the sound horizon. This was equivalent to measuring the Hubble diagram, because that is also the relation between the distance of an object and its radial velocity (or z-value). If feasible such a study would give you a completely independent confirmation of the Hubble diagram out to much more distant redshifts than was achieved by the supernova studies. Then one would test whether the present-day assumptions about dark energy are right.

Figure 17.18 contains four diagrams, each at a different interval in z, where the sound horizon is present in the data. The figure may be compared to Fig. 16.4 in Chap. 16.

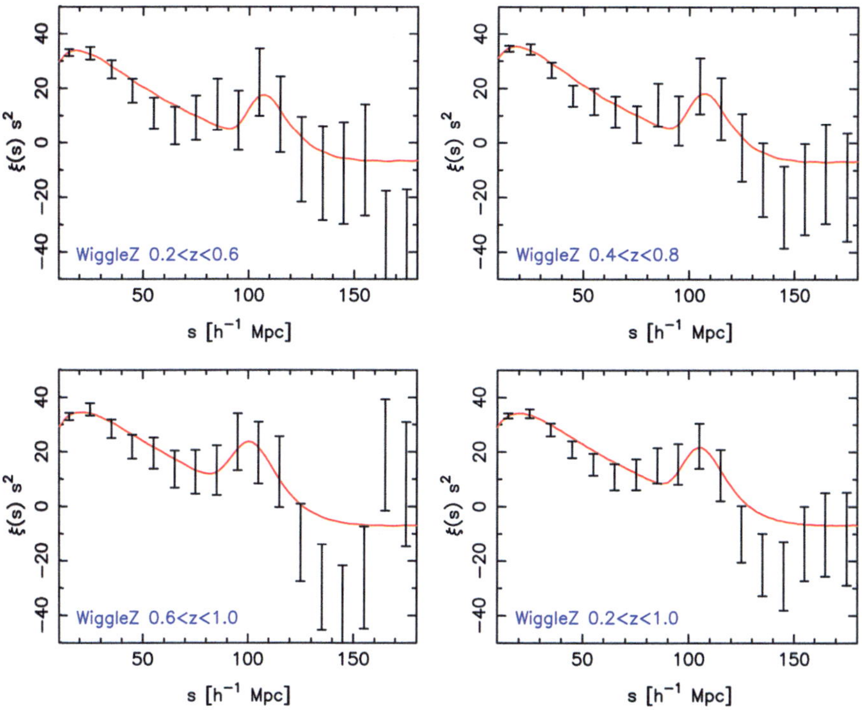

Fig. 17.18 Four correlation diagrams at three different intervals of z. The fourth diagram is the sum of the other 3. In each diagram the signature of the sound horizon is present. In the horizontal direction the co-variant distance is shown: this implies that the change in the scale of the Universe is taken care of (Blake et al. 2011)

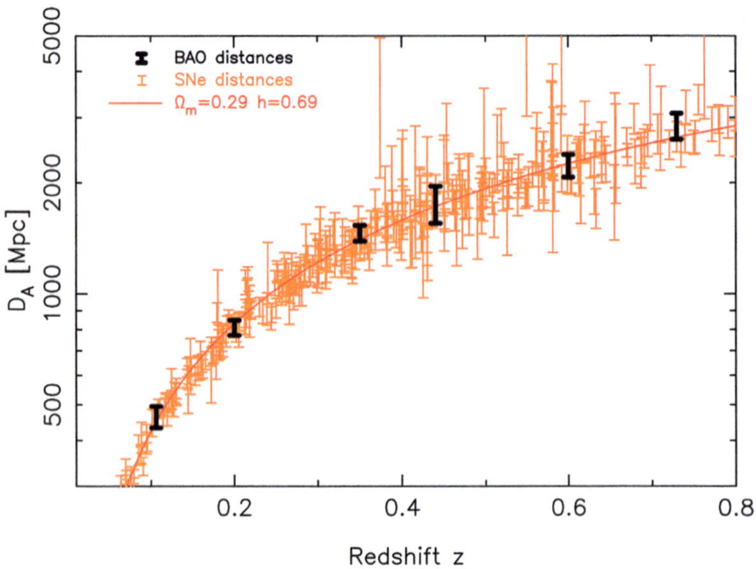

Fig. 17.19 Using the known length of the cosmic sound horizon makes it possible to derive, in a totally independent way, the relation between distance and radial velocity (Hubble relation). In this diagram the velocity is in the horizontal direction (expressed in redshift parameter) and the distance is in the vertical direction. The orange points are derived from measurements of supernovae Ia, the black points are from the analysis of the sound horizon. The full curve is that of the concordant model of the Universe (Blake et al. 2011)

The recognition of the sound horizon in the three diagrams in Fig. 17.18 leads to an independent determination of the relation between distance and redshift (the Hubble relation) and is shown in Fig. 17.19. There is a high degree of correlation between the curve from the analysis of the sound horizon and the curve found by the supernova studies. The curve drawn through the points is that of the concordant model of the Universe.

References

Abell, G.O.: Astrophys. J. Suppl. **3**, 211 (1958)

Bañados, E., Venemans, B.P., et al.: Astrophys. J. Suppl. **227**, 27 (2016)

Biviano, A.: Constructing the Universe with Clusters of Galaxies. In: Durret, F., Gerbal, D. (eds.). IAP 2000 meeting, Paris, France (2000)

Blake, C., Kazin, E.A., et al.: Mon. Not. R. Astron. Soc. **418**, 1707 (2011)

Bouwens, R.J., Illingworth, G.D., et al.: Astrophys. J. **709**, L133 (2010)

Bouwens, R.J., Illingworth, G.D., et al.: Nature. **469**, 504 (2011)

Carlstrom, J.E., Ade, P.A.R., et al.: Publ. Astron. Soc. Pac. **123**, 569 (2011)

Chambers, K.C., Miley, G.K., van Breugel, W.J.M.: Astrophys. J. **363**, 21 (1990)

Clowe, D., Bradač, M., et al.: Astrophys. J. **648**, L109 (2006)
Cole, S., Percival, W.J., et al.: Mon. Not. R. Astron. Soc. **362**, 505 (2005)
Eisenstein, D.J., Zehavi, I., et al.: Astrophys. J. **633**, 560 (2005)
Gunn, J.E., Gott III, J.R.: Astrophys. J. **176**, 1 (1972)
Kormendy, J., Kennicutt Jr., R.C.: Annu. Rev. Astron. Astrophys. **42**, 603 (2004)
Kosowsky, A.: NewAR. **47**, 939 (2003)
Kriek, M., van Dokkum, P.G., et al.: Astrophys. J. **705**, L71 (2009)
Lilly, S.J.: Astrophys. J. **333**, 161 (1988)
Lilly, S.J., Tresse, L., et al.: Astrophys. J. **455**, 108 (1995)
Madau, P., Dickinson, M.: Annu. Rev. Astron. Astrophys. **52**, 415 (2014)
Maiolino, R., Cox, P., et al.: Astron. Astrophys. **440**, L51 (2005)
Markevitch, M., Gonzalez, A.H., et al.: Astrophys. J. **567**, L27 (2002)
Morales, M.F., Wyithe, J.S.B.: Annu. Rev. Astron. Astrophys. **48**, 127 (2010)
Mortlock, D.J., Warren, S.J., et al.: Nature. **474**, 616 (2011)
Mortlock, D.J.: ASSL. **423**, 187 (2016)
Oesch, P.A., Brammer, G., et al.: Astrophys. J. **819**, 129 (2016)
Omont, A., Petitjean, P., et al.: Nature. **382**, 428 (1996)
Pettini, M.: arXiv: astro-ph/0303272 (2003)
Planck Collaboration XXVII: Astron. Astrophys. **594**, A27 (2016)
Pritchard, J.R., Loeb, A.: RPPh. **75**, 6901 (2012)
Somerville, R.S., Dave, R.: Annu. Rev. Astron. Astrophys. **53**, 51 (2015)
Springel, V., White, S.D.M., et al.: Nature. **435**, 629 (2005)
Steidel, C. C., Giavalisco, M. et al.: Astrophys. J., **462** L17-L21 (1996)
Steidel, C.C., Adelberger, K.L., et al.: Astrophys. J. **519**, 1 (1999)
Steidel, C.C.: Proc. Natl. Acad. Sci. U. S. A. **96**, 4232 (1999)
Sunyaev, R.A., Zeldovich, Y.B.: Astron. Astrophys. **20**, 189 (1972)
van Breugel, W., De Breuck, C., et al.: Astrophys. J. **518**, L61 (1999)
Venemans, B.P., Walter, F. et al.: Astrophys. J. **851** L8 (2017)
Whitaker, K.E., van Dokkum, P.G., et al.: Astrophys. J. **754**, L29 (2012)

Epilogue

I do not want to summarize this book. A very short epilogue should be enough.

The detection of radiation ranging from TeV energies to long radio waves, supported and stimulated by theoretical concepts, has opened the history of our Universe; it is a story that is convincing, complex and yet also simple and straightforward. It is the story that, timewise, precedes and completes the geological and biological history of the Earth. Mankind now knows its full history. Having said this, it is a sobering thought that we, of course, do not know the story of our future. Astronomers can look backward over 13.6 billion years and not look forward by one week.

The path of our history is at times very small and leads along several deep abysses. To name the three greatest threats: dark matter, dark energy and inflation–major hypotheses that had to be made and that remain unproven. At some time in the future it will become clear whether we can keep walking along our path or whether we will glide sideways in one of these abysses. Let us, however, be optimistic and enjoy what we have obtained: for the time the story looks very good.

H. J. Habing, *The Birth of Modern Astronomy*, Historical & Cultural Astronomy,
https://doi.org/10.1007/978-3-319-99082-8

Index

H. J. Habing, *The Birth of Modern Astronomy*, Historical & Cultural Astronomy,
https://doi.org/10.1007/978-3-319-99082-8

T

Printed by Printforce, the Netherlands